AF886444

New Trends In Biomarkers and Diseases Research: An Overview

Edited by

Juan Antonio Vílchez

Clinical Analysis Department, Santa Lucía General University Hospital, Cartagena, Spain

Co-Edited by

María Dolores Albaladejo-Otón

Head of Service of Clinical Analysis Department, Santa Lucía General University Hospital, Cartagena, Spain

General:

1. Any dispute or claim arising out of or in connection with this License Agreement or the Work (including non-contractual disputes or claims) will be governed by and construed in accordance with the laws of the U.A.E. as applied in the Emirate of Dubai. Each party agrees that the courts of the Emirate of Dubai shall have exclusive jurisdiction to settle any dispute or claim arising out of or in connection with this License Agreement or the Work (including non-contractual disputes or claims).
2. Your rights under this License Agreement will automatically terminate without notice and without the need for a court order if at any point you breach any terms of this License Agreement. In no event will any delay or failure by Bentham Science Publishers in enforcing your compliance with this License Agreement constitute a waiver of any of its rights.
3. You acknowledge that you have read this License Agreement, and agree to be bound by its terms and conditions. To the extent that any other terms and conditions presented on any website of Bentham Science Publishers conflict with, or are inconsistent with, the terms and conditions set out in this License Agreement, you acknowledge that the terms and conditions set out in this License Agreement shall prevail.

Bentham Science Publishers Ltd.
Executive Suite Y - 2
PO Box 7917, Saif Zone
Sharjah, U.A.E.
Email: subscriptions@benthamscience.org

BENTHAM SCIENCE

CONTENTS

DEDICATION

This book is dedicated to all who love laboratory medicine and especially to each of the authors who have participated in this exciting project.

Thank you all

FOREWORD

The concept of biomarker applies to any biological characteristic that can be objectively measured and evaluated to improve the clinical ability to evaluate different aspects of health or disease status.

Referring to a disease, a biomarker can help to evaluate the susceptibility of developing a disease in which case the biomarker is a risk factor; to recognize its existence both in the subclinical (screening biomarker) or clinical (diagnostic biomarker) stage, to control the progression or regression (staging biomarker), to prognoses the disease (prognostic biomarker) or its response to therapy (monitoring biomarker). Furthermore, a biomarker could be the measure of a biological variable, such as blood pressure, the result of a scanning technique as an X-ray or a biological variable measured in the clinical laboratory. And it is in this latter type of variables where the term of biomarker is more widely applied.

The literature is full of new biomarkers which are claimed to add value to clinical practice and, hence to be a part of the clinical laboratory portfolio. The aim of the current book is to summarize the existing evidence on the role that classical and some novel biomarkers can play in the most prevalent diseases.

Jordi Ordoñez-Llanos, MD, Ph.D.
Senior Consultant, Hospital de Sant Pau. Professor Clinical Biochemistry
IIB-Sant Pau Biomedical Research Institute, Hospital Santa Creu i Sant Pau and
Department of Biochemistry and Molecular Biology
Universitat Autònoma, Barcelona,
Spain

PREFACE

Biomarkers, whether measured in peripheral fluids (blood, serum or urine) may improve our knowledge of the pathophysiology of many diseases. Moreover, and most importantly, biomarkers could help in the assessment of diagnosis, prognosis and surviving. The aim of this book was to summarize published data about new emerging and classic biomarkers studied in prevalent diseases, with focus on data from clinical trials and large community based cohorts.

The use of biomarkers has become a fundamental practice in medicine, involving significantly greater scope of laboratory medicine. Biomarkers are measurable characteristics of an organism reflecting a particular physiological state. In medicine, biomarkers considered as compounds isolated from serum, urine, or other fluids, can be used as indicators of the presence or severity of a particular disease state.

Biomarkers can take many different forms including particular proteins or peptides, antibodies, cell types, metabolites, hormones, enzyme levels, compounds related to genomics, *etc*. A biomarker can also be a substance introduced into a patient to assess the internal organ systems role/function. Moreover, biomarkers facilitate the adaptation of treatment to the specific needs of each patient, which is known as personalized medicine. In this way, using biomarkers to monitor a patient's reaction to a particular drug, it is possible to determine whether a treatment is effective for an individual or has to be changed due to toxic adverse effects. Also, this aspect could lead to their convenient use in terms of cost-efficiency. Biomarkers are also important to manage new disease therapies through the use of biomarkers of progression to delineate the development and course of a disease, thus focusing on the risk or progression of a disease or with the susceptibility of a disease to a given treatment.

Nowadays, the way to diagnose general pathologies and diseases has changed importantly due to the development of high throughput technologies, such as microarrays or large-scale studies of proteins (proteomics, genomics, metabolomics *etc*). These investigations have allowed the discovery of new biomarkers of diseases. Also, the use of established biomarkers to other aspects in the progression of diseases, as the inclusion of biomarkers in many stratification schemes of severity to stratify the risk in patient's state.

A single biomarker is often inconclusive or ambiguous because diseases with very different pathophysiological mechanisms can affect the same molecule or marker. Therefore, it is crucial to study and manage patterns of biomarkers with respect to disease state, prognosis and treatment. This complex scenario is the main focus of this book. We tried to summarize, relating to principal groups of pathologies with respect to blood biomarkers, a compilation of the present and upcoming knowledge about biomarkers and diseases. The chapters cover a wide number of areas including those in which biomarkers could mainly add clear information about the illness. This book includes, reviews about sepsis, nutrition, liver state, cancer, cardiology, bone metabolism, *etc*. The chapters are written by experts and specialists in general laboratory medicine.

Dr. Juan Antonio Vílchez
Clinical Analysis Department,
Santa Lucía General University Hospital,
Cartagena,
Spain

List of Contributors

Ana M. Moreno-Fuentes, BSc	Clinical Analysis Department, Santa Lucía General University Hospital, Cartagena, Murcia, Spain
Africa de Béjar-Almira, BSc	Clinical Analysis Department, Santa Lucia General University Hospital, Cartagena, Spain
Anna Hernández-Aguilera, MSc	Unitat de Recerca Biomèdica, and Laboratori de Referència SUD, Hospital Universitari Sant Joan, Institut d'Investigació Sanitària Pere Virgili, Universitat Rovira i Virgili, Reus, Spain Carrer Sant Llorenç 21, 43201-, Reus, Spain
Ana Martínez-Ruiz, Pharm, PhD	Clinical Analysis Department, Reina Sofia General University Hospital, Murcia, Spain
Ana Hernando-Holgado, Pharm	Clinical Analysis Department, Santa Lucia General University Hospital, Cartagena, Spain
Alice Charlotte Viney, Pharm	Pharmacy Department, Santa Lucia General University Hospital, Cartagena, Spain
Carmen Nieto-Sánchez, Pharm. PhD	Clinical Analysis Department, Santa Lucía General University Hospital, Cartagena, Murcia, Spain
Carmen María Puche-Morenilla, PhD	Clinical Analysis Department, Clinical University Hospital Virgen de la Arrixaca, Murcia, Spain
Diana Hernández-Romero, PhD	Department of Cardiology, Virgen de la Arrixaca Clinic University Hospital Instituto de Investigación Biomédica-Virgen de la Arrixaca, Murcia, Spain
Enrique Jiménez-Santos, Chem	Clinical Analysis Department, Santa Lucia General University Hospital, Cartagena, Spain
Francisco Marín, PhD	Department of Cardiology, . Virgen de la Arrixaca Clinic University Hospital, Murcia, Spain. Instituto de Investigación Biomédica-Virgen de la Arrixaca. IMIB-Arrixaca, Murcia, Spain
Iria Cebreiros-López, Pharm, PhD	Clinical Analysis Department, Clinical University Hospital Virgen de la Arrixaca, Murcia, Spain
Irene De Miguel-Elízaga, BSc	Clinical Analysis Department Unilabs Laboratory, Torrevieja Salud Hospital, Alicante, Spain
Isabel Fort-Gallifa, MSc	Unitat de Recerca Biomèdica, and Laboratori de Referència SUD, Hospital Universitari Sant Joan, Institut d'Investigació Sanitària Pere Virgili, Universitat Rovira i Virgili, Reus, Spain Carrer Sant Llorenç 21, 43201-, Reus, Spain
Irene Gutiérrez-García, BSc	Clinical Analysis Department, Santa Lucia General University Hospital, Cartagena, Spain
Iris Muñoz-García, Pharm	Pharmacy Department, Santa Lucia General University Hospital, Cartagena, Spain
Juan Antonio Vílchez, Pharm, PhD	Clinical Analysis Department, Santa Lucía General University Hospital, Murcia, Spain

Jordi Ordóñez-Llanos, MD, PhD	IIB-Sant Pau Biomedical Research Institute, , Hospital Santa Creu i Sant Pau and Department of Biochemistry and Molecular Biology, Universitat Autònoma, Barcelona, Spain
Jorge Joven, MSc	Unitat de Recerca Biomèdica, and Laboratori de Referència SUD, Hospital Universitari Sant Joan, Institut d'Investigació Sanitària Pere Virgili, Universitat Rovira i Virgili, Reus, Spain Carrer Sant Llorenç 21, 43201-, Reus, Spain
Jordi Camps, PhD	Unitat de Recerca Biomèdica, and Laboratori de Referència SUD, Hospital Universitari Sant Joan, Institut d'Investigació Sanitària Pere Virgili, Universitat Rovira i Virgili, Reus, Spain Carrer Sant Llorenç 21, 43201-, Reus, Spain
Jose Pedregosa-Díaz, Chem	Clinical Analysis Department, Santa Lucia General University Hospital, Cartagena, Spain
Luis García de Guadiana-Romualdo, Pharm	Clinical Analysis Department, Santa Lucia General University Hospital, Cartagena, Spain
Natalia Sancho-Rodríguez, Pharm, PhD	Santa Lucía, General University Hospital, Cartagena, Murcia, Spain
Marta Sancho-Rodríguez, Pharm	Pharmaceutical, Production Team in MSD, Oss, The, Netherlands
María Dolores Albaladejo-Otón, PhD	Clinical Analysis Department, Santa Lucia General University Hospital, Cartagena, Spain
María Henar García-Lagunar, Pharm	Pharmacy Department, Santa Lucia General University Hospital, Cartagena, Spain
M. Pilar Gallego-Hernanz, MD, PhD	Hématologie Oncologique et Thérapie cellulaire, Centre Hospitalier Universitaire de Poitiers, France
Miriam Martínez-Villanueva, Pharm, PhD	Clinical Analysis Department, Virgen de la Arrixaca Clinic University Hospital, Murcia, Spain
Marta M. Castañeda San Cirilo, PhD	Clinical Analysis Department, Santa Lucia General University Hospital, Cartagena, Spain
Maria S García-Simón, Pharm	Pharmacy Department, Santa Lucia General University Hospital, Cartagena, Spain
Mariano Valdés, MD, PhD	Department of Cardiology, Virgen de la Arrixaca Clinic University Hospital Instituto de Investigación Biomédica-Virgen de la Arrixaca, Murcia, Spain
Patricia Esteban-Torrella, Pharm	Clinical Analysis Department, Santa Lucia General University Hospital, Cartagena, Spain
Pablo Perez-Cañadas, Pharm	Clinical Analysis Department, Santa Lucía General University Hospital, Murcia, Spain
Silvia Montoro-García, PhD	Department of Cardiovascular Risk, Faculty of Health Sciences, Universidad Católica San Antonio de Murcia (UCAM), Campus de los Jerónimos, Guadalupe, Murcia, Spain
Sara María Sánchez-Martínez, Pharm	Department of Cardiovascular Risk, Faculty of Health Sciences, Universidad Católica San Antonio de Murcia (UCAM), Campus de los Jerónimos, Guadalupe, Murcia, Spain

Vanessa Roldán, PhD	Department of Hematology, Morales Meseguer University Hospital Instituto de Investigación Biomédica-Virgen de la Arrixaca, Murcia, Spain
Vicente Martín-Paredero, MSc	Unitat de Recerca Biomèdica, and Laboratori de Referència SUD, Hospital Universitari Sant Joan, Institut d'Investigació Sanitària Pere Virgili, Universitat Rovira i Virgili, Reus, Spain Carrer Sant Llorenç 21, 43201-, Reus, Spain

CHAPTER 1

Biomarkers in Pre-eclampsia: Is it Possible to Predict it?

Ana Martínez-Ruiz[1,*], **Irene De-Miguel-Elízaga**[2] and **Natalia Sancho-Rodríguez**[1]

[1] *Clinical Analysis Department, Reina Sofia General University Hospital, Murcia, Spain*

[2] *Clinical Analysis Department Unilabs Laboratory, Torrevieja Salud Hospital, Alicante, Spain*

Abstract: pre-eclampsia is a syndrome with high maternal and fetal mortality. The pathophysiology remains unknown. Prediction, diagnosis and management of the disease has allowed the identification of multiple biomarkers, some of which help to predict those at risk. Some of these biomarkers have demonstrated, even in isolation, an effi-ciency of the test that allows to incorporate them into clinical practice. The combination of these biomarkers and clinical factors may help predict pre-eclampsia risk by developing integrated clinical risk models. This chapter aims to delve into the literature related to biomarkers in pre-eclampsia and its possible clinical applications.

Keywords: A disentigrin and Metalloprotease 12, Activin A, Cell-free DNA, Cystatin C, Fetal hemoglobin, First trimester, Inhibin A, Metabolomics, P-selectin, Pentraxin 3, Placental growth factor, Placental protein-13, Preclampsia, Pregnancy-associated plasma protein-A, Proteomics, Renin angiotensin system, Soluble endoglin, Soluble fms-like tyrosine Kinase, Vascular endothelial growth factor, Vifastin.

INTRODUCTION

pre-eclampsia (PE) affects 3-5% of pregnancies. It is diagnosed by an increase in blood pressure and proteinuria [1]. PE has become one of the causes of maternal, fetal and neonatal mortality, especially in countries with medium or low incomes.

The etiology of PE is unclear. In women with PE, placental antiangiogenic factors are up regulated and disrupt the maternal endothelium, leading to an antiangiogenic state which can result in clinical signs of PE [2].

It is a unique disease in several ways: it is one of the rare pathologic conditions

[*] **Corresponding author Ana Martínez-Ruiz:** Clinical Analysis Department, Reina Sofia General University Hospital, Murcia, Spain; Tel/Fax: 0034968359000; E-mail: anamr82@gmail.com

that are specific to pregnancy; it is, by definition, a precursor of a potentially-severe disease (eclampsia) but is lethal by itself. Despite this, it has the same essential treatment (delivery) for hundreds of years; and researchers are unable to know what its fundamental cause is and how to prevent it.

DEFINITION OF PRE-ECLAMPSIA

PE is defined as *de novo* hypertension present after 20 weeks of gestation. It is combined with proteinuria (>300 mg/day), and other maternal dysfunction, just like renal failure, hepatic impairment, uteroplacental dysfunction, growth restriction fetal and other complications at neurological or haematological level [3]. As proteinuria is no longer needed in the new definition, proteinuric and non-proteinuric PE are now two separate categories. We defined hypertension as systolic blood pressure greater than 140 mm Hg or diastolic blood pressure greater than 90 mm Hg on two occasions which are 4-6 h apart [4].

PE can be subdivided into early-onset PE with <34 weeks gestation and late-onset PE with ≥34 weeks. It exists a higher incidence of ad-verse outcome associated with early onset PE.

EPIDEMIOLOGY

Two-thirds of PE cases occur in otherwise healthy, nulliparous women, so there is no single most important recognizable risk factor. However, there is a classic list of conditions that predispose a patient to pre-eclampsia (Table **1**) [5, 6].

Table 1. Risk factors for pre-eclampsia.

	Relative Risk [6]	**Risk [5]**
Nulliparity	3	-
Prior pre-eclampsia	7	-
Advanced maternal age	2	10%-40% (age>40)
Chronic hypertension	-	15%-40%
Chronic renal disease	-	15%-40%
Diabetes	3.5	10%-35%
Obesity	-	10%-15%
Multiple gestation	3	-
Vascular/conective tissue disorder	-	10%-20%
Antiphospholipid antibody syndrome/trombophilia	9	10%-40%
Family history of pre-eclampsia	2-4	10%-15%
Patient born small For gestational age	-	1.5-fold

(Table 1) contd.....

	Relative Risk [6]	Risk [5]
Prior adverse pregnancy outcomes	-	2-fold to 3-fold

ETIOPATHOLOGY OF PRE-ECLAMPSIA

Although PE is a systemic disease, its origin appears to be in the placenta. However, failure in placentation is not enough to explain the endothelial alteration that causes the maternal syndrome. Maternal risk factors for the appearance of PE are related to medical situations that condition predisposition of develop vascular dysfunction, such as chronic hypertension, diabetes mellitus, obesity or thrombophilia. All this points towards a relationship between a deficient placentation and the induction of maternal vascular damage, which could be mediated by factors released into the maternal circulation from an insufficient placenta [7].

The pathogenesis is a result of multifactorial origin, which can grossly be understood under following components:

Uteroplacental pathology: the starting factor in PE would be the reduction of utero-placental perfusion, as result of the abnormal invasion of the spiral arteries by the trophoblast. Invasive trophoblastic cells differ abnormally to syncytium (giant cells), which lose their power of penetration.

Angiogenic factors: in PE, there is an imbalance in the production and release to the maternal circulation of factors regulating angiogenesis from the placenta in the situation of ischemia.

Lipid peroxides: oxidation of lipoproteins is present in normal pregnancy, but is greatly increased in PE.

Inflammation and cytokines: PE is a disease characterized by generalized dysfunction of the endothelial cell, related to several factors: fatty acids, lipoproteins, lipid peroxidation, tumor necrosis factor α (TNFα) and degradation products of fibronectin. All these factors together, result from a generalized intravascular inflammatory response present during pregnancy.

Autoantiobodies: recent studies have shown that women with PE have autoantibodies termed ATI-AAs. These antibodies activate the angiotensin II receptors.

Genetics: PE is a genetic disorder and is influenced by environmental factors. It is found with increased frequency in mothers, daughters, sisters, and grand daughters of women having pre-eclampsia studied.

Immunological factors: the immunization concept is supported by the observation that PE develops more frequently in multiparous women impregnated by a new consort.

PREDICTION OF PRE-ECLAMPSIA

Although perfect prediction of PE has been a noble but hitherto elusive goal, it is possible to distinguish between women who are at low risk and high.

Previous PE or hypertension in pregnancy, chronic kidney disease, hypertension, diabetes (type 1 or type 2), and autoimmune disorders, among which are included systemic lupus erythematosus or antiphospholipid syndrome are considered Strong risk factors [8]. First pregnancy, multiple pregnancy, age 40 years or more, a pregnancy interval greater than 10 years, body-mass index of 35 kg/m^2 or more, polycystic ovarian syndrome, and family history of PE are considered as moderate risk factors [8, 9].

At present, a combination of biomarkers is being studied to predict PE, since the use of a biomarker in isolation does not present a high yield. The combination of these biomarkers with Doppler would allow increased sensitivity and specificity [10 - 12].

The most promising strategies for the prediction of PE involve multiparametric approaches, which use a variety of individual parameters in combination.

In order to identify a high proportion of pregnancies at high risk for early-onset PE a combination of maternal risk factors, the uterine artery pulsatility index (PI), mean arterial pressure (MAP), and maternal serum pregnancy-associated plasma protein-A (PAPP-A), placental growth factor (PlGF), placental protein-13 (PP 13), *etc.* at 11–13 weeks' gestation [13, 14] can be used.

MANAGEMENT OF PRE-ECLAMPSIA

Adequate identification of risk factors is important for PE management [15, 16]. After the diagnosis, the safety of the mother and the fetus must be ensured. The decision between delivery and expectant management depends on fetal gestational age, fetal status, and severity of maternal condition at time of assessment.

Women with mild disease developing at 38 weeks' gestation or longer have in general a similar pregnancy outcome to that seen in normotensive pregnancy [16, 17]. Childbirth is also recommended in those women at week 34 or more of gestation with severe PE.

PREVENTION OF PRE-ECLAMPSIA

Aspirin is the drug of choice for the prevention of PE, based on the results of a meta-analysis of individual patient data showing a moderate benefit of aspirin (RR 0.90, 95% CI 0,84-0.97) [18]. Other studies use heparin and dal-teparin for the prevention of PE. No conclusive results are obtained because of the small sample size [19, 20].

Have also been associated with PE low dietary calcium and low serum calcium concentrations [21]. In women with low calcium intake in the diet, high-dose calcium supplements reduced PE (RR 0.36, 95% CI 0.20-0.65) [22]. Although calcium supplementation is not recommended in women with normal calcium intake in the diet, WHO recommends for women with low energy intake of calcium in the diet the supplementation of calcium (1.5-2 g daily) in the second half of pregnancy [23]. Studies are being conducted in women with previous PE to observe the effectiveness of calcium supplementation in early stages of pregnancy to prevent PE [24].

The risk of PE is not reduced by dietary supplementation with vitamin C and vitamin E (RR 1.00, 95% CI 0.92–1.09) or magnesium (0.87,0.58–1.32) [25, 26]. Vitamin D insufficiency is associated with an increased risk of gestational diabetes, PE, and small size for gestational age, but prophylactic vitamin D supplementation has only been assessed in one randomized controlled trial (0.67, 0.33–1.35) [27, 28].

In a large randomized controlled trial of women at risk for PE, the precursor nitric oxide L-arginine reduced the risk of PD when given in combination with antioxidants (RR 0.17, 95% CI 0.12-0.21) [29], which was subsequently confirmed in a meta-analysis (0.34, 0.21-0.55.55) [30]. It would be interesting to study with pregnant women with low risk of developing PE and in them to observe if the supplementation of L-arginine prevents the development of the disease.

Has been show, according to systematic reviews [31, 32] that risk of PE in pregnant women can be reduced making changes in diet and lifestyle, including women with gestational diabetes although this effect was not confirmed in a more recent randomized controlled trial [33] of diet and lifestyle interventions in pregnant women who are overweight or obese. In summary, treatment with aspirin is the only intervention to prevent PE for which robust evidence exists, but its effect is not large. Except for calcium supplementation in women with low dietary calcium intake, all other preventive interventions need further assessment and should not be prescribed outside the context of clinical trials.

BIOMARKERS

Vascular Endothelial Growth Factor (VEGF) and Placental Growth Factor (PlGF)

Angiogenesis consists of the interaction between VEGF and PlGF factors with their VEGF receptor-1 receptors (VEGFR-1, alternatively called fms-like tyrosine kinase-1 (Flt-1)) and VEGFR-2 [34].

Serum levels of PlGF increased during the first and second trimesters, peaking at week 30 of gestation [35]. PlGF is a member of the vascular endothelial growth factor family and is implicated in angiogenesis and trophoblastic invasion of the maternal spiral arteries.

In normal pregnancies, PlGF peaks at 30 weeks and decreases towards term [36]. In PE, circulating PlGF is decreased, especially in cases of early-onset PE, although there are some discrepant findings [37]. Serum levels of PlGF decreased at the end of the first trimester predict PE cases [38, 39]. Other studies did not find any predictive value early in pregnancy [40, 41], or only for early-onset PE. PlGF is considered a good marker for predicting onset-early PE rather than onset-late PE.

Soluble Flt-1 (sFlt-1)

SFlt-1 is a truncated splice variant of the membrane-bound Flt-1. This splice variant circulates freely in the serum, where it binds and neutralizes VEGF and PlGF. Some studies relate PE to increased levels of sFlt-1 [42]. Increased levels of sFlt-1 have been observed five weeks before the onset of the disease [36]. Maynard *et al.*, [43] reported that the excessive placental production of sFlt-1 (an antagonist of VEGF and PlGF) contributes to the pathogenesis of PE, extensive research has been published showing the usefulness of angiogenic markers in diagnosis and subsequent prediction and management of PE and placenta related disorders. PlGF circulates free or in complexes with sFlt-1.

The levels of sFlt-1 remain constant during the first two quarters of pregnancy and increase in the third trimester in a normal pregnancy. The action of sFlt1 and its interaction with the "pro-angiogenic proteins" VEGF and PlGF during pregnancy is complex [44]. It is believed that this is a "decoy protein" in pregnancy, reducing free concentrations of VEGF and PlGF, thereby reducing the vaso-dilation effect on the endothelium, and at the same time inducing vasoconstriction and contributing to the development of hypertension and proteinuria [45].

In women with PE, increased levels of sFlt-1 were found in the second and third trimesters of gestation [46]. The associations are similar to PlGF, stronger in early-onset and severe PE; However, not all studies agree here to [47]. Mean arterial pressure (MAP) and proteinuria correlate positively with increased levels of sFlt-1 [48].

Early-onset EP correlates with low levels of sFlt-1 in the first trimester and increased levels of sFlt-1 at the end of the first and second-trimester second-stage is related to PE risk [49].

Results from the latter cohort recently showed that low sFlt1 in the first trimester could predict early-onset PE independently of small-for-gestational age (SGA) and also late-onset PE together with SGA [50]. Other studies have not shown a predictive value for sFlt1 in the first trimester [51, 52].

Pregnancy-Associated Plasma Protein-A (PAPP-A)

PAPP-A is a syncytiotrophoblast derived metalloproteinase, which enhances the mitogenic function of the insulin-like growth factors by cleaving the complex formed between such growth factors and their binding proteins [53, 54]. The insulin-like growth factor system seems to have an important role in placental growth and development; because of this, it is not surprising that PAPP-A serum is associated with a higher incidence of PE [55].

Decreased serum levels of PAPP-A in the first and second trimesters of gestation are related to the risk of developing PE. However, measurement of PAPP-A alone is not an effective method of screening for PE because only 8-23% of cases have serum levels below the fifth percentile, which is about 0.4 multiple of the median (MoM). The reported odds ratios for PE varied between 1.5 and 4.6 at the fifth percentile of normal for PAPP-A [56].

Soluble Endoglin (sEng)

SEng is a truncated form of receptor for TGF (transforming growth factor β family) β1 and TGFβ2. SEng is a potential anti-angiogenic factor which interferes with binding of TGFβ1 to its receptor, and which results in the production of nitric oxide (NO), vasodilation, and capillary formation by endothelial cells *in vitro* [57]. Serum levels of sEng decrease in normal pregnancies in the first and second trimester of gestation. However, it has been reported that sEng is elevated in maternal serum in patients who are destined to develop severe PE [58].

Inhibin A and Activin A

Inhibin A and Activin A are glycoproteins and members of the TGF, which are released by the placenta during pregnancy. Inhibin A has an important endocrine role in the negative feedback of gonadotrophins while Activin A is involved in various biological activities [59]. In severe PE, in the third trimester of gestation, the levels of both hormones are increased tenfold compared to normal pregnancies [60]. In PE, there is increased oxidative stress and maternal systemic inflammation. It was documented that oxidative stress stimulates activin A production and its secretion from placental explants and endothelial cells [61].

Other markers (PIGF, PP-13, inhibin-A, sEng, pentraxin-3, and P-selectin) present greater differences in early-onset PE, whereas activin-A in late-onset PE [62]. Inhibin-A and activin-A have been shown to be increased prior to 14 weeks in PE pregnancies [63, 64].

Placental Protein-13 (PP-13)

PP-13 is a relatively small, 32-kDa dimer protein. It is highly expressed in the placenta. It probably has an immunobiological function at the feto-maternal interface and in maternal vascular remodeling. The levels of PP-13 in serum gradually increase in normal pregnancy, but abnormally low levels of PP-13 were detected in first trimester serum samples of women who subsequently developed PE [65]. Furthermore, it was reported that first-trimester serum PP-13 levels may serve as a marker for early onset PE (before 34 weeks of gestation) only, but not for severe PE. Combined measuring of maternal serum PP-13 and median uterine artery pulsatility index by using ultrasound early in pregnancy seems to predict severe PE [66].

Increased shedding of subcelluar necrotic microparticles (STBM) is most likely a source of high concentration of PP13 into maternal blood as PE progresses. The severity of the signs of PE is proportional to the increase of PP13 from first to third trimester [67].

A Disintegrin and Metalloprotease 12 (ADAM 12)

ADAM12 is the protease for insulin-like growth factor binding proteins. Low levels of ADAM12 reflect an increased amount of insulin-like growth factor in the bound state, and this is then unavailable to promote placental growth and development [68]. Studies on ADAM12 and PE are discordant [69, 70]. Spencer *et al.*, demonstrated only a modest predictive efficiency of ADAM12 for PE with an AUC of 0.694 for ADAM12 alone and an AUC of 0.714 when ADAM12 and PAPP-A were combined [70].

Renin Angiotensin System (RAS)

One of the most important regulators to control blood pressure are auto antibodies against angiotensin II Type 1 (AT1) receptor RAS, especially for long term control of blood pressure. In addition, RAS has been implicated in vascular remodelling, inflammation and tumour development [71].

Normal pregnancy is characterized by resistance to the vasoconstrictive effects of angiotensin II. In PE, there is increased sensitivity to angiotensin II as compared to that seen in normotensive pregnant women. The angiotensin receptor, AT1 is a G protein-coupled receptor (GPCR) for angiotensin II, whose signalling leads to strong vasoconstriction [72]. The enhanced activation of AT1 induces hypertension, oedema and proteinuria. There seems to be at least two mechanisms which operate in PE, that accelerate AT1 signalling. These are: (1) formation of AT1-bradykinin B2 heterodimers [73] and (2) agonistic autoimmune antibody against AT1 (AT1-AA). Interestingly, increased levels of AT1-AA are found in PE [74]. AT1-AA may also contribute to the development of hypertension in later life, as its increased levels are observed in some women with a history of PE, even after their deliveries. Stimulation of AT1 receptor of cultured trophoblasts using IgG obtained from women with PE was found to result in the elevation of sFlt1 *in vitro*. Therefore, a close association is thought to exist between accelerated AT1 signalling and sFlt1 production. AT1-AA is detectable in fetal cord blood in PE pregnancies, which is suggestive of its usage as a fetal-side marker for evaluating other fetal conditions [75].

In addition, a new form of oxidized angiotensinogen has been found in the circulation of PE subjects, which enhances the formation of angiotensin [76]. Surprisingly, circulating angiotensin II and aldosterone are suppressed in PE subjects. Studies are needed to evaluate whether this oxidized form of angiotensinogen is altered before clinical disease.

Vifastin

Visfatin is an adipokine which is secreted by adipose tissue and which is involved in the biosynthesis of nicotinamide adenine dinucleotide, as it catalyzes the condensation of nicotinamide with 5-phosphoribosyl-1-pyrophosphate to yield nicotinamide mononucleotide. It is involved in glucose homeostasis. It has been associated with various pathologies such as type-2 diabetes mellitus, obesity, and gestational diabetes mellitus. It is expressed in the placenta and myometrium.

There is a correlation between decreased levels of vifastin and the development of PE as well as with the severity of the disease [77]. Although there are other conflicting studies linking increased levels of vifastin with the development of PE [78]. Therefore, larger scale studies are required to evaluate the role of visfatin as a potential marker for PE.

Cystatin C

Cystatin C is a marker of renal function that increases when glomerular filtration decreases. It is expressed in the placenta. Increased levels of cystatin C have been observed in women with PE [79, 80].

Median cystatin C concentrations in the first trimester of pregnancy are significantly higher in women who subsequently develop PE (median, 0.65 mg/L) compared with those with a normal pregnancy (median, 0.57 mg/L, $p = 0.0001$) [80].

Pentraxin 3

Pentraxin 3 (tumor necrosis factor-stimulated gene-14) belongs to the same family as C-reactive protein and serum amyloid P component. Pentraxin 3 consists of 381 amino acids [81]. Serum levels of Pentraxin 3 are increased in PE due to the inflammatory response that originates in this pathology. It occurs in several tissues (fibroblasts, mononuclear phagocytes, vascular endothelial cells and smooth muscle cells) [82]. The levels of Pentraxin 3 have been found to be increased in patients with PE compared with normal pregnancies. Further longitudinal studies throughout pregnancy may be warranted to determine whether or not pentraxin 3 will be a useful early marker for PE.

P-Selectin

P-selectin is a member of the selectin family of cell surface adhesion molecules. P-selectin is expressed by platelets and endothelial cells upon activation. This cell surface adhesion molecule plays a crucial role both in inflammatory reactions by supporting recruitment and activation of circulating leucocytes and in coagulation through generation of leukocyte-derived “bloodborne” tissue factor [83].

P-selectin is rapidly shed from the cellular membrane of activated platelets, and this release is suggested to contribute to most of the soluble isoform of the molecule that is found in plasma. PE is associated with great platelet activation [84]. P-selectin-exposing micro-particles with procoagulant activity, released from activated platelets, have been detected in the peripheral blood of women with PE [85]. Higher serum levels of P-selectin were observed in patients with PE [86]. Another study did not show significant differences in P-selectin levels in normotensive patients and PE [87]. Therefore, further studies are needed before using P-selectin as a marker predictor of PE.

Fetal Hemoglobin (HbF)

HbF is being considered a new predictor of PE. Centlow *et al.*, found an up regulation of HbF genes and accumulation of extracellular HbF in the vascular lumen in PE placentas [88]. Furthermore, the heme scavenger and antioxidant alpha (1) - microglobulin (A1M) increases in parallel with fetal haemoglobin [89]. Endothelial dysfunction, hypertension and proteinuria occur as a result of a defect in placental haematopoiesis [89].

Anderson *et al.*, observed increased levels of HbF in the first and second trimester of gestation in patients with PE [90].

Cell-Free DNA (cfDNA)

Human fetal DNA provokes *in vitro* activation of NF-κB, with resulting proinflammatory interleukin-6 production in both a human B-cell line and in peripheral blood mononuclear cells from both pregnant and nonpregnant donors. Administration of human fetal (but not adult) DNA into pregnant BALB/c mice provokes increased tumor necrosis factor-α and IL-6 [91].

Fetal DNA originates from the placenta, and placental-specific messenger RNA molecules are also easily detected in maternal plasma. There is a positive correlation between Fetal DNA and the development of PE [92].

Placental DNA present in the maternal circulation could be responsible for the systemic response that originates in PE [93].

Studies are being conducted to evaluate cffDNA as a predictor of PE along with other biomarkers (*e.g.*, P-selectin, PAPP-A, PP-13, sFlt-1, sEng, and PlGF) [94].

Papantoniou *et al.*, reported that cfDNA and free fetal DNA (cffDNA) levels from blood samples obtained at 11-13 weeks of gestation. They were significantly increased in women who developed PE compared to those with uncomplicated pregnancies (median cfDNA: 9402 *vs.* 2698 g/mL; median cFFDNA: 934.5 *vs.* 62 gEq/mL, respectively). Following operating characteristic curve analysis, cutoff values of 7486 g/ml for cfDNA and 512 g/ml for cffDNA were chosen. These provided a sensitivity of 75% and 100% and a specificity of 98% and 100%, respectively, to identify women at risk for PE [95].

PROTEOMICS

The proteome is the total complement of proteins present in any defined biological compartment such as a whole organism, a cell, an organelle, or a fluid such as blood, Misfolded serpin proteins are detected in the placenta of preeclamptic subjects. Women with PE who required delivery exhibited a unique proteomic profile in the urine consisting of nonrandom fragments of SERPINA1 and albumin [96]. This profile was also a better predictor compared with the sFlt1:PlGF ratio and urine protein:creatinine ratio. Chen *et al.*, also identified 31 proteins that are differentially expressed in women with pre-eclampsia and gestational hypertension as compared with normal pregnancy *via* proteomics [97]. These proteins played a role in coagulation, cell adhesion, and immune response. The three most significant markers were angiotensinogen, albumin, and SERPINA1. It is interesting to note that levels of SERPINA1 and albumin are upregulated in women with PE, but they are down regulated in women with gestational hypertension.

This suggests that these proteins may be used to differentiate various hypertensive disorders of pregnancy. On the other hand, urinary angiotensinogen levels are significantly lower in PE and gestational hypertension when compared with normal pregnancy [97]. Carty *et al.*, discovered a urinary proteome consisting of fragments of collagen, fibrinogen, and uromodulin, which predicted better than sFlt-1 and PlGF [98]. However, their 100% sensitivity and specificity occurred only at week 28 of gestation in women with 2 risk factors. It is interesting to note that the uromodulin gene, which encodes for tamm-horsfall protein, the most abundant protein that naturally occurs in the urine, is thought to play a role in protecting against inflammation and infection [99].

Urinary proteomics has been used to identify biomarkers for PE more than 10 weeks before clinical presentation. Two such markers are fragments of SERPINA1 and albumin. SERPINA1 is serine protease inhibitor and is synthesized in many cell types including trophoblasts; increased levels have been found in inflammatory conditions such as vasculitis and cardiovascular disease [100].

METABOLOMICS

Metabolomics is a rapidly growing technology to characterize the complete collection of metabolites or small molecules found in an organism or in its cells, tissues, and biofluids [101]. Odibo *et al.*, described the presence of 4 metabolites (hydroxyhexanoylcarnitine, alanine, phenylalanine, and glutamate) that showed significant increases in PE. However, the ability of individual markers to predict PE was approximately 70% to 80% and this was associated with false-positive rates of almost 20% [102].

A different set of metabolites (citrate, glycerol, hydroxyisovalerate, and methionine) that were identified in the first trimester predicted early PE with a sensitivity of 75% and a false-positive rate of less than 5%. The sensitivity increased to approximately 80% and the false-positive rate decreased to less than 2% when these biomarkers were used in combination with uterine artery Doppler pulsatility index and fetal crown-rump length [103].

DISCUSSION

First Trimester Combined Screening

Of the markers studied to predict PE, PAPP-A and PlGF show a high predictive value of disease [35, 62, 104 - 106]. Three studies, derived from prospective first-trimester screening for adverse obstetric outcomes in the UK by the Fetal Medicine Foundation, have reported the superiority of multiple biomarkers in the prediction of PE. The first study, which combines PAPP-A, PIGF, PP-13, inhibin-A, activin-A, sEng, pentraxin-3, and P-selectin levels, maternal characteristics, MAP, uterine artery pulsatility index (PI), were obtained from case-control studies. Algorithms that combine maternal characteristics and biophysical and biochemical tests at 11–13 weeks' gestation could potentially identify approximately 90%, 80%, and 60% of pregnancies that subsequently develop early (<34 weeks), intermediate (34–36 weeks), and late (≥37 weeks) PE, respectively, at the false-positive rate of 5% [62].

The other two studies combined maternal characteristics, uterine artery PI, MAP, and levels of PAPP-A and PlGF determined at 11-13 weeks of gestation. This combination predicted a large number of patients who developed PE [107, 108].

In the second study, screening by maternal characteristics with the uterine artery PI and MAP detected 90% of PE cases requiring delivery before 34 weeks' gestation and 57% of all PE cases at a fixed false-positive rate of 10% [107]. In the third study, in screening for PE requiring delivery before 34 weeks' gestation, the detection rate at a false-positive rate of 10% was approximately 50% by maternal characteristics, and improved to approximately 90% by the addition of biophysical markers and to approximately 75% by the addition of biochemical markers [108]. This detection rate improved to more than 95% by an algorithm combining maternal factors, biophysical markers, and biochemical markers (PAPP-A and PlGF).

Recents Developments

In the last years, a large number of studies have been carried out to select bio-markers to predict PE. These biomarkers must contradistinguish between the different hypertensive disorders of pregnancy. The utility of angiogenic factors in the prediction of PE has been evaluated. The diagnostic sensitivity and specificity of sFlt-1 for differentiation of PE from gestational and chronic hypertension were 84% and 95% respectively [109].

SFlt-1 measurements in plasma showed 89% diagnostic sensitivity and 90% specificity in early PE (at <34 weeks) as compared to 55% diagnostic sensitivity and 58% specificity in late PE (at> 37 weeks). The sFlt-1/PlGF ratio shows a better predictive value than if both markers were used separately [110].

Further studies are required to develop tools for early detection and management of PE. However, question remains whether a single set of parameters are useful for diagnosing a disease entity which is as complex as PE. Thus, it is necessary to design a set of predictive and feasible tests using multiple biomarkers.

CONCLUSION

In recent years, there has been a decrease in mortality and morbidity caused by PE due to better management and diagnosis of the disease. However, challenge lies in its early diagnosis, when there are no apparent clinical signs. Therefore, an early intervention could be initiated and maternal mortality and morbidity could be reduced substantially, even in the developing countries. Till date, several promising biomarkers have been reported, which could be used to make an early diagnosis. However, large-scale, multicenter, multi-ethnic, prospective trials with women considering different risks of developing PE are required to propose an ideal combination of markers for routine screening.

CONFLICT OF INTEREST

The author (editor) declares no conflict of interest, financial or otherwise.

ACKNOWLEDGEMENTS

Declared none.

REFERENCES

[1] Ananth CV, Keyes KM, Wapner RJ. Pre-eclampsia rates in the United States, 1980-2010: age-perio--cohort analysis. BMJ 2013; 347: f6564.
[http://dx.doi.org/10.1136/bmj.f6564]

[2] Venkatesha S, Toporsian M, Lam C, *et al.* Soluble endoglin contributes to the pathogenesis of pre-eclampsia. Nat Med 2006; 12: 642-9.
[http://dx.doi.org/10.1038/nm1429]

[3] Tranquilli AL, Dekker G, Magee L, *et al.* The classification, diagnosis and management of the hypertensive disorders of pregnancy: A revised statement from the ISSHP. Pregnancy Hypertens 2014; 4: 97-104.
[http://dx.doi.org/10.1016/j.preghy.2014.02.001]

[4] Magee LA, Helewa M, Moutquin JM, Von DP. Diagnosis, evaluation, and management of the hypertensive disorders of pregnancy. J Obstet Gynaecol Can 2008; 30: S1-S48.
[http://dx.doi.org/10.1016/S1701-2163(16)32776-1]

[5] Barton JR, Sibai BM. Prediction and prevention of recurrent pre-eclampsia. Obstet Gynecol 2008; 112: 359-72.
[http://dx.doi.org/10.1097/AOG.0b013e3181801d56]

[6] Duckitt K, Harrington D. Risk factors for pre-eclampsia at antenatal booking: systematic review of controlled studies. BMJ 2005; 330: 565. [http://dx.doi.org/10.1136/bmj.38380.674340.E0]

[7] Sircar M, Thadhani R, Karumanchi SA. Pathogenesis of pre-eclampsia. Curr Opin Nephrol Hypertens 2015; 24: 131-8. [http://dx.doi.org/10.1097/MNH.0000000000000105]

[8] Bramham K, Parnell B, Nelson-Piercy C, Seed PT, Poston L, Chappell LC. Chronic hypertension and pregnancy outcomes: systematic review and meta-analysis. BMJ 2014; 348: g2301. [http://dx.doi.org/10.1136/bmj.g2301]

[9] Roos N, Kieler H, Sahlin L, Ekman-Ordeberg G, Falconer H, Stephansson O. Risk of adverse pregnancy outcomes in women with polycystic ovary syndrome: population based cohort study. BMJ 2011; 343: d6309. [http://dx.doi.org/10.1136/bmj.d6309]

[10] Zhong Y, Tuuli M, Odibo AO. First-trimester assessment of placenta function and the prediction of pre-eclampsia and intrauterine growth restriction. Prenat Diagn 2010; 30: 293-308.

[11] Scazzocchio E, Figueras F. Contemporary prediction of pre-eclampsia. Curr Opin Obstet Gynecol 2011; 23: 65-71. [http://dx.doi.org/10.1097/GCO.0b013e328344579c]

[12] Harrington K. Early screening for pre-eclampsia and intrauterine growth restriction. Ultrasound Obstet Gynecol 2011; 37: 623-4. [http://dx.doi.org/10.1002/uog.9018]

[13] Akolekar R, Syngelaki A, Beta J, Kocylowski R, Nicolaides KH. Maternal serum placental protein 13 at 11-13 weeks of gestation in pre-eclampsia. Prenat Diagn 2009; 29: 1103-8. [http://dx.doi.org/10.1002/pd.2375]

[14] Cuckle HS. Screening for pre-eclampsia--lessons from aneuploidy screening. Placenta 2011; 32 (Suppl.): S42-8. [http://dx.doi.org/10.1016/j.placenta.2010.07.015]

[15] Brown MA, Hague WM, Higgins J, *et al.* The detection, investigation and management of hypertension in pregnancy: executive summary. Aust N Z J Obstet Gynaecol 2000; 40: 133-8. [http://dx.doi.org/10.1111/j.1479-828X.2000.tb01136.x]

[16] Sibai BM. Diagnosis and management of gestational hypertension and pre-eclampsia. Obstet Gynecol 2003; 102: 181-92.

[17] Hauth JC, Ewell MG, Levine RJ, *et al.* Pregnancy outcomes in healthy nulliparas who developed hypertension. Calcium for pre-eclampsia Prevention Study Group. Obstet Gynecol 2000; 95: 24-8.

[18] Askie LM, Duley L, Henderson-Smart DJ, Stewart LA. Antiplatelet agents for prevention of pre-eclampsia: a meta-analysis of individual patient data. Lancet 2007; 369: 1791-8. [http://dx.doi.org/10.1016/S0140-6736(07)60712-0]

[19] Rodger MA, Hague WM, Kingdom J, *et al.* Antepartum dalteparin *versus* no antepartum dalteparin for the prevention of pregnancy complications in pregnant women with thrombophilia (TIPPS): a multinational open-label randomised trial. Lancet 2014; 384: 1673-83. [http://dx.doi.org/10.1016/S0140-6736(14)60793-5]

[20] Dodd JM, McLeod A, Windrim RC, Kingdom J. Antithrombotic therapy for improving maternal or infant health outcomes in women considered at risk of placental dysfunction. Cochrane Database Syst Rev 2013; 7: CD006780.

[21] Ephraim RK, Osakunor DN, Denkyira SW, Eshun H, Amoah S, Anto EO. Serum calcium and magnesium levels in women presenting with pre-eclampsia and pregnancy-induced hypertension: a case-control study in the Cape Coast metropolis, Ghana. BMC Pregnancy Childbirth 2014; 14: 390.

[http://dx.doi.org/10.1186/s12884-014-0390-2]

[22] Hofmeyr GJ, Lawrie TA, Atallah AN, Duley L, Torloni MR. Calcium supplementation during pregnancy for preventing hypertensive disorders and related problems. Cochrane Database Syst Rev 2014; 6: CD001059.
[http://dx.doi.org/10.1002/14651858.CD001059.pub4]

[23] Hofmeyr GJ, Duley L, Atallah A. Dietary calcium supplementation for prevention of pre-eclampsia and related problems: a systematic review and commentary. BJOG 2007; 114: 933-43.
[http://dx.doi.org/10.1111/j.1471-0528.2007.01389.x]

[24] Hofmeyr GJ, Seuc AH, Betran AP, *et al.* The effect of calcium supplementation on blood pressure in non-pregnant women with previous pre-eclampsia: an exploratory, randomized placebo controlled study. Pregnancy Hypertens 2015; 5: 273-9.
[http://dx.doi.org/10.1016/j.preghy.2015.04.001]

[25] Conde-Agudelo A, Romero R, Kusanovic JP, Hassan SS. Supplementation with vitamins C and E during pregnancy for the prevention of pre-eclampsia and other adverse maternal and perinatal outcomes: a systematic review and metaanalysis. Am J Obstet Gynecol 2011; 204: 503-12.
[http://dx.doi.org/10.1016/j.ajog.2011.02.020]

[26] Makrides M, Crosby DD, Bain E, Crowther CA. Magnesium supplementation in pregnancy. Cochrane Database Syst Rev 2014; 4: CD000937.
[http://dx.doi.org/10.1002/14651858.CD000937.pub2]

[27] Aghajafari F, Nagulesapillai T, Ronksley PE, Tough SC, O'Beirne M, Rabi DM. Association between maternal serum 25-hydroxyvitamin D level and pregnancy and neonatal outcomes: systematic review and meta-analysis of observational studies. BMJ 2013; 346: f1169.
[http://dx.doi.org/10.1136/bmj.f1169]

[28] De-Regil LM, Palacios C, Ansary A, Kulier R, Pena-Rosas JP. Vitamin D supplementation for women during pregnancy. Cochrane Database Syst Rev 2012; 2: CD008873.
[http://dx.doi.org/10.1002/14651858.CD008873.pub2]

[29] Dorniak-Wall T, Grivell RM, Dekker GA, *et al.* The role of L-arginine in the prevention and treatment of pre-eclampsia: a systematic review of randomised trials. J Hum Hypertens 2014; 28: 230-5.
[http://dx.doi.org/10.1038/jhh.2013.100]

[30] Vadillo-Ortega F, Perichart-Perera O, Espino S, *et al.* Effect of supplementation during pregnancy with L-arginine and antioxidant vitamins in medical food on pre-eclampsia in high risk population: randomised controlled trial. BMJ 2011; 342: d2901.
[http://dx.doi.org/10.1136/bmj.d2901]

[31] Allen R, Rogozinska E, Sivarajasingam P, Khan KS, Thangaratinam S. Effect of diet- and lifestyle-based metabolic risk-modifying interventions on pre-eclampsia: a meta-analysis. Acta Obstet Gynecol Scand 2014; 93: 973-85.
[http://dx.doi.org/10.1111/aogs.12467]

[32] Thangaratinam S, Rogozinska E, Jolly K, *et al.* Effects of interventions in pregnancy on maternal weight and obstetric outcomes: meta-analysis of randomised evidence. BMJ 2012; 344: e2088.
[http://dx.doi.org/10.1136/bmj.e2088]

[33] Dodd JM, Turnbull D, McPhee AJ, *et al.* Antenatal lifestyle advice for women who are overweight or obese: LIMIT randomised trial. BMJ 2014; 348: g1285.
[http://dx.doi.org/10.1136/bmj.g1285]

[34] Tjwa M, Luttun A, Autiero M, Carmeliet P. VEGF and PlGF: two pleiotropic growth factors with distinct roles in development and homeostasis. Cell Tissue Res 2003; 314: 5-14.
[http://dx.doi.org/10.1007/s00441-003-0776-3]

[35] Tsiakkas A, Duvdevani N, Wright A, Wright D, Nicolaides KH. Serum placental growth factor in the three trimesters of pregnancy: effects of maternal characteristics and medical history. Ultrasound

Obstet Gynecol 2015; 45: 591-8.
[http://dx.doi.org/10.1002/uog.14811]

[36] Levine RJ, Karumanchi SA. Circulating angiogenic factors in pre-eclampsia. Clin Obstet Gynecol 2005; 48: 372-86.
[http://dx.doi.org/10.1097/01.grf.0000160313.82606.d7]

[37] Masuyama H, Segawa T, Sumida Y, *et al.* Different profiles of circulating angiogenic factors and adipocytokines between early- and late-onset pre-eclampsia. BJOG 2010; 117: 314-20.
[http://dx.doi.org/10.1111/j.1471-0528.2009.02453.x]

[38] Kusanovic JP, Romero R, Chaiworapongsa T, *et al.* A prospective cohort study of the value of maternal plasma concentrations of angiogenic and anti-angiogenic factors in early pregnancy and midtrimester in the identification of patients destined to develop pre-eclampsia. J Matern Fetal Neonatal Med 2009; 22: 1021-38.
[http://dx.doi.org/10.3109/14767050902994754]

[39] Romero R, Nien JK, Espinoza J, *et al.* A longitudinal study of angiogenic (placental growth factor) and anti-angiogenic (soluble endoglin and soluble vascular endothelial growth factor receptor-1) factors in normal pregnancy and patients destined to develop pre-eclampsia and deliver a small for gestational age neonate. J Matern Fetal Neonatal Med 2008; 21: 9-23.
[http://dx.doi.org/10.1080/14767050701830480]

[40] Taylor RN, Grimwood J, Taylor RS, McMaster MT, Fisher SJ, North RA. Longitudinal serum concentrations of placental growth factor: evidence for abnormal placental angiogenesis in pathologic pregnancies. Am J Obstet Gynecol 2003; 188: 177-82.
[http://dx.doi.org/10.1067/mob.2003.111]

[41] Powers RW, Roberts JM, Plymire DA, *et al.* Low placental growth factor across pregnancy identifies a subset of women with preterm pre-eclampsia: type 1 *versus* type 2 pre-eclampsia? Hypertension 2012; 60: 239-46.
[http://dx.doi.org/10.1161/HYPERTENSIONAHA.112.191213]

[42] Kar M. Role of biomarkers in early detection of pre-eclampsia. J Clin Diagn Res 2014; 8: BE01-4.
[http://dx.doi.org/10.7860/JCDR/2014/7969.4261]

[43] Maynard SE, Min JY, Merchan J, *et al.* Excess placental soluble fms-like tyrosine kinase 1 (sFlt1) may contribute to endothelial dysfunction, hypertension, and proteinuria in pre-eclampsia. J Clin Invest 2003; 111: 649-58.
[http://dx.doi.org/10.1172/JCI17189]

[44] Staff AC, Benton SJ, Von DP, *et al.* Redefining pre-eclampsia using placenta-derived biomarkers. Hypertension 2013; 61: 932-42.
[http://dx.doi.org/10.1161/HYPERTENSIONAHA.111.00250]

[45] Staff AC, Braekke K, Harsem NK, Lyberg T, Holthe MR. Circulating concentrations of sFlt1 (soluble fms-like tyrosine kinase 1) in fetal and maternal serum during pre-eclampsia. Eur J Obstet Gynecol Reprod Biol 2005; 122: 33-9.
[http://dx.doi.org/10.1016/j.ejogrb.2004.11.015]

[46] Haggerty CL, Seifert ME, Tang G, *et al.* Second trimester anti-angiogenic proteins and pre-eclampsia. Pregnancy Hypertens 2012; 2: 158-63.
[http://dx.doi.org/10.1016/j.preghy.2012.01.005]

[47] Sezer SD, Kucuk M, Yenisey C, *et al.* Comparison of angiogenic and anti-angiogenic factors in maternal and umbilical cord blood in early- and late-onset pre-eclampsia. Gynecol Endocrinol 2012; 28: 628-32.
[http://dx.doi.org/10.3109/09513590.2011.650759]

[48] Aggarwal PK, Chandel N, Jain V, Jha V. The relationship between circulating endothelin-1, soluble fms-like tyrosine kinase-1 and soluble endoglin in pre-eclampsia. J Hum Hypertens 2012; 26: 236-41.
[http://dx.doi.org/10.1038/jhh.2011.29]

[49] Vatten LJ, Eskild A, Nilsen TI, Jeansson S, Jenum PA, Staff AC. Changes in circulating level of angiogenic factors from the first to second trimester as predictors of pre-eclampsia. Am J Obstet Gynecol 2007; 196: 239-6.
[http://dx.doi.org/10.1016/j.ajog.2006.10.909]

[50] Vatten LJ, Asvold BO, Eskild A. Angiogenic factors in maternal circulation and pre-eclampsia with or without fetal growth restriction. Acta Obstet Gynecol Scand 2012; 91: 1388-94.
[http://dx.doi.org/10.1111/j.1600-0412.2012.01516.x]

[51] Myatt L, Clifton RG, Roberts JM, *et al.* First-trimester prediction of pre-eclampsia in nulliparous women at low risk. Obstet Gynecol 2012; 119: 1234-42.
[http://dx.doi.org/10.1097/AOG.0b013e3182571669]

[52] Erez O, Romero R, Espinoza J, *et al.* The change in concentrations of angiogenic and anti-angiogenic factors in maternal plasma between the first and second trimesters in risk assessment for the subsequent development of pre-eclampsia and small-for-gestational age. J Matern Fetal Neonatal Med 2008; 21: 279-87.
[http://dx.doi.org/10.1080/14767050802034545]

[53] Bonno M, Oxvig C, Kephart GM, *et al.* Localization of pregnancy-associated plasma protein-A and colocalization of pregnancy-associated plasma protein-A messenger ribonucleic acid and eosinophil granule major basic protein messenger ribonucleic acid in placenta. Lab Invest 1994; 71: 560-6.

[54] Lawrence JB, Oxvig C, Overgaard MT, *et al.* The insulin-like growth factor (IGF)-dependent IGF binding protein-4 protease secreted by human fibroblasts is pregnancy-associated plasma protein-A. Proc Natl Acad Sci USA 1999; 96: 3149-53.
[http://dx.doi.org/10.1073/pnas.96.6.3149]

[55] Bersinger NA, Smarason AK, Muttukrishna S, Groome NP, Redman CW. Women with pre-eclampsia have increased serum levels of pregnancy-associated plasma protein A (PAPP-A), inhibin A, activin A and soluble E-selectin. Hypertens Pregnancy 2003; 22: 45-55.
[http://dx.doi.org/10.1081/PRG-120016794]

[56] Spencer K, Cowans NJ, Chefetz I, Tal J, Meiri H. First-trimester maternal serum PP-13, PAPP-A and second-trimester uterine artery Doppler pulsatility index as markers of pre-eclampsia. Ultrasound Obstet Gynecol 2007; 29: 128-34.
[http://dx.doi.org/10.1002/uog.3876]

[57] Mutter WP, Karumanchi SA. Molecular mechanisms of pre-eclampsia. Microvasc Res 2008; 75: 1-8.
[http://dx.doi.org/10.1016/j.mvr.2007.04.009]

[58] Robinson CJ, Johnson DD. Soluble endoglin as a second-trimester marker for pre-eclampsia. Am J Obstet Gynecol 2007; 197: 174-5.
[http://dx.doi.org/10.1016/j.ajog.2007.03.058]

[59] Muttukrishna S, North RA, Morris J, *et al.* Serum inhibin A and activin A are elevated prior to the onset of pre-eclampsia. Hum Reprod 2000; 15: 1640-5.
[http://dx.doi.org/10.1093/humrep/15.7.1640]

[60] Muttukrishna S, Knight PG, Groome NP, Redman CW, Ledger WL. Activin A and inhibin A as possible endocrine markers for pre-eclampsia. Lancet 1997; 349: 1285-8.
[http://dx.doi.org/10.1016/S0140-6736(96)09264-1]

[61] Mandang S, Manuelpillai U, Wallace EM. Oxidative stress increases placental and endothelial cell activin A secretion. J Endocrinol 2007; 192: 485-93.
[http://dx.doi.org/10.1677/JOE-06-0061]

[62] Akolekar R, Syngelaki A, Sarquis R, Zvanca M, Nicolaides KH. Prediction of early, intermediate and late pre-eclampsia from maternal factors, biophysical and biochemical markers at 11-13 weeks. Prenat Diagn 2011; 31: 66-74.
[http://dx.doi.org/10.1002/pd.2660]

[63] Salomon LJ, Benattar C, Audibert F, *et al.* Severe pre-eclampsia is associated with high inhibin A levels and normal leptin levels at 7 to 13 weeks into pregnancy. Am J Obstet Gynecol 2003; 189: 1517-22.
[http://dx.doi.org/10.1016/S0002-9378(03)00902-5]

[64] Ong CY, Liao AW, Munim S, Spencer K, Nicolaides KH. First-trimester maternal serum activin A in pre-eclampsia and fetal growth restriction. J Matern Fetal Neonatal Med 2004; 15: 176-80.
[http://dx.doi.org/10.1080/14767050410001668275]

[65] Burger O, Pick E, Zwickel J, *et al.* Placental protein 13 (PP-13): effects on cultured trophoblasts, and its detection in human body fluids in normal and pathological pregnancies. Placenta 2004; 25: 608-22.
[http://dx.doi.org/10.1016/j.placenta.2003.12.009]

[66] Nicolaides KH, Bindra R, Turan OM, *et al.* A novel approach to first-trimester screening for early pre-eclampsia combining serum PP-13 and Doppler ultrasound. Ultrasound Obstet Gynecol 2006; 27: 13-7.
[http://dx.doi.org/10.1002/uog.2686]

[67] Huppertz B, Sammar M, Chefetz I, Neumaier-Wagner P, Bartz C, Meiri H. Longitudinal determination of serum placental protein 13 during development of pre-eclampsia. Fetal Diagn Ther 2008; 24: 230-6.
[http://dx.doi.org/10.1159/000151344]

[68] Shi Z, Xu W, Loechel F, Wewer UM, Murphy LJ. ADAM 12, a disintegrin metalloprotease, interacts with insulin-like growth factor-binding protein-3. J Biol Chem 2000; 275: 18574-80.
[http://dx.doi.org/10.1074/jbc.M002172200]

[69] Laigaard J, Sorensen T, Placing S, *et al.* Reduction of the disintegrin and metalloprotease ADAM12 in pre-eclampsia. Obstet Gynecol 2005; 106: 144-9.
[http://dx.doi.org/10.1097/01.AOG.0000165829.65319.65]

[70] Spencer K, Cowans NJ, Stamatopoulou A. ADAM12s in maternal serum as a potential marker of pre-eclampsia. Prenat Diagn 2008; 28: 212-6.
[http://dx.doi.org/10.1002/pd.1957]

[71] Suzuki Y, Ruiz-Ortega M, Lorenzo O, Ruperez M, Esteban V, Egido J. Inflammation and angiotensin II. Int J Biochem Cell Biol 2003; 35: 881-900.
[http://dx.doi.org/10.1016/S1357-2725(02)00271-6]

[72] Paul M, Poyan MA, Kreutz R. Physiology of local renin-angiotensin systems. Physiol Rev 2006; 86: 747-803.
[http://dx.doi.org/10.1152/physrev.00036.2005]

[73] bdAlla S, Lother H, el MA, Quitterer U: Increased AT(1) receptor heterodimers in pre-eclampsia mediate enhanced angiotensin II responsiveness. Nat Med 2001; 7: 1003-9.
[http://dx.doi.org/10.1038/nm0901-1003]

[74] Wallukat G, Homuth V, Fischer T, *et al.* Patients with pre-eclampsia develop agonistic autoantibodies against the angiotensin AT1 receptor. J Clin Invest 1999; 103: 945-52.
[http://dx.doi.org/10.1172/JCI4106]

[75] Irani RA, Zhang Y, Blackwell SC, *et al.* The detrimental role of angiotensin receptor agonistic autoantibodies in intrauterine growth restriction seen in pre-eclampsia. J Exp Med 2009; 206: 2809-22.
[http://dx.doi.org/10.1084/jem.20090872]

[76] Zhou A, Carrell RW, Murphy MP, *et al.* A redox switch in angiotensinogen modulates angiotensin release. Nature 2010; 468: 108-11.
[http://dx.doi.org/10.1038/nature09505]

[77] Hu W, Wang Z, Wang H, Huang H, Dong M. Serum visfatin levels in late pregnancy and pre-eclampsia. Acta Obstet Gynecol Scand 2008; 87: 413-8.
[http://dx.doi.org/10.1080/00016340801976012]

[78] Fasshauer M, Bluher M, Stumvoll M, Tonessen P, Faber R, Stepan H. Differential regulation of visfatin and adiponectin in pregnancies with normal and abnormal placental function. Clin Endocrinol (Oxf) 2007; 66: 434-9. [http://dx.doi.org/10.1111/j.1365-2265.2007.02751.x]

[79] Kristensen K, Larsson I, Hansson SR. Increased cystatin C expression in the pre-eclamptic placenta. Mol Hum Reprod 2007; 13: 189-95. [http://dx.doi.org/10.1093/molehr/gal111]

[80] Thilaganathan B, Ralph E, Papageorghiou AT, Melchiorre K, Sheldon J. Raised maternal serum cystatin C: an early pregnancy marker for pre-eclampsia. Reprod Sci 2009; 16: 788-93. [http://dx.doi.org/10.1177/1933719109336618]

[81] Souza DG, Soares AC, Pinho V, *et al.* Increased mortality and inflammation in tumor necrosis factor-stimulated gene-14 transgenic mice after ischemia and reperfusion injury. Am J Pathol 2002; 160: 1755-65. [http://dx.doi.org/10.1016/S0002-9440(10)61122-4]

[82] Cetin I, Cozzi V, Pasqualini F, *et al.* Elevated maternal levels of the long pentraxin 3 (PTX3) in pre-eclampsia and intrauterine growth restriction. Am J Obstet Gynecol 2006; 194: 1347-53. [http://dx.doi.org/10.1016/j.ajog.2005.11.018]

[83] Andre P. P-selectin in haemostasis. Br J Haematol 2004; 126: 298-306. [http://dx.doi.org/10.1111/j.1365-2141.2004.05032.x]

[84] Holthe MR, Staff AC, Berge LN, Lyberg T. Different levels of platelet activation in preeclamptic, normotensive pregnant, and nonpregnant women. Am J Obstet Gynecol 2004; 190: 1128-34. [http://dx.doi.org/10.1016/j.ajog.2003.10.699]

[85] Bretelle F, Sabatier F, Desprez D, *et al.* gnat-George F: Circulating microparticles: a marker of procoagulant state in normal pregnancy and pregnancy complicated by pre-eclampsia or intrauterine growth restriction. Thromb Haemost 2003; 89: 486-92.

[86] Aksoy H, Kumtepe Y, Akcay F, Yildirim AK. Correlation of P-selectin and lipoprotein(a), and other lipid parameters in pre-eclampsia. Clin Exp Med 2002; 2: 39-43. [http://dx.doi.org/10.1007/s102380200005]

[87] Nasrollahi S, Hoseini Panah SM, Tavilani H, Tavasoli S, Naderan M, Shoar S. Antioxidant status and serum levels of selectins in pre-eclampsia. J Obstet Gynaecol 2015; 35: 16-8. [http://dx.doi.org/10.3109/01443615.2014.935710]

[88] Centlow M, Carninci P, Nemeth K, Mezey E, Brownstein M, Hansson SR. Placental expression profiling in pre-eclampsia: local overproduction of hemoglobin may drive pathological changes. Fertil Steril 2008; 90: 1834-43. [http://dx.doi.org/10.1016/j.fertnstert.2007.09.030]

[89] Olsson MG, Centlow M, Rutardottir S, *et al.* Increased levels of cell-free hemoglobin, oxidation markers, and the antioxidative heme scavenger alpha(1)-microglobulin in pre-eclampsia. Free Radic Biol Med 2010; 48: 284-91. [http://dx.doi.org/10.1016/j.freeradbiomed.2009.10.052]

[90] Gram M, Anderson UD, Johansson ME, *et al.* The human endogenous protection system against cell-free hemoglobin and heme is overwhelmed in pre-eclampsia and provides potential biomarkers and clinical indicators. PLoS One 2015; 10: e0138111. [http://dx.doi.org/10.1371/journal.pone.0138111]

[91] Scharfe-Nugent A, Corr SC, Carpenter SB, *et al.* TLR9 provokes inflammation in response to fetal DNA: mechanism for fetal loss in preterm birth and pre-eclampsia. J Immunol 2012; 188: 5706-12. [http://dx.doi.org/10.4049/jimmunol.1103454]

[92] Cotter AM, Martin CM, O'Leary JJ, Daly SF. Increased fetal DNA in the maternal circulation in early pregnancy is associated with an increased risk of pre-eclampsia. Am J Obstet Gynecol 2004; 191: 515-

20.
[http://dx.doi.org/10.1016/j.ajog.2004.01.040]

[93] Tjoa ML, Cindrova-Davies T, Spasic-Boskovic O, Bianchi DW, Burton GJ. Trophoblastic oxidative stress and the release of cell-free feto-placental DNA. Am J Pathol 2006; 169: 400-4.
[http://dx.doi.org/10.2353/ajpath.2006.060161]

[94] Grill S, Rusterholz C, Zanetti-Dallenbach R, *et al.* Potential markers of pre-eclampsia--a review. Reprod Biol Endocrinol 2009; 7: 70.
[http://dx.doi.org/10.1186/1477-7827-7-70]

[95] Papantoniou N, Bagiokos V, Agiannitopoulos K, *et al.* RASSF1A in maternal plasma as a molecular marker of pre-eclampsia. Prenat Diagn 2013; 33: 682-7.
[http://dx.doi.org/10.1002/pd.4093]

[96] Buhimschi IA, Zhao G, Funai EF, *et al.* Proteomic profiling of urine identifies specific fragments of SERPINA1 and albumin as biomarkers of pre-eclampsia. Am J Obstet Gynecol 2008; 199: 551-16.
[http://dx.doi.org/10.1016/j.ajog.2008.07.006]

[97] Chen G, Zhang Y, Jin X, *et al.* Urinary proteomics analysis for renal injury in hypertensive disorders of pregnancy with iTRAQ labeling and LC-MS/MS. Proteomics Clin Appl 2011; 5: 300-10.
[http://dx.doi.org/10.1002/prca.201000100]

[98] Carty DM, Siwy J, Brennand JE, *et al.* Urinary proteomics for prediction of pre-eclampsia. Hypertension 2011; 57: 561-9.
[http://dx.doi.org/10.1161/HYPERTENSIONAHA.110.164285]

[99] Kottgen A, Glazer NL, Dehghan A, *et al.* Multiple loci associated with indices of renal function and chronic kidney disease. Nat Genet 2009; 41: 712-7.

[100] Engstrom G, Janzon L, Berglund G, *et al.* Blood pressure increase and incidence of hypertension in relation to inflammation-sensitive plasma proteins. Arterioscler Thromb Vasc Biol 2002; 22: 2054-8.
[http://dx.doi.org/10.1161/01.ATV.0000041842.43905.F3]

[101] Wishart DS, Knox C, Guo AC, *et al.* HMDB: a knowledgebase for the human metabolome. Nucleic Acids Res 2009; 37: D603-10.
[http://dx.doi.org/10.1093/nar/gkn810]

[102] Odibo AO, Goetzinger KR, Odibo L, *et al.* First-trimester prediction of pre-eclampsia using metabolomic biomarkers: a discovery phase study. Prenat Diagn 2011; 31: 990-4.
[http://dx.doi.org/10.1002/pd.2822]

[103] Bahado-Singh RO, Akolekar R, Mandal R, *et al.* Metabolomics and first-trimester prediction of early-onset pre-eclampsia. J Matern Fetal Neonatal Med 2012; 25: 1840-7.
[http://dx.doi.org/10.3109/14767058.2012.680254]

[104] Smith GC, Stenhouse EJ, Crossley JA, Aitken DA, Cameron AD, Connor JM. Early pregnancy levels of pregnancy-associated plasma protein a and the risk of intrauterine growth restriction, premature birth, pre-eclampsia, and stillbirth. J Clin Endocrinol Metab 2002; 87: 1762-7.
[http://dx.doi.org/10.1210/jcem.87.4.8430]

[105] Poon LC, Akolekar R, Lachmann R, Beta J, Nicolaides KH. Hypertensive disorders in pregnancy: screening by biophysical and biochemical markers at 11-13 weeks. Ultrasound Obstet Gynecol 2010; 35: 662-70.
[http://dx.doi.org/10.1002/uog.7628]

[106] Hromadnikova I, Kotlabova K, Doucha J, Dlouha K, Krofta L. Absolute and relative quantification of placenta-specific microRNA in maternal circulation with placental insufficiency-related complications. J Mol Diagn 2012; 14: 160-7.
[http://dx.doi.org/10.1016/j.jmoldx.2011.11.003]

[107] Wright D, Akolekar R, Syngelaki A, Poon LC, Nicolaides KH. A competing risks model in early screening for pre-eclampsia. Fetal Diagn Ther 2012; 32: 171-8.

[http://dx.doi.org/10.1159/000338470]

[108] Akolekar R, Syngelaki A, Poon L, Wright D, Nicolaides KH. Competing risks model in early screening for pre-eclampsia by biophysical and biochemical markers. Fetal Diagn Ther 2013; 33: 8-15. [http://dx.doi.org/10.1159/000341264]

[109] Salahuddin S, Lee Y, Vadnais M, Sachs BP, Karumanchi SA, Lim KH. Diagnostic utility of soluble fms-like tyrosine kinase 1 and soluble endoglin in hypertensive diseases of pregnancy. Am J Obstet Gynecol 2007; 197: 28-6. [http://dx.doi.org/10.1016/j.ajog.2007.04.010]

[110] Levine RJ, Lam C, Qian C, *et al.* Soluble endoglin and other circulating antiangiogenic factors in pre-eclampsia. N Engl J Med 2006; 355: 992-1005. [http://dx.doi.org/10.1056/NEJMoa055352]

CHAPTER 2

Metabolic Syndrome and Inflammation: Interrelated Aspects and Biomarkers Involved

Jose Pedregosa-Díaz[1], **María Henar García-Lagunar**[2] and **María Dolores Albaladejo-Otón**[1,*]

[1] *Clinical Analysis Department, Santa Lucia General University Hospital, Cartagena, Spain*

[2] *Pharmacy Department, Santa Lucia General University Hospital, Cartagena, Spain*

Abstract: Metabolic syndrome is considered a cluster of cardiovascular risk factors that are presented both in a single individual. They include insulin resistance, type 2 diabetes mellitus, dyslipidemia, hypertension and central obesity. Metabolic syndrome is related to systemic alterations that involve several tissues, such as liver, muscle, adipose tissue, stomach and immune system, varying the production of many biomolecules. Due to the main outcome of this clinical state being cardiovascular disease, metabolic syndrome has become a significant clinical priority.This global disturbance can be first detected and monitored later by several plasmatic biomarkers. This chapter intends to review a battery of avant-garde biomarkers related to metabolic syndrome.

Keywords: Adipokines, Adiponectin, Angiotensinogen, Biomarkers, C-reactive Protein, Cytokines, Insulin Resistance, Interleukin 1, Interleukin 18, Interleukin 6, Leptin, Metabolic Syndrome, Obesity, Oxidized Low Density Lipoprotein, Plasminogen Activator Inhibitor 1, Resistin, Tumour Necrosis Factor alpha, Type 2 Diabetes Mellitus, Uric Acid.

INTRODUCTION

Metabolic Syndrome

Metabolic syndrome (MS) is a clinical term used to describe a compendium of interrelated biochemical, physiological, clinical, and metabolic factors which heightens risk of cardiovascular events or diseases, type 2 diabetes mellitus (T2DM) and all-cause mortality [1].

* **Corresponding author María Dolores Albaladejo-Otón:** Clinical Analysis Department, Santa Lucia General University Hospital, Cartagena, Spain; Tel/Fax: +34 968 110536; E-mail: mariad.albaladejo@carm.es

Juan Antonio Vílchez (Ed.)

MS, obesity and T2DM have been described as a chronic inflammatory state which associates, aside from the above-mentioned disorders, altered plasmatic levels of cytokines, acute-phase reactants and other inflammatory cues.

MS was born first more as a concept than a diagnosis [2]. Before it was described in 1988 [3], several investigators and physicians set the fundamental pillars for its definition. In 1920 Kylin, a Swedish physician, established the association of high blood pressure (HBP), hyperglycemia and gout [4]. In the late 40's Vague *et al.* related metabolic dysfunctions found in cardiovascular disease (CVD) and type 2 diabetes mellitus (T2DM) [5]. Twenty years later, Avogaro *et al.* described an entity which comprised obesity, hyperglycemia and high blood pressure [6]. In 1988, Reaven *et al.* described a compendium of risk factors for T2DM and CVD and named it as "Syndrome X" [7]. The greatest contribution of his work was the inclusion of the concept of Insulin resistance. The following year, this entity was renamed as "The Deadly Quartet" by Kaplan *et al.* [8]. It included glucose intolerance, hypertriglyceridemia, high blood pressure and upper body obesity. This name only lasted three years because in 1992 it underwent a new change of name: The Insulin Resistance Syndrome [9].

Definition

Metabolic syndrome is defined as a complex systemic disease. It is considered as a set of cardiovascular risk factors that occur concurrently in the same individual [10]. This cluster includes biochemical, physiological, metabolic and clinical factors, and all of them imply an increased suffering of CVD and T2DM.

In order to define the MS, a number of characteristics are needed and these include hyperglycemia, hypertriglyceridemia, low plasma levels of high-density lipoprotein cholesterol (HDL), high blood pressure and abdominal obesity determined by high waist perimeter. As stated above, many definitions have been proposed in the recent decades. The most important ones are from the World Health Organization (WHO) [11], the European Group for the study of Insulin Resistance (EGIR) [12], the National Cholesterol Education Programme Adult Treatment Panel III (NCEP ATPIII) [13], the American Association of Clinical Endocrinologists (AACE) [14] and the International Diabetes Federation (IDF) [15]. All of these different classifications are listed in the following (Table **1**):

Pathophysiology:

<u>Insulin resistance:</u> Insulin is a hormone synthesized in the pancreas in hyperglycaemia situation. It stimulates glucose use in different ways depending on the tissue.

Table 1. Diagnostic criteria proposed for the clinical diagnosis of the MS.

Diagnostic criteria	WHO, 1998 [11]: Insulin resistance or 2TDM plus ≥ 2	EGIR, 1999 [12]: Hyperinsulinemia plus ≥ 2	ATPPIII, 2001 [13]: ≥ 3	AACE, 2003 [14]: IGT or IFG plus ≥ 1	IDF, 2005 [15]: Obesity plus ≥ 2
Insulin resistance	IGT, IFG, T2DM	Plasma insulin >75th percentile			
Obesity	Waist-to-hip ratio >0.90 (♂); waist-to-hip ratio >0.85 and/or BMI >30kg/m^2 (♀)	WC ≥94cm (♂) or ≥80cm (♀)	WC ≥102cm (♂) or ≥88cm (♀)	BMI >25kg/m^2	Increased WC (population specific) plus any 2 of the following
Lipidic profile	TGs ≥150mg/dL and/or HDL-C <35mg/dL (♂) or <39mg/dL (♀)	TGs ≥150mg/dL and/or HDL-C <39mg/dL (⚥)	TGs ≥150mg/dL and/or HDL-C <40mg/dL (♂) or <50mg/dL (♀)	TGs ≥150mg/dL and/or HDL-C <40mg/dL (♂) or <50mg/dL (♀)	TGs ≥150mg/dL or on TGs Tx; HDL-C <40mg/dL (♂) or <50mg/dL (♀) or on HDL-C Tx
Blood pressure	≥140/90mmHg	≥140/90mmHg or on hypertension Tx	≥130/85mmHg	≥130/85mmHg	≥130mmHg systolic or ≥85mmHg diastolic or on hypertension Tx
Hyperglycemia	IGT, IFG or T2DM	IGT or IFG (but not diabetes)	>110mg/dL (includes diabetes)	IGT or IFG (but not diabetes)	>100mg/dL (includes diabetes)
Other factors	Microalbuminuria: Urinary excretion rate of >20mg/min or albumin/creatinine ratio of >30mg/g			family history of T2DM, POS, sedentary lifestyle, advanced age and ethnic groups susceptible to T2DM	

BMI: Body mass index; HDL-C: High density lipoprotein cholesterol; IFG: Impaired fasting glucose; IGT: Impaired glucose tolerance; T2DM: Type 2 diabetes mellitus;POS: Polycystic ovary syndrome; TGs: Triglycerides; Tx: Under treatment; WC: Waist circumference.

Skeletal muscle, liver and adipose tissue remove glucose from the circulation. In the adipose tissuc and skeletal muscle, insulin promotes glucose consumption by translocation of the Glucose Transporter Type 4 (GLUT4) toward the cell surface.

Insulin promotes the synthesis of glycogen from glucose in the skeletal muscle and liver and, at the same time, hinders glycogenolysis. In the liver, insulin lessens gluconeogenesis as well, avoiding the flow of more molecules of glucose into the blood circulation. In adipose tissue, insulin hinders lipolysis while promotes glucose consumption. The resulting effect of those changes is the augment of glucose demand, the reduction of glucose levels in the bloodstream and an raise of the anabolism of glucose into the storage molecules, either glycogen or fat [16, 17]. When a state of insulin resistance is given, adipose, muscle and liver cells cannot react correctly to this hormone, and glucose plasma levels left high, which leads up to pathological situations.

Insulin-mediated glucose disposal rates may vary in the population by over sixfold. Some of these variations produced by adiposity and fitness, and the others are the result of genetic origin. Insulin resistance occurs when there is a reduction in the response of peripheral tissues (such as skeletal muscle, adipose tissues, and hepatocytes) to the effects of insulin. Insulin resistance may predict T2DM, and hyperinsulinemia is a suitable substitute marker for insulin resistance.

Insulin action begins with the joining of insulin to the insulin receptor, which is a ligand-activated tyrosine kinase. After joint of insulin, a tyrosine phosphorylation of downstream substrates is carried out followed by two different pathways: The mitogen-activated protein (MAP) kinase pathways and the phosphoinositide 3-kinase (PI3K). Tyrosine phosphorylation of insulin receptor substrates (IRS) activates first PI3K, and then the 3-phosphoinositide-dependent protein kinase 1 (PDK1) and Akt kinase. The PI3K-Akt pathway triggers most of the main metabolic effects of insulin. In vascular endothelial cells, Akt kinase phosphorylates and stimulates endothelial nitric oxide synthase (eNOS). In skeletal muscle and adipose cells, Akt kinase promotes as well translocation of the insulin responsive glucose transporter GLUT4 toward the cell surface, increasing glucose consumption [17].

At the same time, tyrosine phosphorylation of the Shc protein activates the GTP exchange factor Sos. This factor promotes the activation of MAP kinase pathway, including Ras, Raf, MAP kinase (MEK) and extracellular regulated kinase (ERK). The MAP kinase pathway stimulates endothelin-1 (ET-1) production, bringing on vasoconstriction; expression of the vascular cell adhesion molecules VCAM-1 and E-selectin. All this leads to more leukocyte-endothelial interactions; such as growth and mitogenesis effects on vascular smooth muscle cells [16].

When insulin resistance is given, this affects PI3K-Akt pathway, while the MAP kinase pathway does not. This causes a change in the balance between these two parallel pathways. Inhibition of the PI3K-Akt pathway leads to a reduction in endothelial nitric oxide (NO) production. This results in endothelial dysfunction, and reduces GLUT4 translocation, leading to decreased skeletal muscle and fat glucose uptake. On the contrary, the MAP kinase pathway is unaffected, so ET-1 production continues. In those ways, insulin resistance results in vascular dysfunctions that predispose to atherosclerosis.

Insulin raises local blood flow in tissues by activating eNOS, producing two separable effects [16 - 18]. Capillary recruitment occurs in a few minutes, since dilation of the larger-resistance vessels may need from 30 minutes up to 2 hours. These effects help to vasodilation and raise delivery of glucose and insulin to tissues.

Accordingly, insulin signaling coordinately affects peripheral glucose use, vascular tone and blood flow. Common mechanisms that contribute to insulin resistance can, thereupon, also affect vascular function. This may include hyperglycemia, advanced glycation products, toxicity from free fatty acids (FFA), dyslipidemia, obesity and other proinflammatory conditions [17, 19].

Diabetes mellitus (DM) is possibly one of the oldest known human diseases. Its first report was in an Egyptian document about 3000 years ago [20]. The distinction between type 1 and T2DM was made in 1935 [21]. T2DM was first described as a component of MS in 1988 [22]. T2DM (previously known as noninsulin dependent DM) is the majority form of DM characterized by hyperglycemia, insulin resistance, and relative insulin deficiency [23]. T2DM is a result from interaction between environmental,behavioural risk factors and genetic [24, 25].

T2DM patients are more vulnerable to different kind of short- and long-term complications, which may result in premature death. This trend of increased morbidity and mortality is seen in patients with T2DM because of the commonness of this type of DM, its insidious onset and late recognition, especially in resource-poor developing countries like in the African continent [26].

T2DM is due primarily to lifestyle factors and genetics [27]. A number of lifestyle factors are known to be important to the development of T2DM. These are sedentary tobacco and generous intake of alcohol [28]. Obesity contributes more than 50% of cases of T2DM [29]. The increased rate of childhood obesity between the 1960s and 40 years later is thought to have resulted in an increase in T2DM young population [30]. Environmental poisons may contribute to the recent increases in the rate of T2DM. Relationship between the concentration in the urine of bisphenol A, a constituent of some plastics, and the incidence of T2DM has been established [31].

T2DM is defined as insulin insensitivity as a result of hormone resistance, with an Insulin production lessening, and even pancreatic beta-cell failure [32, 33]. This may lead to a declining in glucose input into the liver, muscle cells, and fat cells. So, a raise in the breakdown of fat accompanied with hyperglycemia happens. The engagement of damaged alpha-cell function has been admitted in the pathophysiology of T2DM within the recent past [34].

<u>Visceral adiposity:</u> Visceral reduces insulin-mediated glucose intake, and this leads to insulin resistant state. This mechanism probably involves adipokines, which are produced in adipose cells, that modulates interferences between metabolism and vascular function. These comprise tumour necrosis factor α (TNF-α) and interleukin-6 (IL-6), that are both pro-inflammatory and confer to insulin resistance and vascular dysfunction. The renin-angiotensin system is activated in adipose tissue as well, leading to hypertension and insulin resistance. Conversely, adiponectin is considered a protective adipokine that matches insulin sensitivity with energy metabolism. Adiponectin concentrations are depleted in T2DM, obesity and metabolic syndrome. Besides, free fatty acids (FFAs), that are released from visceral fat, and bioactive lipid intermediates play together to damage the PI3K-Akt pathway and increase oxidative stress [35].

<u>Atherogenic dyslipidemia:</u> When an increased level of free fatty acid reaches the liver it triggers araised production of Apo B-containing triglyceride-rich very low-density proteins (VLDL) [36]. The action of insulin on this course is slightly convoluted. Within of insulin resistance state, an overflow of FFAs reaching to the liver increases triglyceride concentrations; nevertheless, under some conditions, insulin hinders secretion of VLDL toward the systemic circulation [37]. This answer is due to insulin-mediated degradation of Apo B [38]. Although, insulin has lipogenic action, raising the transcription and enzyme activity of some genes is related to triglyceride biosynthesis [39]. If this process remains operational in insulin resistance is still unknown. Moreover, it has been seen that insulin resistance might also decrease lipoprotein lipase levels in peripheral tissues [40]. Nevertheless, this exchange in this enzyme seems to promote less to hypertriglyceridemia than the overproduction of VLDL does. However, hypertriglyceridemia is a remarkable reflex of the insulin resistance and is one of the important fact for diagnosis of MS.

The other major lipoprotein disturbance in the metabolic syndrome is a reduction in HDL cholesterol. This reduction is a consequence of changes in HDL composition and metabolism. In the presence of hypertriglyceridemia, a depletion in the cholesterol content of HDL results from decreases in the cholesteryl ester content of the lipoprotein core with variable increases in triglyceride making the particle small and dense, a function in part of cholesteryl ester transfer protein [41]. This change in lipoprotein composition results in an increased clearance of HDL from the circulation as well. The relation of these changes in HDL to insulin resistance are probably indirect, arising in concert with the changes in triglyceride-rich lipoprotein metabolism [42, 43].

Endothelial dysfunction: This situation is the last common routine connection with many cardiovascular risk factors and the development of pathologies such as atherosclerosis [16, 44, 45]. Endothelial cells limit the inner surface of blood vessels and participate in very important mechanical and biological functions. Endothelium can sense and respond to normal and pathological stimulus, by producing vasoactive substances, such as NO, prostacyclin and endothelin. Endothelial expression of cell adhesion molecules rules interactions with white blood cells, especially with monocytes, promoting inflammation, and with platelets, stimulating haemostasis and thrombosis. Endothelium inflects as well the answer of vascular smooth muscle.

Endothelial dysfunction takes place when endothelium is unable to perform its normal physiological and protective mechanisms. Characteristically, it consists of spoilt endothelium-dependent vasodilatation, a decreased arterial accordance, and a quickened progress of atherosclerosis [46]. Several factors such as oxidative stress, hyperglycemia, advanced glycation products, FFAs, inflammatory cytokines, or adipokines may cause an unfitness of endothelium to attend its normal physiological and protective actions. It has seen that immune cells have an important role in all the points of the atherosclerotic process [47]; furthermore, a diminution in NO levels, an important modulator of endothelial homeostasis, and a raised reactive oxygen species lead in an endothelial dysfunction and a proatherogenic vascular bedding [48].

Hypercoagulable condition: A pro-inflammatory state is characterized by elevated plasmatic cytokines and acute-phase reactants, such as C-reactive protein (CRP). Besides, a prothrombotic situation leads to anomalies in the procoagulant factors. These would be: An increase in fibrinogen, factor VII and factor VIII aside from the antifibrinolytic factor (PAI-1), platelet abrasions, and endothelial dysfunctions. Fibrinogen, which acts as an acute phase reactant protein, as it could be CRP, increases its plasmatic levels in response to a highcytokine state. This shows that the prothrombotic and proinflammatory states could be metabolically interconnected [49].

Diet: High dietary fat intake is related to an oxidative stress and an activation of the pro-inflammatory transcription factor, known as nuclear factor kappa-beta(NF κβ) [50]. On the contrary, a diet rich in fruits and fibres has no inflammation-inducing ability if it is compared with a high-fat diet even if it contents the same calories [51].

These pathological mechanisms are reflected on the Fig. (**1**).

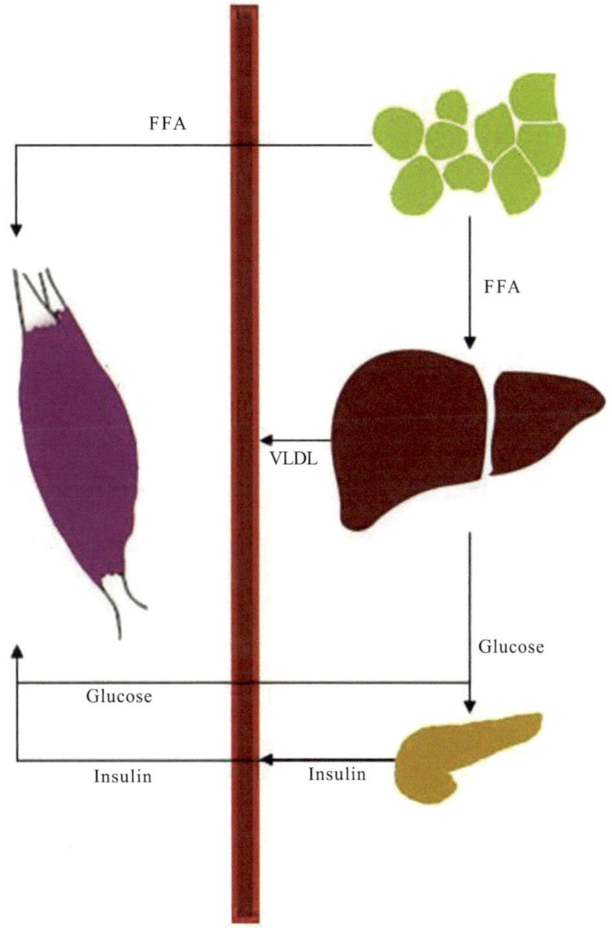

Fig. (1). Mechanism of the metabolic syndrome through the insulin resistance. FFAs are released from the adipose tissue. When the FFAs reach the liver stimulates an increased production of glucose, triglycerides and secretion of very low density lipoproteins (VLDL). FFAs reduce as well insulin sensitivity in skeletal muscle by inhibiting insulin-mediated glucose uptake. When glucose concentration increases in circulating blood it increases pancreatic insulin secretion turning out hyperinsulinemia. This hyperinsulinemia enhances sodium reabsorption and increases sympathetic nervous system (SNS) activity. All this contributes to the hypertension as might increase levels of plasmatic FFAs.

Epidemiology

Prevalence of MS goes from less than 10% to 84% and it depends on the definition of MS used; characteristics of the studied population, such as age, gender or ethnicity; region and if it corresponds to rural or urban area [52].

INFLAMMATION AS A BOND BETWEEN METABOLIC SYNDROME AND TYPE 2 DIABETES MELLITUS

It has been demonstrated that a chronic low-grade inflammation and an activation of the immune system are involved in the pathogenesis of obesity-related insulin resistance and T2DM [10]. Systemic inflammatory markers seem to be factors for the development of T2DM and its macrovascular complications. In presence of obesity adipose tissue, liver, muscle and pancreas are themselves sites of inflammation [53 - 55].

Obesity, particularly visceral adiposity, was associated with insulin resistance, hyperglycemia, dyslipidemia and hypertension, which together are termed metabolic syndrome [56].

Next, the relationship between inflammation and insulin resistance in tissues such as liver, adipose tissue and muscle is going to be explained one by one:

<u>Liver:</u> In contrast to adipose tissue, liver is frequently occupied by macrophages, known as Kupffer cells, which represent about the 10% of total liver cells. Although the number of Kupffer cells does not vary with obesity, their activation state changes [57, 58]. Kupffer cells have been less studied in comparison with adipose tissue macrophages in insulin resistance context but it is known that they clearly promote the production of inflammatory mediators such as TNF-α, IL-6 and IL-1β that contribute to insulin resistance in liver [59]. At the same time, proinflammatory mediators in the portal circulation, mainly produced by abdominal fat, could also initiate hepatic inflammation.

<u>Adipose tissue:</u> There is an increased expression and production of TNF-α in adipose tissue of obese individuals and this induces directly insulin resistance [60]. It has been confirmed that exists a specific up-regulation of genes encoding inflammatory factors and an over-production of several chemokines and cytokines in enlarged adipose tissue [55]. Besides, a decrease in the expression of proinflammatory genes is given [61, 62]. So adipose tissue inflammation is considered as a crucial event leading to metabolic syndrome, T2DM and atherosclerotic cardiovascular diseases. Additionally, adipose tissue has been associated with a marked accumulation of macrophage cells in its stromovascular fraction during obesity [63]. Although it has been demonstrated that enlarged

adipocytes can produce pro-inflammatory cytokines and chemokines by themselves [64], adipose tissue macrophages could be essential for the production of adipose adipocyte-derived pro-inflammatory cytokines. In fact, their recruitment towards the adipose tissue is related to the severity of obesity and is joint to systemic inflammation, insulin resistance [65] and metabolic syndrome [66].

To distinguish macrophages, they are classified into two different subsets:

- M1 macrophages, classically known as activated macrophages: This kind of macrophages secrete pro-inflammatory cytokines like IL-1β, IL-6, TNF-α.
- M2 macrophages, known as alternatively activated macrophages: Produce antiinflammatory cytokines such as IL-10 [67].

This model has been well-established in mice [68, 69], but the existence of different M1 and M2 types of adipose tissue macrophages has not been proved in humans. In order to understand them in human, macrophages are described as mix between M1 and M2 phenotypes [70]. Besides adipose tissue macrophages infiltration, obesity can transform macrophages from the M2 to M1 phenotype, correlating with insulin resistance both in mice and humans [70 - 72]. Signals released from M1 macrophages seem to impair insulin signaling and adipogenesis in adipocytes meanwhile M2 macrophages can protect against obesity-induced insulin resistance [73]. Adaptive immune system can contribute as well to obesity-induced inflammation although macrophages are the most abundant leukocyte adipose tissue.

<u>Muscle:</u> The muscle is another important area of insulin resistance in obesity and T2DM. Similar to adipose tissue, skeletal muscle macrophages belong to pro-inflammatory M1 macrophages [73]. These macrophages promote an increased expression of inflammatory factors in muscle cells [73, 74] contributing locally to insulin resistance. Furthermore, the gene expression of phenotype markers of pro and anti- inflammatory macrophages in human skeletal muscle correlates with insulin sensitivity [74].

BIOMARKERS

The objective of these pages is to review biomarkers that have been suggested in recent years as the link between metabolic syndrome, diabetes and inflammation. Biomarkers have been divided into two groups in order to improve their understanding: Pro-inflammatory and anti-inflammatory molecules.

Proinflammatory Biomarkers

Tumour Necrosis Factor (TNF-α)

Tumour necrosis is one of the main mediating cytokine in immune and inflammatory response [75]. TNF-α plasma levels are increased in obesity. A positive correlation has been proved between TNF-α plasma levels and BMI. Same manner, MS patients presented higher levels of plasmatic TNF-α compared to healthy obese patients [76]. Same study was able as well to stablish a positive correlation between TNF-α plasma levels and insulin resistance measured by means of HOMA-IR index. On the contrary, correlation between TNF-α and HDL cholesterol was negative. As stated before, it has been published that TNF-α is overexpressed in white adipose tissue in insulin resistance and obesity states [77]. TNF-α is a cytokine that associates coronary atheroma. This cytokine is secreted by macrophages, adipose cells, endothelial and muscle cells. Its function is to recruit monocytes towards developing atherosclerotic injury. Association of LDL accumulation in rat arteries with TNF-α expression suggests a role for inflammation in early-stage atherosclerosis [78]. Besides obesity and insulin resistance, TNF-α has been related to infections, cancer, dyslipidemia and arterial hypertension [79].

Blood concentrations of TNF-α are continuously elevated in patients at raised risk for recurrent coronary events [80]. A study demonstrated that weight loss in response to lifestyle modification in obese individuals gets decreased TNF-α levels [81]. Structure of TNF-α is shown in Fig. (**2**).

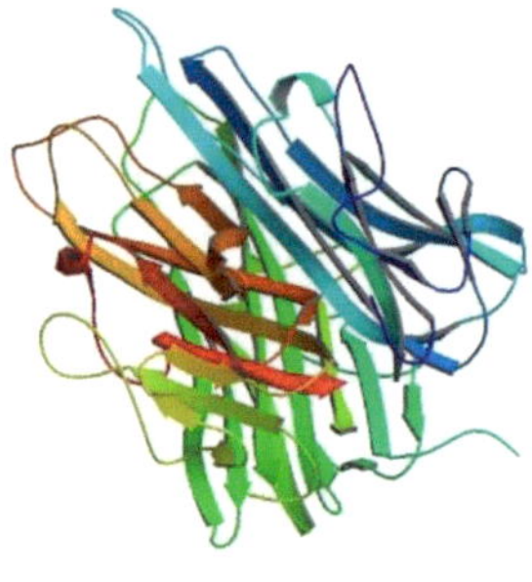

Fig. (2). Structure of tumour necrosis factor-alpha at 2.6 Å resolution [161].

Interleukin-1

IL-1 is a prototypical proinflammatory cytokine and it acts as an early mediator of inflammation. IL-1 is produced by macrophages, monocytes, fibroblasts, dendritic cells, B and NK lymphocytes and epithelial cells [82]. IL-1 increases the expression of adhesion factors on endothelial cells in order to attract macrophages and lymphocytes to the site of infection. IL-1 acts on the hypothalamus, which lets a rise in body temperature. Because IL-1 activates the mechanism of fever is why this cytokine is named an endogenous pyrogen. As well as fever, IL-1 promotes as well hypotension and vasodilation [83].

IL-1 forms a big group of cytokines which includes 11 of them. The most studied ones are IL-1α and IL-1β. Both have a proinflammatory effect. IL-1α and IL-1β are synthesized as precursors (pro-IL-1α and pro-IL-1β), which then undergo proteolytic cleavage by calpain and caspase-1, respectively, to produce the mature active forms. Although pro-IL-1β is biologically inactive, myristoylated membrane-associated pro-IL-1α can signal through IL-1R on adjacent cells to initiate cell-to-cell signalling. In addition, nuclear translocation of pro-IL-1α or the cleaved N-terminal pro-peptide that retains its nuclear localization sequence elicits biological functions [84]. Structure of IL-1 is shown in Figs. (**3** and **4**).

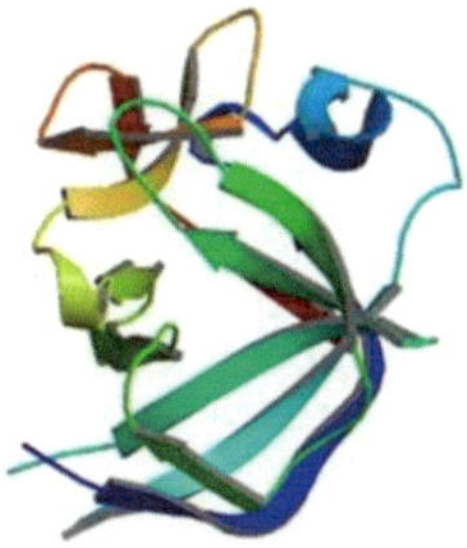

Fig. (3). Structure of interleukin 1α at 2.7-Å resolution [162].

Fig. (4). Crystal structure of recombinant human interleukin 1β at 2.0 Å resolution [163].

Interleukin-6

IL-6 is the main procoagulant cytokine. This interleukin increases plasma levels of CRP, fibrinogen and plasminogen activator inhibitor type I (PAI-1) [85]. Adipocytes secrete about the 30% of total IL-6 and visceral fat contributes more than subcutaneous fat. IL-6 and TNF-α are the best studied cytokines for inflammation and obesity situations [86]. It has been seen that increased concentrations of IL-6 are associated with increased risk of myocardial heart attack in healthy men [87]. Some authors have stablished a relationship between insulin resistance, T2DM and plasma levels of IL-6, being this last one a predictor of future TD2M in several situations [88]. On the contrary, other authors have questioned this fact as after administration of IL-6 no changes have been observed on the glucose metabolism [89].

However, this interleukin has been proposed as a link between stress, obesity, inflammation and heart disease [90]. The role of IL-6 would be implicated in the pathogenesis of heart diseases [91]. Moreover, IL-6 is a good inductor of hepatic acute phase response by stimulating production of several molecules such as CRP, haptoglobin or fibrinogen rebounding on the inflammatory response [92]. Structure of IL-6 is shown in Fig. (**5**).

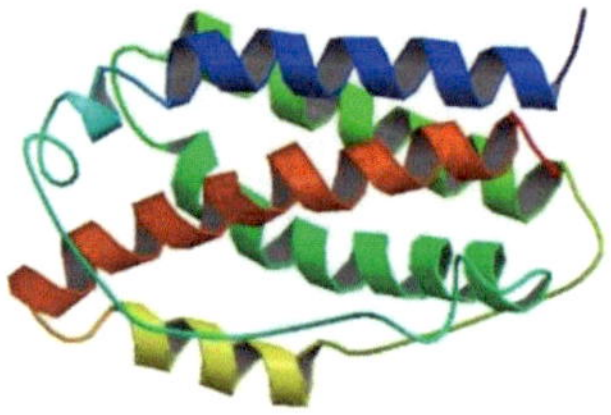

Fig. (5). Human interleukin-6, NMR, minimized average structure [164].

Interleukin-18

Intereleukin-18 is another pro-inflammatory cytokine secreted by adipocytes. It has atherogenic properties by acting on other cytokines such as IL-6, TNF-α and interferon-γ (IFN-γ), which could be an important mediator in subclinical inflammation and obesity. In the same manner, it could be the link between obesity, MS and glucose intolerance, T2DM and cardiovascular diseases [93].

Plasma levels of IL-18 correlate in a positive way with body weight and abdominal fat [94]. It has been found that hyperglycemia causes an increase of plasma levels of this biomarker by means of oxidative stress [94].

T2DM patients have higher plasma levels of this cytokine [95]. Concentrations of IL-18 have been correlated positively with clinical criteria of MS, such as BMI, waist circumference, fasting glucose and insulin concentrations. In the same way, concentrations of IL-18 have been correlated negatively with HDL cholesterol in obese people [96]. Structure of IL-18 is shown in Fig. (**6**).

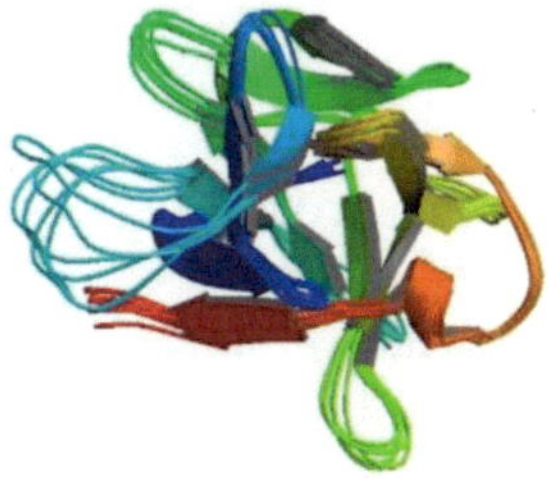

Fig. (6). Solution structure of the human interleukin-18 [165].

Angiotensinogen (AGT)

Angiotensinogen is considered an acute phase reactant protein because its production is increased as a response to inflammation or shock [92]. AGT is part of the renin-angiotensin system [79]. The main role of this system is the regulation of the blood pressure and hypovolemic shock.

It has been seen that there is an overexpression of AGT on visceral adipose tissue, correlating in a positive way with BMI and leptin [97]. Same as TNF-α, AGT has been related to infections, cancer, dyslipidemia and arterial hypertension [79]. Structure of AGT is shown in Fig. (**7**).

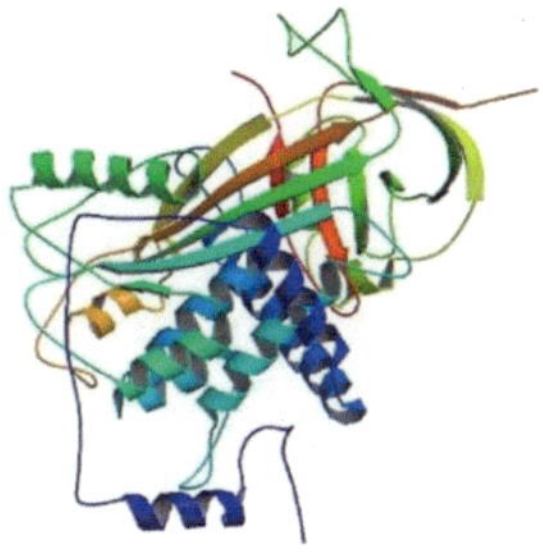

Fig. (7). Crystal structure of human angiotensinogen [166].

Leptin

Leptin is a hormone which is synthetized principally in adipose tissue. It is able to break through the blood-brain barrier towards the hypothalamus. There, it can stimulate anorectic neurones with consequent decreased appetite [98]. Leptin is also able to increase energy use, reducing metabolic effectiveness. Its production is mediated by insulin [98].

Plasmatic levels of leptin are proportional to the amount of body fat. It is considered that circulating leptin acts as a signal indicating adequate levels of energy reserve [99]. On the contrary, plasmatic levels of leptins are reduced in patients who present anorexia. This fact demonstrates that leptin is mediated by nutritional status [100].

Nowadays, it is known that leptin is related to other cytokines, such as IL-6, because they share most of the protein structure. This fact confers a synergy of action between them [99]. Furthermore, not only is known that leptin acts as controller of the body weight, as a matter of fact, it is known that it has a role as a connecting link between neuroendocrine and autoimmune system [101]. Structure of leptin is shown in Fig. (**8**).

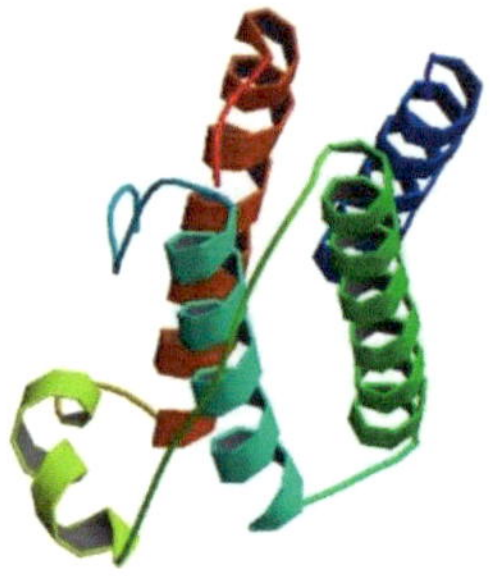

Fig. (8). Crystal structure of the obese protein leptin-E100 [167].

C-Reactive Protein (CRP)

C-reactive protein could be considered the main used biomarker to evaluate systemic inflammation. CRP is synthetized by the liver, some leukocytes, and the adipose tissue in response to multiple signals, particularly increases in interleukin-6 and other systemic inflammatory cytokines [102, 103]. CRP has specific roles including the activation of phagocytic cells—at least in part by binding to the Fc-γ-RIIa receptor [104].

Even though CRP has long been used as a marker of severe inflammation such as it is given in acute infections and chronic inflammatory diseases (with plasmatic concentrations higher than 0.5 mg/dL), it has been strongly demonstrated using high sensitive assays that report hsCRP plasmatic levels that even slight increases in CRP (range 0.2-0.5 mg/dL) are indicative of low-grade inflammation that is not free of clinical relevance in other situations [105]. As proinflammatory state has been associated with visceral adipose tissue, it is logical to assume that hsCRP should be increased when central obesity is set. The level of increase in hsCRP associated with visceral obesity seems to vary by individual [106]; so, once given the correlation between hsCRP and future CVD, evaluating variation of hsCRP could be a good way to differentiate lower risk obese individuals from those at higher risk [107]. Higher plasmatic concentrations of CRP look to be not only a marker of risk but also have an active role in cardiovascular disease. Structure of CRP is shown in Fig. (**9**).

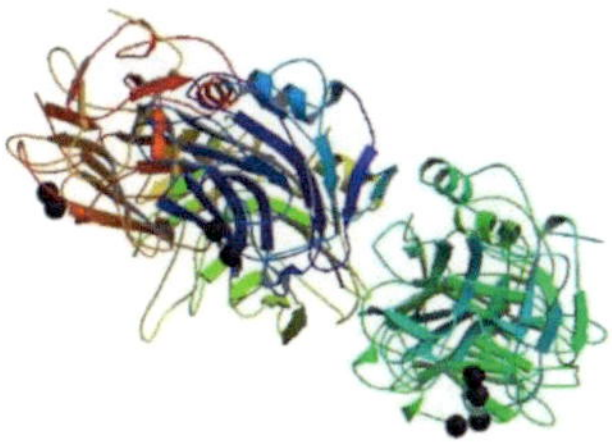

Fig. (9). Three dimensional structure of human C-reactive protein [168].

Uric Acid

Uric acid is a terminal metabolite produced product of purine catabolism, produced in the liver and it is excreted by kidney mainly and intestines secondarily. Uric acid has the most antioxidant power extracellularly, as a matter of fact, it is responsible for two thirds of the whole plasma antioxidant power, where it can chelate metals and clean up oxygen radical species. Nonetheless, intracellularly, it has pro-inflammatory and pro-oxidant activity [108]. It has been seen that uric acid is a biomarker for oxidative damage in conditions like ischemic liver, atherosclerosis, diabetes, and chronic heart failure [109]. As a pro-oxidant, under ischemic conditions or as a result of tissue damage, uric acid oxidizes lipids, which results in inflammation that disrupts reverse cholesterol transport [110]. It also can decrease the availability of NO, which results in less vasodilation and more reactive oxygen species (ROS). This, added to its ability to promote monocytes to produce TNF-α, creates a pro-inflammatory state found in metabolic syndrome. Though its role in pathological diseases is not completely understood, uric acid likely causes systemic inflammation [111].

It is well-known that hyperuricemia means a risk factor for atherosclerotic events like myocardial infarction and stroke, and is associated with other cardiovascular risk factors like hypertension and dyslipidemia. A study have found a positive correlation between uric acid and BMI, blood pressure, and triglycerides, and a negative correlation with HDL-C [112].

Plasmatic levels of uric acid are positively correlated with individual components of the metabolic syndrome, in particular visceral obesity [113, 114]. Most authors have related the presence of hyperuricemia in metabolic syndrome to a secondary response to obesity and hyperinsulinemia. This fact has been attributed to the effects of insulin on proximal tubular urate transport [115], the effect of elevated leptin [116], or altered purine metabolism [117]. Structure of uric acid is shown in Fig. (**10**).

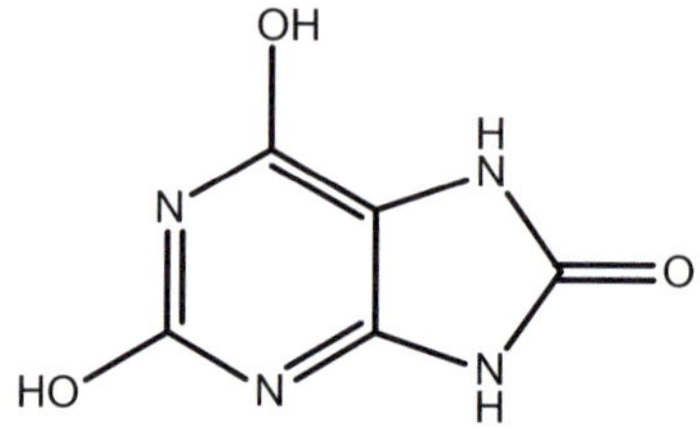

Fig. (10). Chemical structure of Uric Acid.

Plasminogen Activator Inhibitor 1 (PAI-1)

Plasminogen Activator Inhibitor-1 (PAI-1) is the main of four serine peptidase inhibitors that runs to modulate extracellular matrix remodelling and fibrinolysis. It links and deactivates tissue plasminogen (tissue plasminogen activator (tPA) and urokinase plasminogen activator (uPA)). tPA promotes intravascular plasminogen activation, with fibrin regulating its activity, meanwhile uPA activates plasminogen on migrating cells, with the uPA receptor regulating its activity on different cells. Thus, PAI-1 inhibits intravascular fibrinolysis and cell associated proteolysis [118].

The connection between PAI-1 and metabolic syndrome has been long established with elevated levels being strongly correlated, so that the more severe the metabolic syndrome the higher the PAI-1 [119 - 121]. It has been showed that PAI-1 is strongly associated with the components of metabolic syndrome, including BMI, triglycerides and insulin resistance [122]. Furthermore, several groups have found that PAI-1 levels are not associated with dyslipidemia but rather with the distribution phenotype of adipocytes: Visceral adipose tissue primarily and ectopic fat in the liver [123, 124]. Because of this, some authors have suggested PAI-1 could be used as a biomarker for ectopic fat storage. Structure of PAI-1 is shown in Fig. (**11**).

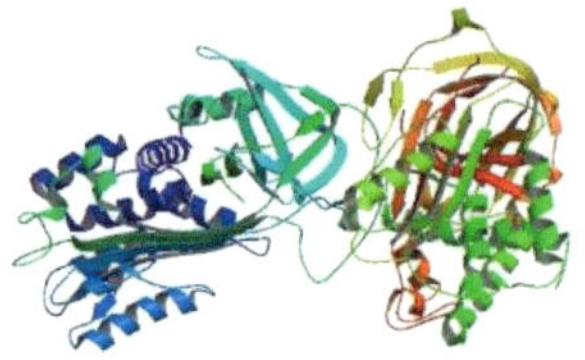

Fig. (11). Crystal structure of native PAI-1 [169].

Oxidized Low Density Lipoprotein (OxLDL)

Oxidized LDL (OxLDL) is the result of lipid oxidation. This product can serve as a marker of oxidative stress. It is known that lipid oxidation promotes the generation of reactive oxygen species (ROS). When ROS reach a molecule of LDL form components of OxLDL. Lipid oxidation products, ROS, and OxLDL in low concentrations can be used as signaling compounds for pathways of cellular antioxidants, such as Heme Oxygenase (HO-1) and glutathione [125]. Nevertheless, if the antioxidant capacity of the cell is altered, as in instance, in metabolic syndrome state, then these compounds lead to an oxidative cascade that, at last, contribute to cell damage and apoptosis [126]. This outspread cell damage and death can result into the vascular dysfunction frequently seen in MS, whilst the dysfunctional OxLDL can further promote dyslipidemia, presenting a risk factor for cardiovascular diseases, which are common comorbidities associated with metabolic syndrome. OxLDL invades and damages the blood vessel endothelium and contributes to atherosclerosis [127]. Additionally, to cardiovascular disease, elevated levels of OxLDL in adults, associate obesity and insulin resistance, two common accompanying components of metabolic syndrome [128].

ANTI-INFLAMMATORY BIOMARKERS

Adiponectin

Adiponectin is a 30 kDa protein. The most important region of this protein is the collagen-like domain. A cysteine residue at this collagenous rod is the main mediator of multimeric complexes, which could represent the most biologically active form of the protein. Adiponectin is structurally related to proteins of the complement system (C1q) and TNF-α [129].

Adiponectin is principally synthetized in white adipose tissue and concretely in mature adipocytes. Three major adipose tissue deposits are recognized as producers of this protein: Subcutaneous, visceral and perivascular [130, 131]. Visceral fat compartments are very important for the secretion of adiponectin, if it is compared to subcutaneous fatty deposits. This fact confers the high association between visceral obesity and metabolic disease; anyway, *in vitro* and *in vivo* studies performed in humans have shown that adipocytes with exhausted lipid storage, filled with fatty acids, and located in the intra-abdominal region can inhibit transcription of the adiponectin gene by secreting inflammatory and angiogenic factors, reducing its plasma levels [132]. The epicardial fat is also a source of adiponectin. Lower levels are synthetized as well in liver, cardiomyocytes, skeletal muscle, colon, salivary gland, placenta and hypophysis, but the contribution of these tissues to the circulating adiponectin is considerably small. Adiponectin can also be detected in breast milk and cerebrospinal fluid, although in low concentrations [133].

Adiponectin works in an autocrine and paracrine manner in the adipose tissue, and in an endocrine manner in distal tissues. The autocrine effects are illustrated by their role in adipocyte differentiation. Induction is greater than 100 times that of adiponectin mRNA during the course of distinction of adipose cells [134]. Similarly, in experiments involving adipocytes overexpressing adiponectin, cell differentiation was accelerated, leading to increased lipid accumulation and insulin-stimulated glucose transport activity in fully differentiated cells. So, adiponectin promotes adipocyte distinction and increased insulin sensitivity [135, 136]. In addition to the regulatory capacity, adiponectin acts as an autocrine and paracrine factor to inhibit the secretion by adipocytes of interleukins 6 and 8, macrophage inflammatory protein 1α/β and monocyte chemotactic protein-1, which, in turn, can inhibit the storage of lipids and insulin sensitivity in adipocytes [137, 138].

Adiponectin can act as an endocrine factor as well. It has been shown that adiponectin levels are responsible for 73% of the variation in insulin sensitivity [139].

It has been shown that Leptin:Adiponectin ratio (LAR) is more beneficial than either alone. In instance, a high LAR is a better biomarker than leptin or adiponectin alone for the diagnosis of metabolic syndrome. One limiting factor with using just adiponectin or leptin is that the difference between adiponectin and leptin tends to be small in the fasting *vs* postprandial state [140]. Furthermore, one of the benefits of using the LAR is that it has the potential to assess insulin sensitivity and metabolic syndrome in the nonfasting state [141]. Structure of adiponectin is shown in Fig. (**12**).

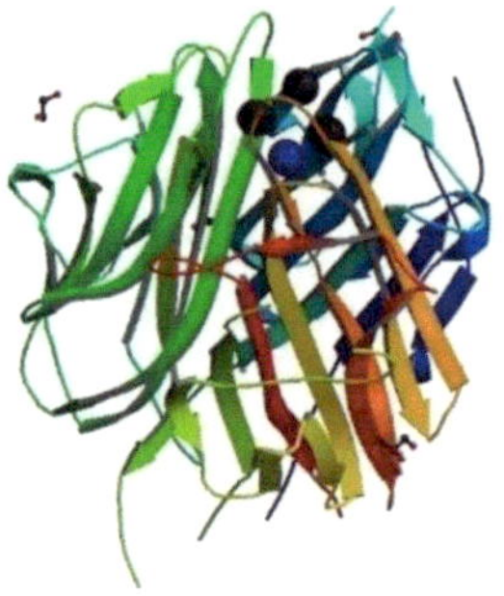

Fig. (12). Crystal structure of a single-chain trimer of human adiponectin globular domain.

Ghrelin

Ghrelin is a neuroendocrine hormone secreted mainly by the stomach. Its function is to stimulate the appetite directly *via* activation of the growth hormone (GH) secretagogue receptor 1a (GHSR-1a) in the hypothalamus, and indirectly by increasing expression of orexigenic peptides, such as neuropeptide Y (NPY) [142, 143]. It might also be protective of vasculature by antagonizing the effects of vasoconstrictors, such as endothelin 1, and promoting the effects of vasodilators, such as NO [144]. Besides, it may help to stimulate lipolysis *via* AMP-activated protein kinase (AMPK) [145]. Nowadays, vasoprotective and lipolytic properties of ghrelin are being studied and represents two mechanisms by which ghrelin could perform a protective effect against MS.

MS presents lower levels of ghrelin, and progressively lower ghrelin levels are associated with increasing metabolic syndrome severity. Ghrelin levels decrease with increasing number of metabolic syndrome items [145 - 147]. This trend is very significant even after adjusting for sex or age, though ghrelin levels have been seen to be higher in females than males [148, 149]. The components of metabolic syndrome including obesity, insulin resistance, and hypertension associate low plasmatic levels of ghrelin [150, 151]. Notwithstanding, the correlation between MS and low circulating ghrelin is likely primarily explained by the relationship to obesity as obese patients with metabolic syndrome have lower ghrelin levels than not obese equivalents [152]. Besides, among obese patients, ghrelin levels are lower in insulin resistant patients if it is compared to insulin sensitive obese patients [153]. Circulating ghrelin is also decreased in the healthy descendant of type 2 diabetes patients. Thus, this fact suggests a genetic component to ghrelin regulation. Structure of ghrelin is shown in Fig. (**13**).

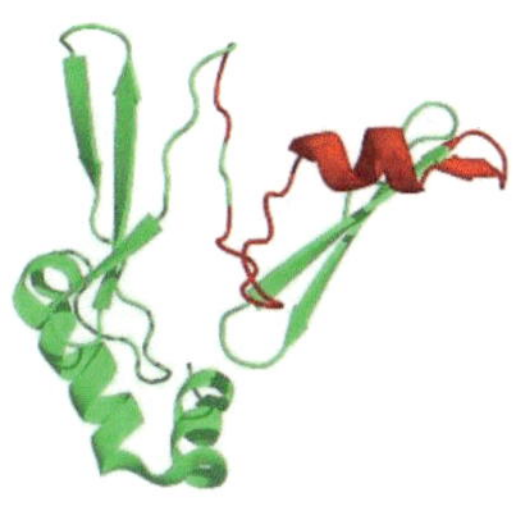

Fig. (13). Theoretical structure proposed for Ghrelin. 1P7X. Umashankar, V. RSCB PDB.

Resistin, a Biomarker to Understand

Up to now, all the biomarkers have been able to be classified by their proinflammatory or anti-inflammatory nature. We have an exceptional case represented by resistin. Until date it has not been able to determine its function regarding inflammation. Resistin is a hormone synthetized by adipocytes. It has a protein structure rich in cysteine amino acids [154]. To date, its role is not understood. Even though resistin has been proposed as a proinflammatory cytokine [155], results in human are still contradictory. Several studies have not found resistin role as a proinflammatory cytokine and few of these conclude that resistin is not related to obesity [156, 157]. Going into details, some studies have not been able to establish the relationship between circulating resistin levels and obesity and insulin resistance, weight loss, insulin resistance, adiposity, plasmatic glucose and some parameters of the lipidic profile but HDL-cholesterol [157]. Nevertheless, other study has related circulating resistin to some factors of metabolic syndrome [158]. About inflammation, plasmatic concentration of resistin has been correlated with plasmatic levels of IL-6, PCR and TNF-α in one study with severe inflammatory state [159]. Structure of resistin is shown in Fig. (**14**).

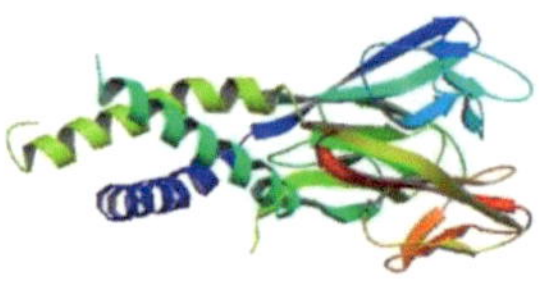

Fig. (14). Crystal Structure of resistin.

Resistin has been related to inflammatory processes, such as atherosclerosis and arthritis [160]. Several authors consider resistin a cardiovascular risk factor, the same way that it is considered as responsible of endothelial diseases.

It is necessary more information than we have nowadays in order to clear up the role of resistin and its suitability as a biomarker for this purpose.

On Fig. (**15**), all the studied biomarkers in this chapter are shown, and the variation of their plasmatic levels in the context of MS:

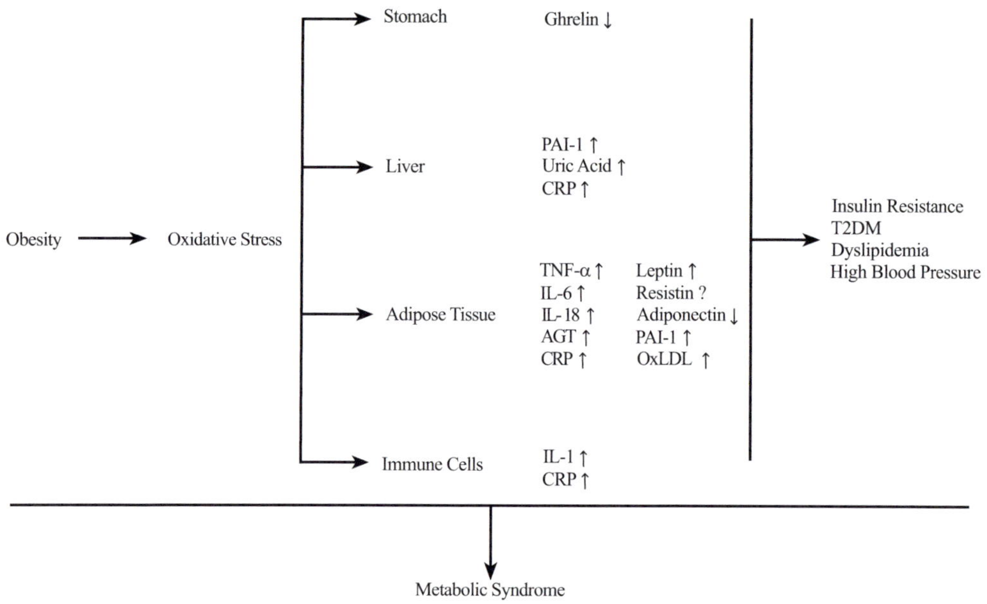

Fig. (15). Variation of the concentration on different biomarkers involved in the metabolic syndrome.

CONCLUSION

Metabolic syndrome is a complex systemic disease. Up to date, its underlying causes are only partway understood. It is very important to decode its causes, mechanisms and relations to other pathologies.

Metabolic syndrome, obesity and T2DM have been described as a chronic inflammatory state which associates altered plasmatic levels of cytokines, acute-phase reactants and other inflammatory cues.

As it has been shown in this chapter, there is an important battery of biomarkers in order to assess these clinical states which are known to develop more severe diseases. Be able to predict the future establishment of more serious diseases could change the natural develop by treating it or changing habits such as lifestyle in order to avoid further complications.

Since the discovery of biomarkers such as adipokines, the view on adipose tissue has changed radically. This kind of tissue will never be considered as just a lipid storage depot anymore. It has been demonstrated that adipose tissue acts as active endocrine organ.

Even there are several biomarkers for this purpose, to date they still cannot be used as biomarkers which involve clinical decisions. Even though some of these biomarkers reflect better than others the clinical state of the patient, no single biomarker has demonstrated to be a good biomarker by itself.

Disposing a battery of biomarkers will improve the clinical practice on obese, metabolic syndrome, T2DM patients. In order to achieve this fact, more studies are necessary so as to assess and discuss their clinical utility since contradictory findings are still in force.

CONFLICT OF INTEREST

The author (editor) declares no conflict of interest, financial or otherwise.

ACKNOWLEDGEMENTS

Declared none.

REFERENCES

[1] Kaur J. A comprehensive review on metabolic syndrome. Cardiol Res Pract 2014; 2014: 1-21. [http://dx.doi.org/10.1155/2014/943162]

[2] Shaw JE, Chisholm DJ. 1: Epidemiology and prevention of type 2 diabetes and the metabolic syndrome. Med J Aust 2003; 179(7): 379-83. [PMID: 14503906]

[3] Schnell M. Dominguez Z a: Aspectos gen ticos , cl nicos y fisiopatol gicos del S ndrome Metab lico Genetical , clinical and pathophysiological aspects of the Metabolic Syndrome An Venez Nutr 2007. [cited 2015 Nov 26]

[4] Kylin E. Studien uber das Hypertonie-Hyperglykamie-hyperurikamiesyndrome. Zentralblatt fuer Inn Medizin 1926; 44: 105-27.

[5] Vague J. La différenciation sexuelle; facteur déterminant des formes de l'obésité. Presse Med 1947; 55(30): 339-40.
[PMID: 18918084]

[6] Avogaro P, Crepaldi G. Essential hyperlipidemia, obesity and diabetes. Diabetologia 1965; 1: 137.

[7] Reaven GM. Banting lecture 1988. Role of insulin resistance in human disease. Diabetes 1988; 37: 1595-607.

[8] Kaplan NM. The deadly quartet. Upper-body obesity, glucose intolerance, hypertriglyceridemia, and hypertension. Arch Intern Med 1989; 149: 1514-20.
[http://dx.doi.org/10.1001/archinte.1989.00390070054005]

[9] Haffner SM, Valdez RA, Hazuda HP, Mitchell BD, Morales PA, Stern MP. Prospective analysis of the insulin-resistance syndrome (syndrome X). Diabetes 1992; 41: 715-22.

[10] Alberti KG, Eckel RH, Grundy SM, *et al.* Harmonizing the metabolic syndrome: A joint interim statement of the international diabetes federation task force on epidemiology and prevention; National heart, lung, and blood institute; American heart association; World heart federation; International. Circulation 2009; 120(16): 1640-5.
[http://dx.doi.org/10.1161/CIRCULATIONAHA.109.192644] [PMID: 19805654]

[11] Alberti KG, Zimmet PZ. Definition, diagnosis and classification of diabetes mellitus and its complications. Part 1: diagnosis and classification of diabetes mellitus provisional report of a WHO consultation. Diabet Med 1998; 15(7): 539-53.
[http://dx.doi.org/10.1002/(SICI)1096-9136(199807)15:7<539::AID-DIA668>3.0.CO;2-S] [PMID: 9686693]

[12] Balkau B, Charles MA. Comment on the provisional report from the WHO consultation. Diabet Med 1999; 16(5): 442-3.
[http://dx.doi.org/10.1046/j.1464-5491.1999.00059.x] [PMID: 10342346]

[13] Cleeman JI. Executive summary of the third report of the National Cholesterol Education Program (NCEP) expert panel on detection, evaluation, and treatment of high blood cholesterol in adults (adult treatment panel III). JAMA 2001; 285(19): 2486-97.
[http://dx.doi.org/10.1001/jama.285.19.2486] [PMID: 11368702]

[14] Einhorn D, Reaven GM, Cobin RH, *et al.* American College of Endocrinology position statement on the insulin resistance syndrome. Endocr Pract 2003; 9(3): 237-52.
[PMID: 12924350]

[15] The IDF consensus worldwide definition of the metabolic syndrome http://www.idf.org/metabolic-syndrome

[16] Kim J-A, Montagnani M, Koh KK, Quon MJ. Reciprocal relationships between insulin resistance and endothelial dysfunction: molecular and pathophysiological mechanisms. Circulation 2006; 113(15): 1888-904.
[http://dx.doi.org/10.1161/CIRCULATIONAHA.105.563213] [PMID: 16618833]

[17] Huang PL. A comprehensive definition for metabolic syndrome. Dis Model Mech 2009; 2(5-6): 231-7.
[http://dx.doi.org/10.1242/dmm.001180] [PMID: 19407331]

[18] Jonk AM, Houben AJ, de Jongh RT, Serné EH, Schaper NC, Stehouwer CD. Microvascular dysfunction in obesity: a potential mechanism in the pathogenesis of obesity-associated insulin resistance and hypertension. Physiology (Bethesda) 2007; 22: 252-60.

[http://dx.doi.org/10.1152/physiol.00012.2007] [PMID: 17699878]

[19] Tidyman WE, Rauen KA. Mutational and functional analysis in human Ras/MAP kinase genetic syndromes. Methods Mol Biol 2010; 661: 433-47. [http://dx.doi.org/10.1007/978-1-60761-795-2_27] [PMID: 20812000]

[20] Ahmed AM. History of diabetes mellitus. Saudi Med J 2002; 23(4): 373-8. [PMID: 11953758]

[21] Diabetes Mellit Hist Anc to Mod times 2011.http://www.diabetes-management-center.com/diabetes-mellitus-history.html

[22] Patlak M. New weapons to combat an ancient disease: treating diabetes. FASEB J 2002; 16(14): 1853. [http://dx.doi.org/10.1096/fj.02-0974bkt] [PMID: 12468446]

[23] Maitra A, Abbas AK. The endocrine system. Philadelphia: Elsevier Saunders 2005; pp. 1155-226.

[24] Chen L, Magliano DJ, Zimmet PZ. The worldwide epidemiology of type 2 diabetes mellitus--present and future perspectives. Nat Rev Endocrinol 2011; 8(4): 228-36. [http://dx.doi.org/10.1038/nrendo.2011.183] [PMID: 22064493]

[25] Leja D, Manolio T. Genetic basis of type 1 and type2 diabetes, obesity, and their complications. Adv Emerg Oppor diabetes Res A Strateg Plan Rep DMICC 2011.

[26] Azevedo M, Alla S. Diabetes in sub-saharan Africa: kenya, mali, mozambique, Nigeria, South Africa and zambia. Int J Diabetes Dev Ctries 2008; 28(4): 101-8. [http://dx.doi.org/10.4103/0973-3930.45268] [PMID: 20165596]

[27] Ripsin CM, Kang H, Urban RJ. Management of blood glucose in type 2 diabetes mellitus. Am Fam Physician 2009; 79: 29-36.

[28] Hu FB, Manson JE, Stampfer MJ, *et al.* Diet, lifestyle, and the risk of type 2 diabetes mellitus in women. N Engl J Med 2001; 345(11): 790-7. [http://dx.doi.org/10.1056/NEJMoa010492] [PMID: 11556298]

[29] Sung RY, Tong PC, Yu C-W, *et al.* High prevalence of insulin resistance and metabolic syndrome in overweight/obese preadolescent Hong Kong Chinese children aged 9-12 years. Diabetes Care 2003; 26(1): 250-1. [7]. [http://dx.doi.org/10.2337/diacare.26.1.250] [PMID: 12502696]

[30] Barlow SE. Expert committee recommendations regarding the prevention, assessment, and treatment of child and adolescent overweight and obesity: summary report. Pediatrics 2007; 120 (Suppl. 4): S164-92. [http://dx.doi.org/10.1542/peds.2007-2329C] [PMID: 18055651]

[31] Lang IA, Galloway TS, Scarlett A, *et al.* Association of urinary bisphenol A concentration with medical disorders and laboratory abnormalities in adults. JAMA 2008; 300(11): 1303-10. [http://dx.doi.org/10.1001/jama.300.11.1303] [PMID: 18799442]

[32] Kahn CR. Banting Lecture. Insulin action, diabetogenes, and the cause of type II diabetes. Diabetes 1994; 43(8): 1066-84. [http://dx.doi.org/10.2337/diab.43.8.1066] [PMID: 8039601]

[33] Robertson RP. Antagonist: diabetes and insulin resistance--philosophy, science, and the multiplier hypothesis. J Lab Clin Med 1995; 125(5): 560-4. [PMID: 7738421]

[34] Fujioka K. Pathophysiology of type 2 diabetes and the role of incretin hormones and beta-cell dysfunction. JAAPA 2007; (Suppl.)3-8. [http://dx.doi.org/10.1097/01720610-200712000-00001] [PMID: 18217245]

[35] Kershaw EE, Flier JS. Adipose tissue as an endocrine organ. J Clin Endocrinol Metab 2004; 89(6): 2548-56. [http://dx.doi.org/10.1210/jc.2004-0395] [PMID: 15181022]

[36] Lewis GF, Uffelman KD, Szeto LW, Weller B, Steiner G. Interaction between free fatty acids and insulin in the acute control of very low density lipoprotein production in humans. J Clin Invest 1995; 95(1): 158-66.
[http://dx.doi.org/10.1172/JCI117633] [PMID: 7814610]

[37] Lewis GF, Steiner G. Acute effects of insulin in the control of VLDL production in humans. Implications for the insulin-resistant state. Diabetes Care 1996; 19(4): 390-3.

[38] Taghibiglou C, Rashid-Kolvear F, Van Iderstine SC, *et al.* Hepatic very low density lipoprotein-ApoB overproduction is associated with attenuated hepatic insulin signaling and overexpression of protein-tyrosine phosphatase 1B in a fructose-fed hamster model of insulin resistance. J Biol Chem 2002; 277(1): 793-803.
[http://dx.doi.org/10.1074/jbc.M106737200] [PMID: 11598116]

[39] Foufelle F, Ferré P. New perspectives in the regulation of hepatic glycolytic and lipogenic genes by insulin and glucose: a role for the transcription factor sterol regulatory element binding protein-1c. Biochem J 2002; 366(Pt 2): 377-91.
[http://dx.doi.org/10.1042/bj20020430] [PMID: 12061893]

[40] Eckel RH, Yost TJ, Jensen DR. Alterations in lipoprotein lipase in insulin resistance. Int J Obes Relat Metab Disord 1995; 19 (Suppl. 1): S16-21.
[PMID: 7550532]

[41] Murakami T, Michelagnoli S, Longhi R, *et al.* Triglycerides are major determinants of cholesterol esterification/transfer and HDL remodeling in human plasma. Arterioscler Thromb Vasc Biol 1995; 15(11): 1819-28.
[http://dx.doi.org/10.1161/01.ATV.15.11.1819] [PMID: 7583561]

[42] Brinton EA, Eisenberg S, Breslow JL. Increased apo A-I and apo A-II fractional catabolic rate in patients with low high density lipoprotein-cholesterol levels with or without hypertriglyceridemia. J Clin Invest 1991; 87(2): 536-44.
[http://dx.doi.org/10.1172/JCI115028] [PMID: 1899429]

[43] Eckel RH, Grundy SM, Zimmet PZ. The metabolic syndrome. Lancet 2005; 365(9468): 1415-28.
[http://dx.doi.org/10.1016/S0140-6736(05)66378-7] [PMID: 15836891]

[44] Gimbrone MA Jr, Topper JN, Nagel T, Anderson KR, Garcia-Cardeña G. Endothelial dysfunction, hemodynamic forces, and atherogenesis. Ann N Y Acad Sci 2000; 902: 230-9.
[http://dx.doi.org/10.1111/j.1749-6632.2000.tb06318.x] [PMID: 10865843]

[45] Huang PL. Unraveling the links between diabetes, obesity, and cardiovascular disease. Circ Res 2005; 96(11): 1129-31.
[http://dx.doi.org/10.1161/01.RES.0000170705.56583.45] [PMID: 15947251]

[46] Kraemer-Aguiar LG, Laflor CM, Bouskela E. Skin microcirculatory dysfunction is already present in normoglycemic subjects with metabolic syndrome. Metabolism 2008; 57(12): 1740-6.
[http://dx.doi.org/10.1016/j.metabol.2008.07.034] [PMID: 19013299]

[47] Hansson GK. Mechanisms of disease: Inflammation, atherosclerosis, and coronary artery disease. N Engl J Med 2005; 352: 1685-95.

[48] Harrison DG. Cellular and molecular mechanisms of endothelial cell dysfunction. J Clin Invest 1997; 100(9): 2153-7.
[http://dx.doi.org/10.1172/JCI119751] [PMID: 9410891]

[49] Grundy SM. Obesity, metabolic syndrome, and cardiovascular disease. J Clin Endocrinol Metab 2004; 89(6): 2595-600.
[http://dx.doi.org/10.1210/jc.2004-0372] [PMID: 15181029]

[50] Aljada A, Mohanty P, Ghanim H, *et al.* Increase in intranuclear nuclear factor kappaB and decrease in inhibitor kappaB in mononuclear cells after a mixed meal: evidence for a proinflammatory effect. Am J Clin Nutr 2004; 79(4): 682-90.

[PMID: 15051615]

[51] Dandona P, Aljada A, Chaudhuri A, Mohanty P, Garg R. Metabolic syndrome: a comprehensive perspective based on interactions between obesity, diabetes, and inflammation. Circulation 2005; 111(11): 1448-54.
[http://dx.doi.org/10.1161/01.CIR.0000158483.13093.9D] [PMID: 15781756]

[52] Desroches S, Lamarche B. The evolving definitions and increasing prevalence of the metabolic syndrome. Appl Physiol Nutr Metab 2007; 32(1): 23-32.
[http://dx.doi.org/10.1139/h06-095] [PMID: 17332782]

[53] Cameron AJ, Shaw JE, Zimmet PZ. The metabolic syndrome: Prevalence in worldwide populations. Endocrinol Metab Clin North Am 2004; 33: 351-75.
[http://dx.doi.org/10.1016/j.ecl.2004.03.005]

[54] Wilson PW, Kannel WB, Silbershatz H, D'Agostino RB. Clustering of metabolic factors and coronary heart disease. Arch Intern Med 1999; 159: 1104.
[http://dx.doi.org/10.1001/archinte.159.10.1104]

[55] Palaniappan L, Carnethon MR, Wang Y, Hanley AJ, Fortmann SP, Haffner SM. Predictors of the Incident Metabolic Syndrome in Adults: The Insulin Resistance Atherosclerosis Study. Diabetes Care 2004; 27: 788-93.
[http://dx.doi.org/10.2337/diacare.27.3.788]

[56] Shoelson SE, Lee J, Goldfine AB. Inflammation and insulin resistance. J Clin Invest 2006; 116(7): 1793-801.
[http://dx.doi.org/10.1172/JCI29069] [PMID: 16823477]

[57] Donath MY, Shoelson SE. Type 2 diabetes as an inflammatory disease. Nat Rev Immunol 2011; 11(2): 98-107.
[http://dx.doi.org/10.1038/nri2925] [PMID: 21233852]

[58] Chawla A, Nguyen KD, Goh YP. Macrophage-mediated inflammation in metabolic disease. Nat Rev Immunol 2011; 11(11): 738-49.
[http://dx.doi.org/10.1038/nri3071] [PMID: 21984069]

[59] Ouchi N, Parker JL, Lugus JJ, Walsh K. Adipokines in inflammation and metabolic disease. Nat Rev Immunol 2011; 11(2): 85-97.
[http://dx.doi.org/10.1038/nri2921] [PMID: 21252989]

[60] Esser N, Legrand-Poels S, Piette J, Scheen AJ, Paquot N. Inflammation as a link between obesity, metabolic syndrome and type 2 diabetes. Diabetes Res Clin Pract 2014; 105(2): 141-50.
[http://dx.doi.org/10.1016/j.diabres.2014.04.006] [PMID: 24798950]

[61] Cai D, Yuan M, Frantz DF, Melendez PA, Hansen L, Lee J. Local and systemic insulin resistance resulting from hepatic activation of IKK-beta and NF-kappaB. 2005; 11: 183-90.

[62] Huang W, Metlakunta A, Dedousis N, *et al.* Depletion of liver Kupffer cells prevents the development of diet-induced hepatic steatosis and insulin resistance. Diabetes 2010; 59(2): 347-57.
[http://dx.doi.org/10.2337/db09-0016] [PMID: 19934001]

[63] Hotamisligil GS, Arner P, Caro JF, Atkinson RL, Spiegelman BM. Increased adipose tissue expression of tumor necrosis factor-alpha in human obesity and insulin resistance. J Clin Invest 1995; 95(5): 2409-15.
[http://dx.doi.org/10.1172/JCI117936] [PMID: 7738205]

[64] Lowe G, Woodward M, Hillis G, Rumley A, Li Q, Harrap S. Circulating inflammatory markers and the risk of vascular complications and mortality in people with type 2 diabetes and cardiovascular disease or risk factors: the ADVANCE Study 2014; 63(3): 1115-23.
[http://dx.doi.org/10.2337/db12-1625]

[65] Weisberg SP, McCann D, Desai M, Rosenbaum M, Leibel RL, Ferrante AW Jr. Obesity is associated with macrophage accumulation in adipose tissue. J Clin Invest 2003; 112(12): 1796-808.

[http://dx.doi.org/10.1172/JCI200319246] [PMID: 14679176]

[66] Skurk T, Alberti-Huber C, Herder C, Hauner H. Relationship between adipocyte size and adipokine expression and secretion. J Clin Endocrinol Metab 2007; 92(3): 1023-33. [http://dx.doi.org/10.1210/jc.2006-1055] [PMID: 17164304]

[67] Esser N, L'homme L, De Roover A, *et al.* Obesity phenotype is related to NLRP3 inflammasome activity and immunological profile of visceral adipose tissue. Diabetologia 2013; 56(11): 2487-97. [http://dx.doi.org/10.1007/s00125-013-3023-9] [PMID: 24013717]

[68] Cancello R, Henegar C, Viguerie N, *et al.* Reduction of macrophage infiltration and chemoattractant gene expression changes in white adipose tissue of morbidly obese subjects after surgery-induced weight loss. Diabetes 2005; 54(8): 2277-86. [http://dx.doi.org/10.2337/diabetes.54.8.2277] [PMID: 16046292]

[69] Bruun JM, Helge JW, Richelsen B, Stallknecht B. Diet and exercise reduce low-grade inflammation and macrophage infiltration in adipose tissue but not in skeletal muscle in severely obese subjects. Am J Physiol Endocrinol Metab 2006; 290(5): E961-7. [http://dx.doi.org/10.1152/ajpendo.00506.2005] [PMID: 16352667]

[70] Lumeng CN, Bodzin JL, Saltiel AR. Obesity induces a phenotypic switch in adipose tissue macrophage polarization. J Clin Invest 2007; 117(1): 175-84. [http://dx.doi.org/10.1172/JCI29881] [PMID: 17200717]

[71] Fujisaka S, Usui I, Bukhari A, *et al.* Regulatory mechanisms for adipose tissue M1 and M2 macrophages in diet-induced obese mice. Diabetes 2009; 58(11): 2574-82. [http://dx.doi.org/10.2337/db08-1475] [PMID: 19690061]

[72] Wentworth JM, Naselli G, Brown WA, *et al.* Pro-inflammatory CD11c+CD206+ adipose tissue macrophages are associated with insulin resistance in human obesity. Diabetes 2010; 59(7): 1648-56. [http://dx.doi.org/10.2337/db09-0287] [PMID: 20357360]

[73] Nguyen MT, Favelyukis S, Nguyen AK, *et al.* A subpopulation of macrophages infiltrates hypertrophic adipose tissue and is activated by free fatty acids *via* Toll-like receptors 2 and 4 and JNK-dependent pathways. J Biol Chem 2007; 282(48): 35279-92. [http://dx.doi.org/10.1074/jbc.M706762200] [PMID: 17916553]

[74] Fink LN, Oberbach A, Costford SR, *et al.* Expression of anti-inflammatory macrophage genes within skeletal muscle correlates with insulin sensitivity in human obesity and type 2 diabetes. Diabetologia 2013; 56(7): 1623-8. [http://dx.doi.org/10.1007/s00125-013-2897-x] [PMID: 23595247]

[75] Vincent HK, Taylor AG. Biomarkers and potential mechanisms of obesity-induced oxidant stress in humans. Int J Obes 2006; 30(3): 400-18. [http://dx.doi.org/10.1038/sj.ijo.0803177] [PMID: 16302012]

[76] Xydakis AM, Case CC, Jones PH, *et al.* Adiponectin, inflammation, and the expression of the metabolic syndrome in obese individuals: the impact of rapid weight loss through caloric restriction. J Clin Endocrinol Metab 2004; 89(6): 2697-703. [http://dx.doi.org/10.1210/jc.2003-031826] [PMID: 15181044]

[77] Xu H, Barnes GT, Yang Q, *et al.* Chronic inflammation in fat plays a crucial role in the development of obesity-related insulin resistance. J Clin Invest 2003; 112(12): 1821-30. [http://dx.doi.org/10.1172/JCI200319451] [PMID: 14679177]

[78] Niemann-Jönsson A, Dimayuga P, Jovinge S, *et al.* Accumulation of LDL in rat arteries is associated with activation of tumor necrosis factor-α expression. Arterioscler Thromb Vasc Biol 2000; 20(10): 2205-11. [http://dx.doi.org/10.1161/01.ATV.20.10.2205] [PMID: 11031205]

[79] Cottam DR, Mattar SG, Barinas-Mitchell E, *et al.* The chronic inflammatory hypothesis for the morbidity associated with morbid obesity: implications and effects of weight loss. Obes Surg 2004;

14(5): 589-600.
[http://dx.doi.org/10.1381/096089204323093345] [PMID: 15186624]

[80] Ridker PM, Rifai N, Pfeffer M, Sacks F, Lepage S, Braunwald E. Elevation of tumor necrosis factor-α and increased risk of recurrent coronary events after myocardial infarction. Circulation 2000; 101(18): 2149-53.
[http://dx.doi.org/10.1161/01.CIR.101.18.2149] [PMID: 10801754]

[81] Monzillo LU, Hamdy O, Horton ES, *et al.* Effect of lifestyle modification on adipokine levels in obese subjects with insulin resistance. Obes Res 2003; 11(9): 1048-54.
[http://dx.doi.org/10.1038/oby.2003.144] [PMID: 12972674]

[82] Contassot E, Beer H-D, French LE. Interleukin-1, inflammasomes, autoinflammation and the skin. Swiss Med Wkly 2012; 142: w13590.
[PMID: 22653747]

[83] Dinarello CA. Interleukin-1 in the pathogenesis and treatment of inflammatory diseases. Blood 2011; 117(14): 3720-32.
[http://dx.doi.org/10.1182/blood-2010-07-273417] [PMID: 21304099]

[84] Risbud MV, Shapiro IM. Role of cytokines in intervertebral disc degeneration: pain and disc content. Nat Rev Rheumatol 2014; 10(1): 44-56.
[http://dx.doi.org/10.1038/nrrheum.2013.160] [PMID: 24166242]

[85] Willerson JT, Ridker PM. Inflammation as a cardiovascular risk factor. Circulation 2004; 109(21) (Suppl. 1): II2-II10.
[PMID: 15173056]

[86] Fantuzzi G. Adipose tissue, adipokines, and inflammation. J Allergy Clin Immunol 2005; 115(5): 911-9.
[http://dx.doi.org/10.1016/j.jaci.2005.02.023] [PMID: 15867843]

[87] Ridker PM, Rifai N, Stampfer MJ, Hennekens CH. Plasma concentration of interleukin-6 and the risk of future myocardial infarction among apparently healthy men. Circulation 2000; 101(15): 1767-72.
[http://dx.doi.org/10.1161/01.CIR.101.15.1767] [PMID: 10769275]

[88] Vozarova B, Fernández-Real J-M, Knowler WC, *et al.* The interleukin-6 (-174) G/C promoter polymorphism is associated with type-2 diabetes mellitus in Native Americans and Caucasians. Hum Genet 2003; 112(4): 409-13.
[PMID: 12589429]

[89] Krogh-Madsen R, Plomgaard P, Møller K, Mittendorfer B, Pedersen BK. Influence of TNF-α and IL-6 infusions on insulin sensitivity and expression of IL-18 in humans. Am J Physiol Endocrinol Metab 2006; 291(1): E108-14.
[http://dx.doi.org/10.1152/ajpendo.00471.2005] [PMID: 16464907]

[90] Goyenechea E, Parra MD, Martínez Hernández JA. [Role of IL-6 and its -174G>C polymorphism in weight management and in the metabolic comorbidities associated with obesity]. An Sist Sanit Navar 2005; 28(3): 357-66.
[PMID: 16421614]

[91] Yudkin JS, Kumari M, Humphries SE, Mohamed-Ali V. Inflammation, obesity, stress and coronary heart disease: is interleukin-6 the link? Atherosclerosis 2000; 148(2): 209-14.
[http://dx.doi.org/10.1016/S0021-9150(99)00463-3] [PMID: 10657556]

[92] Engström G, Stavenow L, Hedblad B, *et al.* Inflammation-sensitive plasma proteins and incidence of myocardial infarction in men with low cardiovascular risk. Arterioscler Thromb Vasc Biol 2003; 23(12): 2247-51.
[http://dx.doi.org/10.1161/01.ATV.0000102924.11767.8D] [PMID: 14672880]

[93] Skurk T, Kolb H, Müller-Scholze S, Röhrig K, Hauner H, Herder C. The proatherogenic cytokine interleukin-18 is secreted by human adipocytes. Eur J Endocrinol 2005; 152(6): 863-8.

[http://dx.doi.org/10.1530/eje.1.01897] [PMID: 15941925]

[94] Esposito K, Nappo F, Marfella R, *et al.* Inflammatory cytokine concentrations are acutely increased by hyperglycemia in humans: role of oxidative stress. Circulation 2002; 106(16): 2067-72. [http://dx.doi.org/10.1161/01.CIR.0000034509.14906.AE] [PMID: 12379575]

[95] Aso Y, Okumura K, Takebayashi K, Wakabayashi S, Inukai T. Relationships of plasma interleukin-18 concentrations to hyperhomocysteinemia and carotid intimal-media wall thickness in patients with type 2 diabetes. Diabetes Care 2003; 26(9): 2622-7. [http://dx.doi.org/10.2337/diacare.26.9.2622] [PMID: 12941729]

[96] Hung J, McQuillan BM, Chapman CM, Thompson PL, Beilby JP. Elevated interleukin-18 levels are associated with the metabolic syndrome independent of obesity and insulin resistance. Arterioscler Thromb Vasc Biol 2005; 25(6): 1268-73. [http://dx.doi.org/10.1161/01.ATV.0000163843.70369.12] [PMID: 15790931]

[97] Schorr U, Blaschke K, Turan S, Distler A, Sharma AM. Relationship between angiotensinogen, leptin and blood pressure levels in young normotensive men. J Hypertens 1998; 16(10): 1475-80. [http://dx.doi.org/10.1097/00004872-199816100-00011] [PMID: 9814618]

[98] Solomon A, Martínez JA. Participación del sistema nervioso y del tracto gastrointestinal en la homeostasis energética. Rev Med Unav 2006; 50: 27-37.

[99] Otero M, Nogueiras R, Lago F, Dieguez C, Gomez-Reino JJ, Gualillo O. Chronic inflammation modulates ghrelin levels in humans and rats. Rheumatology (Oxford) 2004; 43(3): 306-10. [http://dx.doi.org/10.1093/rheumatology/keh055] [PMID: 14623951]

[100] Grinspoon S, Gulick T, Askari H, *et al.* Serum leptin levels in women with anorexia nervosa. J Clin Endocrinol Metab 1996; 81(11): 3861-3. [PMID: 8923829]

[101] Fantuzzi G, Faggioni R. Leptin in the regulation of immunity, inflammation, and hematopoiesis. J Leukoc Biol 2000; 68: 437-6.

[102] de Ferranti S, Mozaffarian D. The perfect storm: obesity, adipocyte dysfunction, and metabolic consequences. Clin Chem 2008; 54(6): 945-55. [http://dx.doi.org/10.1373/clinchem.2007.100156] [PMID: 18436717]

[103] Calabro P, Chang DW, Willerson JT, Yeh ET. Release of C-reactive protein in response to inflammatory cytokines by human adipocytes: linking obesity to vascular inflammation. J Am Coll Cardiol 2005; 46(6): 1112-3. [http://dx.doi.org/10.1016/j.jacc.2005.06.017] [PMID: 16168299]

[104] Stein MP, Edberg JC, Kimberly RP, *et al.* C-reactive protein binding to FcgammaRIIa on human monocytes and neutrophils is allele-specific. J Clin Invest 2000; 105(3): 369-76. [http://dx.doi.org/10.1172/JCI7817] [PMID: 10675363]

[105] Ridker PM, Hennekens CH, Buring JE, Rifai N. C-reactive protein and other markers of inflammation in the prediction of cardiovascular disease in women. N Engl J Med 2000; 342(12): 836-43. [http://dx.doi.org/10.1056/NEJM200003233421202] [PMID: 10733371]

[106] Wee CC, Mukamal KJ, Huang A, Davis RB, McCarthy EP, Mittleman MA. Obesity and C-reactive protein levels among white, black, and hispanic US adults. Obesity (Silver Spring) 2008; 16(4): 875-80. [http://dx.doi.org/10.1038/oby.2008.7] [PMID: 18379563]

[107] Wildman RP, Muntner P, Reynolds K, McGinn AP, Rajpathak S, Wylie-Rosett J. The obese without cardiometabolic risk factor clustering and the normal weight with cardiometabolic risk factor clustering: prevalence and correlates of 2 phenotypes among the US population NHANES 1999 2004. Arch Intern Med 2008; 168: 1617-24.

[108] Danenberg HD, Szalai AJ, Swaminathan RV, *et al.* Increased thrombosis after arterial injury in human C-reactive protein-transgenic mice. Circulation 2003; 108(5): 512-5.

[http://dx.doi.org/10.1161/01.CIR.0000085568.13915.1E] [PMID: 12874178]

[109] Datta S, Iqbal Z, Prasad KR. Comparison Between Serum hsCRP and LDL Cholesterol for Search of a Better Predictor for Ischemic Heart Disease. Indian J Clin Biochem 2011; 26(2): 210-3. [http://dx.doi.org/10.1007/s12291-010-0100-4] [PMID: 22468052]

[110] Srikanthan K, Feyh A, Visweshwar H, Shapiro JI, Sodhi K. Systematic Review of Metabolic Syndrome Biomarkers: A Panel for Early Detection, Management, and Risk Stratification in the West Virginian Population. Int J Med Sci 2016; 13(1): 25-38. [http://dx.doi.org/10.7150/ijms.13800] [PMID: 26816492]

[111] McGillicuddy FC, de la Llera Moya M, Hinkle CC, *et al.* Inflammation impairs reverse cholesterol transport *in vivo*. Circulation 2009; 119(8): 1135-45. [http://dx.doi.org/10.1161/CIRCULATIONAHA.108.810721] [PMID: 19221221]

[112] Ishizaka N, Ishizaka Y, Toda E, Nagai R, Yamakado M. Association between serum uric acid, metabolic syndrome, and carotid atherosclerosis in Japanese individuals. Arterioscler Thromb Vasc Biol 2005; 25(5): 1038-44. [http://dx.doi.org/10.1161/01.ATV.0000161274.87407.26] [PMID: 15746438]

[113] Matsuura F, Yamashita S, Nakamura T, *et al.* Effect of visceral fat accumulation on uric acid metabolism in male obese subjects: visceral fat obesity is linked more closely to overproduction of uric acid than subcutaneous fat obesity. Metabolism 1998; 47(8): 929-33. [http://dx.doi.org/10.1016/S0026-0495(98)90346-8] [PMID: 9711987]

[114] Hikita M, Ohno I, Mori Y, Ichida K, Yokose T, Hosoya T. Relationship between hyperuricemia and body fat distribution. Intern Med 2007; 46(17): 1353-8. [http://dx.doi.org/10.2169/internalmedicine.46.0045] [PMID: 17827832]

[115] Reaven GM. The kidney: an unwilling accomplice in syndrome X. Am J Kidney Dis 1997; 30(6): 928-31. [http://dx.doi.org/10.1016/S0272-6386(97)90106-2] [PMID: 9398143]

[116] Fruehwald-Schultes B, Peters A, Kern W, Beyer J, Pfützner A. Serum leptin is associated with serum uric acid concentrations in humans. Metabolism 1999; 48(6): 677-80. [http://dx.doi.org/10.1016/S0026-0495(99)90163-4] [PMID: 10381138]

[117] Hayden MR, Tyagi SC. Uric acid: A new look at an old risk marker for cardiovascular disease, metabolic syndrome, and type 2 diabetes mellitus: The urate redox shuttle. Nutr Metab (Lond) 2004; 1(1): 10. [http://dx.doi.org/10.1186/1743-7075-1-10] [PMID: 15507132]

[118] Masuo K, Kawaguchi H, Mikami H, Ogihara T, Tuck ML. Serum uric acid and plasma norepinephrine concentrations predict subsequent weight gain and blood pressure elevation. Hypertension 2003; 42(4): 474-80. [http://dx.doi.org/10.1161/01.HYP.0000091371.53502.D3] [PMID: 12953019]

[119] Nakanishi N, Okamoto M, Yoshida H, Matsuo Y, Suzuki K, Tatara K. Serum uric acid and risk for development of hypertension and impaired fasting glucose or Type II diabetes in Japanese male office workers. Eur J Epidemiol 2003; 18(6): 523-30. [http://dx.doi.org/10.1023/A:1024600905574] [PMID: 12908717]

[120] Binder BR, Christ G, Gruber F, *et al.* Plasminogen activator inhibitor 1: physiological and pathophysiological roles. News Physiol Sci 2002; 17: 56-61. [PMID: 11909993]

[121] Juhan-Vague I, Alessi M-C, Mavri A, Morange PE. Plasminogen activator inhibitor-1, inflammation, obesity, insulin resistance and vascular risk. J Thromb Haemost 2003; 1(7): 1575-9. [http://dx.doi.org/10.1046/j.1538-7836.2003.00279.x] [PMID: 12871293]

[122] Bilgili S, Celebiler AC, Dogan A, Karaca B. Inverse relationship between adiponectin and plasminogen activator inhibitor-1 in metabolic syndrome patients. Endocr Regul 2008; 42(2-3): 63-8.

[PMID: 18624608]

[123] Vague P, Juhan-Vague I, Aillaud MF, *et al.* Correlation between blood fibrinolytic activity, plasminogen activator inhibitor level, plasma insulin level, and relative body weight in normal and obese subjects. Metabolism 1986; 35(3): 250-3. [http://dx.doi.org/10.1016/0026-0495(86)90209-X] [PMID: 3081778]

[124] Kraja AT, Province MA, Arnett D, *et al.* Do inflammation and procoagulation biomarkers contribute to the metabolic syndrome cluster? Nutr Metab (Lond) 2007; 4: 28. [http://dx.doi.org/10.1186/1743-7075-4-28] [PMID: 18154661]

[125] Sakkinen PA, Wahl P, Cushman M, Lewis MR, Tracy RP. Clustering of procoagulation, inflammation, and fibrinolysis variables with metabolic factors in insulin resistance syndrome. Am J Epidemiol 2000; 152(10): 897-907. [http://dx.doi.org/10.1093/aje/152.10.897] [PMID: 11092431]

[126] Alessi M-C, Juhan-Vague I. PAI-1 and the metabolic syndrome: links, causes, and consequences. Arterioscler Thromb Vasc Biol 2006; 26(10): 2200-7. [http://dx.doi.org/10.1161/01.ATV.0000242905.41404.68] [PMID: 16931789]

[127] Folsom AR, Qamhieh HT, Wing RR, *et al.* Impact of weight loss on plasminogen activator inhibitor (PAI-1), factor VII, and other hemostatic factors in moderately overweight adults. Arterioscler Thromb 1993; 13(2): 162-9. [http://dx.doi.org/10.1161/01.ATV.13.2.162] [PMID: 8427853]

[128] Kruszynska YT, Yu JG, Olefsky JM, Sobel BE. Effects of troglitazone on blood concentrations of plasminogen activator inhibitor 1 in patients with type 2 diabetes and in lean and obese normal subjects. Diabetes 2000; 49(4): 633-9. [http://dx.doi.org/10.2337/diabetes.49.4.633] [PMID: 10871202]

[129] Landar A, Zmijewski JW, Dickinson DA, Le Goffe C, Johnson MS, Milne GL, *et al.* Interaction of electrophilic lipid oxidation products with mitochondria in endothelial cells and formation of reactive oxygen species. Am J Physiol - Hear Circ Physiol 2006; 290: H1777-87. [http://dx.doi.org/10.1152/ajpheart.01087.2005]

[130] Pohjantähti-Maaroos H, Palomäki A, Kankkunen P, Laitinen R, Husgafvel S, Oksanen K. Circulating oxidized low-density lipoproteins and arterial elasticity: comparison between men with metabolic syndrome and physically active counterparts. Cardiovasc Diabetol 2010; 9: 41. [http://dx.doi.org/10.1186/1475-2840-9-41] [PMID: 20727144]

[131] Turer AT, Scherer PE. Adiponectin: mechanistic insights and clinical implications. Diabetologia 2012; 55(9): 2319-26. [http://dx.doi.org/10.1007/s00125-012-2598-x] [PMID: 22688349]

[132] Balsan GA, Vieira JL da C, Oliveira AM, de , Portal VL. Relationship between adiponectin, obesity and insulin resistance. Rev da Assoc Médica Bras 1992; 61: 72-80. [http://dx.doi.org/10.1590/1806-9282.61.01.072]

[133] Han SH, Quon MJ, Kim JA, Koh KK. Adiponectin and cardiovascular disease: response to therapeutic interventions. J Am Coll Cardiol 2007; 49(5): 531-8. [http://dx.doi.org/10.1016/j.jacc.2006.08.061] [PMID: 17276175]

[134] Bruun JM, Lihn AS, Verdich C, *et al.* Regulation of adiponectin by adipose tissue-derived cytokines: *in vivo* and *in vitro* investigations in humans. Am J Physiol Endocrinol Metab 2003; 285(3): E527-33. [http://dx.doi.org/10.1152/ajpendo.00110.2003] [PMID: 12736161]

[135] Vaiopoulos AG, Marinou K, Christodoulides C, Koutsilieris M. The role of adiponectin in human vascular physiology. Int J Cardiol 2012; 155(2): 188-93. [http://dx.doi.org/10.1016/j.ijcard.2011.07.047] [PMID: 21907426]

[136] Yamauchi T, Nio Y, Maki T, *et al.* Targeted disruption of AdipoR1 and AdipoR2 causes abrogation of adiponectin binding and metabolic actions. Nat Med 2007; 13(3): 332-9.

[http://dx.doi.org/10.1038/nm1557] [PMID: 17268472]

[137] Awazawa M, Ueki K, Inabe K, *et al.* Adiponectin enhances insulin sensitivity by increasing hepatic IRS-2 expression *via* a macrophage-derived IL-6-dependent pathway. Cell Metab 2011; 13(4): 401-12. [http://dx.doi.org/10.1016/j.cmet.2011.02.010] [PMID: 21459325]

[138] Fu Y, Luo N, Klein RL, Garvey WT. Adiponectin promotes adipocyte differentiation, insulin sensitivity, and lipid accumulation. J Lipid Res 2005; 46(7): 1369-79. [http://dx.doi.org/10.1194/jlr.M400373-JLR200] [PMID: 15834118]

[139] Lara-Castro C, Fu Y, Chung BH, Garvey WT. Adiponectin and the metabolic syndrome: mechanisms mediating risk for metabolic and cardiovascular disease. Curr Opin Lipidol 2007; 18(3): 263-70. [http://dx.doi.org/10.1097/MOL.0b013e32814a645f] [PMID: 17495599]

[140] Sell H, Dietze-Schroeder D, Eckardt K, Eckel J. Cytokine secretion by human adipocytes is differentially regulated by adiponectin, AICAR, and troglitazone. Biochem Biophys Res Commun 2006; 343(3): 700-6. [http://dx.doi.org/10.1016/j.bbrc.2006.03.010] [PMID: 16563350]

[141] Bacha F, Saad R, Gungor N, Arslanian SA. Adiponectin in youth: relationship to visceral adiposity, insulin sensitivity, and β-cell function. Diabetes Care 2004; 27(2): 547-52. [http://dx.doi.org/10.2337/diacare.27.2.547] [PMID: 14747242]

[142] Falahi E, Khalkhali Rad AH, Roosta S. What is the best biomarker for metabolic syndrome diagnosis? Diabetes Metab Syndr 2015; 9(4): 366-72. [http://dx.doi.org/10.1016/j.dsx.2013.06.014] [PMID: 25470629]

[143] Finucane FM, Luan J, Wareham NJ, *et al.* Correlation of the leptin:adiponectin ratio with measures of insulin resistance in non-diabetic individuals. Diabetologia 2009; 52(11): 2345-9. [http://dx.doi.org/10.1007/s00125-009-1508-3] [PMID: 19756488]

[144] Cong WN, Golden E, Pantaleo N, White CM, Maudsley S, Martin B. Ghrelin receptor signaling: a promising therapeutic target for metabolic syndrome and cognitive dysfunction. CNS Neurol Disord Drug Targets 2010; 9(5): 557-63. [http://dx.doi.org/10.2174/187152710793361513] [PMID: 20632971]

[145] Varela L, Vázquez MJ, Cordido F, *et al.* Ghrelin and lipid metabolism: key partners in energy balance. J Mol Endocrinol 2011; 46(2): R43-63. [PMID: 21169422]

[146] Tesauro M, Schinzari F, Rovella V, *et al.* Ghrelin restores the endothelin 1/nitric oxide balance in patients with obesity-related metabolic syndrome. Hypertension 2009; 54(5): 995-1000. [http://dx.doi.org/10.1161/HYPERTENSIONAHA.109.137729] [PMID: 19786648]

[147] Pulkkinen L, Ukkola O, Kolehmainen M, Uusitupa M. Ghrelin in diabetes and metabolic syndrome. Int J Pept 2010; 2010: 248948.

[148] Santaniemi M, Kesäniemi YA, Ukkola O. Low plasma adiponectin concentration is an indicator of the metabolic syndrome. Eur J Endocrinol 2006; 155(5): 745-50. [http://dx.doi.org/10.1530/eje.1.02287] [PMID: 17062891]

[149] Ukkola O. Ghrelin and metabolic disorders. Curr Protein Pept Sci 2009; 10(1): 2-7. [http://dx.doi.org/10.2174/138920309787315220] [PMID: 19275667]

[150] Tschöp M, Weyer C, Tataranni PA, Devanarayan V, Ravussin E, Heiman ML. Circulating ghrelin levels are decreased in human obesity. Diabetes 2001; 50(4): 707-9. [http://dx.doi.org/10.2337/diabetes.50.4.707] [PMID: 11289032]

[151] Bacha F, Arslanian SA. Ghrelin suppression in overweight children: a manifestation of insulin resistance? J Clin Endocrinol Metab 2005; 90(5): 2725-30. [http://dx.doi.org/10.1210/jc.2004-1582] [PMID: 15728212]

[152] McLaughlin T, Abbasi F, Lamendola C, Frayo RS, Cummings DE. Plasma ghrelin concentrations are

decreased in insulin-resistant obese adults relative to equally obese insulin-sensitive controls. J Clin Endocrinol Metab 2004; 89(4): 1630-5.
[http://dx.doi.org/10.1210/jc.2003-031572] [PMID: 15070922]

[153] St-Pierre DH, Karelis AD, Coderre L, *et al.* Association of acylated and nonacylated ghrelin with insulin sensitivity in overweight and obese postmenopausal women. J Clin Endocrinol Metab 2007; 92(1): 264-9.
[http://dx.doi.org/10.1210/jc.2006-1603] [PMID: 17062757]

[154] Arner P. Resistin: yet another adipokine tells us that men are not mice. Diabetologia 2005; 48(11): 2203-5.
[http://dx.doi.org/10.1007/s00125-005-1956-3] [PMID: 16193286]

[155] Pang SS, Le YY. Role of resistin in inflammation and inflammation-related diseases. Cell Mol Immunol 2006; 3(1): 29-34.
[PMID: 16549046]

[156] Iqbal N, Seshadri P, Stern L, *et al.* Serum resistin is not associated with obesity or insulin resistance in humans. Eur Rev Med Pharmacol Sci 2005; 9(3): 161-5.
[PMID: 16080635]

[157] Chen C-C, Li T-C, Li C-I, Liu C-S, Wang H-J, Lin C-C. Serum resistin level among healthy subjects: relationship to anthropometric and metabolic parameters. Metabolism 2005; 54(4): 471-5.
[http://dx.doi.org/10.1016/j.metabol.2004.10.015] [PMID: 15798953]

[158] Norata GD, Ongari M, Garlaschelli K, Raselli S, Grigore L, Catapano AL. Plasma resistin levels correlate with determinants of the metabolic syndrome. Eur J Endocrinol 2007; 156(2): 279-84.
[http://dx.doi.org/10.1530/eje.1.02338] [PMID: 17287419]

[159] Stejskal D, Adamovská S, Bartek J, Juráková R, Prosková J. Resistin - concentrations in persons with type 2 diabetes mellitus and in individuals with acute inflammatory disease. Biomed Pap Med Fac Univ Palacky Olomouc Czech Repub 2003; 147(1): 63-9.
[http://dx.doi.org/10.5507/bp.2003.009] [PMID: 15034607]

[160] Lehrke M, Reilly MP, Millington SC, Iqbal N, Rader DJ, Lazar MA. An inflammatory cascade leading to hyperresistinemia in humans. PLoS Med 2004; 1(2): e45.
[http://dx.doi.org/10.1371/journal.pmed.0010045] [PMID: 15578112]

[161] Eck M, Sprang SR. The structure of tumor necrosis factor-alpha at 2.6 a resolution. Implications for receptor binding. J Biol Chem 1989; 264: 17595-605.

[162] Graves B, Hatada M, Hendrickson W, Miller J, Madison V, Satow Y. Structure of interleukin 1 alpha at 2.7-A resolution. Biochemistry 1990; 29: 2679-84.

[163] Finzel B, Clancy L, Holland D, Muchmore S, Watenpaugh K, Einspahr HM. Crystal structure of recombinant human interleukin-1 beta at 2.0 A resolution. J Mol Biol 1989; 209: 779-91.

[164] Xu G, Yu H, Hong J, Stahl M, McDonagh T, Kay L, *et al.* Solution structure of recombinant human interleukin-6. J Mol Biol 1997; 268: 468-81.

[165] Kato Z, Jee J, Shikano H, Mishima M, Ohki I, Ohnishi H, *et al.* The structure and binding mode of interleukin-18. Nat Struct Biol 2003; 10: 966-71.
[http://dx.doi.org/10.1038/nsb993]

[166] Zhou A, Carrell R, Murphy M, Wei Z, Yan Y, Stanley P, *et al.* A redox switch in angiotensinogen modulates angiotensin release. Nature 2010; 468: 108.
[http://dx.doi.org/10.1038/nature09505]

[167] Zhang F, Basinski M, Beals J, Briggs S, Churgay L, Clawson D, *et al.* Crystal structure of the obese protein leptin-E100. Nature 1997; 387: 206-9.

[168] Shrive A, Cheetham G, Holden D, Myles D, Turnell W, Volanakis J, *et al.* Three dimensional structure of human C-reactive protein. Nat Struc tBiol 1996; 3: 346-54.

[169] Nar H, Bauer M, Stassen J, Lang D, Gils A, Declerck PJ. Plasminogen activator inhibitor 1. Structure of the native serpin, comparison to its other conformers and implications for serpin inactivation. J Mol Biol 2000; 297: 683-95.

CHAPTER 3

Liver Fibrosis Biomarkers

Iria Cebreiros-López*

Clinical Analysis Department, Clinical University Hospital Virgen de la Arrixaca, Murcia, Spain

Abstract: Chronic liver diseases of differing etiologies are among the leading causes of mortality and morbidity worldwide. Establishing accurate staging of liver disease is very important for enabling both therapeutic decisions and prognostic evaluations. A liver biopsy is considered the gold standard for assessing the stage of hepatic fibrosis, but it has many limitations. During the last decade, several noninvasive markers for assessing the stage of hepatic fibrosis have been developed. Some have been well validated and are comparable to liver biopsy. This chapter will focus on the various noninvasive biochemical markers used to stage liver fibrosis.

Keywords: AAR, APRI, Biomarkers, Collagen, Cirrhosis, Chronic liver diseases, ELF, FIB-4, Fibrosis, FibroIndex, FibroMeter, FibroQ, FibroTest, Fibrospect, Forns, HepaScore, PGA, SHASTA, YKL-40, Hyaluronic acid, Laminin, Liver biopsy, Metalloproteinases, Procollagen, Transforming growth factor.

INTRODUCTION

Chronic liver diseases of differing etiologies are among the leading causes of morbidity and mortality worldwide. Chronic liver disease progresses through different pathological stages that vary from mild hepatic inflammation without fibrosis to advanced hepatic fibrosis and cirrosis [1, 2]. Assessment of the stage of liver disease is important for diagnosis, treatment, and follow-up both during treatment and after cessation of treatment. A liver biopsy is the oldest and most accurate method used to evaluate liver histology and the progression of chronic liver disease. Furthermore, different histological scoring systems have been developed and modified. A liver biopsy is considered the gold standard for assessing liver histology [1, 3].

During the pathological progression of liver fibrosis, excessive amounts of extracellular matrix build up; furthermore, serum levels of various biomarkers change, in addition to the appearance of new biomarkers in the serum during

* **Corresponding author Iria Cebreiros-López:** Clinical Analysis Department, Clinical University Hospital Virgen de la Arrixaca, Murcia, Spain; E-mail: iriacebreiros@hotmail.com

different stages of fibrosis [1, 4].

Recently many noninvasive markers for assessing liver fibrosis have been developed, and they are frequently used in clinical practice. They have been validated in different studies, and some were found to be highly accurate in the assessment of liver fibrosis compared with liver biopsies, which have always been used as the standard reference method for evaluating the accuracy of noninvasive methods.

THE BIOLOGY OF LIVER FIBROSIS

The most important cellular player in the production of extracellular matrix is the myofibroblast (MF). A wide array of cells of different origins can be converted into fibrogenic MFs, including portal MFs and bone marrow-derived mesenchymal stem cells. Some epithelial cells including hepatocytes and biliary epithelial cells (cholangiocytes) could be activated to function as myofibroblasts through the process of Epithelial-Mesenchymal Transition (EMT). However, the predominant MF producing liver cells are quiescent hepatic stellate cells (HSC), also known as Ito cells or perisinusoidal cells residing in the space of Disse and storing retinoid [5, 6].

Non-MF cells actively participate in the process of fibrogenesis. For example, hepatocytes can respond to this damage in multiple ways, including production of reactive oxygen species (ROS) and apoptosis, while the resident liver macrophages called Kupffer cells elicit a massive immune response resulting in the recruitment of other inflammatory cells to the site of injury [7]. Attracted to the chemokines produced by the Kupffer cells, the leukocytes exit out of the vasculature towards the injury site and contribute to the release of additional pro-inflammatory and pro-fibrotic mediators, including cytokines such as tumor necrosis factor alpha (TNF-α) and various interleukins. Reactive oxygen and nitrogen species, proteases, and lipid metabolites such as prostaglandins and thromboxane are also released. As a result of this response, quiescent HSCs are converted to activated myofibroblasts and, in turn, contribute to the chemotaxis of leukocytes as well as their own chemotaxis through the production of chemokines and cytokines such as monocyte chemotactic protein-1 (MCP-1) [8]. As a result, activated HSCs start expressing the Platelet Derived Growth Factor (PDGF) receptor and Transforming Growth Factor (TGF) receptor. TGF-β is the central mediator of fibrogenesis, while PGDF stimulates proliferation of the HSCs. Activation of HSCs is associated with gradual replacement of basement membrane-like extracellular matrix (ECM) within the space of Disse by the collagen rich fibers and the production of fibrous bands. In advanced stages of fibrosis, the liver contains approximately six times more ECM components than normal, including collagens (I, III, and IV), fibronectin, undulin, elastin, laminin, hyaluronan, and proteoglycans [8].

Taking into account the pathophysiological process involved in the development of fibrosis, several molecules could be used as biomarkers of fibrosis. One can look for biomarkers among molecules that change expression in the process of HSC activation or at the methods that enumerate MFs or their products. Another approach can look at direct quantification of ECM within the liver, or the particular molecules involved in the process of profibrogenic inflammation within the liver parenchyma. To prove the clinical utility of any of these biomarkers, it is essential to perform a correct validation. In addition, as the mechanisms involved in the development of fibrosis as well as their distribution in the liver, vary depending on the etiologic agent, each biomarker should be evaluated in several patient cohorts.

Types and Causes of Acquired Liver Fibrosis

The development of fibrosis is related to various etiological agents, such as alcoholic liver disease, cholestasis, viral hepatitis, non-alcoholic fatty liver disease (NAFLD), hepatic copper or iron overload. These etiologic agents may be present in an isolated or combined manner, thereby having a cumulative toxic effect. It should be taken into account that the presence of liver fibrosis has a different clinical meaning depending on the etiological context, so it should always be evaluated in an individualized way. For example, in the case of hepatitis C the presence of fibrotic changes could argue for the need for antiviral treatment, while in the case of NAFLD it does not change the therapeutic options, but may provide important prognostic information.

Alcoholic Liver Disease (ALD)

Excessive and chronic alcohol consumption is one of the main causative agents of liver fibrosis and cirrhosis. Metabolization of ethanol produced acetaldehyde and reactive oxygen species (ROS), both with profibrotic action. In hepatocytes, the primary site for alcohol metabolism, acetaldehyde and ROS are produced in abundance, then they diffuse outside and enter HSCs. Acetaldehyde directly up-regulates the transcription of collagen I and the synthesis of transforming growth factor-beta 1 (TGF-β1). Exposure to ROS sensitizes HSCs to various pro-inflammatory factors and elicits the production of inflammatory mediators that contribute to the fibrotic changes in the liver. In ALD, the fibrotic changes in the liver start in the pericentral and perisinusoidal areas [9, 10].

Chronic Viral Hepatitis

Chronic viral infections such as hepatitis B (HBV) or hepatitis C (HCV) viruses pose an important risk for the development of liver fibrosis. The pathophysiological mechanism involved in the development of fibrosis is less clear in the case of viral hepatitis than in chronic non-viral liver disease. In chronic viral hepatitis, fibrosis is usually initiated in the portal area [9]. The pathogenesis is probably multifactorial, resulting from the interaction between virus-specific factors with specific factors of the host, including oxidative stress, hepatic steatosis, increased iron stores, and increased rate of hepatocyte apoptosis, under the pressure of the viral proteins and viral replication.

In chronic HCV infection, the viral core, NS5 and NS3 proteins have been demonstrated to initiate a cascade of molecular events that can eventually lead to fibrosis. HCV proteins appear to affect both lipid accumulation and degradation, with the consequent disruption of the normal process of lipid compartmentalization and metabolism, skewing towards ROS production. In the case of HBV infection, studies have shown that the X protein of HBV (HBx) directly induces TGF-β secretion by hepatocytes and, thus, contributes to the paracrine activation of HSC's [11]. Both HIV-HBV and HIV-HCV coinfected patients are at increased risk for progression of their liver disease as compared to patients who are mono-infected with HCV or HBV [12].

Non-alcoholic Fatty Liver Disease (NAFLD)

The term "non-alcoholic fatty liver disease" (NAFLD) encompasses a broad spectrum of lesions ranging from simple hepatic steatosis, steatohepatitis (NASH) associated with a variable degree of fibrosis, which can progress to liver cirrhosis and even hepatocellular carcinoma. This type of liver disease is usually seen in subjects who have the alterations associated with the metabolic syndrome (MS) such as obesity, type- 2 diabetes (DM), dyslipidemia, and insulin resistance. To date, the pathogenesis of NASH-related liver fibrosis is not entirely well understood [4]. Evidence provided by numerous studies links obesity, insulin resistance and the progression of fibrosis together in one vicious circle [13]. The same factors are also known for their association with hepatic fibrosis. For example, leptin, the well-know adipokine produced proportionally to the mass of the visceral adipose compartment, also augments fibrogenesis by stimulating phagocytic activity and cytokine secretion by Kupffer cells and macrophages as well as the proliferative and ROS generating activities of the endothelial cells [14, 15]. Another adipokine, resistin, exerts pro-inflammatory actions in HSC by increasing the expression of both MCP-1 and interleukin-8 (IL-8) as well as activating the transcription factor, NFkB [16]. Therefore, taking these data into account, it can be deduced that the development of fibrosis in NAFLD is related to factors produced as a result of excessive visceral adipose and on a skewed distribution of the soluble fat particles in the bloodstream.

Cholestatic Liver Diseases

Cholestasis (reduced bile duct excretion) is another well-known cause of liver fibrosis. Cholestasis triggers the proliferation of the cholangiocyte lining of the intrahepatic and extrahepatic bile duct systems through a complex regulatory milieu that involves both autocrine and paracrine factors [17]. The activation of biliary proliferation is known as ductular reaction. Proliferating bile duct epithelial cells produce the profibrogenic connective tissue growth factor (CTGF) that stimulates myofibroblast generation through Epithelial-Mesenchymal Transition (EMT) and collagen deposition [18]. In the fibrotic reaction that takes place in cholestasis the inflammatory response triggered by neutrophils and resulting from oxidative stress plays a major role. Fibrosis begins in the hepatic portal area, where there are normally large numbers of fibroblasts, which then become myofibroblasts [9, 10]. The hepatocytes adjacent to these fibroblasts are destroyed rapidly, which causes the increase of the portal field and the activation of the portal fibroblasts [10].

DIAGNOSTICS OF LIVER FIBROSIS

The great advances that have occurred in recent years in the treatment of chronic liver diseases has generated the need for diagnostic methods that allow to evaluate fibrosis accurately. The main limitation for the clinical management of these patients, as well as to advance in the field of research, is the absence of reproducible and easily applied methods for assessment of hepatic fibrosis.

Liver Biopsy Scoring Techniques

Liver biopsy is considered the gold standard for the diagnosis and staging of fibrosis. In addition to fibrosis, it allows to obtain much information about other processes of liver damage, such as inflammation, necrosis or steatosis. However, the biopsy presents several important limitations, such as sampling error, inter- and intraobserver variability, and high economic cost [19]. It is also an invasive technique and therefore there is a small but real risk of associated morbidity and mortality. Complications associated include pain, bleeding, infections, and hepato-biliary fistulae. Studies reveal that the risk for hospitalization after liver biopsy is 1-5%, the risk for severe complications is 0.57%, and mortality rates vary from 0.009% to 0.12%. These reasons justify that many patients refuse to be tested, which could imply serious prognostic consequences [20, 21].

These limitations and complications associated with biopsy have led to the development of new non-invasive alternative methods that accurately detect and quantify fibrosis. These alternative methods include laboratory biomarkers and imaging studies.

However, the diagnostic performance of all noninvasive methods should be assessed with reference to liver biopsy, which at present, despite its limitations, remains the reference technique.

The rate of adoption of non-invasive diagnostics for liver fibrosis differs from country to country, but remains limited. At present, the use of non-invasive procedures could be recommended as pre-screening tools, which may allow physicians to narrow down the patients' population before definitive testing of liver fibrosis by biopsy of the liver.

Imaging Techniques

Ultrasonography (US)

US is an innocuous technique, inexpensive and of great diagnostic value in abdominal pathology. Therefore, after anamnesis and physical examination, it is the technique of choice in patients with suspected liver disease, simultaneously with the analytical tests aimed at establishing the cause of the process.

US findings that suggest progression of fibrosis in patients with chronic liver disease include altered parenchymal echogenicity with coarsened echotexture and surface nodularity that reflects the presence of regenerative nodules and fibrous septa. As cirrhosis progresses, characteristic hypertrophy of the caudate and lateral segment with volume loss of the right lobe of the liver is observed, while in the advanced phase liver atrophy is complete. These findings may lack high sensitivity and specificity, and liver morphology may be normal in the early stage of cirrhosis [22].

However, correlation between US and histologic diagnosis has not been fully investigated in large series of patients [23].

Elastography and Fibroscan

It has been reported that hepatic stiffness proportionately increases with fibrosis progression. Elastography has been studied as an assessment tool for the staging of liver fibrosis. Comparison of elastic ratio with various markers and scores showed that elastography appears to be a better technique than Forns score, hepascore, fibro index and serum markers like hyaluronic acid. The role of

Transient Elastography (Fibroscan®) in measuring fibrosis seems promising, but there are certain limitations. It is not practical for patients who have ascites because the elastic waves are unable to propagate through liquids. It is not only a cost-intensive procedure, but also Elastography has limited significance in patients with narrow intercostal spaces and in those who are obese [24].

Acoustic Radiation Force Impulse (ARFI) Imaging Sonoelastography

ARFI is an imaging technique which uses sonography for assessment of liver fibrosis. Unlike Fibroscan and Elastography, ARFI can be performed on obese patients and on patients with narrow intercostals spaces. The diagnostic accuracy of ARFI when expressed as area under receiver operating curve (AUROC) showed validity of 90.2% for significant fibrosis, being higher with progressing fibrosis. The results showed overlap between fibrosis stages, especially at early stages, hence the role in diagnosing early fibrosis could not be established [24].

Biomarkers for Assessing Liver Fibrosis

In recent years, interest in the identification of serological biomarkers as diagnostic methods for hepatic fibrosis has been increasing. In addition to its non-invasive nature, sampling error and variability in relation to the analyzer is practically null, so they are a good alternative to biopsy. On the other hand, the determinations can be performed repeatedly over time, thereby allowing a dynamic control of the disease. These are the main reasons that biomarkers are presented as a good tool for monitoring the progression of chronic liver disease, either as part of monitoring the natural history of the disease or to control treatment regimens.

The Ideal Biomarker of Liver Fibrosis

The diagnostic value of serum markers of liver fibrosis has been investigated in numerous studies. Based on clinical and research needs, the ideal marker for liver fibrosis would have the following characteristics:

- Be highly sensitive and specific to identify different stages of fibrosis.
- Be readily available, safe, inexpensive and reproducible.
- Be applicable to the monitoring of disease progression or regression as a part of natural history of liver disease or treatment regimens.
- Not be susceptible to false positive results, for example, in individuals with inflammation related to other diseases.

Although no single ideal marker exists, several markers have been identified as possible useful indicators of fibrosis when used in conjunction with each other.

Noninvasive biomarkers for assessing liver fibrosis are grouped into two main categories: class 1 fibrosis markers, or direct biomarkers, and class 2 fibrosis markers, or indirect biomarkers. The direct markers directly correlate with or are parts of the liver matrix produced by the Hepatic stellate cells during ECM turnover in the fibrosis process. In contrast, the indirect markers reflect changes in liver functions and are molecules released into the blood due to liver inflammation, but they do not correlate with ECM turnover [1, 25, 26].

Direct Biomarkers

(i) Direct Markers Linked to Matrix Deposition

- ***Procollagen Type 1 and Type III***

Procollagen is a collagen precursor. It is cleaved by two different enzymes at its carboxy-terminal (type 1 (PC1CP)) and amino-terminal (type III (PCIIINP)), leading to the production of collagen. Mature collagen integrates into ECM [1, 19].

a. The PCICP terminal peptide is major component of the connective tissue. It has a higher upper limit of normal limit in male compared to females 202 μg and 170 μg, respectively. It is normal in patients with mild liver disease but increases in patients with moderate to severe cirrhosis [27].
b. PCIIINP or PIIINP is another major component of connective tissue that has been extensively studied. Serum levels of PCIIINP reflect the stage of hepatic fibrosis. During cirrhosis, PCIIINP serum levels correlate with serum bilirubin. An upper limit of normal for PIIINP was defined by *Gallorini et al.*, as 0.8 U/Ml [27]. Available data on PCIIINP in viral hepatitis and alcoholic disease show that it is increased and that its levels correlate with the severity of liver disease. Furthermore, a reduction in PCIIINP correlates with the response of chronic hepatitis C (CHC) patients to treatment with interferon [27]. The main limitation of using PCIIINP is that it is not specific to hepatic fibrosis and that it can be detected in other conditions. Furthermore, it shows lower efficacy compared with type IV collagen and hyaluronic acid.

- ***Type IV Collagen***

Is a component of ECM that was investigated as a surrogate marker of liver fibrosis [19]. It has three different regions, an amino-terminal domain, a central helix domain, and a carboxy-terminal domain. Type IV collagen has been studied extensively in liver diseases of different etiologies. It is increased in patients with liver diseases and its levels correlate significantly with the extent of hepatic fibrosis [28, 29]. At a cutoff level of ≥5.0 ng/mL type VI collagen had an AUC of 0.82 and negative predictive value of 83,6% for the detection of severe fibrosis in non alcoholic fatty liver disease (NAFLD) [28]. *Walsh et al.*, had shown that type VI collagen is elevated in hepatitis C patients compared to control (median of 127,1 ng/mL, range 17,7–317,4 and median 61,3 ng/mL, range 11,5–102,3), respectively [29].

- ***Hyaluronic Acid (HA)***

Is a glycosaminoglycan, and it is a component of the ECM that is produced by hepatic stellate cells. The liver is the main tissue responsible for the synthesis and elimination of HA, is synthesized primarily by hepatic stellate cells and serves as an integral component of ECM, and is degraded by sinusoidal endothelial cells. The serum concentration of HA depends mainly on the morphological changes that accompany hepatic sinusoidal capillarization in the evolution from chronic liver diseases to cirrhosis. It is for this reason that although their concentrations increase as fibrosis progresses, the higher levels are observed in cirrhotic patients [30]. An upper limit of normal range was defined by *Gallorini et al.*, as 98 µg/L [27]. Variable cutoff point had been defined by different authors. HA has been studied in CHC, NAFLD, alcoholic liver disease, and chronic hepatitis B (CHB), but it has been more extensively studied in the former two diseases. HA has been of great value in detecting advanced fibrosis. HA shows a negative predictive value of 98–100% for cirrhosis and is of great value in excluding cirrhosis. In treated CHC patients, the response to treatment was reported to be associated with a reduction in serum HA levels [30].

- ***Laminin***

Is a noncollagenous glycoprotein component of the ECM that is produced by hepatic stellate cells. It is deposited in the basement membrane of the liver [31]. Serum levels of laminin are elevated above the upper limit of normal range (0.59–1.4 U/mL) or (9.74–2.46) as defined by different authors. Furthermore *Kropf et al.*, [32] had proposed a cutoff value of 1.45 for laminin for the detection of both liver fibrosis and cirrhosis in patients with chronic liver disease, and they correlate with the degree of perisinusoidal fibrosis. It showed an accuracy of 77% in the detection of significant fibrosis in CHC. Laminin has also been found to be of prognostic value, with a diagnostic accuracy of 70% for predicting the risk of variceal bleeding. The data on basement membrane related to direct noninvasive markers showed that PICP, PIIINP, type VI collagen, and laminin levels decrease during abstinence from alcohol intake.

- ***YKL-40 Chondrex***

Is a member of the chitinase family, and it is involved in the remodeling and degradation of the ECM. Increased serum levels of YKL-40 chondrex to (330 μg/L), have been shown to correlate with the degree of fibrosis in all forms of liver disease and similar observations were made for cirrhosis (425 μg/L) as compared to age matched normal controls (102 μg/L). *Saitou et al.*, had defined different cutoff levels for fibrosis and cirrhosis 186.4 and 284.8 μg/L, respectively. YKL-40 levels have also been observed to correlate with HA levels. The serum level of YKL-40 during post interferon therapy for CHC significantly decreased in both responder and non responder patients [33, 34].

(ii) Direct Markers Linked to Matrix Degradation

Degradation of the EMC is an action primarily of the family of metalloproteinase enzymes (MMPs). Tissue inhibitors of matrix metalloproteinases (TIMPs) interfere with MMP functions and lead to the inhibition of ECM degeneration. The balance between both determines the rate of replenishment of the ECM.

- ***Metalloproteinases (MMPs)***

MMPs are enzymes that are produced intracellularly and are secreted as proenzymes, which require a process of cleavage by mechanisms of the cell surface to have functional activity. These enzymes participate directly in the elimination of ECM, as they are the main responsible for the degradation of collagen [35, 36]. The three most commonly studied human metalloproteinases are MMP-2 (gelatinase-A), MMP-3 (stromelysin), and MMP-9 (gelatinase-B). MMP-2 is secreted by activated HSCs; elevated levels of MMP-2 and its proenzyme have been observed in various liver diseases [37]. During hepatic fibrogenesis, the expression of MMP-2 is markedly increased. The potential for MMP-2 for predicting liver fibrosis remains unclear as some contradictory data have been reported by studies performed so far [38, 39]. In contrast to MMP-2, MMP- 9 levels show their value primarily in the diagnosis of hepatocellular carcinoma [40]. In one of the studies, MMP-9 levels were negatively correlated to the histological severity of the liver disease in patients with chronic hepatitis C [41].

- ***Tissue Inhibitors of Matrix Metalloproteinases (TIMPs)***

TIMPs are secreted proteins that interact with MMPs and modulate their activation and functioning. TIMP-1 controls activity of most MMPs, whereas TIMP-2 specifically inhibits MMP-2. TIMPs-dependent inhibition of ECM degradation may promote liver fibrosis; elevation of TIMPs' levels has been observed in chronic liver disease. For example, chronic hepatitis C causes the elevation of both TIMP-1 and TIMP-2 in corollary with fibrosis progression [39]. A recent study of the relationship between serum MMP-9, TIMP-1 and fibrosis in 50 patients with various chronic liver disease showed that serum levels of MMP-9 in chronic hepatitis patients were low as compared to the controls [41]. Moreover, serum MMP-9 levels decrease as chronic hepatitis progresses to cirrhosis, while TIMP-1 levels increase along with an increase of the degree of fibrosis. These findings prompt using serum TIMP-1 as a non-invasive assay in liver fibrosis [41].

(iii) Cytokines and Chemokines Linked to Liver Fibrosis

▪ ***Transforming Growth Factor-β1 (TGF-β1)***

TGF-β1 is a pleiotropic cytokine involved in tissue growth, differentiation, ECM production and the immune response. Three isoforms (β1, β2 and β3) of this cytokine have been identified, but only TGF-β1 is linked to liver fibrogenesis. TGF- β1 is also commonly accepted as a central component of fibrogenic response to wounding and is up-regulated in a variety of different diseases. A correlation between TGF-β1 levels and the rate of fibrosis progression is widely accepted [42, 43].

▪ ***Transforming Growth Factor Alpha (TGF-α)***

Transforming growth factor alpha (TGF-α) was found to enhance the proliferation of hepatic stellate cells by inducing the entry of hepatic stellate cells into S-phase. In patients with liver disease, TGF-α was found to correlate with the progression of liver disease, Child-Pugh classification, and it is increased in patients with hepatocellular carcinoma [44].

Limitations of Direct serum biomarkers of fibrosis

- They reflect the rate of matrix turnover (not only deposition) and have a tendency to be more elevated when associated with high inflammatory activity. As a consequence, extensive matrix deposition might not be detected in the presence of minimal inflammation.
- They are not liver-specific and their serum levels may be elevated in the presence of concomitant sites of inflammation.
- Serum levels of markers depend on their clearance rates, which are influenced by the dysfunction of endothelial cells, impaired biliary excretion or renal function [45].

Indirect Biomarkers

Most indirect biomarkers of fibrosis are integrated with one or more fibrosis predicting biomarker panels.

These indices or mathematical algorithms have resulted from several clinical studies, using statistical models, such as multiple logistic regression, the parameters have been selected offered the best results for the detection and classification of hepatic fibrosis. In Table **1**, the main mathematical indices available today can be observed.

The following describes in more detail some of the most commonly used biomarkers and algorithms:

- ***Serum Alanine Aminotransferase (ALT)***

ALT is one of the oldest markers used to assess liver disease. *Pradat et al.* have shown that serum ALT is beneficial to measure due to its high sensitivity and specificity (2.25-fold greater than the normal levels predicts liver histology) [46]. However, serum ALT levels are affected by many factors, including gender, body mass index, and the use of hepatotoxic medications.

- ***The Aspartate Aminotransferase (AST)/ALT (AAR)***

The predictive value of the AST/ALT ratio has been validated in non-alcoholic liver disease, chronic viral hepatitis, primary sclerosing cholangitis, and primary biliary cirrhosis [47]. In many forms of acute and chronic liver injury or steatosis (fatty infiltration of the liver), this ratio is less than or equal to 1, while in alcoholic hepatitis, an AST/ALT ratio is often greater than 2. While these ratios are suggestive of certain etiology of liver conditions, there is too much overlap between groups to rely on AST/ALT exclusively when making a diagnosis - for example, in patients with both hepatitis C and history of alcohol abuse [48].

- ***The AST/Platelet Ratio (APRI)***

This ratio was developed by *Wai et al.*, [50] in 2003 and is calculated as (AST/upper limit of normal range)/platelet count (10^9/L) × 100. This index has previously been validated as a surrogate marker of significant hepatic fibrosis in HIV/HCV-coinfected patients, and has recently been used to determine advanced fibrosis in HIV-monoinfected patients [51]. However, recent large meta-analysis suggested that APRI can identify hepatitis C-related fibrosis with only a moderate degree of accuracy [52].

Table 1. Main mathematical algorithms for detecting liver fibrosis are illustrated. Adapted from *Gressner et al.*, [49]. Abbreviations: GGT, γ-glutamyltransferase; PIIINP, N-terminal propeptide of type III-procollagen; TIMP, tissue inhibitors of metalloproteinases;MMP,matrix metalloproteinases; β-NAG, N-acetyl-β-glucosaminidase; AST, aspartate-aminotransferase; ALT, alanine aminotransferase; INR, international normalized ratio.

Index	Parameters
PGAA index	Prothrombin time, γGT, apolipoprotein A1, α2
Bonacini index	ALT/AST-ratio, INR, platelet count
Sheth index	AST/ALT (De Ritis)
PGA index	Prothrombin time, γGT, apolipoprotein A1
Fortunato score	Fibronectin, prothrombin time, PCHE, ALT, Mn-SOD, β-NAG
Fibrotest	Haptoglobin, α2-macroglobulin, apolipoprotein A1, γGT, bilirubin
Pohl score	AST/ALT-ratio, platelet count
Actitest	Fibrotest + ALT
Forns index	Age, platelet count, γGT, cholesterol
APRI	AST, platelet count
ELF score	PIIINP, hyaluronan, TIMP
FibroSpect	Hyaluronan, TIMP-1, α2-macroglobulin
Sud-index	Age, AST, cholesterol, insulin resistance (HOMA), past alcohol intake
Leroy score	PIIINP, MMP
Fibrometer test	Platelet count, prothrombin index, AST, α2-macroglobulin, hyaluronan, urea, age
Hepascore	Bilirubin, cGT, hyaluronan, α2-macroglobulin, age, gender
Testa index	Platelet count/spleen diameter-ratio
FIB-4	Platelet count, AST, ALT, age
FibroIndex	Platelet count, AST, γ-globulin

▪ *The Forns Index*

This index was described by *Forns et al.*, [53] in 2002. It is calculated based on the age of the patient and three routine laboratory tests, namely, platelet count, cholesterol level, and γ glutamyl transferase (GGT). At a cut of value of 6.9, it was noted to be of value in differentiating mild fibrosis from severe fibrosis, but it is less accurate in the differentiation of from.

▪ *The PGA Index*

Combines the measurement of the Prothrombin Index, γ glutamyl transferase levels and apolipoprotein-A1. It was subsequently modified to the PGAA index by the addition of α2- macroglobulin, which resulted in marginal if any improvement in its performance. In chronic liver diseases, the PGA index has a relationship to both the inflammation and the fibrosis [54, 55].

- ***The FIB-4 Score***

This score is calculated based on age, platelet count, AST, and ALT. It was first developed by *Sterling et al.*, [56] to assess fibrosis in HIV/HCV coinfected patients at a cutoff value of 3.25; 87% of patients were correctly classified, with an area under the receiver operating curve (AUC in the ROC) of 0.765 for significant fibrosis. The Fib-4 score was subsequently validated for detection of the monoinfections HCV and HBV [57, 58]. It showed AUCs of 0.85 and 0.81 for the detection of severe fibrosis, for isolated HCV and HBV infection, respectively.

- ***The FibroQ Test***

This test was proposed by *Hsieh et al.*, [59] in 2009. It is calculated based on age, AST, prothrombin time, platelet count, and ALT. In that study, using a cutoff value of 1.6 the AUC for the detection of significant fibrosis was 0.783, and the negative predictive value was 100% for the exclusion of cirrhosis. These values were both higher than those obtained when using the APRI and AAR in the same cohort [59, 60].

- ***The FibroTest and FibroSure***

These tests are identical marketed under different names in Europe and America for the assessment of fibrosis and necroinflammatory activity. The FibroTest score is computed by accessing a proprietary website and entering the patient's age, sex, and results for serum haptoglobin, α2-macroglobulin, apolipoprotein-A1, γ-glutamyltransferase, and bilirubin analyses [61]. It generates a score that is correlated with the degree of liver damage in people with a variety of liver diseases. Due to the variability of component of assays and analyzers, FibroTest assays can only be performed in validated laboratories [62].

- ***The FibroIndex***

This index was developed by *Koda et al.*, [63] for liver fibrosis in chronic hepatitis C. This test relies on platelet count, AST and serum IgG. FibroIndex showed high predictive values for significant fibrosis, including in a subgroup of HCV cases with normal alanine aminotrasferase [64]. The sensitivity and specificity of FibroIndex for detecting fibrosis in patients with HCV were 78% and 74%. In a comparative study, the validated AUC of the FibroIndex for predicting significant fibrosis was found to be 0.83 and 0.82 [65].

Combination of Direct and Indirect Biomarkers

- ***The FibroMeter***

Is a combination of the platelet count, prothrombin index, AST, α2 macroglobulin, hyaluronate, blood urea nitrogen and age. The good performance and applicability of FibroMeter was validated in a number of chronic liver diseases, including chronic viral hepatitis B or C, alcoholic liver disease and nonalcoholic fatty liver disease. An important feature of the FibroMeter is that it presents the amount of liver fibrosis as a percentage of fibrous tissue within the liver. Another significant feature of FibroMeter is that it validates the results through an expert system that detects erroneous results. FibroMeter has two main diagnostic targets - fibrosis stage corresponding to the histological staging system Metavir and the amount of fibrosis which corresponds to morphometric determinations of the fibrotic area [66].

- ***Fibrospect II Test***

The index combines three parameters: hyaluronic acid, TIMP-1, and *α2* macroglobulin. At a cutoff value of 42, it can differentiate mild from severe fibrosis [67]. It was validated in CHC patients, and it showed an AUC of 0.831 for the detection of significant fibrosis [68]. Furthermore a similar AUC 0.83 for detection of advanced fibrosis was obtained by *Jeffers et al.*, in a study of 145 CHB and CHC patients [69].

- ***SHASTA Index***

Is based on serum hyaluronic acid, AST, and albumin. In a study of 95 HIV/HCV coinfected patients, an index of 0.3 showed a sensitivity of >88% and a negative predictive value of >94%, and a level of 0.8 showed a specificity of 100% and a positive predictive value of 100% for detection of severe fibrosis or more. Using this index only, 42% of patients were correctly classified, whereas the remaining 58% showed values between 0.3 and 0.8 [70].

- ***The HepaScore***

This index combines age, gender, bilirubin, γ.glutamyl transferase, hyaluronic acid, and α2-macroglobulin into a score from 0.00 to 1.00. In 512 chronic HCV patients, automated HepaScores showed good predictive performances for significant fibrosis (AUC=0.81), severe fibrosis (AUC=0.82), and cirrhosis (AUC=0.88). Importantly, HepaScore test can be automated using a single analyzer [71].

▪ ***European Liver Fibrosis Panel (ELF) Test***

ELF test was proposed by the ELF panel. Its calculation is based on age, hyaluronic acid, amino-terminal properties of type III collagen (PIIINP), and the tissue inhibitor of matrix metalloproteinase 1 [72]. In the original calculation, age was included and the value was called the OELF, but the calculation was then simplified to a set of parameters that did not include age. The sensitivity of ELF for the detection of severe fibrosis was 90%. The ELF has been found to be of value for assessing fibrosis in chronic viral hepatitis, autoimmune liver disease, ALD, and NAFLD because the AUC in different studies has ranged from 0.773 for CHC to 0.98 for NAFLD [72 - 74].

DISCUSSION

Techniques of non invasive assessment of liver fibrosis have evolved tremendously in recent years. The diagnosis and monitoring of liver diseases has long relied on biopsy, but today it is increasingly clear that the value of this technique to assess the severity of liver disease, and especially for monitoring progression disease, is very limited.

However, the main problems in this area include the lack of a standard real gold to validate the evidence and steady growth in the number of techniques and diagnostic algorithms, which are not always properly validated so do not just join the daily clinical practice.

The most comprehensive understanding of the pathogenesis of hepatic fibrosis has been responsible for the development of new non-invasive biomarkers of fibrosis.

Currently available direct and indirect serum biomarkers, and mathematical algorithms that combine several of them, should be used with caution because neither single nor panel markers fulfill the requirements of an ideal non-invasive biomarker of fibrosis which is analytical simplicity allowing performance in any laboratory, standardization of the test system and calibrators allowing comparison between the laboratories over a long period, cost-effectiveness, specificity for the liver and the disease, clear association with the stage of fibrosis or grade of fibrogenesis and independency of the etiology of the fibrosis.

Powerful research studies that show new mechanisms related to the pathogenesis of fibrosis, as well as advances in analytical techniques, will allow the development of new biomarkers useful for the monitoring of fibrosis from the clinical laboratory. The following describes briefly a selection of newly developed or proposed future candidate biomarkers of non-invasive diagnosis and follow-up of liver fibrogenesis.

The most important molecular counterpart is the bone morphogenetic protein (BMP)-7, also belonging to the TGF-β superfamily. BMP-7 inhibits EMT (epithelial–mesenchymal transition), and can also induce a mesenchymal–epithelial transition (reverse EMT=MET). It has anti-apoptotic properties and anti-inflammatory and proliferation-stimulating effects. BMP-7 inhibits TGF-β signaling via Smads. The determination of the TGF-β/BMP-7 ratio in serum or plasma is potentially promising, since this ratio might reflect the process of EMT and, thus, at least partially the rate of progression of fibrosis. A decrease of this ratio might indicate those patients with slow progression (slow fibroser) and an increase, a fast progression (rapid fibroser). However, some limitations must be considered: the proportion of circulating cytokines does not always reflect their activity in epithelial cells and fibroblasts, and in addition the main fraction of these cytokines may be inactive. Thus, the protein ratio does not necessarily mimic the diagnostically important activity ratio of these mediators [49].

The determination of connective tissue growth factor (CTGF) in serum or plasma is suggested as a further innovative parameter of fibrogenesis, since this modulator protein is strongly up-regulated in the fibrotic liver, synthesized and secreted by parenchymal and non-parenchymal cells and since the action of the pro-fibrogenic TGF-β is stimulated but that of the anti-fibrogenic BMP-7 is inhibited. Preliminary studies point to significantly enhanced concentrations of CTGF in blood of patients with active liver fibrogenesis in contrast to advanced cirrhosis with low concentration of active fibrogenesis, which is reflected by a relative decrease of serum CTGF [49].

Further successful developments could emerge from serum proteome profiling and from total serum protein glycomics, which is the pattern of N-glycans. It was reported that a unique serum proteomic fingerprint is identified in the sera of patients with fibrosis, which enables differentiation between different stages of fibrosis and a prediction of fibrosis and cirrhosis in patients with a chronic hepatitis B infection [75]. Supplementation of all these laboratory tests by modern high resolution or molecular imaging analyses would be extremely helpful in the consolidation of objective and valid non-invasive biomarkers of diagnosis and follow-up of fibrogenic liver diseases.

The flow cytometric detection of circulating fibrocytes in blood or in buffy coat leucocytes by using CD34+, CD45+ and collagen I positivities as identifying markers might be a way for the evaluation of their diagnostic potential. Alternatively, these antigens might be detected by amplifying their mRNA using a quantitative PCR approach [49].

These new biomarkers are shown as promising options, however, despite the good results obtained, implementation in daily clinical practice would require additional studies, which be validated its usefulness in independent patient cohorts.

Currently, fibrosis biomarkers most commonly used in clinical practice are indirect biomarkers, in isolation or combined using algorithms such as those already described. Although these biomarkers have less specificity to assess fibrosis than direct biomarkers, they have the enormous advantage of being magnitudes available in most clinical laboratories, routinely performed in monitoring patients with chronic liver disease. Initially lies most useful in screening for liver disease, and later in their ability to rule out cirrhosis. In this regard, it would be interesting a greater integration into clinical practice of direct biomarkers, whose diagnostic accuracy has proven to be higher in many studies. Its limited use is due to their reduced availability because they generally are made by non-automated techniques, as well as its higher cost, and in some cases the absence of more rigorous studies to support their use.

On the other hand, despite the good results provided by fibrosis biomarkers, the latest trends betting on the combination of these with other non-invasive techniques, such as radio diagnostic techniques as the best diagnostic approach in the assessment of liver fibrosis. The combination of two different non-invasive methods involves a broader and probably more appropriate approach, since when trying to limit the number of biopsies performed, the fact of using together two complementary methods with different biological basis, is an advantage against the use of individual techniques and increases diagnostic confidence. Some of the experimental serum markers, especially those that are liver-specific, combined with novel imaging and physical techniques could create a nearly biopsy-free scenario in the near future.

Continuation of research in this area will give us the opportunity to offer patients non-invasive diagnostic tools more accurate. Liver biopsy will remain part of clinical practice in the coming years, but advances in biomedicine will challenge many of the assumptions made for decades and probably change the current approach of liver disease.

CONFLICT OF INTEREST

The author (editor) declares no conflict of interest, financial or otherwise.

ACKNOWLEDGEMENTS

Declared none.

REFERENCES

[1] Baranova A, Lal P, Birerdinc A, Younossi ZM. Non-invasive markers for hepatic fibrosis. BMC Gastroenterol 2011; 11: 91.
[http://dx.doi.org/10.1186/1471-230X-11-91]

[2] Bravo AA, Sheth SG, Chopra S. Liver biopsy. N Engl J Med 2001; 344: 495-500.
[http://dx.doi.org/10.1056/NEJM200102153440706]

[3] Castera L. Assessing liver fibrosis. Expert Rev Gastroenterol Hepatol 2008; 2: 541-52.
[http://dx.doi.org/10.1586/17474124.2.4.541]

[4] Bataller R, Brenner DA. Liver fibrosis. J Clin Invest 2005; 115: 209-18.
[http://dx.doi.org/10.1172/JCI24282]

[5] Brenner DA. Molecular pathogenesis of liver fibrosis. Trans Am Clin Climatol Assoc 2009; 120: 361-8.

[6] Rygiel KA, Robertson H, Marshall HL, *et al.* Epithelial-mesenchymal transition contributes to portal tract fibrogenesis during human chronic liver disease. Lab Invest 2008; 88: 112-23.
[http://dx.doi.org/10.1038/labinvest.3700704]

[7] Malhi H, Guicciardi ME, Gores GJ. Hepatocyte death: a clear and present danger. Physiol Rev 2010; 90: 1165-94.
[http://dx.doi.org/10.1152/physrev.00061.2009]

[8] Jiao J, Friedman SL, Aloman C. Hepatic fibrosis. Curr Opin Gastroenterol 2009; 25: 223-9.
[http://dx.doi.org/10.1097/MOG.0b013e3283279668]

[9] Fontana L, Jerez D, Rojas-Valencia L, Solis-Herruzo JA, Greenwel P, Rojkind M. Ethanol induces the expression of alpha 1(I) procollagen mRNA in a co-culture system containing a liver stellate cell-line and freshly isolated hepatocytes. Biochim Biophys Acta 1997; 1362: 135-44.
[http://dx.doi.org/10.1016/S0925-4439(97)00056-2]

[10] Purohit V, Brenner DA. Mechanisms of alcohol-induced hepatic fibrosis: a summary of the Ron Thurman Symposium. Hepatology 2006; 43: 872-8.
[http://dx.doi.org/10.1002/hep.21107]

[11] Martin-Vilchez S, Sanz-Cameno P, Rodriguez-Munoz Y, *et al.* The hepatitis B virus X protein induces paracrine activation of human hepatic stellate cells. Hepatology 2008; 47: 1872-83.
[http://dx.doi.org/10.1002/hep.22265]

[12] Lin W, Wu G, Li S, *et al.* HIV and HCV cooperatively promote hepatic fibrogenesis *via* induction of reactive oxygen species and NFkappaB. J Biol Chem 2011; 286: 2665-74.
[http://dx.doi.org/10.1074/jbc.M110.168286]

[13] Bertolani C, Marra F. The role of adipokines in liver fibrosis. Pathophysiology 2008; 15: 91-101.
[http://dx.doi.org/10.1016/j.pathophys.2008.05.001]

[14] Diehl AM. Nonalcoholic steatosis and steatohepatitis IV. Nonalcoholic fatty liver disease abnormalities in macrophage function and cytokines. Am J Physiol Gastrointest Liver Physiol 2002; 282: G1-5.
[http://dx.doi.org/10.1152/ajpgi.00384.2001]

[15] Rahmouni K, Haynes WG. Endothelial effects of leptin: implications in health and diseases. Curr Diab Rep 2005; 5: 260-6.
[http://dx.doi.org/10.1007/s11892-005-0020-5]

[16] Bertolani C, Sancho-Bru P, Failli P, *et al.* Resistin as an intrahepatic cytokine: overexpression during chronic injury and induction of proinflammatory actions in hepatic stellate cells. Am J Pathol 2006; 169: 2042-53.
[http://dx.doi.org/10.2353/ajpath.2006.060081]

[17] Glaser SS, Gaudio E, Miller T, Alvaro D, Alpini G. Cholangiocyte proliferation and liver fibrosis. Expert Rev Mol Med 2009; 11: e7.
[http://dx.doi.org/10.1017/S1462399409000994]

[18] Magness ST, Bataller R, Yang L, Brenner DA. A dual reporter gene transgenic mouse demonstrates heterogeneity in hepatic fibrogenic cell populations. Hepatology 2004; 40: 1151-9.
[http://dx.doi.org/10.1002/hep.20427]

[19] Sebastiani G, Alberti A. Non invasive fibrosis biomarkers reduce but not substitute the need for liver biopsy. World J Gastroenterol 2006; 12: 3682-94.
[http://dx.doi.org/10.3748/wjg.v12.i23.3682]

[20] Terjung B, Lemnitzer I, Dumoulin FL, *et al.* Bleeding complications after percutaneous liver biopsy. An analysis of risk factors. Digestion 2003; 67: 138-45.
[http://dx.doi.org/10.1159/000071293]

[21] West J, Card TR. Reduced mortality rates following elective percutaneous liver biopsies. Gastroenterology 2010; 139: 1230-7.
[http://dx.doi.org/10.1053/j.gastro.2010.06.015]

[22] Lurie Y, Webb M, Cytter-Kuint R, Shteingart S, Lederkremer GZ. Non-invasive diagnosis of liver fibrosis and cirrhosis. World J Gastroenterol 2015; 21: 11567-83.
[http://dx.doi.org/10.3748/wjg.v21.i41.11567]

[23] Shen L, Li JQ, Zeng MD, Lu LG, Fan ST, Bao H. Correlation between ultrasonographic and pathologic diagnosis of liver fibrosis due to chronic virus hepatitis. World J Gastroenterol 2006; 12: 1292-5.
[http://dx.doi.org/10.3748/wjg.v12.i8.1292]

[24] Hassan S, Syed S, Kehar SI. Review of diagnostic techniques of hepatic fibrosis. J Pak Med Assoc 2014; 64: 941-5.

[25] Castera L. Invasive and non-invasive methods for the assessment of fibrosis and disease progression in chronic liver disease. Best Pract Res Clin Gastroenterol 2011; 25: 291-303.
[http://dx.doi.org/10.1016/j.bpg.2011.02.003]

[26] Pinzani M, Rombouts K, Colagrande S. Fibrosis in chronic liver diseases: diagnosis and management. J Hepatol 2005; 42 (Suppl.): S22-36.
[http://dx.doi.org/10.1016/j.jhep.2004.12.008]

[27] Gallorini A, Plebani M, Pontisso P, *et al.* Serum markers of hepatic fibrogenesis in chronic hepatitis type C treated with alfa-2A interferon. Liver 1994; 14: 257-64.
[http://dx.doi.org/10.1111/j.1600-0676.1994.tb00085.x]

[28] Sakugawa H, Nakayoshi T, Kobashigawa K, *et al.* Clinical usefulness of biochemical markers of liver fibrosis in patients with nonalcoholic fatty liver disease. World J Gastroenterol 2005; 11: 255-9.
[http://dx.doi.org/10.3748/wjg.v11.i2.255]

[29] Walsh KM, Fletcher A, MacSween RN, Morris AJ. Basement membrane peptides as markers of liver disease in chronic hepatitis C. J Hepatol 2000; 32: 325-30.
[http://dx.doi.org/10.1016/S0168-8278(00)80079-3]

[30] Rostami S, Parsian H. Hyaluronic Acid: from biochemical characteristics to its clinical translation in assessment of liver fibrosis. Hepat Mon 2013; 13: e13787.
[http://dx.doi.org/10.5812/hepatmon.13787]

[31] Rosa H, Parise ER. Is there a place for serum laminin determination in patients with liver disease and cancer? World J Gastroenterol 2008; 14: 3628-32.
[http://dx.doi.org/10.3748/wjg.14.3628]

[32] Kropf J, Gressner AM, Negwer A. Efficacy of serum laminin measurement for diagnosis of fibrotic liver diseases. Clin Chem 1988; 34: 2026-30.

[33] Saitou Y, Shiraki K, Yamanaka Y, *et al.* Noninvasive estimation of liver fibrosis and response to interferon therapy by a serum fibrogenesis marker, YKL-40, in patients with HCV-associated liver disease. World J Gastroenterol 2005; 11: 476-81.
[http://dx.doi.org/10.3748/wjg.v11.i4.476]

[34] Johansen JS, Christoffersen P, Moller S, *et al.* Serum YKL-40 is increased in patients with hepatic fibrosis. J Hepatol 2000; 32: 911-20.
[http://dx.doi.org/10.1016/S0168-8278(00)80095-1]

[35] Lieber CS, Weiss DG, Paronetto F. Value of fibrosis markers for staging liver fibrosis in patients with precirrhotic alcoholic liver disease. Alcohol Clin Exp Res 2008; 32: 1031-9.
[http://dx.doi.org/10.1111/j.1530-0277.2008.00664.x]

[36] Sun J. Matrix metalloproteinases and tissue inhibitor of metalloproteinases are essential for the inflammatory response in cancer cells. J Signal Transduct 2010; 2010: 985132.

[37] Takahara T, Furui K, Yata Y, *et al.* Dual expression of matrix metalloproteinase-2 and membrane-type 1-matrix metalloproteinase in fibrotic human livers. Hepatology 1997; 26: 1521-9.
[http://dx.doi.org/10.1002/hep.510260620]

[38] Murawaki Y, Ikuta Y, Idobe Y, Kawasaki H. Serum matrix metalloproteinase-1 in patients with chronic viral hepatitis. J Gastroenterol Hepatol 1999; 14: 138-45.
[http://dx.doi.org/10.1046/j.1440-1746.1999.01821.x]

[39] Walsh KM, Timms P, Campbell S, MacSween RN, Morris AJ. Plasma levels of matrix metalloproteinase-2 (MMP-2) and tissue inhibitors of metalloproteinases -1 and -2 (TIMP-1 and TIMP-2) as noninvasive markers of liver disease in chronic hepatitis C: comparison using ROC analysis. Dig Dis Sci 1999; 44: 624-30.
[http://dx.doi.org/10.1023/A:1026630129025]

[40] Hayasaka A, Suzuki N, Fujimoto N, *et al.* Elevated plasma levels of matrix metalloproteinase-9 (92-kd type IV collagenase/gelatinase B) in hepatocellular carcinoma. Hepatology 1996; 24: 1058-62.
[http://dx.doi.org/10.1002/hep.510240513]

[41] Badra G, Lotfy M, El-Refaie A, *et al.* Significance of serum matrix metalloproteinase-9 and tissue inhibitor of metalloproteinase-1 in chronic hepatitis C patients. Acta Microbiol Immunol Hung 2010; 57: 29-42.
[http://dx.doi.org/10.1556/AMicr.57.2010.1.3]

[42] Kanzler S, Baumann M, Schirmacher P, *et al.* Prediction of progressive liver fibrosis in hepatitis C infection by serum and tissue levels of transforming growth factor-beta. J Viral Hepat 2001; 8: 430-7.
[http://dx.doi.org/10.1046/j.1365-2893.2001.00314.x]

[43] Nelson DR, Gonzalez-Peralta RP, Qian K, *et al.* Transforming growth factor-beta 1 in chronic hepatitis C. J Viral Hepat 1997; 4: 29-35.
[http://dx.doi.org/10.1046/j.1365-2893.1997.00124.x]

[44] Harada K, Shiota G, Kawasaki H. Transforming growth factor-alpha and epidermal growth factor receptor in chronic liver disease and hepatocellular carcinoma. Liver 1999; 19: 318-25.
[http://dx.doi.org/10.1111/j.1478-3231.1999.tb00056.x]

[45] Grigorescu M. Noninvasive biochemical markers of liver fibrosis. J Gastrointestin Liver Dis 2006; 15: 149-59.

[46] Pradat P, Alberti A, Poynard T, *et al.* Predictive value of ALT levels for histologic findings in chronic hepatitis C: a European collaborative study. Hepatology 2002; 36: 973-7.
[http://dx.doi.org/10.1053/jhep.2002.35530]

[47] Haukeland JW, Schreiner LT, Lorgen I, *et al.* ASAT/ALAT ratio provides prognostic information independently of Child-Pugh class, gender and age in non-alcoholic cirrhosis. Scand J Gastroenterol 2008; 43: 1241-8.
[http://dx.doi.org/10.1080/00365520802158614]

[48] Giboney PT. Mildly elevated liver transaminase levels in the asymptomatic patient. Am Fam Physician 2005; 71: 1105-10.

[49] Gressner OA, Gao C. Monitoring fibrogenic progression in the liver. Clin Chim Acta 2014; 433: 111-22.
[http://dx.doi.org/10.1016/j.cca.2014.02.021]

[50] Wai CT, Greenson JK, Fontana RJ, *et al.* A simple noninvasive index can predict both significant fibrosis and cirrhosis in patients with chronic hepatitis C. Hepatology 2003; 38: 518-26.
[http://dx.doi.org/10.1053/jhep.2003.50346]

[51] DallaPiazza M, Amorosa VK, Localio R, *et al.* Prevalence and risk factors for significant liver fibrosis among HIV-monoinfected patients. BMC Infect Dis 2010; 10: 116.
[http://dx.doi.org/10.1186/1471-2334-10-116]

[52] Lin ZH, Xin YN, Dong QJ, *et al.* Performance of the aspartate aminotransferase-to-platelet ratio index for the staging of hepatitis C-related fibrosis: an updated meta-analysis. Hepatology 2011; 53: 726-36.
[http://dx.doi.org/10.1002/hep.24105]

[53] Forns X, Ampurdanes S, Llovet JM, *et al.* Identification of chronic hepatitis C patients without hepatic fibrosis by a simple predictive model. Hepatology 2002; 36: 986-92.
[http://dx.doi.org/10.1053/jhep.2002.36128]

[54] Naveau S, Poynard T, Benattar C, Bedossa P, Chaput JC. Alpha-2-macroglobulin and hepatic fibrosis. Diagnostic interest. Dig Dis Sci 1994; 39: 2426-32.
[http://dx.doi.org/10.1007/BF02087661]

[55] Poynard T, Aubert A, Bedossa P, *et al.* A simple biological index for detection of alcoholic liver disease in drinkers. Gastroenterology 1991; 100: 1397-402.
[http://dx.doi.org/10.1016/0016-5085(91)90795-M]

[56] Sterling RK, Lissen E, Clumeck N, *et al.* Development of a simple noninvasive index to predict significant fibrosis in patients with HIV/HCV coinfection. Hepatology 2006; 43: 1317-25.
[http://dx.doi.org/10.1002/hep.21178]

[57] Mallet V, Dhalluin-Venier V, Roussin C, *et al.* The accuracy of the FIB-4 index for the diagnosis of mild fibrosis in chronic hepatitis B. Aliment Pharmacol Ther 2009; 29: 409-15.
[http://dx.doi.org/10.1111/j.1365-2036.2008.03895.x]

[58] Vallet-Pichard A, Mallet V, Nalpas B, *et al.* FIB-4: an inexpensive and accurate marker of fibrosis in HCV infection. comparison with liver biopsy and fibrotest. Hepatology 2007; 46: 32-6.
[http://dx.doi.org/10.1002/hep.21669]

[59] Hsieh YY, Tung SY, Lee IL, *et al.* FibroQ: an easy and useful noninvasive test for predicting liver fibrosis in patients with chronic viral hepatitis. Chang Gung Med J 2009; 32: 614-22.

[60] Hsieh YY, Tung SY, Lee K, *et al.* Routine blood tests to predict liver fibrosis in chronic hepatitis C. World J Gastroenterol 2012; 18: 746-53.
[http://dx.doi.org/10.3748/wjg.v18.i8.746]

[61] Imbert-Bismut F, Messous D, Thibault V, *et al.* Intra-laboratory analytical variability of biochemical markers of fibrosis (Fibrotest) and activity (Actitest) and reference ranges in healthy blood donors. Clin Chem Lab Med 2004; 42: 323-33.
[http://dx.doi.org/10.1515/cclm.2004.42.6.681]

[62] Friedrich-Rust M, Rosenberg W, Parkes J, Herrmann E, Zeuzem S, Sarrazin C. Comparison of ELF, FibroTest and FibroScan for the non-invasive assessment of liver fibrosis. BMC Gastroenterol 2010; 10: 103.
[http://dx.doi.org/10.1186/1471-230X-10-103]

[63] Koda M, Matunaga Y, Kawakami M, Kishimoto Y, Suou T, Murawaki Y. FibroIndex, a practical index for predicting significant fibrosis in patients with chronic hepatitis C. Hepatology 2007; 45: 297-

306.
[http://dx.doi.org/10.1002/hep.21520]

[64] Sebastiani G, Vario A, Guido M, Alberti A. Performance of noninvasive markers for liver fibrosis is reduced in chronic hepatitis C with normal transaminases. J Viral Hepat 2008; 15: 212-8.
[http://dx.doi.org/10.1111/j.1365-2893.2007.00932.x]

[65] Pilette C, Rousselet MC, Bedossa P, *et al.* Histopathological evaluation of liver fibrosis: quantitative image analysis vs semi-quantitative scores. Comparison with serum markers. J Hepatol 1998; 28: 439-46.
[http://dx.doi.org/10.1016/S0168-8278(98)80318-8]

[66] Cales P, Boursier J, Oberti F, *et al.* FibroMeters: a family of blood tests for liver fibrosis. Gastroenterol Clin Biol 2008; 32: 40-51.
[http://dx.doi.org/10.1016/S0399-8320(08)73992-7]

[67] Patel K, Gordon SC, Jacobson I, *et al.* Evaluation of a panel of non-invasive serum markers to differentiate mild from moderate-to-advanced liver fibrosis in chronic hepatitis C patients. J Hepatol 2004; 41: 935-42.
[http://dx.doi.org/10.1016/j.jhep.2004.08.008]

[68] Patel K, Nelson DR, Rockey DC, *et al.* Correlation of FIBROSpect II with histologic and morphometric evaluation of liver fibrosis in chronic hepatitis C. Clin Gastroenterol Hepatol 2008; 6: 242-7.
[http://dx.doi.org/10.1016/j.cgh.2007.11.009]

[69] Jeffers LJ, Cortes RA, Bejarano PA, *et al.* Prospective Evaluation of FIBROSpect II for Fibrosis Detection in Hepatitis C and B Patients Undergoing Laparoscopic Biopsy. Gastroenterol Hepatol (N Y) 2007; 3: 367-76.

[70] Kelleher TB, Mehta SH, Bhaskar R, *et al.* Prediction of hepatic fibrosis in HIV/HCV co-infected patients using serum fibrosis markers: the SHASTA index. J Hepatol 2005; 43: 78-84.
[http://dx.doi.org/10.1016/j.jhep.2005.02.025]

[71] Guechot J, Lasnier E, Sturm N, Paris A, Zarski JP. Automation of the Hepascore and validation as a biochemical index of liver fibrosis in patients with chronic hepatitis C from the ANRS HC EP 23 Fibrostar cohort. Clin Chim Acta 2010; 411: 86-91.
[http://dx.doi.org/10.1016/j.cca.2009.10.011]

[72] Rosenberg WM, Voelker M, Thiel R, *et al.* Serum markers detect the presence of liver fibrosis: a cohort study. Gastroenterology 2004; 127: 1704-13.
[http://dx.doi.org/10.1053/j.gastro.2004.08.052]

[73] Guha IN, Parkes J, Roderick P, *et al.* Noninvasive markers of fibrosis in nonalcoholic fatty liver disease: Validating the European Liver Fibrosis Panel and exploring simple markers. Hepatology 2008; 47: 455-60.
[http://dx.doi.org/10.1002/hep.21984]

[74] Vizzutti F, Arena U, Nobili V, *et al.* Non-invasive assessment of fibrosis in non-alcoholic fatty liver disease. Ann Hepatol 2009; 8: 89-94.

[75] Poon TC, Hui AY, Chan HL, *et al.* Prediction of liver fibrosis and cirrhosis in chronic hepatitis B infection by serum proteomic fingerprinting: a pilot study. Clin Chem 2005; 51: 328-35.
[http://dx.doi.org/10.1373/clinchem.2004.041764]

CHAPTER 4

Iron Metabolism: New Biomarkers Implicated

Ana Hernando- Holgado[1] and **M. Pilar Gallego-Hernanz**[2,*]

[1] *Clinical Analysis Department, Santa Lucia General University Hospital, Cartagena, Spain*

[2] *Hématologie Oncologique et Thérapie cellulaire. Centre Hospitalier Universitaire de Poitiers, France*

Abstract: Human organisms require sufficient amounts of iron to satisfy metabolic needs and to accomplish specialized functions such as erythropoiesis, cellular immune response and oxidative metabolism. Despite its abundance, iron physico-chemical properties (practically insoluble at neutral pH under aerobic conditions) hinder its availability and thus specific ligands have evolved for absorption, transport and storage. Iron plays a corner-stone role both in erythropoiesis and inflammation. Iron deficit leads to anemia, but this deficit can be functional (with normal storage) or a true lack of iron: the accuracy of diagnosis results in a completely different treatment. Correct assessment of iron metabolism allows to diagnose and treat (not only in treatment decision-making, but. also predicting and assessing response to treatment) anemia.

In this chapter, we review principal biomarkers related to iron metabolism as hemoglobin (whose levels are used as a measure of anemia), ferritin (related to iron storage and inflammation), hepcidin (related mainly to inflammation), transferrin and its receptors and other proteins involved in absorption and transport.

Finally we review the phases of iron deficiency and its main clinical manifestation: anemia, both ferropenic and inflammation anemia, as well as clinical implication of iron overload.

Keywords: Anemia, Cardiovascular risk, DMT1, Ferroportin, Ferritin, Haptoglobin, Hemoglobin, Hemochromatosis, Hephaestin, HFE, Iiron status, Iron-deficit, Immature hepcidin, Inflammation, Iron overload, Percentage of hypochromic cells, Reticulocytes per-centage, sTfR-F index, Transferrin, Transferrin receptor, Transferrin saturation.

INTRODUCTION

Human organisms require sufficient amounts of iron to satisfy metabolic needs

* **Corresponding author M. Pilar Gallego-Hernanz:** Hématologie Oncologique et Thérapie cellulaire. Centre Hospitalier Universitaire de Poitiers, France; Tel/Fax: +33(0)549444444; E-mail: maria-pilar.gallego-hernanz@c-u-poitiers.fr

Juan Antonio Vílchez (Ed.)

and to accomplish specialized functions such as erythropoiesis, cellular immune response and oxidative metabolism [1]. Iron importance lies in its ability to mediate electron transfer in many enzymatic reactions (reduction and oxidation): in the ferrous state, iron acts as an electron donor, while in the ferric state as an acceptor.

Despite its abundance, iron physico-chemical properties (practically insoluble at neutral pH under aerobic conditions) hinder its availability and thus specific ligands have evolved for transport and storage.

Absorption of dietary intake plays a critical role in iron homeostasis, as human body lacks iron excretion mechanisms and its overload may cause cell death through free radical formation and lipid peroxidation [2].

Free Fe (II) catalyze the conversion of hydrogen peroxide into free radicals, that can damage cellular structures and ultimately kill the cell, especially in the presence of dioxygen.

As prevention, iron is bound to proteins allowing cells to benefit from iron while also limiting its ability to do harm [3].

In healthy adults, the normal iron content is 3 to 4 grams (50mg/kg body weight), distributed as follows [4, 5]:

- 65% Hemoglobin in circulating red cells – Approximately 2.5 grams
- 10% Iron-containing proteins other than hemoglobin (*e.g.*, myoglobin, cytochromes, catalase) – 400 mg
- Iron bound to transferrin in plasma – 3 to 7 mg
- Storage in the form of ferritin or hemosiderin. Adult men have approximately 1 g of iron storage (mostly in liver, spleen, and bone marrow). Comparatively, adult women have less iron storage, which is highly variable depending upon the extent of menses, pregnancies, deliveries, lactation, and iron intake, and some may have no stores.

Clinical assessment of iron status is essentially based on the level of iron stores, reflected by ferritin concentration in serum. However, stored iron is useless for metabolism: the real clinical issue lies on removing iron from storage for its use, mainly in hematopoiesis.

Recently, many new proteins connected with iron metabolism - as well as new functions of already known proteins- have been identified, deeping our knowledge of the physiopathology of iron [6].

IRON METABOLISM (SEE FIG. (1))

Absorption

The iron homeostasis is regulated mainly by the controlled absorption by enterocytes of dietary iron to face the uncontrolled losses from epithelial cells, sweat, injuries and blood losses [3]. But the continuous recycling of senescent erythrocytes provide most of the iron delivered daily to plasma [7].

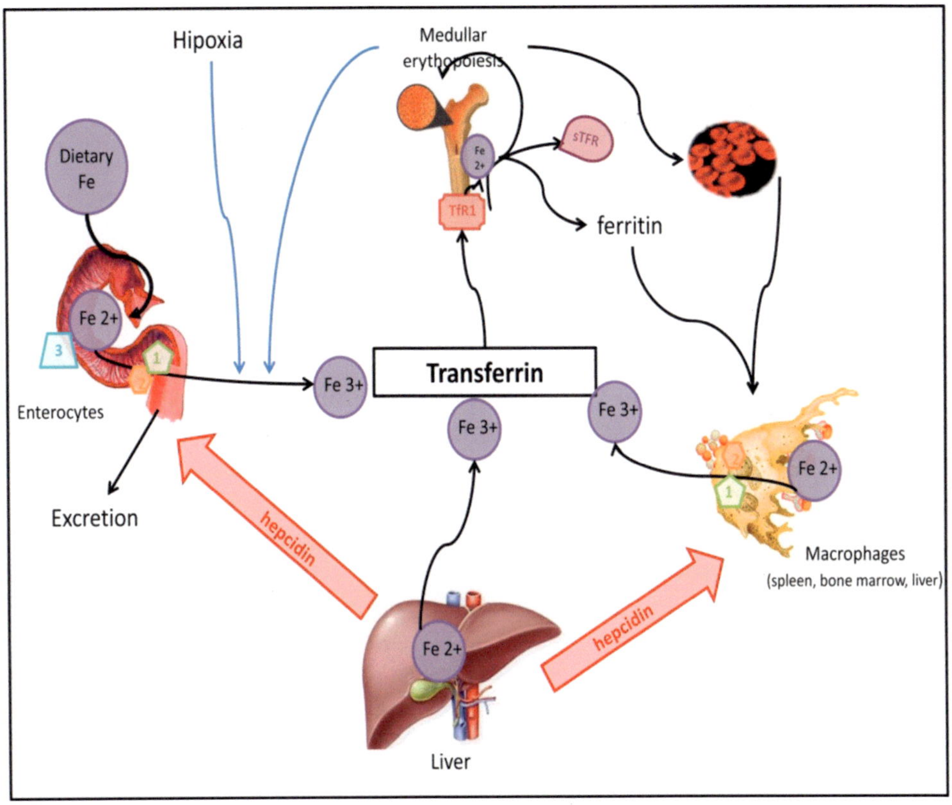

Fig. (1). Iron metabolism (1, ferroportin; 2, hephaestin; 3: divalent metal transporter 1 (DMT1); TfR1, Transferrin Receptor 1; sTfR, soluble transferrin receptor). Adapted from Muñoz *et al.* [5] and Goodnough *et al.* [21].

Iron absorption responds to a variety of interdependent factors, including total iron stores, production of new red blood cells, concentration of hemoglobin and the oxygen content of the blood. Only a small proportion of ingested iron is absorbed, through a very variable process ranging from 5 to 35%, depending on the circumstances and the source. Many external factors influence dietary iron absorption, increasing (ascorbate, citrate, carbohydrates and amino acids) or inhibiting (plant phytates, phosphates and tannins) it [8].

Gastric acid lowers the pH, enhancing the solubility and uptake of ferric iron. The majority of the iron (35% of intake) is absorbed by enterocytes of the duodenal coating, which have particular molecules allowing them to move iron into the body. The dietary iron intake comes either in the form of heme iron (best absorbed form) or as inorganic iron (mechanism less efficient as depending on the ferrous Fe^{2+} form, and thus susceptible to pH variations) [8].

Absorption of intestinal iron is regulated by four different mechanisms:

The iron amount recently ingested: the enterocyte becomes resistant to iron after absorbing it for days.

The iron status: the crypt of the enterocytes acts as a sensor reserve and modulate the behavior of enterocytes of the villus apex. The absorption of inorganic iron can increase up to double or triple in case of deficiency, whereas that it can be greatly reduced in cases of overload [9].

Erythropoiesis: An increased erythropoiesis needs higher absorption. However, not all diseases with increased erythropoiesis determine high intestinal absorption: ineffective erythropoiesis(as in thalassemia,congenital dyserythropoietic anemias, and sideroblastic anemias) carry a cell destruction within bone marrow, allowing iron recycling. On the contrary, in hemolytic anemias (such as spherocytosis or autoimmune hemolysis), where the destruction occurs away from the bone marrow, intestinal absorption is increased. This fact suggests that the factor causing increased absorption occurs in the more immature erythropoietic cells [3].

Ultimately, hypoxia increases the intestinal absorption, although through an unknown mechanism [3]. On the other hand, Hindering is a ferroxidase that can oxidize Fe^{2+} to Fe^{3+}, helping iron transfer across the basolateral end of the intestine cells through ferroportin. This ferroportin is post-translationally repressed by hepcidin [10].

Transport

Once absorbed and oxidized, iron passes to the portal system and is distributed to the organism. It can be trapped by the mononuclear phagocyte system, which is responsible for iron reuse, but finally, it access the main destination: erythropoiesis.

In plasma, iron can travel in various ways [11]:

- in the form of low-mass complexes molecules, which are highly toxic; these are formed only in appreciable concentrations in the advanced stages of the hereditary hemochromatosis.

- joined to ferritin

- bounded to transferrin (the most frequent): different transferrins exist and their ratio depends on the individual's iron reserves: apotransferrin predominate in states of iron deficiency, and ditransferrin in the ferric overload [11].

Cellular Uptake

Once out of the bloodstream, the cellular uptake of iron can occur in two ways: depending or Independent of transferrin.

Transferrin-Dependent Pathway

The interaction with its cell surface receptor leads to the internalization of the iron-transferrin-transferrin receptor complex, *via* endocytosis [12] and is then transported to a endosomal compartment. An acid pump hydrogen (coupled to DMT1) changes pH of the interior: this acidification (pH lower than 5.6) causes conformational change of the transferrin receptor allowing the release of iron in few minutes [13]. Subsequently, iron passes through the membrane endosome to exit to the cytoplasm. Iron can then be used (release it into the body by the ferroportin) or bound by ferritin (for cellular storage). In other cases, a ferric reductase enzyme on the surface of the enterocytes besides a duodenal cytochrome B reduces Fe^{3+} to Fe^{2+}, and then the DMT1 transports iron across the enterocytes membrane into the cell [14, 15].

Meanwhile the complex formed by transferrin and its receptor is transported back to the plasma membrane *via* endocytic recycling [16]. The receptor gets dephosphorylated, and sent outside the cell, truncated [17], resulting in the appearance of a cleaved monomer, called soluble transferrin receptor (sTfR): increased serum concentrations of sTfR are thus a marker of either iron deficiency or stimulated erythropoiesis [12].

But ditransferrin binds also - although with low affinity- to nonspecific points of the cell surface (*i.e.* non mediate by TfR), to be internalized in order to capture iron [18].

Transferrin-Independent Pathway

Transferrin-independent route is the least known, and thought to be responsible for removing from circulation iron complexes in low molecular weight [19].

Recently it has been established a role in cellular iron uptake by the ferroxidase. Such capture is performed by a multi-enzyme complex present in the surface of the mammalian cells. A ferrireductase, not yet characterized, reduces iron (III) to iron (II) releasing the iron from complexes forms; ferroxidase subsequently oxidizes to iron (III) using ascorbate as a reducing agent, and providing appropriate substrate to a specific transporter of trivalent cations, which transfers iron into the cytosol. This transporter is only synthesized in cases of iron deficiency, and explains why ferroxidase increases the cellular uptake of iron only in iron deficiency situations [10].

Other proteins seem to be involved in uptake processes of transferrin-independent iron as the specific stimulator of Fe transport, a membrane protein binding specific to Fe, which facilitates the uptake of iron (II) and iron (III) free [20].

Excretion

The mechanisms of iron excretion are quantitatively unimportant, but qualitatively is the only way to eliminate excess of this potentially toxic metal.

Excretion has a modulator role in iron metabolism, adapting to the physiological needs. If an individual has normal iron stores its absorption is not increased nor diminished; if the stores are low, he has an increased absorption, and conversely, if there exist an overload of this ion, the excretion increases and absorption decreases [9].

The main mechanism takes place in the enterocyte; if the body stores are high, iron becomes part of ferritin as entering the enterocyte, and the peel is lost in faeces. Iron excretion also takes place through the epithelial desquamation, urine, bile and sweat [9].

CLINICAL IMPLICATIONS

Any condition of Iron-Out-of-Balance can cause many vague symptoms or health issues including fatigue, bone or joint disease (osteoarthritis, osteoporosis), shortness of breath, irregular heartbeat, liver trouble, diabetes, infertility, impotence, depression, mood or mental disorders, poor cognitive skills or neurodegenerative diseases [3]. Diseases might occur due to both overload -such as hemochromatosis- and deficiency -anemia [5].

Overload

Excess iron in vital organs, even in mild cases of iron overload, increases the risk for organ failure. The liver is the main iron-storage organ, and the generation of free radicals and lipid peroxidation products in iron-overload states may result in progressive hepatic tissue injury and eventually cirrhosis or hepatocellular carcinoma [5].

Iron storage in cardiomyocytes is also of great interest, as cardiac failure is the leading cause of death among patients with untreated hereditary hemochromatosis or iron overload associated with chronic transfusion. In cardiac cells, excess of iron may result in oxidative stress and alteration of myocardial function due to DNA damage by hydrogen peroxide [5].

Iron overload can be inherited or acquired by blood transfusions, iron injections, or consuming high levels of supplemental iron. Some of the genetic disorders that result in iron overload include hereditary hemochromatosis (all types), African iron overload, sickle cell disease, thalassemia, X-linked sideroblastic anemia, enzyme deficiencies (pyruvate kinase; glucose-6-phosphate dehydrogenase) and very rare protein transport disorders such as aceruloplasminemia and atransferrinemia [22].

Deficit

Lack of replenishment of iron lost may compromise its supply to the erythroid marrow and lead to impaired hemoglobin production. Iron-deficit is the main cause of anemia all around the world [23]. However, not all anemic people are deficient in iron, and this deficiency may initially occur without anemia.

In fact, anemia only appears in the last phases of iron depletion. There are no rates for prevalence of iron disorders in the world, but we can have an idea if we look at the prevalence estimates of anemia in the world of the World Health Organization (WHO) [24].

More recent studies show a significant decrease in prevalence from 1990 to 2010, both in male and females. However anemia was responsible for 68.3 million years of life lived with disability in 2010 (8.8% of global total for all conditions), and iron-deficiency remains the main cause of anemia [25], affecting over 40% of young (15-59 year-old) women in developing countries and 10 to 45% of elderly in developed *vs* developing countries (according to WHO data) [24].

But anemia can also appear because of an impaired use of iron storage, as seen in functional iron deficiency (FID), even with adequate iron supplies. FID is usually defined by the presence of stainable iron in the bone marrow together with a serum ferritin value within normal limits. It encompasses the partial block in iron transport to the erythroid marrow in subjects with infectious, inflammatory or malignant diseases [26].

Iron-deficit anemia will respond to iron treatment, but this supplementation will be useless in FID. Accuracy of diagnosis allows to correctly treat patients, and not giving supplemental iron to every anemic patient.

Other Clinical Implications

Infection

Proteins related to iron metabolism may have effect over infections. While the role of hepcidin in iron regulation is well established (hepcidin inhibits the transport of iron out of enterocytes and macrophages), its contribution to host defense is emerging as complex and multifaceted. In fact, hepcidin production during infection causes depletion of extracellular iron, which is thought to be a general defense mechanism against many infections by withholding iron from invading pathogens. It remains unclear whether hepcidin exerts any iron-independent effects on host defenses [27]. There is a role of systemic and cellular iron-regulating mechanisms in protecting hosts from infection, probably by decreasing the iron supply to invading microbes [27, 28].

On the other hand, iron levels also modulate host defense, as iron content in macrophages regulates their cytokine production [28]. Moreover, iron supplementation during infections in experimental conditions has shown to have deleterious effects [29, 30].

Inflammatory States

Serum ferritin relates faithfully to iron stores. However, it is also an acute phase protein, and thus rises during inflammation. Therefore, the conventional threshold to indicate iron deficiency does no longer apply [23].

The hepatic synthesis of hepcidin is induced by systemic iron levels but also by inflammatory stimuli [27], as interleukin 6 and lipopolysaccharide. During inflammation, iron absorption is enhanced (by stimulation of DMT1 as discussed further on this chapter) but kidnapped inside macrophages [11].

But the influence of inflammation, even at very low levels has also been observed in healthy subjects, where low grade inflammation still leads to a decrease in hemoglobin (although not reaching "anemia" levels) [31].

Endocrine Alterations

Plasma hepcidin levels rise under secondary iron overload caused by chronic transfusion, resulting in ferroportin degradation. The precipitation of different forms of iron in these tissues may lead to tissue necrosis and destruction of endocrine cells. Iron overload in endocrine glands result in insufficiency including pituitary, thyroid, parathyroid, and adrenal glands. Al- Hakeim *et al.,* studied some endocrine and biochemical parameters to determinate factors that mainly affect hepcidin in patients with Thalassemia major. Changes in the level of cortisol between thalassemic and control groups revealed unique results describing the involvement of the adrenal cortex. The results of blood transfusion also showed that the thyroid and adrenal cortex gland functions are correlated with the number of transfusions and subsequent iron overload state. In Al-Hakeim *et al.* study, serum thyroid-stimulating hormone positively correlated with hepcidin, iron, and ferritin ($p < 0.05$). T4 hormone was correlated with ferritin; ferritin and, subsequently, estimated iron body storage showed a significant positive correlation with insulin, and insulin showed a significant positive correlation with hepcidin [32].

The KORA F4 study showed an association of several markers of iron metabolism with hyperglycemia and insulin resistance, suggesting that iron stores as well as iron-related metabolic pathways contribute to the pathogenesis of impaired glucose metabolism (IGM) and type 2 diabetes mellitus (T2DM). Ferritin (odds ratio (OR)=2.08 [95% CI 1.43-3.04]) and transferrin (OR=1.89 [95% CI 1.32-2.70]) were positively associated with IGM; with T2DM (ferritin OR=1.98 [(95% CI 1.22-3.22] and transferrin OR=2.42 [95% CI 1.54-3.81]). Transferrin saturation (OR=0.55 [95% CI 0.34-0.88]) and iron (OR=0.61 [95% CI 0.38-0.97]) were inversely associated with T2DM; sTfR-F-index was inversely associated with IGM (OR=0.67 [95% CI 0.48-0.95]). Moreover, transferrin saturation levels are decreased in T2DM patients [33].

Cardiovascular Risk

Iron balance may influence atherosclerotic events and different studies support the association of high body iron status and coronary heart disease. First, the increase with aged in serum ferritin levels correlates with the increased cardiovascular risk. Moreover, in young patients, the male/female ratio for serum ferritin levels correlates with the increased risk for heart disease in males [34]. Age at first atherosclerotic event is significantly lower in men *versus* women, and this fact is not supported by any hormonal effect (randomized trials have shown inconsistent and pejorative results of hormone therapy [35 - 38]). Therefore monthly iron loss has recently been postulate to contribute to this advantage [39]. Recently, women in transition to menopause were studied with a magnetic resonance imaging of arterial wall, and this marker was correlated to some well-known inflammatory biomarkers, such as protein C reactive as well as with an increase in iron balance in women transitioning through menopause [40].

Other experimental evidence support iron's role in promoting the atheroma progression as some animal studies revealed that iron chelators prevent the oxidation of low-density lipoproteins and vessel-wall damage [34, 41]. Free Fe catalyzes the formation of oxygen free radicals, and oxidized low-density lipoproteins are a well-established risk factor for vascular damage.

Moreover, the Iron in Atherosclerosis Study (FeAST) showed that ferritin reduction with statin therapy -rather than lipid changes- was responsible for the reduction in non-fatal myocardial infarction and stroke in patients with established peripheral arterial disease [42 - 45].

However the meta-analysis by Das De *et al.* found a negative association of transferrin saturation levels and coronary heart disease with high transferrin saturations being associated with a lower risk of cardiovascular disease and myocardial infarction [46]. Nevertheless there are some limitations in terms of reverse causality bias and residual confounders [46] as well of the different role in pathogenesis of stored iron, free Fe and mobilized protein-bound iron that must be solved with more studies.

Another axis of iron cardiovascular toxicity is the formation of homocysteine from methionine, S-adenosylhomocysteine and cystathionine (catalyzed by non-protein-bound iron, or free Fe). Thus an elevated amount of free Fe increases circulating homocysteine. Hyperhomocysteinemia has been regarded as an independent risk factor for cardiovascular disease (CVD) as a large proportion of patients with CVD have hyperhomocysteinemia. Therefore, circulating homocysteine concentrations are thought to be in part a surrogate marker for another, yet-to-be-identified risk factor(s) for CVD, although more studies are needed [47].

Nevertheless all this facts point towards a deep interrelation between inflammatory state and cardiovascular risk in which iron plays a central role.

Others

Environmental risk factors, such as smoking status and air pollution, interact with genes to produce pathology and this interaction is responsible for the wide range of presentation seen in patients with the same diagnosis. Inhalational iron may be a source of environmental variation but it is also found in cigarette smoke, the strongest causative link to pulmonary pathology. The role of iron in pulmonary pathology has been well established, both from an environmental viewpoint as well as in terms of genetic susceptibility [48]. Disruption of cellular iron homeostasis appears to have an adverse effect on the lung. Conditions such as emphysema and lung cancer, which have a predilection for the upper lobes of the lungs, may be due to regional variation of iron. Fluid from the upper lobes of smokers had significantly higher concentrations of extracellular ferritin-bound iron and less transferrin, which may contribute to the pathogenesis of emphysema and lung cancer *via* oxidative stress [49].

RELATED BIOMARKERS

Accurate assessment of iron status requires the combined use of several laboratory tests, mainly in "grey zone" or overlapping situations.

Proteins Function In Iron Metabolism

Ferritin

With a half-life of 60 hours [50], ferritin is the major protein involved in the storage of iron and became therefore the standard test for iron stores assessment, as it is the most sensitive and specific single test to identify isolated iron deficiency [21].

The protein consists of an outer polypeptide shell (also termed apoferritin) composed of 24 symmetrically polypeptide subunits interconnected by non-covalent links, resulting in spheroidal form, with a space inside which is fitted up to 4500 iron atoms deposited as ferric oxyhydroxyhydroxide. The ratio of iron to polypeptide is not constant, because the protein has the ability to catch and release iron through channels from the surface according to physiological needs [51].

Ferritin consists of two different types of chains, H subunit (heavy; 178 amino acids, 21 kDa) and L (light, 171 amino acids, 19.7 kDa), encoded respectively on chromosomes 11 and 19 that provide various isoprotein forms. H subunits predominate in nucleated blood cells and heart and L subunits in liver and spleen. *In vitro*, H-rich ferritins take up iron faster than L-rich and may function more in iron detoxification rather than in storage [52].

Synthesis of both subunits is regulated mainly by the concentration of free intracellular iron as well as by cytokines: interleukin-1 increases synthesis [53]; and so its concentration in plasma rises in inflammatory states [54].

The main storage of iron occurs in hepatocytes, reticuloendothelial cells and skeletal muscle. When iron is in excess, the storage capacity of newly synthesized apoferritin is increased, leading to iron deposition adjacent to ferritin spheres, in an amorphous way called hemosiderin (leading to a clinical condition termed as hemosiderosis) [55].

Toxic potential of iron (II) is due to its reaction with reactive oxygen, forming free radicals that affect membranes, proteins and nucleic acids. Ferritin captures iron (II) in the cytosol cells to avoid its toxicity. Later the H subunit oxidizes the iron and this one is deposited inside the structure as a storage to be used where appropriate. Thus, ferritin has a cytoprotective and backup iron role [56].

Ferritin levels might vary with age (see Table **1**) and sufficient stores to supply erythropoiesis are unfortunately not related to ferritin levels: arised ferritin levels may reflect iron excess in the liver and spleen, and none available for erythropoiesis. The serum ferritin concentration is unhelpful in predicting response to erythropoiesis stimulating agents (ESA) therapy in cancer-related anemia [26].

Table 1. Physiological changes of markers during life. Taken of Choi *et al.*, 1999 [63].

Age group	Fe (μg/L)	CTS(μg/L)	TSI (%)	Ferritin (μg/L)	STfR (mg/L)
0-5 months	2015±484	2859±783	69.2±15.3	178.5±72.1	4.95±1.24
4-24 months	732±181	3164±325	23±7.1	23.5±10.4	4.51±1.12
3-7 years	773±298	2623±561	29.4±11.4	25.7±9.7	3.02±0.76
14-16 years	905±387	3141±432	28.3±10.5	39.1±10.4	2.86±0.74
17-19 years	834±292	3125±46.3	27.1±13.2	47.3±26.2	2.09±0.55
23-62 years	1391±354	3107±312	44.5±12.9	54.1±15.6	2.13±0.51

In the other hand, ferritin levels are optimal in assessing and monitoring iron overload in hemochromatosis and polytransfused patients. However, it rises in inflammatory diseases (responds as an acute phase reactant), acute leukemia, Hodgkin's disease and breast carcinoma, and might be misleading under those circumstances. It is decreased in iron deficiency, whose screening and diagnosis is the main clinical usefulness of ferritin (Table **3**) [11].

Transferrin (Tf)

Structure:

Also called siderofilin, transferrin is one single chain glycoprotein of 679 amino acids (molecular mass of 79.6 kDa) with an ellipsoidal shape. Three variations (D1, D2 and DCH1) because of changes in the amino sequence are described, but with no differences in iron affinity. The encoding gene is situated on chromosome 3, near the corresponding to the Tf receptor [57].

It has two sugar residues terminating in sialic acid molecule [58]. For the relation to the polypeptide chain, there is a high coincidence in the amino acid sequence of both ends. The molecule has two reversible binding points for two iron (III) ions, one in the carboxy terminus and one on the amino-terminal [59]. His electrophoretic migration corresponds to the zone β1-globulin [57].

Synthesis:

With a half-life of 8 days, it is synthesized in the liver, in an inversely proportional amount to the concentration of hepatocyte intracellular ferritin [60].

Reticulum endoplasmatic and the Golgi apparatus in the hepatocytes hold a reservation form of Tf. This pool is liberated when its synthesis is stimulated by bleeding [57].

The Tf plasma concentration rises in iron deficiency, pregnancy and treatment oral contraceptives, since estrogen increases their synthesis [61].

It is diminished in congenital deficiency of this protein (a very rare disease), any chronic inflammation or neoplasia and infections; also in states of accelerate protein catabolism or loss, such as malnutrition and nephrotic syndrome; in the states in which the body has high oncotic pressure as multiple myeloma or hepatocellular diseases, and in the states of iron overloads [61].

Function:

Tf transports iron from either the intestinal absorption, the catabolism of hemoglobin (by the mononuclear phagocyte system) or tanks tissue, to reticulocytes and erythroblasts for hemoglobin synthesis (or other cells synthesizing molecules containing this cation), or for deposits mainly in the liver (Table **3**).

Tf has also a protective role since its union with free Fe prevents the adverse oxide-reduction effects. Tf is also related with the transport of Zn (II), Mn (II), Cr (III), Cu (III), and perhaps holds a detoxifying role of these metals [62].

Besides it has also bacteriostatic action by limiting the access of iron in bacterial growth [62].

Some additional laboratory measures are related to its content in iron (or saturation capacity) [63]:

Total Capacity of Transferrin Saturation (CTS):

The Tf plasma concentration can be expressed indirectly as the mass of iron that can bind a plasma volume. This magnitude is expressed as the substance concentration or mass of iron that the sample can contain and there exist physiological changes of this magnitude along life (Table **2**) [64].

Table 2. Hemoglobin levels for anemia, adapted from Hemoglobin concentrations for the diagnosis of anemia and assessment of severity. WHO/NMH/NHD/MNM/11.1 www.who.int/vmnis/indicators /haemoglobin.pdf.

	Anemia degree		
	low	**mild**	**severe**
6 to 59 month-old infants	10.0 - 10.9 g/dl	7.0 - 9.9 g/dl	< 7.0 g/dl
5 to 11 year-old infants	11.0 - 11-4 g/dl	8.0 - 10.9 g/dl	< 8.0 g/dl
12 to 14 year-old	11.0 - 11.9 g/dl	8.0 - 10.9 g/dl	< 8.0 g/dl
Women over 15 years	11.0 - 11.9 g/dl	8.0 - 10.9 g/dl	< 8.0 g/dl
Pregnancy	10.0 - 10.9 g/dl	7.0 - 9.9 g/dl	< 7.0 g/dl
Men	11.0 - 12.9 g/dl	8.0 - 10.9 g/dl	< 8.0 g/dl

The CTS is a reflection of plasma Tf, increasing and decreasing for the same reasons except as happens in hereditary hemochromatosis in which the CTS is high and transferrin decreased [65].

CTS (μmol Iron/L) = Tf (μmol Iron/L)*2

CTS (μmol Iron/L) = Tf (g/L)*25.1

Percentage Saturation of Transferrin:

Reflects iron in transport (not in stores) and may be used to monitor response to erythropoietin stimulating agents and/or iron therapy in chronic kidney disease. However, used as single determination, it has poor sensitivity and specificity, although improves accuracy when combined [26]:

- to serum ferritin concentration or percentage of hypochromic red cells: useful in functional iron deficit diagnosis

- to soluble transferring receptor: predicting response to intravenous iron therapy

Transferrin Saturation Index (TSI):

This index expresses the percentage of iron present in the plasma in relation to the totality iron that can assume this system. It also suffers physiological changes during life (as shown on Table **1**), although in physiological conditions 30% of available iron binding sites are occupied [12].

TSI= Iron (μmol/L)/ CTS (μmol Iron/L)*100

TSI= Iron (μmol/L)/ Transferrin (μmol Iron/L)*50

TSI= Iron (μmol/L)/ Transferrin (g/L)*4

Major factors influencing transferrin saturation are dietary iron absorption, recycled iron released by reticuloendothelial macrophages and iron used.

It is elevated in hereditary hemochromatosis, ingestion excessive iron, thalassemia, vitamin deficiency B6, aplastic anemia, sideroblastic anemias and it is decreased in iron deficiency erythropoiesis, the malignant diseases of the stomach and small intestine, and pregnancy. Its main application is as screening test in the diagnosis of hereditary hemochromatosis, with a sensitivity of 100 and a specificity of 97% [66].

Another utility of this marker is the differential diagnosis between ferropenic anemia and anemia of chronic disease.

Transferrin carbohydrate deficient plasma is also used as a marker for diagnosing and monitoring the excessive consumption of alcohol [67].

Transferrin Receptor

The gene encoding the transferrin receptor (TfR) is in the chromosome 3 near the transferrin one [57]. TfRs are homodimeric trans membrane glycoproteins, with 760 amino acids and a molecular mass of 90 kDa, linked by two disulfide bridges forming 3 domains: trans membrane, cytoplasmic, and extracellular, which is the largest one and contains the Tf binding site [68]. TfRs are a member of the family of tyrosine kinase-linked receptors, specifics for Fe-loaded Tf, and possess an intrinsic tyrosine kinase involved in signaling pathways [69, 70].

TfR is present on the surface of most cells in the body, although in higher concentration in those cells with a higher iron requirement or increased proliferation rate. For instance, erythroid lineage stem cells in bone marrow have the higher concentration of this protein in their membrane [57].

Two TfRs have been described: TfR1, with high-affinity uptake for Fe-loaded transferrin, and primary target of transferrin in the iron transport system. Compared with TfR1, the role of TfR2 in the uptake of transferrin-bound iron into the cells is considered less significant, as TfR2 has a 25-fold lower affinity for iron loaded transferrin. In addition, TfR2 shows much higher expression levels in liver, compared to other tissues (and in fact, mutations in the TfR2 gene causes hereditary hemochromatosis type 3) [12].

However, transferrin affinity to TfR depends on the iron status of transferrin: non-iron-loaded transferrin presents almost no affinity to its receptor, whereas diferric transferrin presents a 30-fold higher affinity to TfR1 compared to monoferric transferrin [12]. At low Tf concentrations (<0.3μmol/L), TfR1 mediates Tf internalization, but at high Tf concentrations, there is a low affinity for Fe-loaded Tf and it is not mediated by TfR. TfR2 is involved in this low affinity uptake, expressed in only a few organs, and other proteins are responsible for binding and internalize Tf with low affinity, such as the proteoglycans in hepatocytes and glyceraldehyde- 3-phosphate dehydrogenase (GAPDH) on the macrophage cell surface [71].

TfR has an intraindividual variation coefficient of 12%. It is decreased in aplasic anemia, anemia chronic renal failure, and increased in iron deficiency anemia, hemolytic, and in ineffective erythropoiesis such as sideroblastic anemias, myelodysplastic syndromes and megaloblastic anemias (Table **3**) [72].

Table 3. Summary of main biomarkers and variation according to the underlying pathology.

Laboratory test (and normal values)	**Iron deficiency anemia**	**Anemia of inflammation**	**Mixed anemia**	**Functional iron deficiency (resistance to ESA therapy)**	**Iron-loading anemias (ineffective erythropoiesis)**	**Chronic transfusion**	**Hemochro-matosis**
Serum iron (50-180μg/dl)	low						
Ferritin (30-300ng/ml)	Low (<30ng/ml)	Normal to high (>100ng/ml)	Low to normal (<100ng/ml)	variable	high	high	
Transferrin (200-360 mg/dl)	high	low					low
Capacity of transferrin saturation (20-50%)	Low (<20%)	low	low	low	high	high	high
Transferrin saturation index	low				high		high
Transferrin receptor	high	low			high		
Soluble Transferrin receptor (0.76-1.76mg/l)	high	normal	Normal to high				low
sTfR/log ferritin index (<1)			>2				

(Table 3) contd.....

Laboratory test (and normal values)	**Iron deficiency anemia**	**Anemia of inflammation**	**Mixed anemia**	**Functional iron deficiency (resistance to ESA therapy)**	**Iron-loading anemias (ineffective erythropoiesis)**	**Chronic transfusion**	**Hemochro-matosis**
Hepcidin	low	high	normal	high	low	Normal to high	
Hemoglobin W 12-16g/dl M 13-17g/dl	low	low	Low to normal	low	low	Low to normal	Normal to High
Percentage of hypochromic red cells (<5%)	High (>5%)		High (>5%)		Low		
Reticulocytary hemoglobin (28-35pg)	Low (<25pg)			Low (<29pg)	Normal to high		

Soluble Transferrin Receptor (sTfR)

The soluble transferrin receptor (sTfR) is a truncated form of the intact trans membrane receptor, lacking the cytoplasmic and trans membrane domains. Almost all circulating transferrin receptor is the truncated monomeric form, and plasma concentration is in equilibrium with the membrane-bound receptor [73]. The quantity of soluble transferrin receptor depends on the patient's iron homeostasis; in the absence of iron deficiency anemia, there are predominantly two receptor molecules attached to two transferrin, whereas in iron deficiency anemia cases, two receptors bound to only one transferrin [74]. The ratio of transferrin not bound to iron (apotransferrin), attached to an iron ion (III) (monotransferrin) and two ions or iron (III) (ditransferrin) depends on iron reserves. 80% of sTfR comes from erythroblasts [75].

Measurement of sTfR in clinical laboratories is usually realized through latex-enhanced immunonephelometric/turbidimetric assays, with poor agreement among different kits and thus, lack of international standards [12]. However, sTfR has been proposed as a marker of either iron deficiency or erythropoietic activity (or both). The major clinical role of soluble transferring receptor detection is in differentiating iron-deficit anemia (sTfR increased in plasma) from anemia of inflammation, which has no effect on sTfR values [26], allowing to identify iron deficit in patients with anemia and a concurrent inflammatory disease [12].

Although sTfR levels vary with age (as shown on Table **1**), they are decreased in situations of erythroid hypoplasia (such as hyper transfusions, chronic renal failure, or intensive chemotherapy). On the contrary, they are increased when erythropoiesis is enhanced, as in hemolytic anemias, megaloblastic or iron deficiency anemia and secondary polycythemia. It could be therefore used to assess erythropoietic response to erythropoiesis stimulating agents, or even the recovery of erythropoiesis after renal transplantation [12].

The combination of sTfR and ferritin plasma concentrations can be used both to decide the treatment of anemia associated with chronic renal failure, and for monitoring: high plasma concentrations of sTfR associated with normal or low ferritin indicate iron treatment while low sTfR plasma concentration with normal or high ferritin predict increased erythropoiesis after treatment with erythropoietin (Table **3**); the sTfR can be used as a marker of erythroblastic mass [72].

The [soluble transferrin receptor/log (ferritin)] index relates both measures (sTfR and ferritin) as a ratio of the ferritin concentration to that of soluble transferring receptor. The sTfR-F index predicts response to intravenous iron in renal patients under long term erythropoietin stimulating agents therapy [26].

It also offers the most accurate detection of iron deficiency anemia: serum ferritin usually allows to establish the differential diagnosis between iron deficiency anemia and anemia of chronic disease. However in the ferritin grey zone (normal values, between 10-100ng/ml) sTfR/log ferritin index enable the most accurate diagnosis of iron deficiency anemia [76, 77].

Hemosiderin

Hemosiderin is a form of ferritin aggregated and partially deproteinized. In iron overload context, most of this metal is stored as hemosiderin, which released iron much more slowly than ferritin because of a bigger aggregation of this metal inside this protein [78]. Pearls staining reveals the presence of hemosiderin [79].

Hemosiderin is the preferred form of iron reserve in bone marrow macrophages, spleen cells and Kupfer; on the contrary, it is lower in the liver [11].

HFE

The protein encoded by HFE gene (formerly known as HLA-H) is a membrane protein similar to major histocompatibility complex class I proteins but without function in the immune system [80]. The HFE gen product competes with transferrin itself to associate with transferrin receptors (TfR 1 and 2), and therefore, the HFE-TfR2 complex modulate hepcidin transcription [5].

The transferrin receptor fixed on the membrane is attached to a molecule of the HFE protein. It is a trans membrane molecule composed of 343 amino acids, with two polypeptide chains: one "heavy", joined to the cytoplasmic membrane, with three extracellular domains, and other "light", which is a β2-microglobulin molecule. The heavy chain has an α3 domain, similar to the constant domains of immunoglobulin chains, by which it is bound by non-covalent binding to the light chain [80].

HFE expression has alternative splicing. The predominant HFE full-length transcript has approximately 4.2 kb. Alternative HFE splicing variants serve as iron regulatory mechanisms in certain cells or tissues [81].

HFE is present mainly in small intestinal absorptive cells, gastric epithelial cells, macrophages, and in monocytes and granulocytes. It is also located in the syncytiotrophoblast, an iron transport tissue in the placenta [82].

HFE binds to the transferrin receptor in crypt enterocytes and placenta. This binding does not produce a conformational change in HFE but it does in the receptor, which suffers from changes throughout the dimer interface to the cytoplasmic end. The final effect is to reduce the passage of iron through the membrane holding the transferrin receptor. The mechanism of this movement remains unknown: some postulate that HFE decreases the affinity for the transferrin receptor to iron-bound transferrin; others, that HFE decreases the number of functional receptor molecules at the cell surface and it does not stimulate the internalization in cycling endosomal; and finally, that the conformational changes in HFE make difficult to export the iron [80].

Divalent Metal Transporter 1 (DMT1 or Nramp 2)

The divalent metal transporter 1 (DMT1), also known as natural resistance-associated macrophage protein 2 (NRAMP 2), is a protein encoded by the SLC11A2 gene (solute carrier family 11, member 2) [83].

DMT1 binds different divalent metals as cadmium (Cd^{2+}), copper (Cu^{2+}), and zinc (Zn^{2+}), but is better known for its role in transporting ferrous iron (Fe^{2+}) although it also mediates the absorption and transport of manganese (Mn^{2+}) [84].

The transport of iron and other divalent metal ions by DMT1 is pH dependent, but the exact way in which pH performs its effect is unknown [85].

DMT1 mobilizes the iron-bound transferrin present in the endosomes in order to transfer it to the enterocyte basolateral membrane thanks to an active transporter [85].

Different isoforms are expressed depending on the tissues. Intracellular iron status is the leading signal for DMT1 expression, although other yet undefined signals may also contribute. DMT1 gene expression is regulated by body iron stores, to keep the iron homeostasis [86].

Interestingly, DMT1 function differ between humans and other species: the sDMT1 mutation found in mice and rats limits iron uptake in the intestine and iron utilization in red cell precursors, whereas regulating iron utilization by erythroid cells in humans. However, the importance of DMT1 function and its regulation mechanisms on a molecular scale are still barely understood [86].

Ferroportin 1

Ferroportin 1, also called IREG1 (iron -regulated transporter 1) or solute carrier family 40 member 1 (SLC40A1), is formed by 571 amino acid residues, with a highly conserved histidine at residue position 32, and at least 10 trans membrane domains [87]. The mutated form shows a lower activity in its iron transport role [88].

Expressed mainly in the duodenum (especially in the basolateral membrane of enterocytes), placenta, liver, kidney and macrophages [87], ferroportin 1 is the single iron exporter in mammals and therefore considered a key coordinator of the iron balance between intracellular and systemic iron homeostasis [89]. When iron stores diminish, it might transport iron from enterocytes to the blood.

It also plays a role in iron transport across the yolk sac to the embryo as well as from maternal plasma to embryo [87].

Hephaestin

The hephaestin is another trans membrane protein with a single domain within the membrane. Its interaction with ferroportin 1 allows iron passage through the spleen-lateral membrane of enterocytes. It has a high homology with the plasmatic ferroxidase, and possesses the same enzymatic activity as a cofactor, having several copper ions [90].

It is highly expressed in the small intestine (although practically nothing in the crypts) and colon. This protein links the metabolism of iron and copper explaining why copper deficiency can lead to ferropenia (the hephaestin needs copper to be functional in iron absorption), while iron deficiency does not result in copper deficiency [90].

Haptoglobin

Haptoglobin is a glycoprotein of 85 kDa synthesized mostly in the liver, although other tissues such as lung, skin and kidney can also produced small amounts; 19% of its weight corresponds to carbohydrates and its electrophoretic migration corresponds to the region α-2-globulin. It has a variable tetrameric structure, with two β chains and two α chains united by disulfide bridges, presenting different genetic polymorphism as well as three different phenotypes with a different electrophoretic mobility [62].

Haptoglobin has a high affinity for free hemoglobin, joining it in a molar ratio 1:1. The formed complex allows degradative enzymes (mononuclear phagocytic system of the body) to destroy it. Given the big size of the complex, hemoglobin cannot be filtered by the glomerulus. Hemoglobin not joined to haptoglobin binds to hemopexin, albumin and other proteins. Free hemoglobin that has not been captured by any proteins rapidly is going to be filtered by the kidneys [60].

In clinical settings, the haptoglobin assay is used to diagnose and monitor intravascular hemolytic anemia: in intravascular hemolysis, free hemoglobin is liberated into circulation, joining the haptoglobin (leading to an acute reduction of haptoglobin levels) (Table **3**). Conversely, in extravascular hemolysis, the reticuloendothelial system, especially splenic monocytes, phagocytose the erythrocytes and hemoglobin is not released into circulation. However, in massive hemolysis situations, some hemoglobin might be released, decreasing haptoglobin levels [91].

Moreover, haptoglobin is considered as an acute-phase reactant and thus, levels can also become elevated in infection and inflammation [92]. Therefore haptoglobin is not always a very reliable marker to differentiate between intravascular and extravascular hemolysis.

Hemopexin

Haemopexin is a single chain glycoprotein of 57 kDa molecular weight; 20% of its weight is formed by carbohydrates [93]. Electrophoretic migration of hemopexin corresponds to the area of β-globulins [94]. It is mainly expressed in liver, and is considered as an acute phase reactant.

It is the plasma protein with the highest binding affinity to heme group. Heme protein is potentially toxic due to its ability to intercalate into lipid membrane and produce free hydroxyl radicals. The strong bond between heme and hemopexin, and the existence of a specific heme-hemopexin receptor able to catabolize the complex and to induce intracellular antioxidant activities, suggest that hemopexin is the major vehicle for the transportation of heme in the plasma. Rather than participate in iron metabolism, hemopexin might act as an antioxidant after blood heme overload [94].

This protein is, after haptoglobin, the second line of defense against hemoglobin-mediated oxidative damage during intravascular hemolysis. A decrease in plasma hemopexin concentration reflects a recent release of heme compounds in the extracellular compartment. Heme-hemopexin complexes are delivered to hepatocytes by receptor-mediated endocytosis after which hemopexin is recycled to the circulation [95].

Ferroxidase

Ferroxidase, previously called ceruloplasmin, with a molecular weight of 132 kDa, is a polypeptide with various carbohydrate moieties containing six copper ions that are essential for the stability of the molecule; it is synthesized in the liver [10].

It exerts its oxidizing function on the iron (II), which is absorbed by the intestinal microvilli. On the other hand, ferroxidase is responsible for maintaining iron homeostasis in the liver during periods of iron deficiency [10]. This protein may also transfer copper to enzymes requiring it [10].

Hepcidin

Hepcidin (or LEAP 1, liver-expression Antimicrobial peptide), a 25-amino acid peptide synthesized by the liver, is the main iron regulatory hormone [96]. Apart from hepatocytes which are the main source of circulating hepcidin, other cells such as macrophages [97] and adipocytes [98] also express hepcidin mRNA, but at a much lower level. The relevance of the extra hepatic hepcidin production remains unclear, but it could have a role in local regulation of iron fluxes [99].

When bound to its receptor (the cellular iron exporter protein ferroportin), both are internalized, and degradate, leading to the sequestration of iron (keeping off its release from the reticuloendothelial system to the blood). Thus hepcidin has been postulate a marker of iron supply and retention, rather than of restricted erythropoiesis [12].

Moreover, hepcidin reduces iron absorption in the duodenum [31], decreases plasma concentrations, and impairs the accessibility of stored iron from macrophages [100].

The iron content of the blood regulates the production of hepcidin: more hepcidin is produced by hepatocytes when iron is abundant, limiting further iron absorption and release from stores. When iron is deficient, hepatocytes produce less or no hepcidin, allowing more iron to enter the plasma. Both diferric plasma transferrin and stored iron in hepatocytes can stimulate hepcidin synthesis [101].

Hepcidin interacts with villous enterocytes to regulate the absorption of iron.

Increased expression of hepcidin has a double effect: it inhibits intestinal iron absorption and blocks iron reserve within the macrophages [102]; with both actions hepcidin appears to be dedicated to prevent iron overload. Serum levels of hepcidin are usually undetectable in conditions causing iron overload, such as hereditary hemochromatosis.

However, inflammation increases the expression of hepcidin (it behaves as an acute phase reactant type II) regardless of iron status [26, 102]. Inflammatory cytokines (such as interleukine 6 which is produced in early phases of acute inflammation, or tumor necrosis factor) up regulate hepcidin transcription. Stress pathways signaling through the cellular endoplasmic reticulum response have also been found to induce hepcidin expression [5]. Thus, serum levels of hepcidin are elevated in acute and chronic inflammatory states, such as infection, rheumatoid arthritis, inflammatory bowel disease and chronic kidney disease, decreasing iron absorption in such clinical conditions (and causing the so called inflammatory anemia, or anemia of chronic states) (Table **3**).

Lactoferrin

Lactoferrin, also known as lactotransferrin, is a globular glycoprotein of 703 amino acids and a molecular mass of about 80 kDa. It belongs to the transferrin family, also composed by the plasma transferrin, ovotransferrin, and melanotransferrin. They regulate the concentration of free iron in the system. It is widely represented in various secretory fluids, such as milk, saliva, tears, and nasal secretions and stored in polymorph nuclear cells secondary granules and secreted by some acinar cells [103].

Lactoferrin belongs to the basic proteins, with an isoelectric point of 8.7. Two forms exist, with different tertiary structures: an iron-free one -apolactoferrin- characterized by "open" conformation of the N-lobe and the "closed" conformation of the C-lobe, and an iron-rich one -hololactoferrin- with both lobes closed [104]. There has been identified three isoforms and receptors have been isolated for lactoferrin in the intestine, monocytes, neutrophils, lymphocytes, platelets, and some bacteria. Each lactoferrin molecule can bind two ions of iron, zinc, copper or other metals [105].

Lactoferrin affinity for iron is 300 times higher than iron to transferrin [106], and its increased in weakly acidic medium. This facilitates the transfer of iron from transferrin to lactoferrin during inflammation, when the pH of tissues decreases due to accumulation of lactic acid among others [107].

Titer of lactoferrin in the blood corresponds to this particular "transition concentration" and thus lactoferrin in the blood could be in both forms, as a monomer and tetramer [108]. But lactoferrin plays also an immunoregulatory function, by decreasing the release of interleukins 1 and 2, and tumor-necrosis factor by monocytes, as well as increasing the cytotoxic activity of natural killer and monocytes [108].

Moreover, it has also some antimicrobial activity (bactericide, fungicide) and is part of the innate defense, mainly at mucoses [103], through the kidnapping of iron. This protein has also a direct bactericidal effect on gram negative bacteria by damaging their membrane external [109].

It has a high affinity for the lipid A, a fraction of lipopolysaccharide which it is the endotoxin of gram negative. When it binds to the lipid A, this interaction inhibits the join of this lipid with antigen CD14 monocytes; thus lactoferrin neutralizes the toxic effect of endotoxin [109].

Hemoglobin

The globin superfamily of heme proteins offers a large set of proteins whose function is tightly related to small gaseous non polar ligand (mainly O_2 but also NO and CO) affinity and reactivity [110, 111]. But hemoglobin has functions other than storage and transport, including enzymatic and sensing functions [111].

Structure:

Hemoglobin comprises 4 subunits each having one polypeptide chain and one heme group (with a ferrous ion linked inside), equals in pairs. The polypeptide chains might vary, conforming the over 200 hemoglobin variants described to date [112], which differ by the length of the protein and the number of alpha helices structures. The tetrahedral assemblage of the subunits forms its quaternary structure, with hydrogen bridges stabilizing the internal proteins, keeping its specific ellipsoidal shape [11]. In healthy adults, the most common hemoglobin, called hemoglobin A, consists of two α and two β chains, made of 141 and 146 amino acid residues respectively (form denoted therefore as α2β2) non-covalently bounded [112]. In infants, the hemoglobin molecule is made up of 2 α chains and 2 γ chains. The gamma chains are gradually replaced by β chains as the infant grows.

A heme group consists of an iron ion hold by a heterocyclic ring, designated as a porphyrin. This porphyrin ring consists of four pyrrole molecules cyclically linked together with the iron ion bound in the center. In this structure, iron molecule presents 6 valences, each of one binds a globin chain, leaving 2 free valences to bind oxygen. A sixth position can reversibly bind oxygen by a coordinate covalent bond completing the octahedral group of six ligands. However, union to oxygen is only possible if the iron molecule is in its reduce form (Fe^{2+}) as oxide iron (Fe^{3+}) is unable to bond oxygen [11].

The ferrous iron atom present in each heme group can bind reversibly oxygen [112]. Therefore hemoglobin can bind up to four oxygen molecules [113] and transport it in the red blood cells from lungs to the rest of the body.

Synthesis:

Through 2 different metabolic pathways, hemoglobin is synthesized in a combined stepwise procedure: the heme part is synthesized by immature red blood cells both in the mitochondria and the cytosol, while the globin protein parts are synthesized by ribosomes in the cytosol. Production of hemoglobin continues in the cell throughout its early development in the bone marrow from the proerythroblast to the reticulocyte. At this point, the nucleus is lost, but residual ribosomal RNA allows the synthesis of hemoglobin until the reticulocyte loses its RNA (this hemoglobin-synthetic RNA gives the reticulocyte its reticulated appearance in optical microscopy and name) turning into a mature red cell [11, 114].

Function:

Hemoglobin function as a two-way respiratory carrier is to reversibly fix oxygen for transport from the lungs to the tissues, and reversely, with dioxide carbon. Thanks to the cooperative binding of oxygen, hemoglobin saturation changes very quickly with oxygen partial pressure; this fact turns hemoglobin into a highly efficient oxygen carrier and justifies its binding at high oxygen partial pressure (in the lungs) and its release at lower pressures (as in the other tissues) [11, 112].

Even though carbon dioxide is carried by hemoglobin, it does not compete with oxygen for the iron-binding positions because it is bounded to the protein chains of the structure. However carbon monoxide binds to hemoglobin more efficiently than oxygen, and therefore, once bound, carbon monoxide cannot be removed by oxygen [112].

Hemoglobin is also present in other cells such as the dopaminergic neurons in the substantia nigra, macrophages, alveolar cells, and mesangial cells in the kidney. In these tissues, hemoglobin acts as an antioxidant and as regulator of iron metabolism [115]. Its buffering capacity allows hemoglobin to participate in blood pH regulation [11].

Hemoglobin levels are used as a measure of anemia, with a threshold varying according to sex and age (but needing also adjustment for altitude, smoking and ethnic group), as shown below on Table **2**.

Biological Determination of Erythropoiesis

Red Blood Cells Variables

Mean cell volume (MCV) and mean cell hemoglobin (MCH) values are useful for anemia diagnosis, and reflect trends over large periods: they do not reflect acute changes (secondary to therapy with erythropoiesis stimulating agents or to acute bleeding events) and do not allow therefore assessing early response to iron therapy [26]. However, MCV is used for the traditional classification of anemia (and first step in differential diagnosis) in microcytic (low mean cell volume) or macrocytic (high cell volume).

Percentage of Hypochromic Red Cells

It is the most accurate parameters to assess functional iron deficit, both in erythrocytes and reticulocytes. However it have technical limitations (in terms of sample stability or equipment availability), it is expected to be solved within the next few years. These measures are effectives in both detection and monitoring of functional iron deficit secondary to ESA therapy (Table **3**) [26].

As erythrocytes lifespan is about 120 days, hypochromic red cells constitute a late indicator of iron-restricted erythropoiesis, providing information over a 3-month period. In contrast, reticulocytes have a much shorter lifespan, of less than 2 days: reticulocyte hemoglobin content has emerged as an early marker of functional iron deficit, allowing an almost "real time" control of iron [12].

Reticulocytary Hemoglobin

The great increase in precision of the automated reticulocyte count and their hemoglobin content allows now assessing the effects of fluctuations in iron status on this transient red cell population.

Low reticulocytary hemoglobin (Ret-He) predicts functional iron deficit in patients under erythropoiesis stimulating agents (Ret-He<29pg), as well as response to intravenous iron therapy in chronic kidney disease patients on hemodialysis (Ret-He<30.6pg) (Table **3**) [26].

Immature Reticulocyte Fraction

Immature reticulocytes are released during periods of intense erythropoietic stimulation, such as after important blood loss (bleeding or hemolysis), or in early response to ESA [26].

DISCUSSION

Ferropenic Deficiency

In some physiological situations, such as growth, pregnancy and breastfeeding, contributions of iron might not be adequate, resulting in a reduction of iron deposits to release and supply these needs. Moreover, in these situations, enterocytes increase the synthesis of molecules that absorb iron. This phase is the phase I of iron deficiency, and clinical laboratory levels show reduced plasma concentrations of ferritin and a slight increase of soluble transferrin receptor is observed [116].

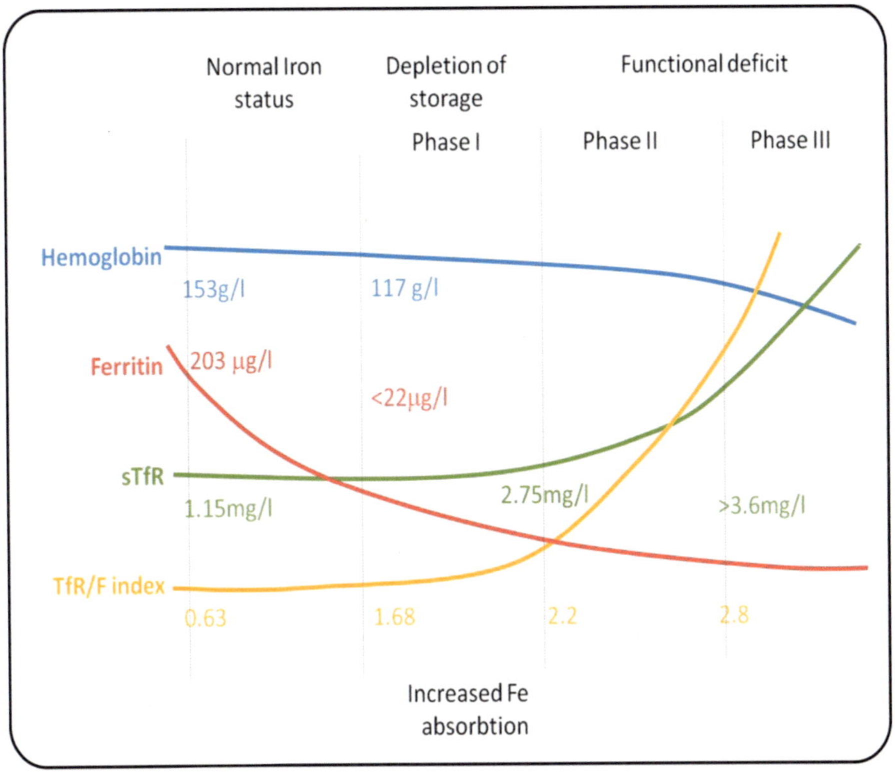

Fig. (2). Phases of iron deficiency. Modified from Suominen *et al.* [116].

When stockage can no longer cope with the needs of the organism, the iron transported by transferrin diminishes also. Consequently, in the bone marrow, the ferrochelatase of erythroblasts mitochondria drops the incorporation of iron and erythropoiesis is defective, having a functional iron deficiency [11]. In the laboratory, we observe a decrease of hemoglobin content in the reticulocytes (indicating a ferropenic erythropoiesis started in the previous two days), and a rise in hypochromic erythrocyte count (indicating the beginning of iron deficiency erythropoiesis more than two months ago); these quantities can be estimated by measuring analyzers erythrocytes on an individual basis, rather than averages indices [117].

In the second stage of iron deficiency, the plasma concentration of soluble transferrin receptor increases such as the indexes that relate these levels to plasma concentration levels of ferritin (still decreasing); transferrin saturation begins also to decrease, as does the medium level of corpuscular volume of erythrocytes (globin synthesis in impaired, and smaller erythrocyte are caused by lack of these proteins) [118]. The blood concentration of hemoglobin begins to decrease, but over the anemia threshold. In phase three of iron deficiency, anemia appears and variations of the mentioned biochemical markers are more pronounced.

In summary, the process has several phases: iron deficiency (phase I), ferropenic erythropoiesis or functional iron deficiency (phase II) and ferropenic anemia (phase III), as shown in Fig. (**2**).

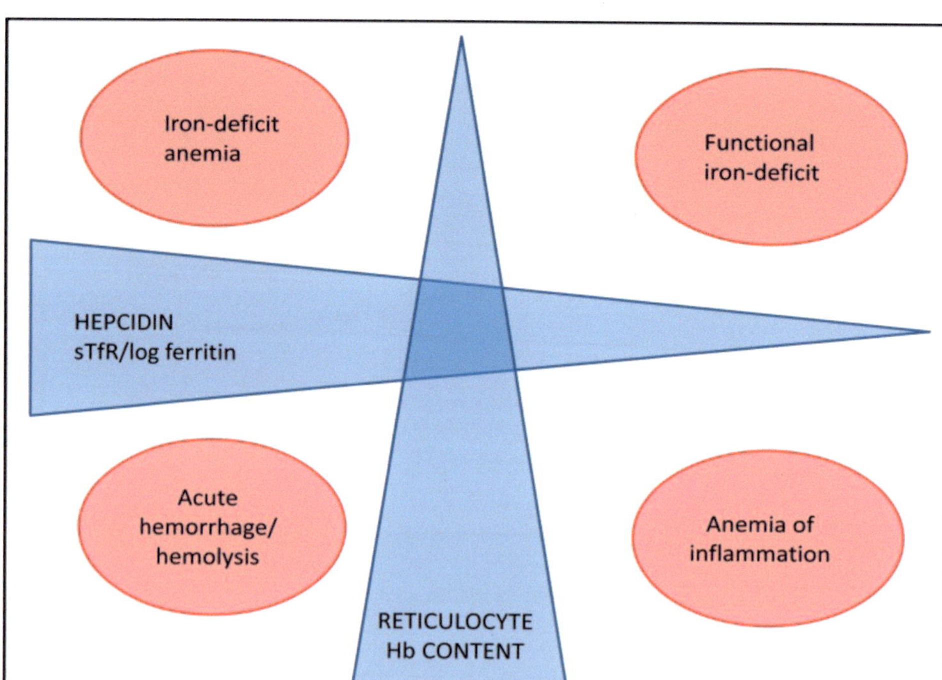

Fig. (3). Use of hepcidin or soluble transferrin receptor values and reticulocytes hemoglobin content for anemia diagnostic assessing.

Anemia of Chronic Diseases

Chronic infections, inflammation, and neoplasias often lead to anemia through inflammation context, in the so called anemia of chronic disease, or anemia of chronic disorders. These processes share an inflammatory nature, which does not have to be chronic, and thus in the light of actual knowledge, is rather called anemia of inflammation [119]. Defined by low serum iron despite adequate systemic iron stores, it usually present as anemia with low ferremia, low transferrin saturation and presence of Prussian-blue-stainable iron in bone marrow macrophages [119].

Bone marrow biopsy is no longer considered as in diagnosis of iron status in anemia states because of the invasive nature of the sampling procedure, and the qualitative nature of the reading, than in addition is evaluator-dependent. It has been replaced by serum ferritin determination: low levels (less than 15ng/ml) are highly specific of iron deficiency and rule out anemia of inflammation as serum ferritin is increased by inflammation [119].

In the physiopathology of anemia several mechanisms are involved, initiated by immune response activation to a foreign antigen or a tumor. Inflammation or infections lead to accumulation of iron in macrophages, which explains the hypoferremia in those situations [7]. This induces the secretion of cytokines: in rheumatoid arthritis, tumor necrosis factor alpha cause apoptosis of erythroid precursors and inhibits their proliferation; lack of precursors leads to anemia. Interleukin-1 has a similar effect [11].

Another mechanism is the alteration of iron metabolism: iron concentration and plasma transferrin decrease. This action is probably mediated by interleukin 6 [120].

Intestinal iron absorption also decreases, and iron is locked in macrophages which leads to ferropenic erythropoiesis. In iron deficiency anemia the number of sideroblasts is reduced, while in the anemia of chronic disease is normal or increased; however, the iron cannot be transferred to the erythroblasts. These actions are probably mediated by hepcidin [121]; interleukin-1 stimulates the activity of polymorphonuclear neutrophils: released lactoferrin (from leukocyte's granules) binds to iron (II); the complex formed is captured by the mononuclear phagocytic system [11].

Moreover, in the anemia of chronic disease the tumor necrosis factor, interleukin-1, and interferon-γ inhibit the effect that low iron availability has on transferrin receptor mRNA transcription, so their synthesis does not increase. In consequence, erythroblasts do not express the receptor on their surface cellular and capture less iron. This mechanism adds to the previous one, and iron deficiency erythropoiesis and anemia occur, despite normal (or even high) body iron content [122].

Biochemical findings are decreased transferrin concentration and transferrin saturation capacity, and an increased plasma concentration of ferritin and transferrin saturation index; all occurs with little or no increase in plasma concentration soluble transferrin receptor [3].

Anemia of chronic disease and iron deficiency anemia are the most prevalent forms of anemia and may often occur concurrently. Standard test of iron status used in differential diagnosis are affected by inflammation, hidering clinical interpretation. In contrast, sTfR indicates iron deficiency and is unaffected by inflammation. Use of sTfR and sTfR Index improves the detection of iron deficiency anemia, particularly in situations where routine markers provide equivocal results [123]. Differential diagnosis between main causes of anemia (iron-deficit, functional iron-deficit or anemia of inflammation) might be guided by biomakers expression, as reflected on Fig. (**3**).

The relationship between functional iron deficit and iron supply to erythropoiesis can be represented in a diagnostic plot as suggested by Thomas *et al.* [117, 124]. In iron-deficit anemia, the iron supply depends on the amount of iron stores, whereas in anemia of chronic disease the supply depends on its mobilization rate, with functional iron-deficit occurring even in the presence of large iron stores (because iron release is impaired). Therefore the relationship between iron status and sTfR levels in patients with inflammation depends on the severity of inflammation, the degree of anemia, the adequacy of erythropoietin production, and the effect of cytokines on marrow activity [12]. Hepcidin has been proposed to be the pathogenic feature of anemia in inflammation, the same as it get bound to ferroportin, inducing its degradation. Iron exportation is kept off, and thus iron is trapped inside enterocytes, macrophages and hepatocytes [12].

Iron Overload

Iron overload occur in normal erythropoiesis situation but with an excess of iron in relation to transferrin capacity. The iron surplus can deposit on organs, affecting to their function. Overload can also appear as a result of increased catabolism of erythrocytes. This iron cumulates in macrophages of the mononuclear phagocytic system and is therefore much better tolerated. Only when their reserve capacity is surpassed, the excess of iron goes to the liver. In the first of these cases we are dealing with a hereditary hemochromatosis, in the second to other ferric overload [3].

Hereditary Hemochromatosis

Hereditary hemochromatosis appeared after a mutation about 2000 years ago, and its consequences should give some advantage to carriers when iron deficiency was prevalent in all age groups by the introduction of agriculture. Currently in the first world, where the diet contains the bioavailable iron, there are no such benefits [125].

In 1976 it was discovered that the genetic disposition of hereditary hemochromatosis is inherited together with the HLA.A3 complex. Twenty years later Feder *et al.* found that in the HLA.A3 complex chromosome 6p was the gene coding for the HFE protein and it is transmitted in an autosomal recessive manner [126].

The C282Y mutation, which is the most important, consists in changing cysteine for a tyrosine in domain 3; this change prevents the formation of a disulfide bridge involving the folding of HFE that interfere with the binding of β2-microglobulin to HFE and also not expressed in the membrane. It was later shown that disruption of the HFE gene in the mouse originated the synthesis of a nonfunctional protein HFE, with the development of hereditary hemochromatosis [3].

H63D, the second most important mutation is a change from a histidine to an aspartate, in this case, HFE is formed and it can join β2-microglobulin and this complex to transferrin receptor, but these connections have no effect on the subsequent binding with transferrin [80].

Another mutation is the S65C, it is given by changing a serine by a cysteine, this change affects the region involved in the binding of HFE with transferrin receptor [127].

Patients affected by these genotypes absorb iron chronically, with independence of the iron stores, hypoxia or erythropoietic activity. The absorption is two to three times higher than in normal subjects and it occurs because the HFE protein does not exert its effect on transferrin receptor. Therefore the concentration of cytosolic iron in crypt enterocyte is low (although the body iron stores are high), and this signal causes high expression of the transferrin receptor, DMT1, ferroportin 1 and hephaestina and low ferritin; as the signal always persists, iron absorption is preserved [128].

This iron plasma exceeds the capacity of protection by binding to proteins such as transferrin and ferritin, and travels as complexes of a low molecular mass, that can react with various reactive species and form oxygen radicals. In the early stages of hereditary hemochromatosis, iron bounded to transferrin it is taken up by type 2 transferrin receptor of hepatocytes; this capture occurs indefinitely. As the disease progresses, excess iron is deposited in tissues, creating fibrosis: in the liver, leading to cirrhosis and a high risk of hepatocellular carcinoma (liver damage is the leading cause of death of hereditary hemochromatosis) [55]. The iron accumulation in the heart, pancreas, pituitary, parathyroid gland leads to cardiomyopathy, diabetes mellitus, hypopituitarism, hypogonadism and hypoparathyroidism [3].

Until recently, iron overload confirmation was made by liver biopsy and measurement of the concentration of iron per gram of dry liver tissue. Nowadays only biopsies for prognosis of some patients are performed. Detection of homozygous C282Y mutation genotypes or C282Y / H63D, C282Y / S65C, by DNA amplification by polymerase chain reaction followed by restriction enzyme, and evidence of iron overload confirm the diagnosis of hereditary hemochromatosis [99].

Other Ferric Overloads

All non ferropenic anemia course with a high plasma concentrations of ferritin [129], but only when iron is deposited locally or generally over a normal way we call it overload; this is what happens in some of these anemias, which have an increased erythropoiesis and/or intestinal absorption (such as thalassemia and overall hemolytic anemias, anemia congenital dyserythropoietic, sideroblastic anemia, or porphyria cutanea tarda, in all of them there is an ineffective erythropoiesis). In some of these diseases the mechanism of why there is an increased intestinal absorption is unknown and in other hepcidin is high [102]. For instance, the African iron overload, siderosis transfusion, and alcoholic liver disease, present also ferric overload. This group has also been called acquired [22] or secondary hemochromatosis [11].

CONCLUSION

Iron metabolism influences many diseases. Accurate diagnosis of its disorders allows to prevent or reverse the associated damage. Some biomarkers guide the diagnosis and even allow to predict and/or assess response to a specific treatment.

In a patient with microcytic anemia, we need to differentiate anemia of chronic disease or inflammation from iron deficiency anemia, which are the most prevalent forms and may often occur concurrently. In most cases, serum ferritin determination allows to diagnose iron-deficit anemia as low levels (less than 15ng/ml) are highly specific of iron deficiency. However, they are affected by inflammation, hindering clinical interpretation. In contrast, sTfR indicates iron deficiency and is unaffected by inflammation and thus should be measured when iron deficit is suspected and ferritin levels are elevated.

High hepcidin levels show an inflammatory state and are thus predictive of no response to oral iron treatment. In such situation, if sTfR levels indicate iron deficiency, intravenous treatment is needed. However, inflammatory states can also lead to anemia without a real iron deficit (sTfR levels would be normal) and thus only the treatment of subjacent disease will improve the anemia.

On the other side of iron metabolism, hemochromatosis diagnosis is realized through the determination of HFE and genetic testing. Once diagnose, ferritin levels allow to assess treatment response.

CONFLICT OF INTEREST

The author (editor) declares no conflict of interest, financial or otherwise.

ACKNOWLEDGEMENTS

Declared none.

REFERENCES

[1] Wang J, Pantopoulos K. Regulation of cellular iron metabolism. Biochem J 2011; 434(3): 365-81. [http://dx.doi.org/10.1042/BJ20101825]

[2] Brown E, Aisen P, Fielding J, Crichton R. Proteins of Iron Metabolism. New York 1977.

[3] Andrews NC. Disorders of iron metabolism. N Engl J Med 1999; 341(26): 1986-95. [http://dx.doi.org/10.1056/NEJM199912233412607]

[4] Cook JD, Flowers CH, Skikne BS. The quantitative assessment of body iron. Blood 2003; 101(9): 3359-64. [http://dx.doi.org/10.1182/blood-2002-10-3071]

[5] Munoz M, Garcia-Erce JA, Remacha AF. Disorders of iron metabolism. Part 1: molecular basis of iron homoeostasis. J Clin Pathol 2011; 64(4): 281-6. [http://dx.doi.org/10.1136/jcp.2010.079046]

[6] Perez-Surribas D. Proteínas relacionadas con el metabolismo del hierro. Química Clínica 2005; 24(1): 5-40.

[7] Ganz T, Nemeth E. Iron sequestration and anemia of inflammation. Semin Hematol 2009; 46(4): 387-93. [http://dx.doi.org/10.1053/j.seminhematol.2009.06.001]

[8] Bridges KR. Anemias with disturbed iron metabolism. In: Hill MG, Ed. Harrisons's Principles of Internal Medicine. 13th ed. New York: Mc Graw Hill 1994; pp. 1717-21.

[9] Finch C. Regulators of iron balance in humans. Blood 1994; 84(6): 1697-702.

[10] Attieh ZK, Mukhopadhyay CK, Seshadri V, Tripoulas NA, Fox PL. Ceruloplasmin ferroxidase activity stimulates cellular iron uptake by a trivalent cation-specific transport mechanism. J Biol Chem 1999; 274(2): 1116-23. [http://dx.doi.org/10.1074/jbc.274.2.1116]

[11] Vives Corrons JL, Sans-Sabrafen J. Introducción al estudio de la patología eritrocitaria. In: SA EE, Ed. Hematología Clínica. 4th ed. Madrid, España: Mosby Doyma 2001; pp. 81-106.

[12] Harms K, Kaiser T. Beyond soluble transferrin receptor: old challenges and new horizons. Best Pract Res Clin Endocrinol Metab 2015; 29(5): 799-810. [http://dx.doi.org/10.1016/j.beem.2015.09.003]

[13] Gudjoncik A, Guenancia C, Zeller M, *et al.* Iron, oxidative stress, and redox signaling in the cardiovascular system. Mol Nutr Food Res 2014; 58(8): 1721-38. [http://dx.doi.org/10.1002/mnfr.201400036]

[14] McKie AT, Barrow D, Latunde-Dada GO, *et al.* An iron-regulated ferric reductase associated with the absorption of dietary iron. Science 2001; 291(5509): 1755-9. [http://dx.doi.org/10.1126/science.1057206]

[15] Andrews NC, Levy JE. Iron is hot: an update on the pathophysiology of hemochromatosis. Blood 1998; 92(6): 1845-51.

[16] Olivares M, Walter T, Cook JD, Hertrampf E, Pizarro F. Usefulness of serum transferrin receptor and serum ferritin in diagnosis of iron deficiency in infancy. Am J Clin Nutr 2000; 72(5): 1191-5.

[17] Daniels TR, Bernabeu E, Rodriguez JA, *et al.* The transferrin receptor and the targeted delivery of therapeutic agents against cancer. Biochim Biophys Acta 2012; 1820(3): 291-317. [http://dx.doi.org/10.1016/j.bbagen.2011.07.016]

[18] Sciot R, van Eyken P, Desmet VJ. Transferrin receptor expression in normal and iron overloaded liver. APMIS Supplementum 1991; 23: 21-31.

[19] Kaplan J, Jordan I, Sturrock A. Regulation of the transferrin-independent iron transport system in cultured cells. J Biol Chem 1991; 266(5): 2997-3004.

[20] Remacha AF, Altes A. Nuevos avances en la patología de las alteraciones del metabolismo del hierro. Haematologica ed esp 2000; 85: 25-31.

[21] Goodnough LT, Nemeth E, Ganz T. Detection, evaluation, and management of iron-restricted erythropoiesis. Blood 2010; 116(23): 4754-61. [http://dx.doi.org/10.1182/blood-2010-05-286260]

[22] Powell LW, Isselbacher KJ, Braunwald E, *et al.* Hemochromatosis. In: Hill MG, Ed. Harrisons's principles of internal medicine. 13th ed. New York 1994; pp. 1717-21.

[23] World Health Organization Department of Nutrition for Health and Development. World Health Organization Department of Nutrition for Health and Development. Assessing the iron status of populations. Geneva, Switzerland 2004.

[24] World Health Organization. World Health Organization. Iron deficiency anaemia: assessment, prevention and control 2001.

[25] Kassebaum NJ, Jasrasaria R, Naghavi M, *et al.* A systematic analysis of global anemia burden from 1990 to 2010. Blood 2014; 123(5): 615-24. [http://dx.doi.org/10.1182/blood-2013-06-508325]

[26] Thomas DW, Hinchliffe RF, Briggs C, *et al.* Guideline for the laboratory diagnosis of functional iron deficiency. Br J Haematol 2013; 161(5): 639-48. [http://dx.doi.org/10.1111/bjh.12311]

[27] Michels K, Nemeth E, Ganz T, Mehrad B. Hepcidin and host defense against infectious diseases. PLoS Pathog 2015; 11(8): e1004998. [http://dx.doi.org/10.1371/journal.ppat.1004998]

[28] Ganz T, Nemeth E. Iron homeostasis in host defence and inflammation. Nat Rev Immunol 2015; 15(8): 500-10. [http://dx.doi.org/10.1038/nri3863]

[29] Schaible UE, Collins HL, Priem F, Kaufmann SH. Correction of the iron overload defect in beta--microglobulin knockout mice by lactoferrin abolishes their increased susceptibility to tuberculosis. J Exp Med 2002; 196(11): 1507-13.
[http://dx.doi.org/10.1084/jem.20020897]

[30] Holbein BE. Enhancement of Neisseria meningitidis infection in mice by addition of iron bound to transferrin. Infect Immun 1981; 34(1): 120-5.

[31] Kotze SR, Pedersen OB, Petersen MS, *et al.* Low-grade inflammation is associated with lower haemoglobin levels in healthy individuals: results from the Danish blood donor study. Vox Sang 2016; 111(2): 144-50.
[http://dx.doi.org/10.1111/vox.12396]

[32] Al-Hakeim HK, Al-Khakani MM, Al-Kindi MA. Correlation of hepcidin level with insulin resistance and endocrine glands function in major thalassemia. Adv Clin Exp 2015; 24(1): 69-78.
[http://dx.doi.org/10.17219/acem/38158]

[33] Huth C, Beuerle S, Zierer A, *et al.* Biomarkers of iron metabolism are independently associated with impaired glucose metabolism and type 2 diabetes: the KORA F4 study. Eur J Endocrinol 2015; 173: 643-53.
[http://dx.doi.org/10.1530/EJE-15-0631]

[34] Sullivan JL. Iron and the sex difference in heart disease risk. Lancet 1981; 1(8233): 1293-4.
[http://dx.doi.org/10.1016/S0140-6736(81)92463-6]

[35] Manson JE, Allison MA, Rossouw JE, *et al.* Estrogen therapy and coronary-artery calcification. N Engl J Med 2007; 356(25): 2591-602.
[http://dx.doi.org/10.1056/NEJMoa071513]

[36] Heiss G, Wallace R, Anderson GL, *et al.* Health risks and benefits 3 years after stopping randomized treatment with estrogen and progestin. JAMA 2008; 299(9): 1036-45.
[http://dx.doi.org/10.1001/jama.299.9.1036]

[37] Wassertheil-Smoller S, Hendrix SL, Limacher M, *et al.* Effect of estrogen plus progestin on stroke in postmenopausal women: the Women's Health Initiative: a randomized trial. JAMA 2003; 289(20): 2673-84.
[http://dx.doi.org/10.1001/jama.289.20.2673]

[38] Rossouw JE, Anderson GL, Prentice RL, *et al.* Risks and benefits of estrogen plus progestin in healthy postmenopausal women: principal results From the Women's Health Initiative randomized controlled trial. JAMA 2002; 288(3): 321-33.
[http://dx.doi.org/10.1001/jama.288.3.321]

[39] Mozaffarian D, Benjamin EJ, Go AS, *et al.* Heart disease and stroke statistics--2015 update: a report from the American Heart Association. Circulation 2015; 131(4): e29-e322.
[http://dx.doi.org/10.1161/CIR.0000000000000152]

[40] Raman SV, Sharkey-Toppen TP, Tran T, *et al.* Iron, inflammation and atherosclerosis risk in men *vs.* perimenopausal women. Atherosclerosis 2015; 241(1): 249-54.
[http://dx.doi.org/10.1016/j.atherosclerosis.2015.03.032]

[41] Howes PS, Zacharski LR, Sullivan J, Chow B. Role of stored iron in atherosclerosis. J Vascular Nur 2000; 18(4): 109-16.
[http://dx.doi.org/10.1067/mvn.2000.111614]

[42] Hansson GK. Atherosclerosis--an immune disease: The Anitschkov Lecture 2007. Atherosclerosis 2009; 202(1): 2-10.
[http://dx.doi.org/10.1016/j.atherosclerosis.2008.08.039]

[43] Nagy E, Eaton JW, Jeney V, *et al.* Red cells, hemoglobin, heme, iron, and atherogenesis. Arterioscler Thromb Vasc Biol 2010; 30(7): 1347-53.
[http://dx.doi.org/10.1161/ATVBAHA.110.206433]

[44] Zacharski LR, DePalma RG, Shamayeva G, Chow BK. The statin-iron nexus: anti-inflammatory intervention for arterial disease prevention. Am J Public Health 2013; 103(4): e105-12. [http://dx.doi.org/10.2105/AJPH.2012.301163]

[45] Mascitelli L, Goldstein MR. Might the beneficial effects of statin drugs be related to their action on iron metabolism? QJM : monthly journal of the Association of Physicians 2012; 105(15): 1225-9. [http://dx.doi.org/10.1093/qjmed/hcs204]

[46] Das De S, Krishna S, Jethwa A. Iron status and its association with coronary heart disease: systematic review and meta-analysis of prospective studies. Atherosclerosis 2015; 238(2): 296-303. [http://dx.doi.org/10.1016/j.atherosclerosis.2014.12.018]

[47] Baggott JE, Tamura T. Homocysteine, iron and cardiovascular disease: a hypothesis. Nutrients 2015; 7(2): 1108-18. [http://dx.doi.org/10.3390/nu7021108]

[48] Khiroya H, Turner AM. The role of iron in pulmonary pathology. Multidiscip Respir Med 2015; 10: 34. [http://dx.doi.org/10.1186/s40248-015-0031-2]

[49] Nelson ME, O'Brien-Ladner AR, Wesselius LJ. Regional variation in iron and iron-binding proteins within the lungs of smokers. Am J Respir Crit Care Med 1996; 153(4 Pt 1): 1353-8. [http://dx.doi.org/10.1164/ajrccm.153.4.8616566]

[50] McLaren GD, Nathanson MH, Jacobs A, Trevett D, Thomson W. Control of iron absorption in hemochromatosis. Mucosal iron kinetics *in vivo*. Ann N Y Acad Sci 1988; 526: 185-98. [http://dx.doi.org/10.1111/j.1749-6632.1988.tb55505.x]

[51] Theil EC. Ferritin: structure, gene regulation, and cellular function in animals, plants, and microorganisms. Annu Rev Biochem 1987; 56: 289-315. [http://dx.doi.org/10.1146/annurev.bi.56.070187.001445]

[52] Levi S, Luzzago A, Cesareni G, *et al.* Mechanism of ferritin iron uptake: activity of the H-chain and deletion mapping of the ferro-oxidase site. A study of iron uptake and ferro-oxidase activity of human liver, recombinant H-chain ferritins, and of two H-chain deletion mutants. J Biol Chem 1988; 263(34): 18086-92.

[53] Dinarello CA. Interleukin-1. Rev Infect Dis 1984; 6(1): 51-95. [http://dx.doi.org/10.1093/clinids/6.1.51]

[54] Ponka P, Beaumont C, Richardson DR. Function and regulation of transferrin and ferritin. Semin Hematol 1998; 35(1): 35-54.

[55] Fleming RE, Migas MC, Holden CC, *et al.* Transferrin receptor 2: continued expression in mouse liver in the face of iron overload and in hereditary hemochromatosis. Proc Natl Acad Sci USA 2000; 97(5): 2214-9. [http://dx.doi.org/10.1073/pnas.040548097]

[56] Levi S, Corsi B, Bosisio M, *et al.* A human mitochondrial ferritin encoded by an intronless gene. J Biol Chem 2001; 276(27): 24437-40. [http://dx.doi.org/10.1074/jbc.C100141200]

[57] Huebers HA, Finch CA. The physiology of transferrin and transferrin receptors. Physiol Rev 1987; 67(2): 520-82.

[58] MacGillivray RT, Mendez E, Shewale JG, *et al.* The primary structure of human serum transferrin. The structures of seven cyanogen bromide fragments and the assembly of the complete structure. J Biol Chem 1983; 258(6): 3543-53.

[59] Van der Heul C, Kroos MJ, Van Noort WL, Van Eijk HG. No functional difference of the two iron-binding sites of human transferrin *in vitro*. Clin Sci (Lond) 1981; 60(2): 185-90. [http://dx.doi.org/10.1042/cs0600185]

[60] J.D. B; L.A. K; A.J. P. Hemoglobina, porfirina y metabolismo del hierro. Química clínica Técnicas de laboratorio, fisiopatología, métodos de análisis Buenos Aires: Panamericana 1986; pp. 721-73.

[61] Wu AHB. Tietz Clinical Guide to Laboratory Tests 4. Saunders Elsevier 2006.

[62] Aguzzi F, Merlini G. Aspectos clínicos del análisis de las proteínas plasmáticas. Milan: Fabio Mainardi 1990.

[63] Choi JW, Pai SH, Im MW, Kim SK. Change in transferrin receptor concentrations with age. Clin Chem 1999; 45(9): 1562-3.

[64] Gambino R, Desvarieux E, Orth M, *et al.* The relation between chemically measured total iron-binding capacity concentrations and immunologically measured transferrin concentrations in human serum. Clin Chem 1997; 43(12): 2408-12.

[65] Lentjes EG, Lindeman JH, van de Bent W, Berger HM. Measured *versus* calculated latent iron binding capacity in plasma of newborns. Ann Clin Biochem 1995; 32(Pt 5): 478-81. [http://dx.doi.org/10.1177/000456329503200507]

[66] González-Revaldería J, de Paula Ruiz M, Pascual Durán T, Cabezas Martinez A, Miravalles González E. Utilidad de la medida del contenido del hierro del hígado en el diagnóstico de la hemocromatosis. Química Clínica 2001; 20: 227-31.

[67] Anton RF. Carbohydrate-deficient transferrin for detection and monitoring of sustained heavy drinking. What have we learned? Where do we go from here? Alcohol 2001; 25(3): 185-8. [http://dx.doi.org/10.1016/S0741-8329(01)00165-3]

[68] El Hage Chahine JM, Hemadi M, Ha-Duong NT. Uptake and release of metal ions by transferrin and interaction with receptor 1. Biochim Biophys Acta 2012; 1820(3): 334-47. [http://dx.doi.org/10.1016/j.bbagen.2011.07.008]

[69] Bou-Abdallah F, Terpstra TR. The thermodynamic and binding properties of the transferrins as studied by isothermal titration calorimetry. Biochim Biophys Acta 2012; 1820(3): 318-25. [http://dx.doi.org/10.1016/j.bbagen.2011.07.013]

[70] Feelders RA, Kuiper-Kramer EP, van Eijk HG. Structure, function and clinical significance of transferrin receptors. Clin Chem Lab Med 1999; 37(1): 1-10. [http://dx.doi.org/10.1515/CCLM.1999.001]

[71] Anderson GJ, Vulpe CD. Mammalian iron transport. Cellular and molecular life sciences. Cell Mol Life Sci 2009; 66(20): 3241-61. [http://dx.doi.org/10.1007/s00018-009-0051-1]

[72] Skikne BS. Circulating transferrin receptor assay--coming of age. Clin Chem 1998; 44(1): 7-9.

[73] Young SP, Bomford A, Williams R. The effect of the iron saturation of transferrin on its binding and uptake by rabbit reticulocytes. Biochem J 1984; 219(2): 505-10. [http://dx.doi.org/10.1042/bj2190505]

[74] Kato J, Kobune M, Kohgo Y, *et al.* Ratio of transferrin (Tf) to Tf-receptor complex in circulation differs depending on Tf iron saturation. Clin Chem 2002; 48(1): 181-3.

[75] Raya G, Henny J, Steinmetz J, Herbeth B, Siest G. Soluble transferrin receptor (sTfR): biological variations and reference limits. Clin Chem Lab Med 2001; 39(11): 1162-8. [http://dx.doi.org/10.1515/CCLM.2001.183]

[76] Shin DH, Kim HS, Park MJ, Suh IB, Shin KS. Utility of Access Soluble Transferrin Receptor (sTfR) and sTfR/log Ferritin Index in Diagnosing Iron Deficiency Anemia. Ann Clin Lab Sci 2015; 45(4): 396-402.

[77] Punnonen K, Irjala K, Rajamaki A. Serum transferrin receptor and its ratio to serum ferritin in the diagnosis of iron deficiency. Blood 1997; 89(3): 1052-7.

[78] Fairbanks V, Klee G, Burtis CA, Ashwood ER. Biochemical aspects of hematology. In: Saunders , Ed.

Tietz textbook of clinical chemistry 3rd ed. Philadelphia: Saunders 1999.

[79] JD.B; AC S-n; L.J.. Panamericana , Ed. Investigación de laboratorio de la hemoglobina.. Métodos y diagnósticos del laboratorio clínico 8th ed. Buenos Aires: Panamericana 1983.

[80] Bennett MJ, Lebron JA, Bjorkman PJ. Crystal structure of the hereditary haemochromatosis protein HFE complexed with transferrin receptor. Nature 2000; 403(6765): 46-53. [http://dx.doi.org/10.1038/47417]

[81] Martins R, Silva B, Proenca D, Faustino P. Differential HFE gene expression is regulated by alternative splicing in human tissues. PLoS One 2011; 6(3): e17542. [http://dx.doi.org/10.1371/journal.pone.0017542]

[82] Griffiths WJ, Kelly AL, Smith SJ, Cox TM. Localization of iron transport and regulatory proteins in human cells. Int J Med 2000; 93(9): 575-87. [http://dx.doi.org/10.1093/qjmed/93.9.575]

[83] Vidal S, Belouchi AM, Cellier M, Beatty B, Gros P. Cloning and characterization of a second human NRAMP gene on chromosome 12q13 Mammalian genome: official journal of the International Mammalian Genome Society. 1995; 6: pp. (4)224-30.

[84] Gunshin H, Mackenzie B, Berger UV, *et al.* Cloning and characterization of a mammalian proton-coupled metal-ion transporter. Nature 1997; 388(6641): 482-8. [http://dx.doi.org/10.1038/41343]

[85] Han O, Fleet JC, Wood RJ. Reciprocal regulation of HFE and NNamp2 gene expression by iron in human intestinal cells. J Nutr 1999; 129(1): 98-104.

[86] Mims MP, Prchal JT. Divalent metal transporter 1. Hematology 2005; 10(4): 339-45. [http://dx.doi.org/10.1080/10245330500093419]

[87] Donovan A, Brownlie A, Zhou Y, *et al.* Positional cloning of zebrafish ferroportin1 identifies a conserved vertebrate iron exporter. Nature 2000; 403(6771): 776-81. [http://dx.doi.org/10.1038/35001596]

[88] Zohn IE, De Domenico I, Pollock A, *et al.* The flatiron mutation in mouse ferroportin acts as a dominant negative to cause ferroportin disease. Blood 2007; 109(10): 4174-80. [http://dx.doi.org/10.1182/blood-2007-01-066068]

[89] Le Gac G, Ka C, Joubrel R, *et al.* Structure-function analysis of the human ferroportin iron exporter (SLC40A1): effect of hemochromatosis type 4 disease mutations and identification of critical residues. Hum Mutat 2013; 34(10): 1371-80. [http://dx.doi.org/10.1002/humu.22369]

[90] Vulpe CD, Kuo YM, Murphy TL, *et al.* Hephaestin, a ceruloplasmin homologue implicated in intestinal iron transport, is defective in the sla mouse. Nat Genet 1999; 21(2): 195-9. [http://dx.doi.org/10.1038/5979]

[91] Dobryszycka W. Biological functions of haptoglobin--new pieces to an old puzzle European journal of clinical chemistry and clinical biochemistry journal of the Forum of European Clinical Chemistry Societies 1997; 35(9): 647-54.

[92] Crichton RR. Cellular Iron Uptake and Export in Mammals.Iron metabolism: from molecular mechanisms to clinical consequences. 3rd ed. UK: John Wiley & Sons; 2009; p. 171. [http://dx.doi.org/10.1002/9780470010303.ch5]

[93] Baker HM, Anderson BF, Baker EN. Dealing with iron: common structural principles in proteins that transport iron and heme. Proc Natl Acad Sci USA 2003; 100(7): 3579-83. [http://dx.doi.org/10.1073/pnas.0637295100]

[94] Tolosano E, Altruda F. Hemopexin: structure, function, and regulation. DNA Cell Biol 2002; 21(4): 297-306. [http://dx.doi.org/10.1089/104454902753759717]

[95] Delanghe JR, Langlois MR. Hemopexin: a review of biological aspects and the role in laboratory medicine. Elsevier 2001; 312(1-2): 13-23. [http://dx.doi.org/10.1016/S0009-8981(01)00586-1]

[96] Park CH, Valore EV, Waring AJ, Ganz T. Hepcidin, a urinary antimicrobial peptide synthesized in the liver. J Biol Chem 2001; 276(11): 7806-10. [http://dx.doi.org/10.1074/jbc.M008922200]

[97] Liu XB, Nguyen NB, Marquess KD, Yang F, Haile DJ. Regulation of hepcidin and ferroportin expression by lipopolysaccharide in splenic macrophages. Blood Cells Mol Dis 2005; 35(1): 47-56. [http://dx.doi.org/10.1016/j.bcmd.2005.04.006]

[98] Bekri S, Gual P, Anty R, *et al.* Increased adipose tissue expression of hepcidin in severe obesity is independent from diabetes and NASH. Gastroenterology 2006; 131(3): 788-96. [http://dx.doi.org/10.1053/j.gastro.2006.07.007]

[99] Nicolas G, Bennoun M, Devaux I, *et al.* Lack of hepcidin gene expression and severe tissue iron overload in upstream stimulatory factor 2 (USF2) knockout mice. Proc Natl Acad Sci USA 2001; 98(15): 8780-5. [http://dx.doi.org/10.1073/pnas.151179498]

[100] Ganz T, Nemeth E. Hepcidin and iron homeostasis. Biochim Biophys Acta 2012; 1823(9): 1434-43. [http://dx.doi.org/10.1016/j.bbamcr.2012.01.014]

[101] Ramos E, Kautz L, Rodriguez R, *et al.* Evidence for distinct pathways of hepcidin regulation by acute and chronic iron loading in mice. Hepatology 2011; 53(4): 1333-41. [http://dx.doi.org/10.1002/hep.24178]

[102] Gehrke SG, Kulaksiz H, Herrmann T, *et al.* Expression of hepcidin in hereditary hemochromatosis: evidence for a regulation in response to the serum transferrin saturation and to non-transferrin-bound iron. Blood 2003; 102(1): 371-6. [http://dx.doi.org/10.1182/blood-2002-11-3610]

[103] Sanchez L, Calvo M, Brock JH. Biological role of lactoferrin. Arch Dis Child 1992; 67(5): 657-61. [http://dx.doi.org/10.1136/adc.67.5.657]

[104] Jameson GB, Anderson BF, Norris GE, Thomas DH, Baker EN. Structure of human apolactoferrin at 2.0 A resolution. Refinement and analysis of ligand-induced conformational change. addendum. Acta Crystallogr D Biol Crystallogr 1999; 55(5): 1108. [http://dx.doi.org/10.1107/S0907444999009431 7]

[105] Levay PF, Viljoen M. Lactoferrin: a general review. Haematologica 1995; 80(3): 252-67.

[106] Mazurier J, Spik G. Comparative study of the iron-binding properties of human transferrins. I. Complete and sequential iron saturation and desaturation of the lactotransferrin. Biochim Biophys Acta 1980; 629(2): 399-408. [http://dx.doi.org/10.1016/0304-4165(80)90112-9]

[107] Broc JHk. M DS Iron in immunity, cancer, and inflammation. New York: Wiley 1989.

[108] Actor JK, Hwang SA, Kruzel ML. Lactoferrin as a natural immune modulator. Curr Pharm Des 2009; 15(17): 1956-73. [http://dx.doi.org/10.2174/138161209788453202]

[109] Caccavo D, Pellegrino NM, Altamura M, *et al.* Antimicrobial and immunoregulatory functions of lactoferrin and its potential therapeutic application. J Endotoxin Res 2002; 8(6): 403-17.

[110] Hsia CC. Respiratory function of hemoglobin. N Engl J Med 1998; 338(4): 239-47. [http://dx.doi.org/10.1056/NEJM199801223380407]

[111] Vinogradov SN, Tinajero-Trejo M, Poole RK, Hoogewijs D. Bacterial and archaeal globins - a revised perspective. Biochim Biophys Acta 2013; 1834(9): 1789-800. [http://dx.doi.org/10.1016/j.bbapap.2013.03.021]

[112] Marengo-Rowe AJ. Structure-function relations of human hemoglobins. Proc Bayl Univ Med Cent 2006; 19(3): 239-45.

[113] Costanzo LS, Hagerstwon MD. Physiology. In: Wilkins LW, Ed. Lippincott Williams & Wilkins. 2007; p. 512.

[114] Burka ER. Characteristics of RNA degradation in the erythroid cell. J Clin Invest 1969; 48(7): 1266-72.
[http://dx.doi.org/10.1172/JCI106092]

[115] Biagioli M, Pinto M, Cesselli D, *et al.* Unexpected expression of alpha- and beta-globin in mesencephalic dopaminergic neurons and glial cells. Proc Natl Acad Sci USA 2009; 106(36): 15454-9.
[http://dx.doi.org/10.1073/pnas.0813216106]

[116] Suominen P, Punnonen K, Rajamaki A, Irjala K. Serum transferrin receptor and transferrin receptor-ferritin index identify healthy subjects with subclinical iron deficits. Blood 1998; 92(8): 2934-9.

[117] Thomas C, Thomas L. Biochemical markers and hematologic indices in the diagnosis of functional iron deficiency. Clin Chem 2002; 48(7): 1066-76.

[118] Han AP, Yu C, Lu L, *et al.* Heme-regulated eIF2alpha kinase (HRI) is required for translational regulation and survival of erythroid precursors in iron deficiency. EMBO J 2001; 20(23): 6909-18.
[http://dx.doi.org/10.1093/emboj/20.23.6909]

[119] Nemeth E, Ganz T. Anemia of inflammation. Hematol Oncol Clin North Am 2014; 28(4): 671-81. [vi.].
[http://dx.doi.org/10.1016/j.hoc.2014.04.005]

[120] Brittenham GM, Farrell DE, Harris JW, *et al.* Magnetic-susceptibility measurement of human iron stores. N Engl J Med 1982; 307(27): 1671-5.
[http://dx.doi.org/10.1056/NEJM198212303072703]

[121] Brugnara C. Iron deficiency and erythropoiesis: new diagnostic approaches. Clin Chem 2003; 49(10): 1573-8.
[http://dx.doi.org/10.1373/49.10.1573]

[122] Means RT Jr, Krantz SB. Progress in understanding the pathogenesis of the anemia of chronic disease. Blood 1992; 80(7): 1639-47.

[123] Skikne BS, Punnonen K, Caldron PH, *et al.* Improved differential diagnosis of anemia of chronic disease and iron deficiency anemia: a prospective multicenter evaluation of soluble transferrin receptor and the sTfR/log ferritin index. Am J Hematol 2011; 86(11): 923-7.
[http://dx.doi.org/10.1002/ajh.22108]

[124] Thomas C, Kirschbaum A, Boehm D, Thomas L. The diagnostic plot: a concept for identifying different states of iron deficiency and monitoring the response to epoetin therapy. Med Oncol 2006; 23(1): 23-36.
[http://dx.doi.org/10.1385/MO:23:1:23]

[125] Hunt JR. Bioavailability of iron, zinc, and other trace minerals from vegetarian diets. Am J Clin Nutr 2003; 78(3) (Suppl.): 633S-9S.

[126] Feder JN, Gnirke A, Thomas W, *et al.* A novel MHC class I-like gene is mutated in patients with hereditary haemochromatosis. Nat Genet 1996; 13(4): 399-408.
[http://dx.doi.org/10.1038/ng0896-399]

[127] Lyon E, Frank EL. Hereditary hemochromatosis since discovery of the HFE gene. Clin Chem 2001; 47(7): 1147-56.

[128] Zoller H, Pietrangelo A, Vogel W, Weiss G. Duodenal metal-transporter (DMT-1, NRAMP-2) expression in patients with hereditary haemochromatosis. Lancet 1999; 353(9170): 2120-3.
[http://dx.doi.org/10.1016/S0140-6736(98)11179-0]

[129] Wallach JB. Interpretation of diagnostics tests. Boston: Little, Brown and Company 1996.

CHAPTER 5

Equations of Glomerular Filtration Rate and Biomarkers to Evaluate Renal Function: A Perspective

Irene Gutiérrez-García and **Juan Antonio Vílchez***

Clinical Analysis Department, Santa Lucia General University Hospital, Cartagena, Spain

Abstract: Early kidney disease is asymptomatic and strongly associated with morbidity and mortality, hence an accurate estimation of kidney function is crucial for its diagnosis. In turn, an appropriate diagnosis and treatment of chronic and acute kidney disease depends on early detection. Currently markers like blood concentrations of creatinine and Cystatin-C are used to estimate glomerular filtration rate, and others as proteinuria help to assess renal damage, but all of these have different limitations indicating the need for new biomarkers that can discriminate and detect damage, especially in the most critical situation during the development of kidney disease. Given that kidneys are the most important excretory organ, changes in blood concentration levels of a wide variety of metabolites can be a reflection of reduction of kidney function and renal damage. An ideal biomarker for renal disease has to be able to predict early diagnosis of kidney disease, allowing for identification of the type and etiology of the lesion, as well as initiation and monitoring of treatments. In this chapter, we describe the most relevant eGFR equation and biomarkers for kidney diseases.

Keywords: ADMA, AKI, BTP, CKD, Creatinine, Cystatin-C, FGF-23,GFR equations, IL-18,KIM-1, Kidney disease, L-FABP, Microalbumin, NAG, NGAL, Uric acid.

INTRODUCTION

Acute and chronic kidney diseases are worldwide public health problems and are associated with an increased risk of kidney failure, complications and mortality [1]. More than 10% of the population worldwide are affected by Chronic kidney disease (CKD) [2].This prevalence increases with age, surpassing 20% in individuals older than 60 years and 35% in those older than 70 years [3].

The international organization Kidney Disease Improving Global Outcomes

* **Corresponding author Juan Antonio Vílchez:** Clinical Analysis Department, Santa Lucia General University Hospital, Cartagena, Spain; E-mail: juananv81@hotmail.com

(KDIGO) guidelines uses glomerular filtration rate (GFR) and albuminuria as the principal kidney measures to define and stage acute and chronic kidney diseases [4 - 6]. Proteinuria and estimated glomerular filtration rate (eGFR) are sensitive markers of CKD progression in clinical practice, but they have limitations, indicating the need for other biomarkers. In cases of pre-renal azotemia without tubular injury, the serum creatinine levels can be increased. In addition, several non-renal factors like muscle metabolism, total body volume, body weight and some drugs, can affect levels of creatinine. Furthermore, serum concentration does not rise until the damage progresses to a critical threshold and there are patients with renal injury with no change in serum creatinine levels due to renal reserve or tubular secretion of creatinine [7].

Acute kidney injury (AKI) and CKD have similarities and share common grounds [8], and for this reason they share biomarkers. AKI can cause end-stage renal disease directly, furthermore AKI increases the risk of developing CKD. In turn, CKD is among the most important risk factors for development of AKI [8].

Accurate estimated kidney function is crucial for prevention, diagnosis, prognosis and appropriate treatment decisions. In this chapter, we describe the main tools and biomarkers used to evaluate kidney function in clinical practice.

Acute Kidney Injury

Acute kidney injury (AKI) is often secondary to a lesion that causes functional or structural damage in the kidney.The main causes of AKI are decreased renal or intrarenal perfusion, a toxic or obstructive injury to the renal tubule, tubulointerstitial inflammation and edema, or primary reduction in the filtering capacity of the glomerulus [9]. The majority of AKI cases are related to ischemia and toxins, frequently in patients with sepsis, hematologic cancers or the acquired immunodeficiency syndrome. And about 19-33% of AKI hospital admission cases are related to drug nephrotoxicity [10] (antibiotics, chemotherapeutic agents, immunosuppressives, environmental contaminants, heme pigments and myeloma light-chain proteins) [11].

Table 1. KDIGO criteria for acute kidney injury [4].

Stage	Serum creatinine	Urine Output
1	Increase in 1.5-1.9 times baseline or ≥0.3 mg/dl (≥26.5 µmol/L) increase	<0.5ml/Kg/h 6-12h
2	Increase in 2.0-2.9 times baseline	<0.5ml/Kg/h ≥12h
3	Increase in 3.0 times baseline or serum creatinine ≥4.0 mg/dl (≥353.6 µmol/L) increase or initiation of renal replacement therapy or in patient <18 years, decrease in eGFR to <35 ml/min/1.73m^2	<0.3ml/Kg/h ≥24h or anuria for ≥12h

AKI is defined as a rise in serum creatinine within 2-7 days or oliguria. AKI is classified by the magnitude of the peak rise in serum creatinine concentrations or nadir decline in urine output (Table **1**) [4].

Acute Kidney Diseases and Disorders

Acute kidney diseases and disorders (AKD) refer to less than 3 months of having decreased kidney function or the presence of a marker of kidney damage including AKI [4].

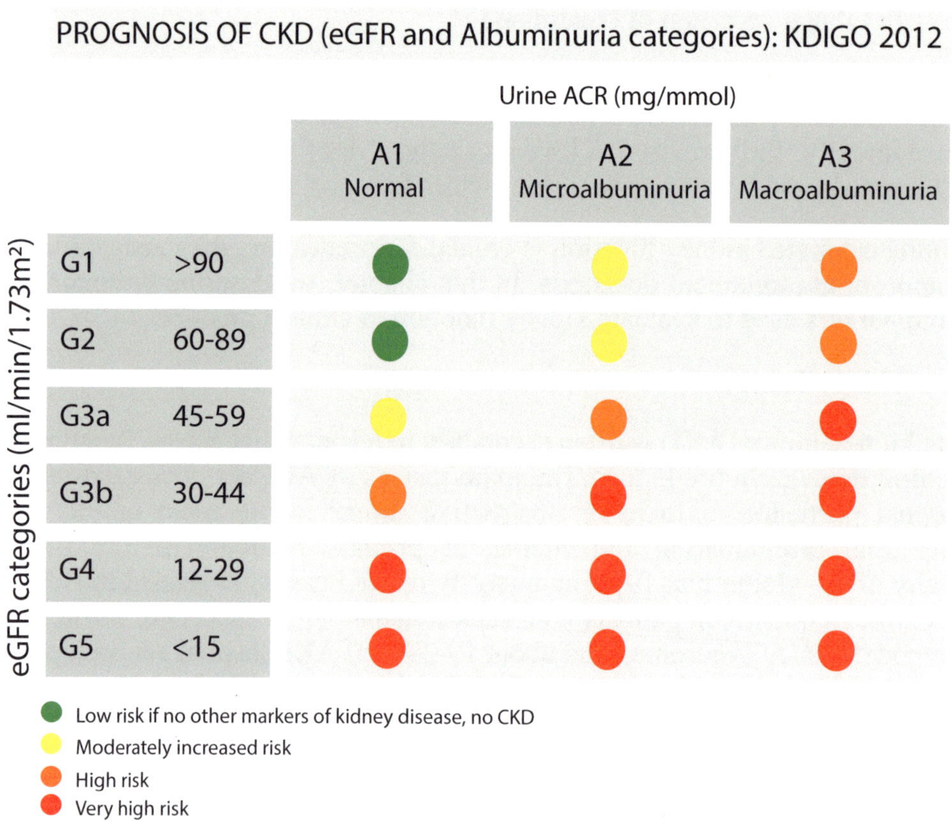

Fig. (1). Prognostic risk and classification of CKD based oneGFR (ml/min/1.73 m^2) and albuminuria (mg/mmol) adapted from KDIGO clinical guidelines, 2012 [5]. eGFR: estimated glomerular filtration rate, ACR: albumin-creatinine ratio.

Chronic Kidney Disease

CKD refers to any extended situation that alter kidney structure and function.

Early kidney disease is asymptomatic, thus biomarkers are very important in its diagnosis, evaluation and treatment. It is important to take into account that declining kidney function occurs during the aging process. For this reason, it is crucial to be able to differentiate between patients with impaired renal function due to disease from patients that are age-related.

CKD was first defined by having more than 3 months of decreased GFR (<60 ml/min/1.73 m^2) or evidence of kidney damage [5]. Kidney damage refers to abnormalities found by biopsy or imaging, alterations in urinary sediment or proteinuria (proteinuria/creatinuria >200mg/g, albuminuria/creatinuria>30mg/g). Hence CKD was classified into five stages according to cause and level of GFR and albuminuria [5]. Subsequently KDIGO added staging by levels of albuminuria in addition to level of eGFR (Fig. **1**).

EQUATION OF GLOMERULAR FILTRATION RATE

GFR is the best method to evaluate kidney function and is defined by the volume of plasma that is cleared of a substance by the kidneys per unit time. Urinary inulin clearance is the "gold-standard" method for GFR measurement butitis complicated to measure in clinical practice and is rarely used. GFR can be estimated by using equations based on plasma concentrations of creatinine and cystatin-C.In the clinical practice, estimated GFR (eGFR) is a useful alternative to formal GFR measurement (mGFR) and provides an easy way of calculating kidney function.But it is important to take into account that all these equations are severely sensitive to the characteristics of the population from which they are originated [12]. The aim of this section is to describe and summarize the main equations for estimating GFR.

Cockcroft and Gault Equation [13]: The first equation published was Cockcroft and Gaul equation which is still used in some settings, particularly for drug dosing recommendations [14] although its use is not for advice [5]. It is based on serum creatinine concentration (SCr) and adjusted for other factors such as age, weight and sex. The disadvantage of this equation is the need to know the weight of the patients. It does not take into account the variability of creatinine production [15, 16] and overestimates the GFR in obese or edematous patients [16].

$$\frac{[((140 - age)x\ weight)\ x\ 1.23]}{(SCr)}\ x\ (0.85\ (if\ female))$$

Serum creatinine (SCr)is given µmol/l, age in years, and weight in Kg [17].

The Modification of Diet in Renal Disease (MDRD) Study Equation (ID-MS traceable) [18]: MDRD equation estimates GFR considering body surface area (1.73 m^2), and was simplified and re-expressed to allow its use with internationally standardized creatinine assays and reference methodology [19].

MDRD equation has been shown to lose accuracy in several groups of patients no included in the MDRD study population, such as elderly, patients with type 1 diabetes, patients with normal renal function and kidney transplant recipients [20, 21].This equation is better at values of GFR less than 60ml/min/1.73m^2 [22].

MDRD equations were developed in people with CKD, and for this reason, one of twofits major limitations are systematic underestimation of measured GFR at higher levels of kidney function leading to an overestimation of the prevalence of stage 3 CKD [23]. The other important limitation of this equation is the imprecision, because the measurement of serum creatinine was performed using a Jaffe method [18]. On the other hand, it does not perform well in Asian and African populations [22].

$$175\ x\ SCr^{-1.154}\ x\ age^{-0.203}\ x\ 0.742(if\ female)x\ 1.210(if\ black)$$

Serum creatinine (SCr)is given μmol/l and age in years [17].

The Chronic Kidney Disease Epidemiology (CKD-EPI$_{\text{creatinine}}$) Collaboration Equation [24]: This equation estimates GFR in based on plasma creatinine concentration and adjusted for sex, age and race. It had the same accuracy as the MDRD equation at GFR less than 60ml/min/1.73m^2 but was especially improved for individuals with GFR higher than 60ml/min/1.73m^2 [22]. It does not perform well in Asian and African population [22].

$$141\ x\ \min\left(\frac{SCr}{\kappa},1\right)^{\alpha}\ x\ \max\left(\frac{SCr}{\kappa},1\right)^{-1.209}\ x\ 0.993^{Age}\ x\ 1.018(if\ female)x\ 1.159\ (if\ black)$$

Serum creatinine (SCr) is given in μmol/l and Age in years. κ is 0.7 for women and 0.9 for men, α is -0.329 for women and -0.411 for men, min indicates the minimum of SCr/κ or 1, and max indicates the maximum of SCr/κ or 1 [17].

CKD-EPI$_{\text{Cystatin}}$ Collaboration Equation [25]: This equation is based on plasma concentration of cystatin-C (Cys-C), a small protein which plasma levels are inversely related to GFR. It is less influenced by non-renal factors, including muscle mass, diet or ethnicity [15]. According to KDIGO, equations based on Cys-C can be used to confirm a diagnosis of CKD in patients without albuminuria and creatinine-based eGFR of 45-59 ml/min/1.73m^2 [5].

$$133\ x\ \min\left(\frac{Scys}{0.8},1\right)^{-0.499}\ x\ \max\left(\frac{Scys}{0.8},1\right)^{-1.328}\ x\ 0.996^{Age}\ x\ 0.932(if\ female)$$

Serum cystatin-C (Scys) is given in mg/land Age in years.Min indicates the minimum of Scys/0.8 or 1and max indicates the maximum of Scys/0.8 or 1 [17].

CKD-EPI$_{\text{Creatinine-Cystatin}}$ Collaboration Equation [25]: This equation is based on plasma concentration of creatinine and Cys-C and it showed better precision and accuracy than that based on creatinine alone or Cys-C alone [25].

$$135\ x\ \min\left(\frac{SCr}{\kappa},1\right)^{\alpha}\ x\ \max\left(\frac{SCr}{\kappa},1\right)^{-0.601}\ x\ \min\left(\frac{Scys}{0.8},1\right)^{-0.375}\ x\ \max\left(\frac{Scys}{0.8},1\right)^{-0.711}\ x$$

Serum creatinine (SCr) is given in µmol/l, serum cystatin-C (Scy) in mg/land Age in years.

κ is 0.7 for women and 0.9 for men, α is -0.248 for women and -0.207 for men, min indicates the minimum of SCr/κ or 1, and max indicates the maximum of SCr/κ or 1 [17].

Berlin Initiative Study (BIS1 and BIS2) Equations: BIS1 and BIS2 are novel equations based on standardized creatinine and Cys-C methodology for application in older white people [26].

BIS1 is based on simple determination of serum creatinine level and was found to be the most reliable for estimating renal function in older white patients, especially in those with CKD stages from 1 to 3 [27], possibly because of the loss of muscle mass secondary to aging [28].

$$3736\ x\ (SCr\ x\ 88.4)^{-0.87}\ x\ age^{-0.95}\ x\ 0.82(if\ female)$$

Age is given in years and serum creatinine (SCr) in µmol/l [17].

And BIS 2 is based on both, creatinine and Cys-C,

$$767\ x\ Scys^{-0.61}\ x\ SCr^{-0.40}age^{-0.57}\ x\ 0.87(if\ female)$$

Age is given in years, serum creatinine (SCr) in µmol/l and serum cystatin-C (Scy) in mg/l [17].

BIOMARKERS TO EVALUATE RENAL FUNCTION

Measurements of blood concentration of creatinineandCys-C are nowadays used to estimate GFR. The combination of serum creatinine, Cys-C and urinary albumin to creatinine ratio allows for risk stratification, predicts renal disease progression, as well as mortality. But creatinine and Cys-C have been shown to be inaccurate, in several situations, particularly in elderly patients, subjects with extreme BMI and in early stages of CKD [29]. Furthermore, the circulating levels of creatinine do not rise until damage progresses to a critical threshold, in particular, after 50% loss of kidney function and for this reason, other early biomarkers are needed [30].

To date, numerous new biomarkers have been studied, some of them are very sensitive and specific, allowing earlier diagnosis of kidney disease. In spite of this, all of them need larger studies before using them in clinical practice.

Biomarkers of Acute Kidney Injury

Currently, more than 30 candidate biomarkers have been studied and investigated. Table **2** shows the most relevant biomarkers for AKI. In this section, the main biomarkers will be described. Amongst all of them, Neutrophil Gelatinase-Associated Lipocalin (NGAL), Cystatin-C, KIM-1 and IL-18 are the most promising biomarkers of AKI. The main advantages and disadvantages of the biomarkers summarized are shown in Table **3**. Fig. (**2**) shows origin AKI biomarkers within a single nephron, adapted from Ostermann *et al.* [31].

Cystatin-C (Cys-C)

Cys-C is a small protein (13 kDa) produced by all nucleated cells. It belongs to the cystatin superfamily, which are the most important endogenous inhibitor of cysteine proteinases, specifically cathepsin H, B, L and calpains [33]. Serum Cys-C concentration appears to be independent of age, muscle mass and sex. Cys-C is s freely filtered by the glomerulus owing to it is not bound to plasma proteins, reabsorbed and degraded, but it is not secreted into the urine. For this reason, an increasedCys-C concentration in urine denotes a reduction of its uptake by damaged proximal tubules.

Urine Cys-C concentrations increase 12-24h post-renal injury [31]. Cys-C can be detected in urine of patients with tubular diseases [34] and in the serum of dialysis patients [35]. It has been found that Cys-C is a sensitive marker of GFR and a stronger predictor than serum creatinine of risk of death and cardiovascular events in older patients [36]. Compared with serum Cys-C, urine Cys-C is an earlier and more sensitive marker in AKI [37]. But it is unknown if Cys-C concentrations are generalizable to all forms of AKI or if they are only specific to particular populations [38].

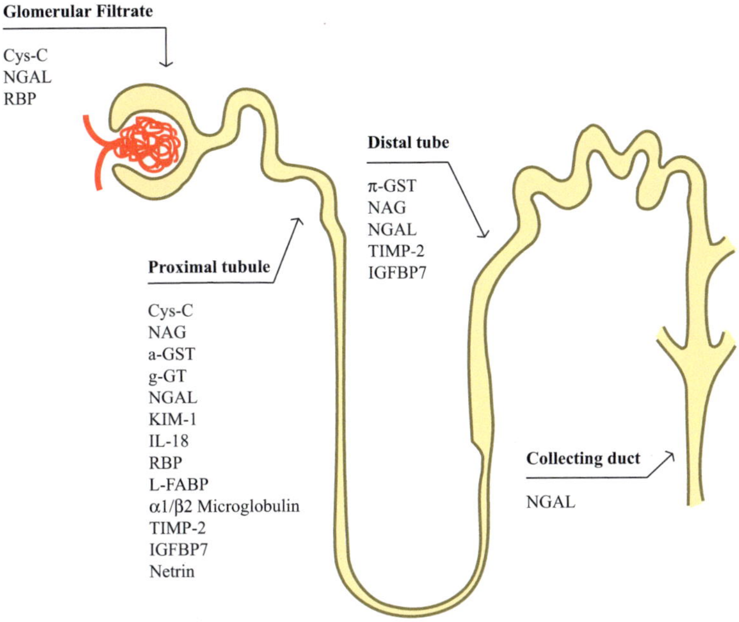

Fig. (2). Origin of AKI biomarkers within a single nephron, adapted from Ostermann *et al.* [31].

Table 2. Biomarker for AKI [32].

Functional biomarkers	Tubular enzymes	Up-regulated proteins
Creatinine Cystatin-C β2-microglobulin α1-microglobulin Retinol-binding-protein Microalbumin	Alanine aminopeptidase Alkaline phosphatase A-glutathione-S-transferase γ-glutamyltranspeptidase N-acetyl-β-glucosaminidase	KIM-1 Clusterin Neutrophil gelatinase-associated lipocalin IL-18 Cysteine-rich-protein Osteopontin FABP Sodium/hydrogen exchanger isoform Exosomalfetuin A

Cys-C concentration appears to be affected by corticosteroids (large doses), inflammation, rheumatoid factor, diabetes, hyperthyroidism, hyperbilirubinemia, hypertriglyceridemia [39, 40] and has not been validated for use in children less than 2 years old [32].

Fatty Acid-Binding Proteins (FABPs)

FABPs are a small cytoplasmic protein (15 kDa), expressed in all tissues with active fatty acid metabolism [41]. FABPs transport free fatty acids in proximal tubules to mitochondria or peroxisomes for metabolism [42]. Two types of FABP have been isolated from the human kidney: liver type FABP (L-FABP) in the proximal tubule and heart-type FABP (H-FABP) in the distal tubule [43, 44].

L-FABP is filtered in glomerulus and reabsorbed in proximal tubular cells. It is expressed in renal tubular epithelial cells and shed into the urine subsequent to tubular cell damage [31]. Urinary L-FABP facilitates the early detection of AKI before an increase in serum creatinine [45]. Urine L-FABP was assessed in various clinical studies with pediatric patients having cardiac surgery [46, 47], adult patients having cardiac surgery [48] and patients with sepsis [49], in which L-FABP appears to be a sensitive and predictive early biomarker of AKI after cardiac surgery [32], especially in combination with other renal biomarkers as NGAL, KIM-1 or NAG.

When using L-FABP as a renal biomarker of AKI, it is important to allow for its expression is up-regulated in patients treated with hypercholesterolemia and hyperlipidemic drugs (statins and fibrates) [49].

Interleukin-18 (IL-18)

IL-18 is a proinflammatory cytokine, member of the IL-1 cytokine superfamily, known as an interferon-γ (IFN-γ)-inducing factor, important regulator of innate and acquired immune responses [50]. It has a potential effector and a regulatory role in several human inflammatory diseases (inflammatory arthritis, multiple sclerosis, inflammatory bowel disease, chronic hepatitis, systemic lupus erythematosis and psoriasis) [50].

The inactive precursor form of IL-18 (24 kDa) is cleaved by IL-1β-converting enzyme to produce active mature 18 kDa IL-18 protein [51]. The source of IL-18 in kidney appears to be the proximal tubule [52]. Compared with normal individuals and patients with other renal diseases, patients with AKI have an increased concentration of urinary IL-18 [53].

Kidney Injury Molecule-1 (KIM-1)

Kidney injury molecule-1 (KIM-1)/T-cell immunoglobulin and mucin domain containing protein-1(Kim-1/TIM-1) is a type I cell transmembrane glycoprotein which contains a six-cysteine immunoglobulin-like domain and a mucin domain in its extracellular region.The function of this protein is still uncertain. KIM-1 expression increases more than any other known gene after kidney injury [54] and its ectodomain is shed from cells *in vitro* [55] and *in vivo* into the urine, in rodents [56] and humans [57], after proximal tubular injury.

Kim-1/TIM1 is expressed in immune cells, also in proliferating and regenerating proximal tubule cells [58]. Kim-1 is a useful biomarker of early detection of AKI within 24 hours after kidney damage. It might be especially beneficial in the diagnosis of ischemic acute tubular necrosis (ATN) [59].

Microalbumin

It is known as urinary albumin concentrations that appear below the threshold of detection by the conventional urinary dipstick (30-300 mg/L). Microalbumin has been recognized as a useful marker of the development and progression of renal disease particularly diabetic nephropathy, signaling altered glomerular structure and function [60]. Urine albumin excretion has been used to identify tubular dysfunction in patients in treatment with cancer chemotherapeutic agents and antibiotics [61, 62]. Recently, it has been found that AKI produces an increased expression of albumin's gene [63]. It is important to take into account that urinary albumin levels can increase in urinary tract infection, hematuria, fever, dehydration, vigorous exercise and poor glycemic control [32]. On the other hand, albuminuria does not permit to identify the site of injury nor to distinguish CKD from AKI.

N-Acetyl-β-glucosaminidase (NAG)

NAG is a proximal tubule lysosomal enzyme that increases its urinary concentration in situations such as proximal tubule injury with loss of lysosomal integrity [64]. NAG is a large protein (~140kDa) and due to its large size, it is not filtrated in glomerulus, and for this reason, a high urinary concentration will more likely be a renal source.

It has been shown that urinary NAG levels (in addition to other tubular enzymes) are highly sensitive in detecting AKI in a population of critically ill adult patients, preceding increases in serum creatinine by 12 h to 4 days [65]. The urinary NAG levels of this kind of patient correlates with poorer outcomes [66].

There are two main advantages of using NAG. Firstly, that any small alterations in the epithelial cells in the brush border of the proximal tubules result in releasing NAG into the urine and its concentration can be directly correlated to tubular injury. Secondly, NAG can be measured in the urine with a simple and reproducible enzymatic assay using a spectrophotometer.

The main disadvantage of NAG is its non-specifity, due to the fact that NAG can be inhibited by endogenous urea [67], as well as heavy metals and nephrotoxicants [68]. Moreover, urinary NAG levels can be increased in a variety of conditions in patients without AKI, such as rheumatoid arthritis [69] impaired glucose tolerance [70] and hyperthyroidism [71]. Furthermore, NAG tends to degrade more easily than other biomarkers, even when stored at -80ºC [32].

Neutrophil Gelatinase-Associated Lipocalin (NGAL)

NGAL is a protein that can be found as monomer (25kDa), homodimer (45kDa) and conjugated to gelatinase in an heterodimeric form (135 kDa) [72]. It is a member of the lipocalin superfamily proteins [72] and contains an α-helix and a β-barrel surrounding a hydrophobic core that binds small lipophilic ligand [73]. NGAL is synthesized during granulocyte maturation (homodimer) in the bone marrow [74] and can be produced in epithelial tissues (monomer and heterodimeric form) throughout the body [75] in a context of inflammation or malignancy, itcan also be released by monocytes/ macrophages and adipocytes [76].

NGAL may perform an important anti-inflammatory function [77], binds siderophore (produced by bacteria) with high affinity, in turn, siderophore traps iron with high affinity [78], inhibits bacterial growth and induces epithelial cell growth. Using a mouse model of AKI, it has been shown, that a single dose of NGAL protects the kidneys and mitigates azotemia [78].

The normal concentrations of NGAL in urine and plasma are very low (~20ng/ml). NGAL is filtered and captured (endocytosis) by the proximal tubule where it is degraded to a 14 kDa fragment in lysosomes [32]. Subsequent to kidney injury, it is upregulated and released into the urine from tubular epithelial cells. In addition, a lesion in proximal tubule alters its reabsorption. NGAL can appear in urine as free form or complexed with matrix metalloproteinase-9 (MMP-9), an endopeptidase that degrades extracellular matrix and is involved in ischemic organ injuries.

NGAL has been proposed as an early and quantitative biomarker for diagnosing AKI in cisplatin nephrotoxicity, induced in the kidney within 3 h of high-dose cisplatin, preceding increase in urinary NAG or serum creatinine [79]. But it is important to know that NGAL is an acute phase reactant that can be released from the liver and neutrophils in systemic infections as well as local urogenital infections [80], and its levels can be influenced by several factors, as baseline renal function, severity of AKI, age [81] and medical conditions such as pre-eclampsia and cancer [82, 83].

α_1-microglobulin(α_1M)

α_1M is a protein synthesized by the liver, with a size of about 27-33 kDa and it is mostly part of IgA.α_1M is filtered by the glomerulus and reabsorbed by proximal tubule cells. It is a marker of tubular proteinuria and proximal tubular dysfunction, even when there is no evidence of histologic damage in the early phase of injury [84]. Its main advantage is its stability. But it is important to take into account that its plasma and serum levels are altered in a variety of conditions such as liver disease, HIV, and mood disorders [85].

Tissue Inhibitor of Metalloproteinases 2 (TIMP-2)

TIMP-2 is a member of the TIMP family of protease inhibitors expressed ubiquitously [86] and induces G1 cell cycle arrest. It is up-regulated after kidney injury.TIMP-2 is found to be a strong predictor of development of KDIGO stage 2 or 3 AKI within 12 hours of sample collection [87].

Insulin-Like Growth Factor-Binding Protein 7 (IGFBP7)

IGFBP7 is a 27 kDa protein expressed ubiquitously in normal human tissues. It has different functions, regulating cells proliferation, apoptosis and senescence. It also has an important role in vascular biology [88]. It is up-regulated after kidney damage. As TIMP-2, it is a solid predictor of moderate or severe AKI within 12 hours of sample collection [87].

Urine TIMP-2 and IGFBP7 have been validated together in more than 1000 critically ill patients at risk for AKI and performed better than any other biomarker reported to date [87]. These proteins are expressed in epithelial cells and act in an autocrine and paracrine manner to arrest cell cycle, which is a mechanism of AKI.Neither TIMP-2 nor IGFBP-7 were elevated in patients without AKI [89].

Retinol Binding Protein (RBP)

RBP is a 21 kDa protein synthesized in the liver and it is a specific carrier for vitamin A from the liver to other tissues. It is filtered by the glomerulus and reabsorbed (but not secreted) and catabolized by the proximal tubule. It was found to be a sensitive indicator of renal tubule dysfunction [90], it is useful to predict AKI in neonates with birth asphyxia [91] and as early diagnostic marker of renal injury in patients with induced nephrotoxicity with cisplatin, lead, mercury, cadmium and cyclosporine [92].

Netrin-1

It is a diffusible laminin-related protein [93]. Its expression increases subsequent to the injury in proximal tubular cells [31]. Inflammatory response of macrophages and neutrophils is regulated by Netrin-1 through suppression of COX-2 mediated PGE2 production [94]. In a study performed in transgenic mice it has been found that overexpression of netrin-1 protects kidney from ischemia reperfusion injury (IRI) by increasing proliferation and suppressing apoptosis. Netrin-1 identifies AKI early and return to baseline within 24h [95].

Table 3. Advantages and disadvantages of the summarized biomarkers for AKI.

Marker	Advantages	Disadvantages
α_1-M	Sensitive marker for tubular dysfunction. Stability	Altered values liver disease, HIV, and mood disorders [85]. Albuminuria increases its excretion due to albumin competes with it for endocytosis in the proximal tubule [99]
Albumin	Useful marker of development and progression of renal disease [60]. Good predictor of CKD progression [100]. Inexpensive	False positive: vigorous exercise, hematuria, urinary tract infection and dehydration, fever and poor glycemic control [32]. Albuminuria does not permit to identify the site of injury nor to distinguish CKD from AKI.
Cys-C	Earlier and sensitive marker of AKI (in urine) [37]. Cys-C is a sensitive biomarker of kidney function in mild-t--moderate kidney disease [97]. Automatic and quantitative assay for detection of Cys-C using and N-latex with a nephelometer [101].	It is unknown if Cys-C concentrations are generalizable to all forms of AKI or if only are specific to particular populations [38].Cys-C concentration appears to be affected by corticosteroids (large doses), inflammation, rheumatoid factor, diabetes, hyperthyroidism, hyperbilirubinemia, hypertriglyceridemia [39, 40] and has not been validated for use in children less than 2 years old [32]. Albuminuria increases its excretion due to albumin competes with it for endocytosis in the proximal tubule [99].
IL-18	Sensitivity and specificity of >90% for the diagnosis of AKI [53].	False positive: inflammation, sepsis and heart failure.
KIM-1	Sensitivity and specificity. Biomarker of early detection of AKI. Marker of AKI to CKD transition [102].	Confounding factors: renal cell carcinoma chronic proteinuria, CKD and sickle cell nephropathy [31].
L-FABP	Urinary L-FABP facilitates the early detection of AKI before an increase in serum creatinine [45]. Useful marker for CKD progression [103]	False positive: CKD, polycystic kidney disease, liver disease, sepsis [31]. L-FABP expression is up-regulated in patients treated with hypercholesterolemia and hyperlipidemic drugs (statins and fibrates) [49].
NAG	Sensitive marker for proximal tubular injury. Simple and reproducible measurement method.	False positive values: rheumatoid arthritis [69] impaired glucose tolerance [70] and hyperthyroidism [71]. Urine NAG is inhibited by urea nephrotoxicants [67] and heavy metals [68]. NAG degrades more easily than other biomarkers, even when stored at -80ºC [32].

(Table 3) contd.....

Marker	Advantages	Disadvantages
NGAL	Early and quantitative biomarker for AKI [79].	Increased in sepsis [80], local urogenital infections, malignancy, chronic obstructive pulmonary disease, endometrial hyperplasia, pancreatitis and CKD [31]. There are different forms of NGAL (monomer, homodimer and heterodimer).
TIMP-2 IGFBP7	Their utility has been validated with the measure of both biomarkers together [87].	The best biomarkers reported to date for predicting the development of moderate or severe AKI (KDIGO stage 2 or 3) within 12hours of sample collection [87].

BIOMARKERS OF CHRONIC KIDNEY DISEASE

Over the last years, one of the main aims of the study of kidney disease has been the identification and characterization of new potential biomarkers of progression and targets for therapy of CKD, especially in initial stages of kidney disease which may be asymptomatic, in which robust biomarkers are crucial to diagnose, evaluate and treat CKD. This chronic disease progresses slowly and its mechanism is not well understood. Delay in diagnosis of CKD may cause deterioration of nephron function which leads to renal damage and end-stage renal disease, increasing morbidity and mortality related to CKD. But CKD includes, not only progression to end-stage renal disease, but also complications related to renal impairment and increased risk of cardiovascular disease. To date, there is evidence that CKD can be detected using laboratory tests and the treatment can prevent complications, delay the progression and reduce the risk of cardiovascular disease [96]. Table **4** shows the most recent findings of biomarkers in CDK. In this section, the main characteristics of the most relevant biomarkers in CDK will be summarized.

Cystatin C

Cys-C is a sensitive biomarker of kidney function in mild-to-moderate kidney disease [97]. It has been shown that Cys-C is a good marker which provides reliable risk prediction for progression of kidney disease in patients with CKD [98].

Microalbumin

In patients with a high risk of CKD, the albumin-creatinine ratio >30 mg/g has been found to be a risk factor of cardiovascular death and all causes mortality, progression of kidney disease, acute kidney injury (as discussed above) and kidney failure [19, 104]. These findings have been reproduced in low risk cohorts [105]. Furthermore it has been shown that microalbuminuria screening in order to detect early-stage CKD is cost-effective in patients with diabetes or hypertension

[106]. On account oftheassociation of albuminuria with the progression of CKD [100], end stage renal disease, and all causes of mortality independent of eGFR, albuminuria staging has been added in the KDIGO guidelines.Combining albuminuria with eGFR improves the prediction of CKD progression [5].

β-Trace Protein (BTP)

BTP is a heterogeneous monomeric glycoprotein belonging to the lipocalin superfamily proteins that bind and transport lipophilic molecules [107]. There are different glycoforms of varying molecular weight (26-29 kDa).

It is also known as "lipocalin prostaglandin D_2 synthase" due to catalyze the conversion of prostaglandin H_2 (PGH_2) to PGD_2. PGD_2 is an eicosanoid involved in a variety of important physiologic processes (nociception, sleep induction, bronchoconstriction, platelet aggregation, vasodilation, inflammation, adipogenesis, skin homeostasis and bone remodeling) [108, 109].

The different glycoforms are found in many human fluids and tissues [110]. The smaller isoform is a component of human cerebrospinal fluid (CSF) and represents almost 3% of the total CSF protein [34]. On the other hand, the larger glycoforms predominates in serum and urine. Albeit, it has been associated with several cardiovascular disorders beyond its role for estimating renal function [111, 112].

Renal handling of BTP is uncertain. Almost all clinical reports show that BTP is freely filtered in glomerulus but it is a protein much larger than creatinine and Cys-C, and even larger than myoglobin, a poorly filtered protein. Ins pite of this, in the only metabolic study of humans,90% to 100% of radiolabelled BTP injected intravenously into 3 patients was recovered in urine indicating near-total renal elimination and little, if anytubularreabsorption or non-renal elimination/metabolism [113]. This study is contradicted with others performed in dogs using a different BTP isoform with different properties [114].

BTP has been studied as an emerging novel marker of glomerular filtration. Some studies have found that BTP concentration is a more sensitive indicator of glomerular filtration than creatinine and Cys-C concentrations, especially to detect reduced GFR (earlier increase in serum concentrations when GFR decreased to <60 ml/min/1.73m^2) [115 - 117], whereas other studies have shown BTB is no better [118 - 120]. On the other hand, several studies show that BTP is a reliable risk predictor for CKD progression [98] but it requires validation in large CKD populations.

More of its characteristics are still unknown and insufficiently studied. For example its serum origin, its non-GFR determinants, extrarenal elimination, effect of medications on serum or urine BTP concentrations and it is not established a reference method [109].

Uric Acid (UA)

UA is the end product of the metabolism of unwanted purine nucleotide and scavenges potential harmful radicals [121]. It is produced endogenously (66%) mainly by liver, muscle, intestine and derived (33%) from diet [122]. Serum UA concentration depends on UA generation, reabsorption and excretion (primarily *via* the kidneys), but may be influenced by other factor such as dietary purine intake, hydration status, and use of diuretics [122]. It is known that only 8-10% of filtered UA is excreted, and 90% is reabsorbed.

Increased serum uric acid concentration is commonly associated with CKD which is supported by many studies, but it is controversial if UA is a true risk factor for the development or exacerbation of CKD [122, 123]. In a recent meta-analysis of 13 observational trials involving over 190,000 patients, the presence of hyperuricemia was an independent predictor of new-onset CKD development [124]. This result was confirmed in patients with and without diabetes. Furthermore, in patients with increased serum UA levels, the risk of developing new-onset CKD increased as the duration of follow-up increased, suggesting that hyperuricemia may play a role in the long-term progression of chronic renal dysfunction [124].Other studies have shown that prevalence of diabetic kidney disease is significantly increased when hyperuricemia exists in patients with diabetes mellitus type 2 [125]. Moreover, mutations in the urate transporter GLUT9 holds a strong predictor power of CKD which may corroborate that the link between UA and CDK is causal [126].

Table 4. Biomarkers of CKD according to the pathophysiological progress [127].

Pathophysiological process	Biomarker
Kidney function (GFR)	Cystatin C B-Trace protein
Tubulointerstitial injury	NGAL KIM-1 NAG L-FABP
Glomerular injury	Podocin Nephrin Podocalyxin
Endothelial dysfunction	ADMA

(Table 4) contd.....

Pathophysiological process	Biomarker
Oxidative stress	Ox-LDL AOPP TBARS Plasma and urinary F2-isoprostanes MDA Protein reduced thiols TAS Protein carbonyls AGE Urinary 8-hydroxydeoxy guanosine 4-hydroxy-nonenal Antioxidantenzymeactivities (superoxidedismutase, glutathioneperoxidase, catalase, *etc.*) GGT
Inflammation	CRP and hs-CRP PTX3 sTNF-RII IL-18 Tenascin TIMP-1
Fibrosis	TGF-β1
Cardiovascular dysfunction	ANP BNP and NT-proBNP cTnT Adrenomedullin
Metabolic disorders	Adiponectin FGF-23 ApoA-IV

Abbreviations: ADMA, asymmetric dimethylarginine; AGE, advanced glycation end product; ANP, atrial natriuretic peptide; AOPP, advanced oxidation protein products;ApoA-IV, apolipoprotein A-IV; BNP, brain natriuretic peptide; CRP, C-reactive protein;cTnT, cardiac troponin T; FGF-23, fibroblast growth factor-23; GFR, glomerular filtration rate; GGT, γ-glutamyltransferase; hs-CRP, high-sen- sitivity-CRP; IL-18,interleukin-18; KIM-1, kidney injury molecule-1; L-FABP, liver-type fatty acid-bindingprotein; MDA, malondialdehyde; NAG, N-acetyl-b-O-glucosaminidase; NGAL, neutrophil gelatinase-associated lipocalin; NT-proBNP, N-terminal brain natriuretic peptide; Ox-LDL, oxidized low-density lipoproteins; PTX3, pentraxin 3; sTNF-RII,soluble tumor necrosis factor receptor II; TAS, total antioxidant status; TBARS,thiobarbituric acid reactive substances; TGF-β1, transforming growth factor-β1;TIMP-1, tissue inhibitor of metallo-proteinases-1.

NGAL

As discussed above, serum and urine NGAL concentrations have been considered as an early biomarker of AKI [79, 128] but its role in CKD has been less studied.

Numerous studies have shown that NGAL is an independent predictor of CKD progression and is a good reflection of the degree of renal disease [129]. In a retrospective clinical study in patients with autosomal dominant polycystic kidney disease it was found that the urine NGAL concentrations was increased in patients whoseCKD progressed rapidly to end-stage kidney disease [130]. Moreover, in kidney tissue from patients with IgA nephropathy and congenital nephron deficit, NGAL expression was elevated in renal tubules and correlated with progressive glomerular mesangial proliferation and tubule-interstitial injury [131]. In an across-sectional study of nondiabetic patients with CKD stages 2-4, serum NGAL levels were increased, reaching the highest levels in patients in stage 4 of CKD [132].

A study of patients affected by CKD with higher baseline NGAL showed a considerably increased risk of worsening residual renal function within 1 year compared with those with lower baseline NGAL values [133].

KIM-1

Several studies suggest that this transmembrane tubular protein may be an indicator of AKI to CKD transition and could be an useful marker for sustained renal injury after AKI [102]. KIM-1 is up-regulated in renal disease and is associated with renal fibrosis and inflammation [134]. In a mouse model of polycystic kidney disease, cystic kidneys had upregulated KIM-1 levels compared with noncystic control kidneys [135]. Furthermore, lower urinary KIM-1 and NAG levels were associated with the regression of microalbuminuria in type 1 diabetes mellitus [136]. Baseline plasma KIM-1 levels correlated with rate of eGFR decline [137]. In spite of these results, long-term studies in larger populations are needed to assess the usefulness of KIM-1 as a CDK biomarker.

NAG

Urine NAG activity is not only an indicator of proximal tubular damage but also a measure of altered function in the renal tubules [138]. Urinary NAG excretion may identify patients with diabetes mellitus type 2 that are susceptible to develop diabetic nephropathy [61]. Also a study of type 1 diabetes mellitus found that lower levels of urinary NAG and KIM-1 were associated with the regression of microalbuminuria [136].

L-FABP

Urinary L-FABP levels may be a useful marker to assess the progression of chronic glomerular disease as it was found in a study of patients with nondiabetic renal disease [103] and appears to be more sensitive than proteinuria [139]. Urinary L-FABP excretion changes with the progression of CKD, increasing with the deterioration of the kidney function [139] and has been correlated with the degree of tubule-interstitial damage and urinary protein excretion, in a study of patients with nondiabetic CKD [140]. Furthermore, high urinary L-FABP was able to predict development and progression of diabetic nephropathy independent of urinary albumin excretion rate and other established risk factors [141].

Fibroblast Growth Factor 23 (FGF-23)

FGF-23 is a protein (32 kDa) produced by osteocytes and osteoblasts, involved in the regulation of phosphate homeostasis. It induces phosphaturia, inhibits the synthesis of 1,25-dihydroxyvitamin D3 and suppress parathyroid hormone [142].This is important considering that one of the main problems in patients with CKD are disorders of calcium and phosphate metabolism. It has an important role in the pathogenesis of Chronic Kidney Disease-Mineral and Bone Disorder (CKD-MBD) and its levels are notably elevated in CKD [143], probably due to chronic hyperphosphatemia and decreased renal clearance. FGF-23 was found to be correlated with progression of kidney disease and mortality in patients with CKD [144 - 146].

Asymmetric Dimethylarginine (ADMA)

It is an aminoacid (202 Da) circulating in the plasma, synthesized intracellularly and eliminated through the urine [147].The amount of ADMA produced depends on the methylation of the arginineend of proteins (mainly histones) as well as on protein kinetics and on the balance of intracellular and extracellular proteins [148, 149].

ADMA has been found to be an inhibitor of the nitric oxide (NO) synthases [150]. It increases in patients with impaired kidney function due to the reduction of enzyme dimethylarginine-dymethyilamino-hydrolase [147].

In patients with CKD,a decrease of NO production or its bioavailability may occur.This situation could be correlated to elevation of blood pressure, CVD and progression of CKD. ADMA has been associated with CKD progression to end-stage renal disease [151] and appears to be a link between CKD and CVD [150].

Tenascin and Tissue Inhibitor of Metalloproteinases 1 (TIMP-1)

Tenascin is an extracellular matrix protein expressed in the medullary interstitium in normal kidney, and ubiquitously present in areas with interstitial fibrosis [152]. TIMP-1 inhibits the degradation of matrix and can promote cell proliferation in different cell types. CKD is characterized by increased synthesis and inhibited destruction of collagenous and non-collagenous matrix protein. It has been shown that serum concentrations of tenascin and TIMP-1 are elevated in patients with CKD compared with healthy controls, although urinary concentrations were lower than serum levels and unrelated to the degree of proteinuria [153].

Nephrin, Podocin and Podocalyxin

Podocyte-associated molecules are new biomarkers of kidney podocyte injury [154, 155]. Urinary nephrin and podocin are elevated in diabetic nephropathy and active lupus nephritis. Urinary podocalyxinincreases in IgA nephropathy, lupus nephritis and post-streptoccocal glomerulonephritis [156].

DISCUSSION

In recent years, numerous studies have developed a large amount of new biomarkers which may improve the information of current markers and facilitate early detection, differential diagnosis and prognosis of kidney diseases, although to date, none of them can replace conventional markers. Furthermore, there have not been enough studies carried out in order to evaluate the usefulness of these biomarkers on treatment decisions.

Biomarkers of AKI

Nowadays the diagnosis of AKI depends on serum creatinine and/or decrease of urine output. Serum creatinine is the most commonly used biomarker to screen and diagnose kidney disease, but it is limited in the early stages of AKI due to its levels can increase in cases of pre-renal azotemia without tubular injury, its concentrations can be influenced by many non-renal factors (muscle metabolism, total body volume and drugs), its circulating levels does not rise after 50% of kidney function loss and in addition, renal injury can exist with no change in serum creatinine. For all of these reasons, serum creatinine should be used only for eGFR and the development of new markers are necessary. Combination of functional and tubular damage biomarkers may contribute and improve the limitations of current markers, but to date, although some of them are available for clinical use around the world, future studies will be necessary.

New biomarkers of AKI, NGAL, Cys-C, KIM-1, IL-18 and recently TIM-2 together with IGFBP7 stand out and are the most promising biomarkers of AKI. However, there have notbeensufficient studies carried out that have evaluated how these new biomarkers can improve decision making.

Biomarkers of CKD

Currently the risk of CKD is established according to eGFR and presence of albuminuria. The pathophysiology of CKD is complex and involves different processes, thus the measurement of newbiomarkersmay improve the diagnosis of CKD especially when it is asymptomatic, in early kidney disease. Although it is improbable that one biomarker can be sufficient.

With regards to eGFR equations, CKD-EPI seems to be more generalized and precise than the MDRD estimating equation. On the other hand, Cys-c based GFR equations are more accurate than the CKD-EPI equation in patients with no change in serum creatinine level such as hospitalized patients, patients with chronic liver disease and cirrhosis, elders and patients with HIV infections and malignancy. Furthermore Cys-c is less influenced by muscle mass than creatinine. But it is important to take into account that Cys-c levels are affected by inflammatory conditions (high levels of CPR), thyroid disease, glucocorticoid use and obesity. Recent eGFR equation combining serum creatinine and Cys-c performed better than equations that used either serum creatinine or Cys-c alone, especially when confirmation of CKD is required.

FGF-23, NGAL and ADMA are emerging biomarkers that can improve prognosis and risk stratification in CKD, but future studies will be necessary before these new biomarkers can be implemented in clinical practice.

Hence, there is a real need for novel and reliable biomarkers for early diagnosis, prediction, detection, progression and treatment decisions. New emerging biomarkers appear to be a powerful tool for a better understanding of patho-physiology and epidemiology of kidney diseases and the links between AKI and the development of CKD, but larger and long term studies are required before applying them in clinical practice.

CONFLICT OF INTEREST

The author (editor) declares no conflict of interest, financial or otherwise.

ACKNOWLEDGEMENTS

Declared none.

REFERENCES

[1] Eckardt KU, Coresh J, Devuyst O, *et al.* Evolving importance of kidney disease: from subspecialty to global health burden. Lancet 2013; 382: 158-69. [http://dx.doi.org/10.1016/S0140-6736(13)60439-0]

[2] James MT, Hemmelgarn BR, Tonelli M. Early recognition and prevention of chronic kidney disease. Lancet 2010; 375: 1296-309. [http://dx.doi.org/10.1016/S0140-6736(09)62004-3]

[3] Coresh J, Selvin E, Stevens LA, *et al.* Prevalence of chronic kidney disease in the United States. JAMA 2007; 298: 2038-47. [http://dx.doi.org/10.1001/jama.298.17.2038]

[4] Kellum JA, Lameire N, Aspelin P, *et al.* Kidney Disease: Improving Global Outcomes (KDIGO) Acute Kidney Injury Work Group. KDIGO Clinical Practice Guideline for Acute Kidney Injury.

Kidney Int 2012; 2: 1.

[5] KDIGO 2012 Clinical Practice Guideline for the Evaluation and Management of Chronic Kidney Disease. Kidney Int Suppl 2013; 3: 4-4.

[6] Levey AS, Becker C, Inker LA. Glomerular filtration rate and albuminuria for detection and staging of acute and chronic kidney disease in adults. JAMA 2015; 313: 837. [http://dx.doi.org/10.1001/jama.2015.0602]

[7] Schrier RW, Wang W, Poole B, Mitra A. Acute renal failure: definitions, diagnosis, pathogenesis, and therapy. J Clin Invest 2004; 114: 5-14. [http://dx.doi.org/10.1172/JCI200422353]

[8] Chawla LS, Kimmel PL. Acute kidney injury and chronic kidney disease: an integrated clinical syndrome. Kidney Int 2012; 82: 516-24. [http://dx.doi.org/10.1038/ki.2012.208]

[9] Thadhani R, Pascual M, Bonventre JV. Acute renal failure. N Engl J Med 1996; 334: 1448-60. [http://dx.doi.org/10.1056/NEJM199605303342207]

[10] Choudhury D, Ahmed Z. Drug-associated renal dysfunction and injury. Nat Clin Pract Nephrol 2006; 2: 80-91. [http://dx.doi.org/10.1038/ncpneph0076]

[11] Vaidya V, Ferguson M, Bonventre J. Biomarkers of acute kidney injury. Annu Rev Pharmacol Toxicol 2008; 48: 1-29. [http://dx.doi.org/10.1146/annurev.pharmtox.48.113006.094615]

[12] Botev R, Mallié J-P, Wetzels JF, Couchoud C, Schück O. The clinician and estimation of glomerular filtration rate by creatinine-based formulas: current limitations and quo vadis. Clin J Am Soc Nephrol 2011; 6: 937-50. [http://dx.doi.org/10.2215/CJN.09241010]

[13] Cockcroft DW, Gault MH. Prediction of creatinine clearance from serum creatinine. Nephron 1976; 16: 31-41. [http://dx.doi.org/10.1159/000180580]

[14] Stevens LA, Nolin TD, Richardson MM, *et al.* Comparison of drug dosing recommendations based on measured GFR and kidney function estimating equations. Am J Kidney Dis 2009; 54: 33-42. [http://dx.doi.org/10.1053/j.ajkd.2009.03.008]

[15] Stevens LA, Schmid CH, Greene T, *et al.* Factors other than glomerular filtration rate affect serum cystatin C levels. Kidney Int 2009; 75: 652-60. [http://dx.doi.org/10.1038/ki.2008.638]

[16] Galteau M-M, Guyon M, Gueguen R, Siest G. Determination of serum cystatin C: biological variation and reference values. Clin Chem Lab Med 2001; 39 [http://dx.doi.org/10.1515/CCLM.2001.141]

[17] Lamb EJ, Stevens PE. Estimating and measuring glomerular filtration rate: methods of measurement and markers for estimation. Curr Opin Nephrol Hypertens 2014; 23: 258-66. [http://dx.doi.org/10.1097/01.mnh.0000444813.72626.88]

[18] Levey AS, Bosch JP, Lewis JB, Greene T, Rogers N, Roth D. A more accurate method to estimate glomerular filtration rate from serum creatinine: a new prediction equation. Modification of Diet in Renal Disease Study Group. Ann Intern Med 1999; 130: 461-70. [http://dx.doi.org/10.7326/0003-4819-130-6-199903160-00002]

[19] Levey AS, Coresh J, Greene T, *et al.* Using standardized serum creatinine values in the modification of diet in renal disease study equation for estimating glomerular filtration rate. Ann Intern Med 2006; 145: 247-54. [http://dx.doi.org/10.7326/0003-4819-145-4-200608150-00004]

[20] Nordén G, Björck S, Granerus G, Nyberg G. Estimation of renal function in diabetic nephropathy. Comparison of five methods. Nephron 1987; 47: 36-42. [http://dx.doi.org/10.1159/000184454]

[21] Stoves J, Lindley EJ, Barnfield MC, Burniston MT, Newstead CG. MDRD equation estimates of glomerular filtration rate in potential living kidney donors and renal transplant recipients with impaired graft function. Nephrol Dial Transplant 2002; 17: 2036-7. [http://dx.doi.org/10.1093/ndt/17.11.2036]

[22] Earley A, Miskulin D, Lamb EJ, Levey AS, Uhlig K. Estimating equations for glomerular filtration rate in the era of creatinine standardization: a systematic review. Ann Intern Med 2012; 156: 785-95. [http://dx.doi.org/10.7326/0003-4819-156-11-201203200-00391]

[23] Stevens LA, Coresh J, Feldman HI, Greene T, Lash JP, Nelson RG, *et al.* Evaluation of the modification of diet in renal disease study equation in a large diverse population. J Am Soc Nephrol 2007; 18: 2749-57. [http://dx.doi.org/10.1681/ASN.2007020199]

[24] Levey AS. A new equation to estimate glomerular filtration rate. Ann Intern Med 2009; 150: 604. [http://dx.doi.org/10.7326/0003-4819-150-9-200905050-00006]

[25] Inker LA, Schmid CH, Tighiouart H, *et al.* Estimating glomerular filtration rate from serum creatinine and Cystatin C. N Engl J Med 2012; 367: 20-9. [http://dx.doi.org/10.1056/NEJMoa1114248]

[26] Schaeffner ES, Ebert N, Delanaye P, *et al.* Two novel equations to estimate kidney function in persons aged 70 years or older. Ann Intern Med 2012; 157: 471. [http://dx.doi.org/10.7326/0003-4819-157-7-201210020-00003]

[27] Koppe L, Klich A, Dubourg L, Ecochard R, Hadj-Aissa A. Performance of creatinine-based equations compared in older patients. J Nephrol 26: 716-23. [http://dx.doi.org/10.5301/jn.5000297]

[28] Garasto S, Fusco S, Corica F, *et al.* Estimating glomerular filtration rate in older people. Biomed Res Int 2014; 916542. [http://dx.doi.org/10.1155/2014/916542]

[29] Shlipak MG, Matsushita K, Ärnlöv J, *et al.* Cystatin C *versus* creatinine in determining risk based on kidney function. N Engl J Med 2013; 369: 932-43. [http://dx.doi.org/10.1056/NEJMoa1214234]

[30] Stevens LA, Coresh J, Greene T, Levey AS. Assessing kidney function--measured and estimated glomerular filtration rate. N Engl J Med 2006; 354: 2473-83. [http://dx.doi.org/10.1056/NEJMra054415]

[31] Ostermann M, Philips BJ, Forni LG. Clinical review: Biomarkers of acute kidney injury: where are we now? Crit Care 2012; 16: 233. [http://dx.doi.org/10.1186/cc11380]

[32] Charlton JR, Portilla D, Okusa MD. A basic science view of acute kidney injury biomarkers. Nephrol Dial Transplant 2014; 29: 1301-11. [http://dx.doi.org/10.1093/ndt/gft510]

[33] Brzin J, Popovic T, Turk V, Borchart U, Machleidt W. Human cystatin, a new protein inhibitor of cysteine proteinases. Biochem Biophys Res Commun 1984; 118: 103-9. [http://dx.doi.org/10.1016/0006-291X(84)91073-8]

[34] Clausen J. Proteins in normal cerebrospinal fluid not found in serum. Proc Soc Exp Biol Med 1961; 107: 170-2. [http://dx.doi.org/10.3181/00379727-107-26569]

[35] Butler EA, Flynn FV. The occurrence of post-gamma protein in urine: a new protein abnormality. J

Clin Pathol 1961; 14: 172-8.
[http://dx.doi.org/10.1136/jcp.14.2.172]

[36] Shlipak MG, Sarnak MJ, Katz R, *et al.* Cystatin C and the risk of death and cardiovascular events among elderly persons. N Engl J Med 2005; 352: 2049-60.
[http://dx.doi.org/10.1056/NEJMoa043161]

[37] Togashi Y, Sakaguchi Y, Miyamoto M, Miyamoto Y. Urinary cystatin C as a biomarker for acute kidney injury and its immunohistochemical localization in kidney in the CDDP-treated rats. Exp Toxicol Pathol Off J Gesellschaft für Toxikologische Pathol 2012; 64: 797-805.

[38] Spahillari A, Parikh CR, Sint K, *et al.* Serum cystatin C- *versus* creatinine-based definitions of acute kidney injury following cardiac surgery: a prospective cohort study. Am J Kidney Dis 2012; 60: 922-9.
[http://dx.doi.org/10.1053/j.ajkd.2012.06.002]

[39] Herget-Rosenthal S, Bökenkamp A, Hofmann W. How to estimate GFR-serum creatinine, serum cystatin C or equations? Clin Biochem 2007; 40: 153-61.
[http://dx.doi.org/10.1016/j.clinbiochem.2006.10.014]

[40] Séronie-Vivien S, Delanaye P, Piéroni L, *et al.* Cystatin C: current position and future prospects. Clin Chem Lab Med 2008; 46: 1664-86.
[http://dx.doi.org/10.1515/CCLM.2008.336]

[41] Glatz JF, van der Vusse GJ. Cellular fatty acid-binding proteins: their function and physiological significance. Prog Lipid Res 1996; 35: 243-82.
[http://dx.doi.org/10.1016/S0163-7827(96)00006-9]

[42] Sweetser DA, Heuckeroth RO, Gordon JI. The metabolic significance of mammalian fatty-aci--binding proteins: abundant proteins in search of a function. Annu Rev Nutr 1987; 7: 337-59.
[http://dx.doi.org/10.1146/annurev.nu.07.070187.002005]

[43] Maatman RG, van de Westerlo EM, van Kuppevelt TH, Veerkamp JH. Molecular identification of the liver- and the heart-type fatty acid-binding proteins in human and rat kidney. Use of the reverse transcriptase polymerase chain reaction. Biochem J 1992; 288(Pt 1): 285-90.
[http://dx.doi.org/10.1042/bj2880285]

[44] Maatman RG, Van Kuppevelt TH, Veerkamp JH. Two types of fatty acid-binding protein in human kidney. Isolation, characterization and localization. Biochem J 1991; 273(Pt 3): 759-66.
[http://dx.doi.org/10.1042/bj2730759]

[45] Kamijo-Ikemori A, Sugaya T, Kimura K. L-type fatty acid binding protein (L-FABP) and kidney disease. Rinsho Byori 2014; 62: 163-70.

[46] Peco-Antić A, Ivanišević I, Vulićević I, *et al.* Biomarkers of acute kidney injury in pediatric cardiac surgery. Clin Biochem 2013; 46: 1244-51.
[http://dx.doi.org/10.1016/j.clinbiochem.2013.07.008]

[47] Krawczeski CD, Goldstein SL, Woo JG, *et al.* Temporal relationship and predictive value of urinary acute kidney injury biomarkers after pediatric cardiopulmonary bypass. J Am Coll Cardiol 2011; 58: 2301-9.
[http://dx.doi.org/10.1016/j.jacc.2011.08.017]

[48] Katagiri D, Doi K, Honda K, *et al.* Combination of two urinary biomarkers predicts acute kidney injury after adult cardiac surgery. Ann Thorac Surg 2012; 93: 577-83.
[http://dx.doi.org/10.1016/j.athoracsur.2011.10.048]

[49] Landrier J-F, Thomas C, Grober J, *et al.* Statin induction of liver fatty acid-binding protein (L-FABP) gene expression is peroxisome proliferator-activated receptor-alpha-dependent. J Biol Chem 2004; 279: 45512-8.
[http://dx.doi.org/10.1074/jbc.M407461200]

[50] Gracie JA, Robertson SE, McInnes IB. Interleukin-18. J Leukoc Biol 2003; 73: 213-24.

[http://dx.doi.org/10.1189/jlb.0602313]

[51] Sugawara I. Interleukin-18 (IL-18) and infectious diseases, with special emphasis on diseases induced by intracellular pathogens. Microbes Infect 2000; 2: 1257-63. [http://dx.doi.org/10.1016/S1286-4579(00)01279-X]

[52] Edelstein CL, Hoke TS, Somerset H, *et al.* Proximal tubules from caspase-1-deficient mice are protected against hypoxia-induced membrane injury. Nephrol Dial Transplant 2007; 22: 1052-61. [http://dx.doi.org/10.1093/ndt/gfl775]

[53] Parikh CR, Jani A, Melnikov VY, Faubel S, Edelstein CL. Urinary interleukin-18 is a marker of human acute tubular necrosis. Am J Kidney Dis 2004; 43: 405-14. [http://dx.doi.org/10.1053/j.ajkd.2003.10.040]

[54] Ichimura T, Bonventre JV, Bailly V, *et al.* Kidney injury molecule-1 (KIM-1), a putative epithelial cell adhesion molecule containing a novel immunoglobulin domain, is up-regulated in renal cells after injury. J Biol Chem 1998; 273: 4135-42. [http://dx.doi.org/10.1074/jbc.273.7.4135]

[55] Bailly V, Zhang Z, Meier W, Cate R, Sanicola M, Bonventre JV. Shedding of kidney injury molecule-1, a putative adhesion protein involved in renal regeneration. J Biol Chem 2002; 277: 39739-48. [http://dx.doi.org/10.1074/jbc.M200562200]

[56] Vaidya VS, Ramirez V, Ichimura T, Bobadilla NA, Bonventre JV. Urinary kidney injury molecule-1: a sensitive quantitative biomarker for early detection of kidney tubular injury. Am J Physiol Renal Physiol 2006; 290: F517-29. [http://dx.doi.org/10.1152/ajprenal.00291.2005]

[57] Han WK, Bailly V, Abichandani R, Thadhani R, Bonventre JV. Kidney Injury Molecule-1 (KIM-1): a novel biomarker for human renal proximal tubule injury. Kidney Int 2002; 62: 237-44. [http://dx.doi.org/10.1046/j.1523-1755.2002.00433.x]

[58] Ichimura T, Asseldonk EJ, Humphreys BD, Gunaratnam L, Duffield JS, Bonventre JV. Kidney injury molecule-1 is a phosphatidylserine receptor that confers a phagocytic phenotype on epithelial cells. J Clin Invest 2008; 118: 1657-68. [http://dx.doi.org/10.1172/JCI34487]

[59] Huang Y, Don-Wauchope AC. The clinical utility of kidney injury molecule 1 in the prediction, diagnosis and prognosis of acute kidney injury: a systematic review. Inflamm Allergy Drug Targets 2011; 10: 260-71. [http://dx.doi.org/10.2174/187152811796117735]

[60] Ferguson MA, Vaidya VS, Bonventre JV. Biomarkers of nephrotoxic acute kidney injury. Toxicology 2008; 245: 182-93. [http://dx.doi.org/10.1016/j.tox.2007.12.024]

[61] Kern W, Braess J, Kaufmann CC, Wilde S, Schleyer E, Hiddemann W. Microalbuminuria during cisplatin therapy: relation with pharmacokinetics and implications for nephroprotection. Anticancer Res 20: 3679-88.

[62] Koch Nogueira PC, Hadj-Aïssa A, Schell M, Dubourg L, Brunat-Mentigny M, Cochat P. Long-term nephrotoxicity of cisplatin, ifosfamide, and methotrexate in osteosarcoma. Pediatr Nephrol 1998; 12: 572-5. [http://dx.doi.org/10.1007/s004670050507]

[63] Ware LB, Johnson AC, Zager RA. Renal cortical albumin gene induction and urinary albumin excretion in response to acute kidney injury. Am J Physiol Renal Physiol 2011; 300: F628-38. [http://dx.doi.org/10.1152/ajprenal.00654.2010]

[64] Waring WS, Moonie A. Earlier recognition of nephrotoxicity using novel biomarkers of acute kidney injury. Clin Toxicol (Phila) 2011; 49: 720-8. [http://dx.doi.org/10.3109/15563650.2011.615319]

[65] Westhuyzen J, Endre ZH, Reece G, Reith DM, Saltissi D, Morgan TJ. Measurement of tubular enzymuria facilitates early detection of acute renal impairment in the intensive care unit. Nephrol Dial Transplant 2003; 18: 543-51. [http://dx.doi.org/10.1093/ndt/18.3.543]

[66] Chew SL, Lins RL, Daelemans R, Nuyts GD, De Broe ME. Urinary enzymes in acute renal failure. Nephrol Dial Transplant 1993; 8: 507-11. [http://dx.doi.org/10.1093/ndt/8.6.507]

[67] Bondiou MT, Bourbouze R, Bernard M, Percheron F, Perez-Gonzalez N, Cabezas JA. Inhibition of A and B N-acetyl-beta-D-glucosaminidase urinary isoenzymes by urea. Clin Chim Acta 1985; 149: 67-73. [http://dx.doi.org/10.1016/0009-8981(85)90274-8]

[68] Vaidya VS, Shankar K, Lock EA, Bucci TJ, Mehendale HM. Role of tissue repair in survival from s-(1,2-dichlorovinyl)-L-cysteine-induced acute renal tubular necrosis in the mouse. Toxicol Sci 2003; 74: 215-27. [http://dx.doi.org/10.1093/toxsci/kfg111]

[69] Iqbal MP, Ali AA, Waqar MA, Mehboobali N. Urinary N-acetyl-beta-D-glucosaminidase in rheumatoid arthritis. Exp Mol Med 1998; 30: 165-9. [http://dx.doi.org/10.1038/emm.1998.24]

[70] Fujita H, Narita T, Morii T, *et al.* Increased urinary excretion of N-acetylglucosaminidase in subjects with impaired glucose tolerance. Ren Fail 2002; 24: 69-75. [http://dx.doi.org/10.1081/JDI-120002662]

[71] Tominaga M, Fujiyama K, Hoshino T, *et al.* Urinary N-acetyl-beta-D-glucosaminidase in the patients with hyperthyroidism. Horm Metab Res 1989; 21: 438-40. [http://dx.doi.org/10.1055/s-2007-1009256]

[72] Kjeldsen L, Johnsen AH, Sengeløv H, Borregaard N. Isolation and primary structure of NGAL, a novel protein associated with human neutrophil gelatinase. J Biol Chem 1993; 268: 10425-32.

[73] Goetz DH, Willie ST, Armen RS, Bratt T, Borregaard N, Strong RK. Ligand preference inferred from the structure of neutrophil gelatinase associated lipocalin. Biochemistry 2000; 39: 1935-41. [http://dx.doi.org/10.1021/bi992215v]

[74] Borregaard N, Sehested M, Nielsen BS, Sengeløv H, Kjeldsen L. Biosynthesis of granule proteins in normal human bone marrow cells. Gelatinase is a marker of terminal neutrophil differentiation. Blood 1995; 85: 812-7.

[75] Cowland JB, Borregaard N. Molecular characterization and pattern of tissue expression of the gene for neutrophil gelatinase-associated lipocalin from humans. Genomics 1997; 45: 17-23. [http://dx.doi.org/10.1006/geno.1997.4896]

[76] Nguyen MT, Devarajan P. Biomarkers for the early detection of acute kidney injury. Pediatr Nephrol 2008; 23: 2151-7. [http://dx.doi.org/10.1007/s00467-007-0470-x]

[77] Nielsen BS, Borregaard N, Bundgaard JR, Timshel S, Sehested M, Kjeldsen L. Induction of NGAL synthesis in epithelial cells of human colorectal neoplasia and inflammatory bowel diseases. Gut 1996; 38: 414-20. [http://dx.doi.org/10.1136/gut.38.3.414]

[78] Mori K, Lee HT, Rapoport D, *et al.* Endocytic delivery of lipocalin-siderophore-iron complex rescues the kidney from ischemia-reperfusion injury. J Clin Invest 2005; 115: 610-21. [http://dx.doi.org/10.1172/JCI23056]

[79] Mishra J, Mori K, Ma Q, Kelly C, Barasch J, Devarajan P. Neutrophil gelatinase-associated lipocalin: a novel early urinary biomarker for cisplatin nephrotoxicity. Am J Nephrol 24: 307-15. [http://dx.doi.org/10.1159/000078452]

[80] Xu SY, Pauksen K, Venge P. Serum measurements of human neutrophil lipocalin (HNL) discriminate between acute bacterial and viral infections. Scand J Clin Lab Invest 1995; 55: 125-31. [http://dx.doi.org/10.3109/00365519509089604]

[81] McIlroy DR, Wagener G, Lee HT. Neutrophil gelatinase-associated lipocalin and acute kidney injury after cardiac surgery: the effect of baseline renal function on diagnostic performance. Clin J Am Soc Nephrol 2010; 5: 211-9. [http://dx.doi.org/10.2215/CJN.04240609]

[82] Smith ER, Zurakowski D, Saad A, Scott RM, Moses MA. Urinary biomarkers predict brain tumor presence and response to therapy. Clin Cancer Res 2008; 14: 2378-86. [http://dx.doi.org/10.1158/1078-0432.CCR-07-1253]

[83] D'Anna R, Baviera G, Giordano D, Todarello G, Corrado F, Buemi M. Second trimester neutrophil gelatinase-associated lipocalin as a potential prediagnostic marker of preeclampsia. Acta Obstet Gynecol Scand 2008; 87: 1370-3. [http://dx.doi.org/10.1080/00016340802464463]

[84] Wolf MW, Boldt J. Kidney-specific proteins: markers for detection of renal dysfunction after cardiac surgery? Clin Res Cardiol Suppl 2007; 2: S103-7. [http://dx.doi.org/10.1007/s11789-006-0043-8]

[85] Penders J, Delanghe JR. Alpha 1-microglobulin: clinical laboratory aspects and applications. Clin Chim Acta 2004; 346: 107-18. [http://dx.doi.org/10.1016/j.cccn.2004.03.037]

[86] Chang C, Werb Z. The many faces of metalloproteases: cell growth, invasion, angiogenesis and metastasis. Trends Cell Biol 2001; 11: S37-43. [http://dx.doi.org/10.1016/S0962-8924(01)02122-5]

[87] Kashani K, Al-Khafaji A, Ardiles T, *et al.* Discovery and validation of cell cycle arrest biomarkers in human acute kidney injury. Crit Care 2013; 17: R25. [http://dx.doi.org/10.1186/cc12503]

[88] Zhu S, Xu F, Zhang J, Ruan W, Lai M. Insulin-like growth factor binding protein-related protein 1 and cancer. Clin Chim Acta 2014; 431: 23-32. [http://dx.doi.org/10.1016/j.cca.2014.01.037]

[89] Wasung ME, Chawla LS, Madero M. Biomarkers of renal function, which and when? Clin Chim Acta 2015; 438: 350-7. [http://dx.doi.org/10.1016/j.cca.2014.08.039]

[90] Bernard AM, Vyskocil AA, Mahieu P, Lauwerys RR. Assessment of urinary retinol-binding protein as an index of proximal tubular injury. Clin Chem 1987; 33: 775-9.

[91] Roberts DS, Haycock GB, Dalton RN, *et al.* Prediction of acute renal failure after birth asphyxia. Arch Dis Child 1990; 65: 1021-8. [http://dx.doi.org/10.1136/adc.65.10_Spec_No.1021]

[92] Garçon G, Leleu B, Zerimech F, *et al.* Biologic markers of oxidative stress and nephrotoxicity as studied in biomonitoring of adverse effects of occupational exposure to lead and cadmium. J Occup Environ Med 2004; 46: 1180-6. [http://dx.doi.org/10.1097/01.jom.0000141665.22881.69]

[93] Barallobre MJ, Pascual M, Del Río JA, Soriano E. The Netrin family of guidance factors: emphasis on Netrin-1 signalling. Brain Res Brain Res Rev 2005; 49: 22-47. [http://dx.doi.org/10.1016/j.brainresrev.2004.11.003]

[94] Ranganathan PV, Jayakumar C, Mohamed R, Dong Z, Ramesh G. Netrin-1 regulates the inflammatory response of neutrophils and macrophages, and suppresses ischemic acute kidney injury by inhibiting COX-2-mediated PGE2 production. Kidney Int 2013; 83: 1087-98. [http://dx.doi.org/10.1038/ki.2012.423]

[95] Wang W, Reeves WB, Pays L, Mehlen P, Ramesh G. Netrin-1 overexpression protects kidney from ischemia reperfusion injury by suppressing apoptosis. Am J Pathol 2009; 175: 1010-8. [http://dx.doi.org/10.2353/ajpath.2009.090224]

[96] Levey AS, Atkins R, Coresh J, *et al.* Chronic kidney disease as a global public health problem: approaches and initiatives - a position statement from Kidney Disease Improving Global Outcomes. Kidney Int 2007; 72: 247-59. [http://dx.doi.org/10.1038/sj.ki.5002343]

[97] Newman DJ, Thakkar H, Edwards RG, *et al.* Serum cystatin C measured by automated immunoassay: a more sensitive marker of changes in GFR than serum creatinine. Kidney Int 1995; 47: 312-8. [http://dx.doi.org/10.1038/ki.1995.40]

[98] Spanaus K-S, Kollerits B, Ritz E, *et al.* Serum creatinine, cystatin C, and beta-trace protein in diagnostic staging and predicting progression of primary nondiabetic chronic kidney disease. Clin Chem 2010; 56: 740-9. [http://dx.doi.org/10.1373/clinchem.2009.138826]

[99] Nejat M, Hill JV, Pickering JW, Edelstein CL, Devarajan P, Endre ZH. Albuminuria increases cystatin C excretion: implications for urinary biomarkers. Nephrol Dial Transplant 2012; 27 (Suppl. 3): 96-103. [http://dx.doi.org/10.1093/ndt/gfr222]

[100] Taal MW, Brenner BM. Renal risk scores: progress and prospects. Kidney Int 2008; 73: 1216-9. [http://dx.doi.org/10.1038/ki.2008.36]

[101] Conti M, Moutereau S, Zater M, *et al.* Urinary cystatin C as a specific marker of tubular dysfunction. Clin Chem Lab Med 2006; 44: 288-91. [http://dx.doi.org/10.1515/CCLM.2006.050]

[102] Ko GJ, Grigoryev DN, Linfert D, *et al.* Transcriptional analysis of kidneys during repair from AKI reveals possible roles for NGAL and KIM-1 as biomarkers of AKI-to-CKD transition. Am J Physiol Renal Physiol 2010; 298: F1472-83. [http://dx.doi.org/10.1152/ajprenal.00619.2009]

[103] Kamijo A, Kimura K, Sugaya T, *et al.* Urinary fatty acid-binding protein as a new clinical marker of the progression of chronic renal disease. J Lab Clin Med 2004; 143: 23-30. [http://dx.doi.org/10.1016/j.lab.2003.08.001]

[104] Gansevoort RT, Matsushita K, van der Velde M, *et al.* Lower estimated GFR and higher albuminuria are associated with adverse kidney outcomes. A collaborative meta-analysis of general and high-risk population cohorts. Kidney Int 2011; 80: 93-104. [http://dx.doi.org/10.1038/ki.2010.531]

[105] Lopez-Giacoman S, Madero M. Biomarkers in chronic kidney disease, from kidney function to kidney damage. World J Nephrol 2015; 4: 57-73. [http://dx.doi.org/10.5527/wjn.v4.i1.57]

[106] Hoerger TJ, Wittenborn JS, Segel JE, *et al.* A health policy model of CKD: 2. The cost-effectiveness of microalbuminuria screening. Am J Kidney Dis 2010; 55: 463-73. [http://dx.doi.org/10.1053/j.ajkd.2009.11.017]

[107] Hoffmann A, Nimtz M, Conradt HS. Molecular characterization of beta-trace protein in human serum and urine: a potential diagnostic marker for renal diseases. Glycobiology 1997; 7: 499-506. [http://dx.doi.org/10.1093/glycob/7.4.499]

[108] Smith WL, Urade Y, Jakobsson P-J. Enzymes of the cyclooxygenase pathways of prostanoid biosynthesis. Chem Rev 2011; 111: 5821-65. [http://dx.doi.org/10.1021/cr2002992]

[109] White CA, Ghazan-Shahi S, Adams MA. β-Trace protein: a marker of GFR and other biological pathways. Am J Kidney Dis 2015; 65: 131-46.

[http://dx.doi.org/10.1053/j.ajkd.2014.06.038]

[110] Urade Y, Hayaishi O. Biochemical, structural, genetic, physiological, and pathophysiological features of lipocalin-type prostaglandin D synthase. Biochim Biophys Acta 2000; 1482: 259-71. [http://dx.doi.org/10.1016/S0167-4838(00)00161-8]

[111] Orenes-Piñero E, Manzano-Fernández S, López-Cuenca Á, Marín F, Valdés M, Januzzi JL. β-Trace protein: from GFR marker to cardiovascular risk predictor. Clin J Am Soc Nephrol 2013; 8: 873-81. [http://dx.doi.org/10.2215/CJN.08870812]

[112] Vílchez JA, Roldán V, Manzano-Fernández S, *et al.* β-Trace protein and prognosis in patients with atrial fibrillation receiving anticoagulation treatment. Chest 2013; 144: 1564-70. [http://dx.doi.org/10.1378/chest.13-0922]

[113] Olsson JE, Link H, Nosslin B. Metabolic studies on 125I-labelled beta-trace protein, with special reference to synthesis within the central nervous system. J Neurochem 1973; 21: 1153-9. [http://dx.doi.org/10.1111/j.1471-4159.1973.tb07570.x]

[114] Li W, Mase M, Inui T, *et al.* Pharmacokinetics of recombinant human lipocalin-type prostaglandin D synthase/beta-trace in canine. Neurosci Res 2008; 61: 289-93. [http://dx.doi.org/10.1016/j.neures.2008.03.006]

[115] Filler G, Priem F, Lepage N, *et al.* Beta-trace protein, cystatin C, beta(2)-microglobulin, and creatinine compared for detecting impaired glomerular filtration rates in children. Clin Chem 2002; 48: 729-36.

[116] Priem F, Althaus H, Birnbaum M, Sinha P, Conradt HS, Jung K. Beta-trace protein in serum: a new marker of glomerular filtration rate in the creatinine-blind range. Clin Chem 1999; 45: 567-8.

[117] Giessing M. Beta-trace protein as indicator of glomerular filtration rate. Urology 1999; 54: 940-1.

[118] Pöge U, Gerhardt TM, Stoffel-Wagner B, *et al.* Beta-Trace protein is an alternative marker for glomerular filtration rate in renal transplantation patients. Clin Chem 2005; 51: 1531-3. [http://dx.doi.org/10.1373/clinchem.2005.048959]

[119] Woitas RP, Stoffel-Wagner B, Poege U, Schiedermaier P, Spengler U, Sauerbruch T. Low-molecular weight proteins as markers for glomerular filtration rate. Clin Chem 2001; 47: 2179-80.

[120] Donadio C, Lucchesi A, Ardini M, Donadio E, Giordani R. Serum levels of beta-trace protein and glomerular filtration rate--preliminary results. J Pharm Biomed Anal 2003; 32: 1099-104. [http://dx.doi.org/10.1016/S0731-7085(03)00215-2]

[121] Hediger MA, Johnson RJ, Miyazaki H, Endou H. Molecular physiology of urate transport. Physiology (Bethesda) 2005; 20: 125-33. [http://dx.doi.org/10.1152/physiol.00039.2004]

[122] Kang D-H, Chen W. Uric acid and chronic kidney disease: new understanding of an old problem. Semin Nephrol 2011; 31: 447-52. [http://dx.doi.org/10.1016/j.semnephrol.2011.08.009]

[123] Johnson RJ, Nakagawa T, Jalal D, Sánchez-Lozada LG, Kang D-H, Ritz E. Uric acid and chronic kidney disease: which is chasing which? Nephrol Dial Transplant 2013; 28: 2221-8. [http://dx.doi.org/10.1093/ndt/gft029]

[124] Li L, Yang C, Zhao Y, Zeng X, Liu F, Fu P. Is hyperuricemia an independent risk factor for new-onset chronic kidney disease?: A systematic review and meta-analysis based on observational cohort studies. BMC Nephrol 2014; 15: 122. [http://dx.doi.org/10.1186/1471-2369-15-122]

[125] Yan D, Tu Y, Jiang F, *et al.* Uric Acid is independently associated with diabetic kidney disease: a cross-sectional study in a Chinese population. PLoS One 2015; 10: e0129797. [http://dx.doi.org/10.1371/journal.pone.0129797]

[126] Testa A, Mallamaci F, Spoto B, *et al.* Association of a polymorphism in a gene encoding a urate transporter with CKD progression. Clin J Am Soc Nephrol 2014; 9: 1059-65.

[http://dx.doi.org/10.2215/CJN.11041013]

[127] Fassett RG, Venuthurupalli SK, Gobe GC, Coombes JS, Cooper MA, Hoy WE. Biomarkers in chronic kidney disease: a review. Kidney Int 2011; 80: 806-21. [http://dx.doi.org/10.1038/ki.2011.198]

[128] Mishra J, Ma Q, Prada A, *et al.* Identification of neutrophil gelatinase-associated lipocalin as a novel early urinary biomarker for ischemic renal injury. J Am Soc Nephrol 2003; 14: 2534-43. [http://dx.doi.org/10.1097/01.ASN.0000088027.54400.C6]

[129] Bolignano D, Lacquaniti A, Coppolino G, *et al.* Neutrophil gelatinase-associated lipocalin (NGAL) and progression of chronic kidney disease. Clin J Am Soc Nephrol 2009; 4: 337-44. [http://dx.doi.org/10.2215/CJN.03530708]

[130] Viau A, El Karoui K, Laouari D, *et al.* Lipocalin 2 is essential for chronic kidney disease progression in mice and humans. J Clin Invest 2010; 120: 4065-76. [http://dx.doi.org/10.1172/JCI42004]

[131] Ding H, He Y, Li K, *et al.* Urinary neutrophil gelatinase-associated lipocalin (NGAL) is an early biomarker for renal tubulointerstitial injury in IgA nephropathy. Clin Immunol 2007; 123: 227-34. [http://dx.doi.org/10.1016/j.clim.2007.01.010]

[132] Malyszko J, Malyszko JS, Bachorzewska-Gajewska H, Poniatowski B, Dobrzycki S, Mysliwiec M. Neutrophil gelatinase-associated lipocalin is a new and sensitive marker of kidney function in chronic kidney disease patients and renal allograft recipients. Transplant Proc 41: 158-61. [http://dx.doi.org/10.1016/j.transproceed.2008.10.088]

[133] Bolignano D, Coppolino G, Lacquaniti A, Nicocia G, Buemi M. Pathological and prognostic value of urinary neutrophil gelatinase-associated lipocalin in macroproteinuric patients with worsening renal function. Kidney Blood Press Res 2008; 31: 274-9. [http://dx.doi.org/10.1159/000151665]

[134] van Timmeren MM, van den Heuvel MC, Bailly V, Bakker SJ, van Goor H, Stegeman CA. Tubular kidney injury molecule-1 (KIM-1) in human renal disease. J Pathol 2007; 212: 209-17. [http://dx.doi.org/10.1002/path.2175]

[135] Kuehn EW, Park KM, Somlo S, Bonventre JV. Kidney injury molecule-1 expression in murine polycystic kidney disease. Am J Physiol Renal Physiol 2002; 283: F1326-36. [http://dx.doi.org/10.1152/ajprenal.00166.2002]

[136] Vaidya VS, Niewczas MA, Ficociello LH, *et al.* Regression of microalbuminuria in type 1 diabetes is associated with lower levels of urinary tubular injury biomarkers, kidney injury molecule-1, and N-acetyl-β-D-glucosaminidase. Kidney Int 2011; 79: 464-70. [http://dx.doi.org/10.1038/ki.2010.404]

[137] Krolewski AS, Niewczas MA, Skupien J, *et al.* Early progressive renal decline precedes the onset of microalbuminuria and its progression to macroalbuminuria. Diabetes Care 2014; 37: 226-34. [http://dx.doi.org/10.2337/dc13-0985]

[138] Bosomworth MP, Aparicio SR, Hay AW. Urine N-acetyl-beta-D-glucosaminidase--a marker of tubular damage? Nephrol Dial Transplant 1999; 14: 620-6. [http://dx.doi.org/10.1093/ndt/14.3.620]

[139] Kamijo A, Sugaya T, Hikawa A, *et al.* Clinical evaluation of urinary excretion of liver-type fatty acid-binding protein as a marker for the monitoring of chronic kidney disease: a multicenter trial. J Lab Clin Med 2005; 145: 125-33. [http://dx.doi.org/10.1016/j.lab.2004.12.003]

[140] Kamijo A, Sugaya T, Hikawa A, *et al.* Urinary excretion of fatty acid-binding protein reflects stress overload on the proximal tubules. Am J Pathol 2004; 165: 1243-55. [http://dx.doi.org/10.1016/S0002-9440(10)63384-6]

[141] Nielsen SE, Sugaya T, Hovind P, Baba T, Parving H-H, Rossing P. Urinary liver-type fatty acid-

binding protein predicts progression to nephropathy in type 1 diabetic patients. Diabetes Care 2010; 33: 1320-4.
[http://dx.doi.org/10.2337/dc09-2242]

[142] Shimada T, Kakitani M, Yamazaki Y, *et al.* Targeted ablation of Fgf23 demonstrates an essential physiological role of FGF23 in phosphate and vitamin D metabolism. J Clin Invest 2004; 113: 561-8.
[http://dx.doi.org/10.1172/JCI200419081]

[143] Diniz H, Frazão JM. The role of fibroblast growth factor 23 in chronic kidney disease-mineral and bone disorder. Nefrol publicación Of la Soc Española Nefrol 2013; 33: 835-44.

[144] Marsell R, Grundberg E, Krajisnik T, *et al.* Fibroblast growth factor-23 is associated with parathyroid hormone and renal function in a population-based cohort of elderly men. Eur J Endocrinol 2008; 158: 125-9.
[http://dx.doi.org/10.1530/EJE-07-0534]

[145] Isakova T, Wahl P, Vargas GS, *et al.* Fibroblast growth factor 23 is elevated before parathyroid hormone and phosphate in chronic kidney disease. Kidney Int 2011; 79: 1370-8.
[http://dx.doi.org/10.1038/ki.2011.47]

[146] Larsson T, Nisbeth U, Ljunggren O, Jüppner H, Jonsson KB. Circulating concentration of FGF-23 increases as renal function declines in patients with chronic kidney disease, but does not change in response to variation in phosphate intake in healthy volunteers. Kidney Int 2003; 64: 2272-9.
[http://dx.doi.org/10.1046/j.1523-1755.2003.00328.x]

[147] Raptis V, Kapoulas S, Grekas D. Role of asymmetrical dimethylarginine in the progression of renal disease. Nephrology (Carlton) 2013; 18: 11-21.
[http://dx.doi.org/10.1111/j.1440-1797.2012.01659.x]

[148] Casellas P, Jeanteur P. Protein methylation in animal cells. I. Purification and properties of S-adenosyl-L-methionine:protein (arginine) N-methyltransferase from Krebs II ascites cells. Biochim Biophys Acta 1978; 519: 243-54.
[http://dx.doi.org/10.1016/0005-2787(78)90077-1]

[149] Casellas P, Jeanteur P. Protein methylation in animal cells. II. Inhibition of S-adenosyl-L-methionine:protein(arginine) N-methyltransferase by analogs of S-adenosyl-L-homocysteine. Biochim Biophys Acta 1978; 519: 255-68.
[http://dx.doi.org/10.1016/0005-2787(78)90078-3]

[150] Ueda S, Yamagishi S-I, Kaida Y, Okuda S. Asymmetric dimethylarginine may be a missing link between cardiovascular disease and chronic kidney disease. Nephrology (Carlton) 2007; 12: 582-90.
[http://dx.doi.org/10.1111/j.1440-1797.2007.00840.x]

[151] Fliser D, Kronenberg F, Kielstein JT, *et al.* Asymmetric dimethylarginine and progression of chronic kidney disease: the mild to moderate kidney disease study. J Am Soc Nephrol 2005; 16: 2456-61.
[http://dx.doi.org/10.1681/ASN.2005020179]

[152] Truong LD, Foster SV, Barrios R, *et al.* Tenascin is an ubiquitous extracellular matrix protein of human renal interstitium in normal and pathologic conditions. Nephron 1996; 72: 579-86.
[http://dx.doi.org/10.1159/000188943]

[153] Hörstrup JH, Gehrmann M, Schneider B, *et al.* Elevation of serum and urine levels of TIMP-1 and tenascin in patients with renal disease. Nephrol Dial Transplant 2002; 17: 1005-13.
[http://dx.doi.org/10.1093/ndt/17.6.1005]

[154] Wang G, Lai FM, Lai K-B, Chow K-M, Li K-T, Szeto C-C. Messenger RNA expression of podocyte-associated molecules in the urinary sediment of patients with diabetic nephropathy. Nephron Clin Pract 2007; 106: c169-79.
[http://dx.doi.org/10.1159/000104428]

[155] Wang G, Lai FM, Tam L-S, *et al.* Messenger RNA expression of podocyte-associated molecules in urinary sediment of patients with lupus nephritis. J Rheumatol 2007; 34: 2358-64.

[156] Kanno K, Kawachi H, Uchida Y, Hara M, Shimizu F, Uchiyama M. Urinary sediment podocalyxin in children with glomerular diseases. Nephron Clin Pract 2003; 95: c91-9. [http://dx.doi.org/10.1159/000074322]

CHAPTER 6

Biomarkers of Bone Turnover: Molecular Approaches and Clinical Relevance

Irene de Miguel-Elízaga[1], **Ana Martínez-Ruíz**[2,*], **Natalia Sancho-Rodríguez**[2] and **Miriam Martínez-Villanueva**[3]

[1] *Clinical Analysis Department Unilabs Laboratory, Torrevieja Salud Hospital, Alicante, Spain*

[2] *Clinical Analysis Department Reina Sofía General University Hospital, Murcia, Spain*

[3] *Clinical Analysis Department, Virgen de la Arrixaca Clinic University Hospital, Murcia, Spain*

Abstract: Osteoporosis, a disease characterized by low bone mass and deterioration of bone microarchitecture, leads to increased fragility and susceptibility to fracture. It is an important public health problem of the elderly and it is expected to rise with an increased life expectancy in the next years. Bone mineral density (BMD) measured by dual-energy X-ray absorptiometry (DXA) is considered as the gold standard of bone status assessment despite the associated limitations. In recent years great progress has been made in the identification and characterization of biomarkers to aid the knowledge of metabolic bone disease and within the last years, a large number of new bone turnover markers (BTM) with increased specificity and sensitivity have been developed. BTM of formation and desorption are released during the process of bone remodeling. Bone biomarkers can be a great tool for the diagnosis of osteoporosis, but their use in clinical practice is limited because there are subjected to various preanalytical and analytical variations. Also, BTM are frequently used to provide valid information about the effectiveness of osteoporosis treatment (antiresorptive or bone formation therapies) although they are not useful alone to estimate bone loss. In this review, conventional and novel BTMs, their use and limitations into clinical practice, available diagnostic methods and current recommendations are discussed.

Keywords: Bone, Bone biomarkers, Bone desorption, Bone disorders, Bone formation, Bone loss, Bone metabolism, Bone mineral density, Bone turnover, Collagen, Fracture, Monitoring of treatment, Osteoblast, Osteoclast, Osteocyte, Osteoporosis, Prediction of fractures, Stability, Standardization, Variability.

* **Corresponding author Ana Martínez-Ruíz:** Clinical Analysis Department Reina Sofía General University Hospital, Murcia, Spain; Tel/Fax: 968128600; E-mail: anamr82@gmail.com

Juan Antonio Vílchez (Ed.)

INTRODUCTION

Bone Properties

Bone is a highly specialized supporting tissue of the body, characterized by its rigidity, hardness, and power of regeneration and repair constantly throughout life. It protects vital organs, provides an environment for bone marrow, acts as a mineral reservoir for calcium, magnesium, phosphorus, sodium, and other ions necessary for homeostatic functions and a reservoir of growth factors and cytokines. Bone also hosts and regulates hematopoiesis by providing niches for hematopoietic cell proliferation and differentiation. The skeleton is highly vascular and receives about 10% of the cardiac output [1].

The basic elements of the bone tissue are cells and extracellular matrix (ECM). The bone matrix consists of an organic part (collagen of type I and other proteins) and nonorganic one (mainly, inorganic calcium and phosphate form hydroxyapatite crystals which mineralize the organic matrix). The ECM supposed approximately 9/10 of its volume, with remaining 1/10 comprising cells and blood vessels [2].

Cellular Component

Three main types of bone cells are involved in the bone metabolism [2, 3]

- Osteoblasts are the bone-forming cells. They secrete osteoid, which forms the bone matrix and also begin mineralization. Osteoblasts arise from osteoprogenitor cells located in the periosteum and the bone marrow. When osteoblasts entombed within the osteoid, become osteocytes, with cytoplasmic processes that communicate with each other. Bone building through increased formation of osteoid is stimulated by the secretion of growth hormone by the pituitary, the thyroid hormone and the sex hormones (estrogens and androgens).
- Osteocytes are the mature osteoblasts that no longer secrete matrix, but being surrounded by it maintain metabolism, and participate in nutrient/waste exchange *via* blood.
- Osteoclasts are huge cells, typically multinucleated, formed by the fusion of many cells derived from circulating monocytes in the blood. These are involved in desorption and degradation of the existing bone; in this role, these are the opposite of osteoblasts. Osteoclasts make and secrete digestive enzymes that dissolve the bone tissue and then take up the bone debris. The collagen is broken down into amino acids, which are recycled to build other proteins, while the calcium and phosphate are released to be used elsewhere in the body.

Extracellular Bone Matrix

Similar to other connective tissues, bone cells are not the primary constituents of bone by weight. Rather, the ECM produced by the osteoblasts makes up most of the dry weight of bone. This bone ECM determines the mechanical properties of the skeleton. The mineralized portion of bone tissue imparts rigidity and hardness to the material, while the organic components of the ECM provide flexibility [4]. The mineralized portion of the ECM is composed largely of calcium-phosphate in the form of hydroxyapatite plus an extensive type I collagen-rich organic ECM [5].

Extracellular components of bone tissue consist of a solid mineral phase closely linked to an organic matrix, of which 90-95% is collagen type I. The non-collagenous portion of the organic matrix is heterogeneous and contains serum proteins like albumin and other proteins produced locally. They play a critical role in bone quality and include adhesion proteins and cell signaling, such as thrombospondin, osteopontin and fibronectin; calcium binding proteins, such as osteocalcin, and proteoglycans such as biglycan and decorin. Some of these proteins organize collagen fibrils; other initiate the mineralization process and fix the mineral phase to the matrix [6]. Osteocalcin and osteopontin are two of the most abundant (and most well-studied) non-collagenous proteins. Both osteopontin and osteocalcin influence in the deposition of mineral within the collagen fibril-rich bone ECM [7].

The end result is a highly organized amalgam of protein, primarily collagen, and mineral, primarily hydroxyapatite, that has sufficient structural integrity to serve the mechanical functions of the skeleton.

The collagen superfamily of proteins constitutes important components of connective tissue throughout the body. Twenty-nine types of collagen have been described, each containing a domain composed of three α-chains intertwined as a triple helix [8]. Type I collagen is a heterotrimer, containing two α1 (I) and one α2 (I) chains intertwined as a triple helix. It is synthesized as a procollagen molecule, with N-terminal and C-terminal globular propeptides flanking the helical domain. The helical domains contain uninterrupted Gly-Xaa-Yaa triplets; Glycine is found at every third position of the triple helix because it is the only amino acid small enough to fit into the center of the helix. X and Y-position amino acids vary more; proline is often present at the X position and hydroxyproline at the Y position. Procollagen chains assemble at their C-propeptides and fold toward the N-terminal. The unfolded chains are subjected to multiple post-translational modifications. Proline and lysine residues along the helical regions of both chains are hydroxylated by prolyl 4-hydroxylase 1 (which acts on the proline ring carbon-4) and Lysol hydroxylase 1 (LH1); hydroxylysine residues can subsequently be glycosylated [9]. Hydroxylation facilitates crosslinking of the α-helices. The ring structure of proline and hydroxyproline residues enables the sharp turns necessary for assembly of the tight triple helix structure. Isomerisation of peptidylprolyl bonds, crucial for proper collagen folding, is catalysed by specific peptidylprolylcis–trans isomerases (PPIase) [10]. After secretion, procollagen molecules undergo an extracellular maturation process, in which the N-propeptides and C-propeptides are removed by specific proteases. After processing, the collagen helices are able to spontaneously assemble into fibrils in tissue, to be further stabilised by crosslinks [11].

Type I collagen is the primary collagen in bone. It is the predominant component of osteoid, the extracellular tissue secreted by osteoblasts during bone formation, growth, and remodeling, and is subsequently mineralized to form bone tissue [12].

Bone Turnover

Bone is a dynamically and metabolically active tissue which undergoes constant remodeling throughout the life span. Bone turnover is balanced with coupling of bone formation and desorption leading to continuous remodeling of bone. Bone remodeling is accomplished by assembly of osteoclasts and osteoblasts into discrete temporary anatomic structures called basic multicellular units (BMUs). Osteoclasts are the first cells observed at the site of bone remodeling and degrade existing bone matrix. Then, osteoblasts migrate to the area of desorption and fill it with a new bone matrix [13]. As old BMUs complete this cycle, new BMUs are forming; the number of BMUs and the rate of bone turnover are usually in a steady state.

Under normal conditions the amount of bone removed is always equal to the amount of newly formed bone and this balance is achieved and regulated by circulating hormones together with locally produced cytokines and growth factors modulate the replication and differentiation of osteoclast and osteoblast progenitors and also govern the lifespan by their effects on apoptosis. The orderly genesis and apoptosis of both osteoclasts and osteoblasts is essential for normal bone homeostasis during bone remodeling [14]. About 10% of the skeleton is remodeled each year, allowing the skeleton to adjust its strength to mechanical stress and to repair any micro damage. Bone remodeling is also necessary for maintaining the metabolic function of the skeleton and calcium homeostasis [15, 16].

However, with natural aging, an imbalance occurs where in bone that is removed during bone desorption is not fully replaced,resulting in age-related bone loss. It could be caused by alteration of bone cell production, and by prolonging osteoclasts lifespan and shortening osteoblast lifespan. Likewise, a reduction in osteoblast function with no change in bone desorption also results in bone loss (as occurs in glucocorticoid induced osteoporosis) [17].

Osteoporosis

Osteoporosis is a major health problem worldwide. It is defined by World Health Organization (WHO) as a systemic disease characterized by a decreased bone mineral density (BMD) and micro architectural deterioration of bone tissue, leading to enhanced bone fragility and consequent increase in fracture risk.

The loss of bone mass and strength can be contributed by: 1) failure to reach an optimal peak bone mass as a young adult; 2) excessive desorption of bone after peak bone mass; 3) an impaired bone formation response during remodeling [3].

Furthermore, osteoporosis may be: 1) Primary (postmenopausal/senile). With advancing age, BMD decreases and prevalence of osteoporosis increases. This problem is accelerated in women by the menopause. The aging process is associated with bone loss even before the menopause sets in, suggesting that factors related to cellular aging, apart from estrogen deficiency, are important [18]. 2) Secondary, caused by nutrition, endocrine factors, drug, malignancy, chronic disease or causes idiopathic.

Fracture risk increases exponentially with age and the number of patients with osteoporosis is and growing due to the general increase in population age [19]. According to recent epidemiologic and demographic data from the Western world, it can be assumed that about 46% of women and 22% of men will incur an osteoporotic fracture beyond 50 years of age [20].

Vertebral and hip fractures result in dramatic restraints in life quality and are associated with an increased rate of morbidity and mortality; also, bone fractures supposed enormous healthcare costs. Therefore, the need to define a diagnosis for osteoporosis, a condition that increases the risk of fracture, has gradually emerged [21]. In suspect bone loss, the main pillars of the diagnostic exploration include detailed anamnesis and risk evaluation, clinical investigations, conventional X-ray of thoracic and lumbar spine, measurement of BMD and laboratory investigations.

Measurements of BMD in the lumbar spine, femoral neck and forearm using dual-energy X-ray absorptiometry (DXA) play a crucial role in diagnosis of osteoporosis [22]. T-scores are calculated by taking the difference between a patient's measured BMD and the mean BMD in healthy young adults, matched for gender and ethnic group, and expressing the difference relative to the young adult population standard deviation (SD). The WHO defines osteoporosis in postmenopausal white women as a BMD T-score at the spine, hip or forearm of ≤2.5 or less [23, 24].

Biochemical markers of bone turnover (BTM) have been developed to provide a non-invasive assessment of bone status. In contrast to DXA, rather static measures, molecular markers of bone metabolism are helpful tools to detect the dynamics of the metabolic imbalance itself [19, 25]. Furthermore, BTM indicate changes in bone metabolism early, before alterations reflected by BMD. Nevertheless, Gold-standard is BMD measurement by DXA.

BONE TURNOVER MARKERS

In metabolic bone diseases such as osteoporosis, bone metabolism is altered, leading to bone loss often accompanied by changes in the microarchitecture, leading to bone fragility. Biochemical markers reflecting enzymatic activities of osteoblasts and osteoclasts or breakdown products of bone tissue have been high valuated to investigate the pathways of bone metabolism and their alterations in bone diseases.

In recent years BTM have been identified and categorized as either biomarkers of bone formation, resorption or osteoclastogenesis. Bone formation markers are produced by osteoblastic cells or derived from collagen metabolism, whereas bone desorption markers are the degradation products of osteoclasts or collagen degradation. In addition, several regulators of bone cells activity may also be used as biomarkers [26]. Rapid, reliable, non-invasive and cost effective assays have been developed with high sensitivity and specificity. BTM offer the potential for screening bone turnover conditions, as well as for monitoring early response to antiresorptive therapies and bone-forming agents [27].

Ideally, BTM should reflect only the activity of osteoblasts or osteoclasts; nevertheless bone formation and bone desorption are usually tightly coupled in time and space and therefore, any marker reflects the overall rate of bone turnover. Certain BTM may reflect different stages of formation and desorption but they cannot reflect disease-specific processes [28]. BTM levels can be measured in serum or urine reflecting the rate of bone remodeling.

Bone Formation Markers

Bone formation markers are products of active osteoblasts expressed during different phases of osteoblast development, function and bone formation. All of them are measured in serum or plasma (Table **1**).

Table 1. Biochemical markers of bone turnover: Formation markers.

Bone Formation markers		
Collagen synthesis by-products	**Source**	**Remarks**
▪ Total N-terminal propeptide of type I procollagen (PINP)	Bone, skin	Precursor molecules of collagen type I synthesised by osteoblasts.
Other unrelated proteins collagen synthesis		
▪ Bone-specific alkaline phosphatase (BALP) ▪ Osteocalcin (OC)	Specific of bone Bone, platelets	Specific product of osteoblasts. Located on the outer cell surface of osteoblasts. Produced by osteoblasts during bone formation. Major non-collagen bone protein.

Bone-Specific Alkaline Phosphatase

Alkaline phosphatases (ALP) belong to a family of ubiquitous, membrane-associated enzymes with a molecular weight of about 140 kDa. There are four isoenzymes of ALP: intestinal, placental, germ cells and a non-specific form expressed in numerous tissues, including bone, kidney and the liver. Three different genes encode the intestine, placental, and germ-line enzymes [29] and tissue-nonspecific ALP are produced by the same gene but there are tissue-specific differences in their posttranslational modification of carbohydrate chain (glycosylation and sialylation) [29, 30]. The bone isoform predominates during childhood and adolescence, whereas bone and liver isoenzymes are the most abundant (95%) in healthy adults and are present in almost equal proportions.

Bone-specific alkaline phosphatase (BALP), a tetrameric glycoprotein specifically produced by osteoblasts is anchored in the plasma membrane of this and is released as a dimer into circulation by phospholipase cleavage during bone formation. It is involved in the mineralization process and its production is correlated positively with bone formation rate [31]. The half-life of BALP is 1 or 2 days, making it less sensitive to circadian variation. In some bone diseases such as Paget's disease or osteomalacia, higher levels of the isoenzyme are observed, as well as after bone fractures or ectopic bone formation. BALP´s values are not influenced by renal dysfunction [29].

Initially the methods developed for the quantification of BALP were not specific for the bone isoenzyme and therefore other isoenzymes such as liver isoenzyme could interfere in the total quantification. Over the years, new methods have been appearing to differentiate BALP from the other isoforms; currently, different enzyme-linked immunosorbent assay (ELISA) are used to measure BALP enzyme activity [32]. Although reproducible and precise, these immunoassays still retain some cross reactivity with the liver isoenzyme. Therefore, BALP results must be carefully interpreted in patients with liver pathology.

Osteocalcin

Osteocalcin (OC) is the most abundant non-collagenous protein of bone. It is a small protein (49 amino acids) expressed by osteoblasts, osteocytes, odontoblasts, and hypertrophic chondrocytes. Its synthesis is dependent upon the presence of active metabolites of vitamin D, especially 1,25-dihydroxyvitamin D, and requires vitamin K for the conversion by carboxylation of three glutamate residues to gamma-carboxyglutamate at positions 17, 21 and 24. These gamma carboxyglutamic acid residues are responsible for the calcium-binding properties of the molecule and it´s critical for the high affinity of osteocalcin to hydroxylapatite [33].

Measurements of serum OC show shigh values in conditions associated with increased bone formation, for example, hyperparathyroidism, hyperthyroidism, Paget's disease, and bone metastases. Thus, OC values are higher in children than in adults, especially during the intensive growth period.Decreased OC values may reflect lower rates of bone formation, as seen, for example, in hypothyroidism, growth hormone deficiency, during estrogen substitution therapy and therapy with biphosphonate and calcitonin. It has been observed that serum levels of OC may decrease substantially in patients with glucocorticosteroids treatment, because these drugs specifically suppress OC synthesis by osteoblasts [34].

The majority of the newly synthesized OC is integrated into the new bone matrix, but 20–30% arrives in circulation where it can be detected. It is metabolized by the liver and cleared by the kidneys. Due to its renal clearance, higher concentrations are observed in renal dysfunction. Serum OC is considered a late specific marker of osteoblast activity and its serum level thus reflect the rate of bone formation [35].

The half-life of OC circulating is short because it is rapidly degraded into fragments.Approximately one-third of serum OC circulating is intact, one-third consists of large (1–43) N-terminal/mid-region fragments, and one-third is small (20–40) mid C-terminal fragments. The smaller OC fragments are thought to be derivatives from the bone matrix during bone desorption [36].

Today, the specific role of the OC is unknown even though this protein has been known for more than 20 years. It seems that OC is involved in the bone remodeling process, more specifically in mineralization in the bone matrix. In vitamin K deficiency a part of OC is imperfectly carboxylated, which increases the risk of hip fractures.Recently an OC knockout mouse model has been developed; these animals exhibited peripheral adiposity, which was unexpected as OC was thought to function exclusively in bone [37]. *In vivo* studies have confirmed a link between bone cells and global metabolism. The hormones insulin and leptin affect postnatal bone formation, whilst OC produced by the osteoblast in turn stimulates insulin secretion by the pancreas [38]. Therefore apparently, there is an endocrine regulation of energy metabolism by the skeleton.

There are different methods to quantify levels of OC: RIA, enzyme-linked immunosorbent assay (ELISA) or chemiluminesence immunoassays. Two-site immunoassays, using antibodies detecting different parts of the OC molecule, have been introduced like immunoassays that detect the intact 1-49 OC molecule. However, only one third of the total OC serum pool represents intact OC, and due to the instability of OC in serum, rapid loss of immunoreactivity is seen with these assays when samples are left for more than 1 hour at room temperature. To circumvent this problem, other assays measure the largest degradation product of OC, the 1-43 (N-terminal/mid-molecule) fragment. This fragment is a result of proteolytic degradation of the intact molecule and may in part be generated by active osteoblasts. In the same way, quick processing of the blood sample is essential for this assays since a loss of reactivity is noted within a few hours at room temperature [39, 40]. Due to sample instability at room temperature, high biological variability and assay variability, use of OC level measurement in clinical practice is limited,despite OC is a sensitive marker of bone formation.

Total N-Terminal Propeptide of Type I Procollagen

The procollagen molecule contains amino- and carboxy-terminal extensions, which are removed by specific proteinases during extracellular processing of newly-synthesized type I collagen. These extension peptides guide the helical folding of the collagen molecule and this cleavage results in the release of procollagentype I N-terminal propeptide (PINP) and procollagen type I C-terminal propeptide (PICP) [41] (Fig. **1**). Both propeptides can be found in the circulation and their concentration reflects the synthesis rate of collagen type I. The analysis of PINP is more extensively described in the literature than that of PICP.

Type I collagen is also a component of several soft tissues such the skin, cartilage and tendons; however the rate of collagen turnover in bone is faster than in other tissues, and therefore, the changes in PINP and PICP are assumed to primarily reflect changes in collagen synthesis in bone [42].

PINP is metabolized by liver endothelial cells. It could be detected in the circulation as the "intact" molecule but it can be present as four different forms:

an intact trimeric form, a dimeric form, a monomeric form and a fragmented form [37]. The intact trimeric form degrades to more stable monomers and fragments because is unstable at body temperature [43, 44].

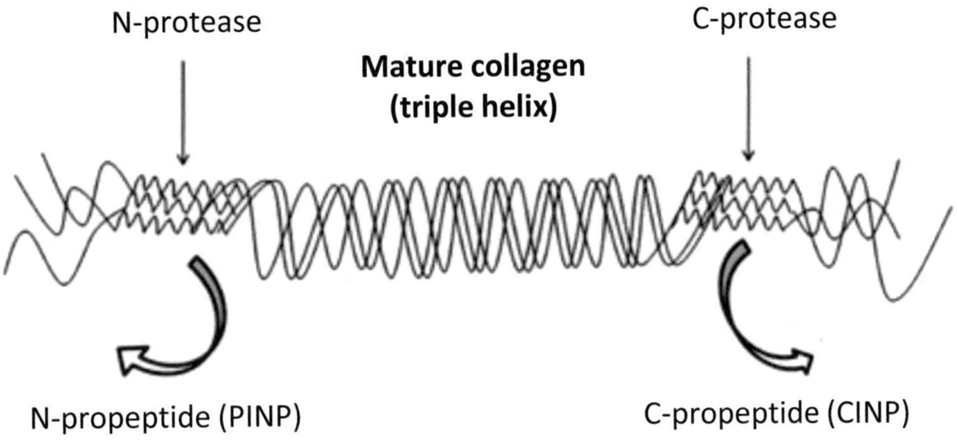

Fig. (1). The catabolism of procollagen extension peptides by proteases during collagen formation leads to PICP and PINP production.

There are two commercially available assays for analysis of serum PINP. The manual competitive radioimmunoassay method (RIA) is specific for the intact trimeric form (intact PINP), whereas the high throughput automated chemiluminesence immunoassays recognizes both the trimeric form and thermal degradation forms of PINP and are thus called 'total' P1NP assays (total P1NP). The automated method is reported to have better analytical reproducibility than the manual RIA [43, 45].

PINP levels are increased during growth and in situations of increased bone formation, such as occur in Paget's disease, and in response to growth hormone. It has been claimed that PINP is considered the most sensitive marker of bone formation rate than other available formation markers, and therefore is particularly useful for monitoring anabolic bone therapy and antiresorptive therapies (such as estrogen and bisphosphonates) [46].

Bone Resorption Markers

The main classic bone desorption markers are degradation products of bone collagen, except tartrate-resistant acid phosphatase (Table **2**).

Table 2. Biochemical markers of bone turnover: Resorption markers.

Bone Resorption markers		
Collagen degradation products	**Source**	**Remarks**
▪ Hydroxyproline (OHP) ▪ Collagen cross-link molecules: Pyridinium crosslinks Pyridinoline (PYD) Deoxypyridinoline (DPD) ▪ Telopeptides of type I collagen: Carboxy-terminal (CTx-; ICTP) Amino-terminal (NTx)	Bone, cartilage, skin Bone, cartilage, tendon,blood vessels Bone,Dentin All tissues containing type I collagen All tissues containing type I collagen	Present in all fibrillar collagens and partly collagenous proteins. Present in newly synthesized and mature collagen. Released when mature collagen breaks and stabilized the molecule Present in mature collagen only. Specific to bone. Collagen type I (highest contribution probably from bone). Isomerization to β aspartyl occurs with ageing of collagen molecule. Collagen type I, with highest contribution from bone. Cleaved from type 1 collagen by cathepsin-K during bone desorption.
Non-collagenous proteins		
▪ Bone sialoprotein (BSP) ▪ Tartrate-resistant acid phosphatase (TRACP-5b) ▪ Cathepsin K	Bone, Dentin, hypertrophic cartilage Bone Blood Osteoclasts	Appears to be associated with osteoclast function. Present in ruffled border of osteoclasts. Reflect osteoclast number. Specific biomarker of osteoclastic activity.

Hydroxyproline

Hydroxyproline (OHP) is a nonessential amino acid derivative formed during post-translational protein modification through hydroxylation of the amino acid proline present in collagen. It was first used in clinical practice in the 1960s as a marker of bone desorption. However, OHP is not specific to bone desorption because it can be found in other tissues such as cartilage and skin and it can also be derived from dietary intake [47].

Ninety percent of the OHP liberated during the degradation of bone collagen is primarily metabolized in the liver and subsequently, it is excreted in the urine where it may be detected. However, ten percent of urinary OHP are derived from the degradation of newly synthesized procollagen peptides. Therefore, urinary hydroxyprolineis considered an unspecific index of collagen turnover. OHP is measured in urine by colorimetric methods or high-performance liquid chromatography (HPLC) [48] although the assay was gradually superseded by more sensitive and specific techniques.

Collagen Cross-Link Molecules: Pyridinoline And Deoxypyridinoline

Collagen fibrils structure is stabilized mechanically by intra- and intermolecular crosslinks. In bone, the predominant crosslinks are Pyridinium crosslinks (pyridinoline (PYD) and deoxypyridinoline (DPD)). Collagen crosslinks are formed during the extracellular maturation of collagen and are released into circulation from the breakdown of mature matrix (Fig. **2**). DPD is considered more specific bone desorption marker than PYD because it is almost solely present in bone and dentine whereas pyridinoline crosslink can be found in other tissues such cartilage, bone or blood vessels [49].

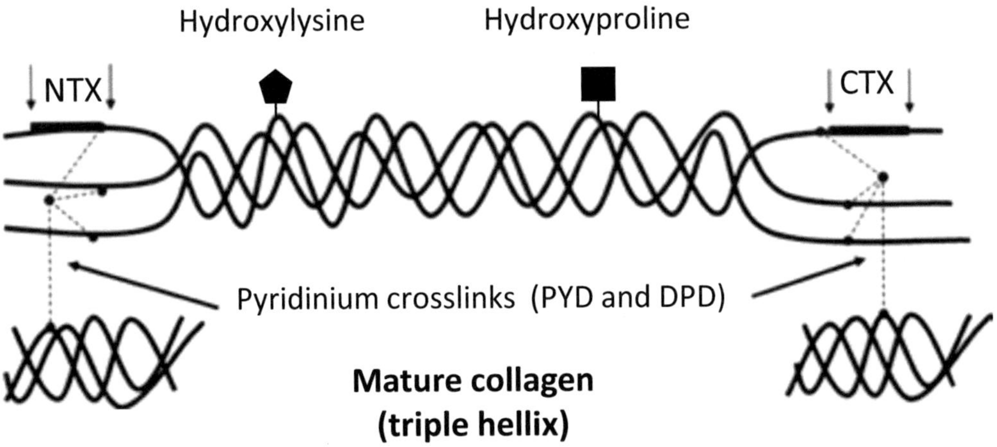

Fig. (2). Mature collagen fibrils with N and C-terminal telopeptides and PYD and DPD crosslinks linking between adjacent fibrils.

Circadian rhythm has been observed in the excretion of PYD and DPD, with a peak excretion early in the morning and their excretion is less influenced by diet. Crosslinks are excreted in urine and it can be measured either as free crosslinks (40%) or conjugated to short collagen peptides (60%) [49].

Assays for the crosslinks were developed twenty years ago and became widely used in the 1990s. A high-performance liquid chromatography (HPLC)-based assay is used to analyze total collagen crosslink concentrations in urine. Free DPD can also be measured by ELISA. Nowadays immunoassays are preferred instead of HPLC [50].

Tartrate-Resistant Acid Phosphatase

Tartrate-resistant acid phosphatase (TRACP) is a member of a heterogeneous group of enzymes secreted from the lysosomes into the resorptive vacuole beneath the ruffled border of the osteoclast. There are two isoforms of TRACP in circulation: TRACP5a and TRACP5b, which are derived by the post-translational modification of a single gene product. TRACP5b is osteoclast-specific and it is formed from two subunits, which are connected by two disulfide bonds [51]. Cathepsin K activates TRACP5b by elimination of a propeptide domain.

TRACP5b is secreted in vesicles by the osteoclast during bone desorption producing reactive oxygen species (ROS) and thus correlates with resorptive behaviour. Therefore TRACP5b has been reported to reflect bone desorption rate but more recent data show that it more accurately reflects the number of osteoclasts rather than the osteoclastic activity [52, 53]. Circulating TRACP5b is hydrolyzed by proteases (about 90% of TRACP is fragmented) and gives rise to small fragments that are metabolized in the liver and excreted in the urine. TRACP5b levels are not affected by the renal function and the effect of food intake is negligible andthus, main indications of TRAP 5b are patients with reduced kidney function, such as dialysis patients. Furthermore, TRACP5b is relatively stable in serum samples [54].

High amounts of TRACP5b are expressed in bone-resorbing osteoclasts states like Paget's disease and bone metastasis. Recently, due to assay evolution, TRACP5b is becoming one of the BTM used for prediction of high bone turnover with a significant correlation with BMD loss [55].

There are enzymatic and immunoassay techniques for the quantitative determination of the active isoform TRACP5B in serum. ELISA assay is a specific method to detect TRACP5b activity freshly liberated from osteoclasts, but not fragments forms. It has been extensively studied and shown to have good specificity, sensitivity, an acceptable intra- and inter-assay coefficient of variation (CV) and good correlations with other desorption markers [56].

There are enzymatic and immunoassay techniques for the quantitative determination of the active isoform TRACP5B in serum. ELISA assay is a specific method to detect TRACP5b activity freshly liberated from osteoclasts, but not fragments forms.

Cross-Linked Telopeptides of Type I Collagen

Theproteolytic degradation of bone matrix forms a variety of type I collagen fragments. This results from the primary structure of protein chains and the characteristic modifications during collagen fibril formation, maturation and ageing of the matrix. Cross-linked telopeptides derived from the enzymatic degradation of the amino-terminal cross-linking telopeptide (NTx) and carboxyterminal collagen cross-linked (CTx) terminal regions of type I collagen. The native α-form of C-Telopeptide undergoes spontaneous and slow β-isomerisation of a peptide-bond which is attributed to protein ageing; analysis of "β-CrossLaps" is specific for elderly tissue and therefore for bone and its proteolytic desorption [57]. CTx is well suited for monitoring the efficacy of antiresorptive drugs including the different bisphosphonates. Currently, β-CrossLaps is perhaps the most frequently used crosslink assay.

These fragments reflect different enzymatic pathways of bone breakdown, by cathepsin K or matrix metalloptoteinase-1 (MMP-1) cleavage of collagen type I, respectively. NTx and CTx are released by cathepsin K cleavage and the carboxyterminal cross-linked telopeptide of type I procollagen (ICTP) is a larger fragment produced by matrix metalloproteinase (MMP-1) (Fig. **2**) [58].

This Cross-linked telopeptides show a circadian rhythm (which is more pronounced for CTx) and the serum levels are influenced by food intake. Therefore the timing of the sample collection is critical: the 24-hour urine collection has the advantage of overcoming circadian changes and is less sensitive to dietary interferences [46, 59, 60]. NTx and CTx also can be measured in serum.

Currently, NTx and CTx are measured using automated ELISA assays, but also there are RIA and electrochemiluminescence assays. Costs and inter-assay variability are similar, and serum samples yield slightly more consistent results than urinary assays [60].

Bone Sialoprotein

Bone sialoprotein (BSP) is a significant component of the bone extracellular matrix and constitute approximately 5-10% of all non-collagenous matrix of bone. This phosphorylated glycoprotein is synthesized by osteoblasts, odontoblasts, osteoclasts and some malignant cells line. Consequently, BSP or its mRNA is detected mainly in mineralized tissue such as bone or dentin. It is thought play an important role in cell-matrix adhesion processes and in the supramolecular organization of the extracellular matrix of mineralized tissues and appears to stimulate osteoclast-mediated bone desorption [39, 61].

RIA or ELISA assays, which are based upon polyclonal antisera, have been designed for the quantitative determination of BSP in serum. Based upon clinical data and the rapid reduction of serum BSP levels following intravenous bisphosphonate treatment, it is assumed that serum BSP reflects processes mainly related to bone desorption.

In the same way, levels of other markers of bone desorption such as PINP or NTX decreased in response to this therapy [39].

Markers of Osteoclastogenesis

Currently we know that osteocytes are one of the principal regulatory cells involved in the response of the bone to mechanical stimulation, modulating the function of osteoclasts and osteoblasts at a molecular level [62]. This control is particularly possible due to a secretion of SOST, an inhibitor of osteoblast activity, and other osteocyte cytokines, stimulators or inhibitors of osteoclast differentiation respectively. A significant role of osteocytes as principal bone mechano-sensitive cells rely on their capability to detect variations in the level of strain and distribute signals leading to adaptive responses. This is possible due to the formation of a functional osteocytes-abundant syncytium and strategic location of these cells within bone [63].

Due to the paramount interest in this field, researches have been encouraged to identify potential BTM (Table **3**). Among these, the following must be mentioned: receptor activator of nuclear-factor kappa-B ligand (RANKL) and osteoprotegerin (OPG), two cytokine of the Tumour Necrosis Factor (TNF) family, and Receptor activator of nuclear-factor kappa-B (RANK), an important member of the Tumour Necrosis Factor Receptor. Proper growth and functioning of osteoclasts is controlled by a pathway in which OPG, RANK and RANKL play the main role [64]. It has become clear that osteoclasts are not simply trench digging cells, but that they have important regulatory functions as immunomodulators in pathologic states and that they may also regulate osteoblast function [65].

Table 3. Biochemical markers of bone turnover: Osteocyte markers and markers of osteoclastogenesis.

Other biomarkers		
Osteocyte markers	**Source**	**Remarks**
▪ Sclerostin (SCL) ▪ Dickkopf-related protein 1 (DKK1) ▪ Dentin matrix protein 1 (DMP1) ▪ Matrix extracellular phodphoglycoprotein (MEPE)	Osteocytes Osteocytes	Inhibitor of wnt signaling and thereby bone formation produced by the osteocytes. Inhibition of Wnt signaling pathway in osteoblasts and decreasing bone formation.
Markers of osteoclastogenesis		
▪ Osteoprotegerin (OPG) ▪ Receptor activator of nuclear factor kappa B (RANK) ▪ Receptor activator of nuclear-factor kappa B Ligand (RANKL)	Osteoblasts Osteoclasts Osteoblasts	Reduces bone desorption by binding to RANK and preventing osteoclastogenesis. Stimulates osteoclastogenesis and bone desorption. Binds to RANK, which is expressed on osteoclasts and their precursors, stimulating their differentiation and activity.

RANK is a type I membrane protein, which is expressed on the surface of osteoclasts. This receptor prods osteoclast precursor cells to develop into fully differentiated osteoclasts when it is activated by its ligand (RANKL) [66]. Therefore RANKL activates osteoclasts and thus it is essential in osteo-clastogenesis and bone desorption [67]. RANKL is one of the key signaling molecules that facilitate cross talk between the osteoblasts and osteoclasts and help coordinate bone remodeling. RANKL also is found on the surface of stromal cells, osteoblasts, and T cells.

OPG is a basic glycoprotein formed from 401 amino-acid residues of the TNF receptor superfamily. It is secreted by osteoblasts and osteogenic stromal stem cells and protects the skeleton from excessive bone desorption by binding to RANKL, *i.e,* OPG inhibit the osteoclasts function [68]. OPG is a RANK homolog that pastes to RANKL on the osteoblast/stromal cells, and acts as a decoy receptor avoiding its interaction with RANK. So, presents the effect of inhibiting the recruitment, proliferation, and activation of osteoclasts through the inactivation of a downstream pathway involving the nuclear translocation of the transcription factor Nuclear Factor-kappa B (NFkB). The RANKL/OPG ratio in bone marrow is an important determinant of bone mass in normal and disease states. OPG levels increase with age; elevated OPG levels has been reported in postmenopausal and elderly women probably as a homeostatic response to limit the bone loss [64, 66].

RANKL/RANK/OPG system regulates the differentiation of precursors into multinucleated osteoclasts as well as osteoclast activation and survival, both normally and in most pathologic conditions associated with increased bone desorption like postmenopausal osteoporosis, Paget's disease, bone loss in metastatic cancers, and rheumatoid arthritis [69]. *Khosla et al.* remarks in his review that the identification of the OPG/RANKL/RANK system as the dominant mediator of osteoclastogenesis represents a major advance in bone biology. The main role played by OPG/RANKL/RANK system in regulating bone metabolism was demonstrated by the findings in RANKL knockout mice which exhibit a reduction of number of osteoclasts along with a phenotype of osteopetrosis (increased bone mass). Also there was demonstrated that overexpression of RANKL is related with many bone diseases such as rheumatoid arthritis and psoriatic arthritis [70]. Also, it was shown that OPG inhibits osteoclast differentiation from precursor cells prevent osteoporosis when administered to ovariectomized rats, and result in osteopetrosis when overexpressed in transgenic mice [68].

The knowledge of RANKL/RANK/OPG system as a pivotal regulator of osteoclast activity helps with the understanding of the bone pathophysiology. It seems that be useful in assessing bone metabolism in pathological bone states and may provide the molecular basis for developing novel strategies targeting bone diseases.

Increased serum levels of OPG and RANKL was observed in postmenopausal and elderly women, probably as a homeostatic response to limit the bone loss [71]. Some studies have demonstrated that OPG and RANKL serum levels are inversely related to BMD and contribute to the development of osteoporosis in postmenopausal women [72, 73]. Nevertheless, *Mezquita-Raya et al.* [74] found that OPG was independently associated with osteoporosis and prevalent vertebral fractures. Therefore, research into the relationship between circulating levels of OPG and RANKL to BMD in postmenopausal osteoporosis are controversial and further studies are needed to confirm the potential clinical value of these novel biochemical markers of bone [75].

Serum levels of RANK-L and OPG are very low and therefore their quantification is complicated, especially RANK-L. It is also unlikely that circulating levels of OPG and RANK-L adequately reflect the bone microenvironment. These limitations may explain the conflicting data on the association of circulating OPG and RANKL with BMD and biochemical markers of bone turnover in postmenopausal women [76]; further commercial assays and identification of the sources of variabilityare needed before quantification of these markers can be applied to clinical practice.

The knowledge of OPG/RANK/RANKL system also has helped to the development of new antiresorptive therapies like Denosumab. Denosumab is a fully human monoclonal IgG antibody with high affinity and specificity for human RANKL approved for the treatment of postmenopausal and osteoporosis. By binding to RANK-L, Denosumab prevents the interaction of RANK-L with RANK and, thus, inhibits bone desorption. Denosumab's specificity for RANKL is far superior to the previously tested recombinant OPG or RANK molecules, resulting in potent and prolonged antiresorptive effect [77, 78]. In the FREEDOM (Fracture Reduction Evaluation of Denosumab in Osteoporosis every 6 Months) study, a clinical trial of postmenopausal women using Denosumab or placebo, there was a suppression of BTM. Denosumab reduced bone desorption by an average of 86% within 1 month, which is a greater reduction seen than with other anti resorptive drugs. The decrease in serum CTx was more pronounced and faster than the decreases in serum PINP and BSP. There was a significant correlation between serum CTx reduction and increased BMD [79].

Osteocyte Markers

In the last years many studies show that osteocytes play a main role in the regulation of bone turnover. They act like bone mechanoreceptors and they can detect changes in bone morphology, particularly micro-fractures [80]. Osteocytes are now known to produce soluble factors that regulate the onset of both bone formation and desorption. Four proteins have been identified as having some role in the mineralization process: sclerostin (SCL), Dickkopf-related protein 1 (DKK1), dentin matrix protein 1 (DMP1) and matrix extracellular phosphoglycoprotein (MEPE) [81]. All these markers can be measured in blood. Also they can regulate bone turnover through direct physical contact with other bone cells.

DKK1 and SCL are secreted by osteocytes and acting as potent inhibitors of the Wnt signaling pathway and are known to have a crucial role in bone remodeling, bone formation and bone growth; expression of both inhibitors is known to decrease bone mass in humans and in animal models [82].

It has been shown that postmenopausal women have higher levels of SCL than premenopausal women resulting in a decrease in bone formation. Therefore, higher serum SCL levels are associated with a greater risk of hip fractures in older women and in addition, the risk of hip fracture is amplified when high SCL levels are combined with lower BMD [83]. Serum SCL levels are regulated by estrogens in women and so estrogen deficiency results in a marked increase in bone turnover and bone loss [84].

Although widely used in research their diagnostic importance remains to be validated due in part to their analytical and biological variability and standardization of SCL assays is necessary before being introduced into general clinical laboratory use [85].

Important Parameters which Regulate Bone Metabolism

Bone carries out some important metabolic functions. The skeleton is also a storehouse for two minerals, calcium and phosphorus, that are essential for the functioning of other body organs, particularly for the nerves and muscle. These bone minerals can be mobilized to maintain systemic mineral homeostasis irrespective of loss of skeletal structural integrity. This metabolic function of bone prevails over its structural function.

The extracellular fluid calcium concentration is tightly controlled by a complex homeostatic mechanism involving fluxes of calcium between the extracellular fluid and the kidney, bone and gut. These fluxes are carefully regulated by a complex system of regulatory hormones and their receptors: parathormone (PTH) and PTH receptor (PTHR) and 1,25(OH)2D and the vitamin D receptor (VDR), as well as serum ionized calcium and the calcium-sensing receptor (CaR) [86].

Blood ionized calcium concentrations are remarkably stable in healthy individuals because of the homeostatic system while allowing the flow of calcium to and from essential stores. CaR is highly expressed in the parathyroid gland, and is activated by extracellular calcium. Activation of CaR regulates calcium homeostasis by suppressing parathyroid hormone (PTH) secretion. Thus, a decrease in serum calcium inactivates the CaR to increase PTH secretion, which acts on the PTHR in kidney to increase tubular calcium reabsorption, and in bone to increase net bone desorption. The increased PTH also stimulates the kidney to increase secretion of 1,25(OH)2D, which activates the VDR in gut to increase calcium absorption, in the parathyroid glands to decrease PTH secretion, and in bone to increase resorption. Thanks to this complex homeostatic mechanism calcium levels are restaured and the negative feedback loop is closed. When serum calcium increase these actions are reversed and integrated hormonal response reduces serum calcium. Therefore, negative feedback mechanisms help to maintain total serum calcium levels in healthy individuals within a narrow physiologic range [87].

Levels of PTH also have important effects in phosphate homeostasis: increased serum PTH increase the loss of calcium in the urine, but more importantly inhibit the loss of phosphate ions.

Phosphate ions will therefore be retained in the plasma where they form insoluble salts with calcium ions, thereby removing them from the ionized calcium pool in the blood. Phosphate ions will therefore be retained in the plasma where they form insoluble salts with calcium ions, thereby removing them from the ionized calcium pool in the blood.

The low levels of PTH also inhibit the formation of calcitriol acting on the enterocytes inhibiting their ability to absorb calcium from the intestinal contents. The low calcitriol levels also act on bone causing the osteoclasts to release less calcium ions into the blood plasma.

SOURCES OF VARIABILITY

Like any other biological markers most BTM exhibit a significant variability. Several factors can impact on the BTM levels, causing variability across samples, which can reduce repeatability and comparability, both within patients and between patients. It can be divided into pre-analytical and analytical sources.

Pre-Analytical Variability

It is one of the limitations affecting the clinical interpretation of bone marker measurement. There are two main sources of pre-analytical variability (Table **4**).

Table 4. Main sources of pre-analytical variability in BTM.

PRE-ANALYTICAL FACTORS	
Technical causes	
▪ Sample requirements (serum/plasma/urine) ▪ Sample collection and transport (on ice/protection from light) ▪ Sample integrity (haemolysis/ thermodegradation) ▪ Storage conditions of samples in lab (freezer temperature)	
Biological causes. Exogenous	
▪ Circadian rhythms ▪ Food, diet ▪ Medication ▪ Others	BTM are highest during night and lowest in afternoon Hydroxyproline increased by collagen-containing foods Ca^{2+} supplements suppress BTM Glucocorticoids increase BTM Antiresorptive drugs decrease bone formation markers Heparin decrease BTM Seasonal variation, exercise...
Biological causes. Endogenous	
▪ Age ▪ Gender ▪ Ethnicity ▪ Pregnancy ▪ Fractures ▪ Immobility ▪ **Diseases**	Children: higher BTM than adults. Adults: age-related increase in BTM until 3rd decade, then stable until menopausal increase in women. Women >aged 50 years have higher BTM than men Increase in BTM, highest in 3rd trimester. Postpartum decrease in BTM BTM increase during fracture healing and remain elevated up to 1 year Inmobility increased bone desorption markers Paget disease, Primary Hyperparathyroidism, Chronic Kidney and Liver disease

Technical Factors

To obtain a reliable test result pre-analytical factors such as the choice of sample (serum/plasma/urine), mode of sample collection (24 hour urine/spot urine), the appropriate preparation of the patient (pre-sampling diet/fasting/exercise) and the correct handling, processing and storage of specimens should always be considered.

As for blood samples, the presence of some anticoagulants in plasma can affect some BTM like OC. Like other chemical analytes, BTM require specific storage precautions, given their sensitivity to thermo degradation and photosensitivity, especially to UV radiation, as well as other ambient influences [88]. OC and TRACP5b are the most difficult to handle and levels can be significantly reduced after only a few hours at room temperature, while collagen peptides and BALP are more resistant to degradation [89]. PYD is light-sensitive and degraded by UV irradiation [90]. Hemolysis is one of the leading causes of unsuitable blood specimens in diagnostic testing. Interference of free hemoglobin and erythrocytes intracellular contents have the potential to jeopardise the accurate measurement of some analytes in serum, plasma or other body fluids resulting in levels lower or higher markers. For example, erythrocytes are known to contain proteases which degrade some BTM like OC [91].

Therefore, sample collection and preparation seem one of the most important factors in successful BTM measurement. This source of variability is controllable and can be minimized with strict standardization of sampling timing and conditions.

Biological Factors

Many biological factors cannot be controlled, such as age, gender, ethnicity, pregnancy or lactation and thus the comparison of the results with reference intervals created for these factors being mandatory. It could be divided into endogenous and exogenous factors.

Highest levels of bone markers are detected during the rapid growth phases such as infancy and puberty when values are 2–10 times higher than in adults [92]. Once the growth is completed, levels of bone markers reaching a plateau around the third decade of life, in both men and women. During menopause, an increase in bone turnover accelerates bone loss due to estrogens deficiency [93, 94].

Calcium requirements are substantially increased during pregnancy and lactation to meet the calcium demands of the fetus and breast milk production. Thus, bone turnover begins to increase from the second trimester of pregnancy resulting in increased levels of both formation (BALP and PINP) and desorption markers (cross-links and telopeptides). Since the placenta produces additional ALP in pregnancy, levels normally increases twice the normal range, especially in late pregnancy. Serum OC concentration is decreased during pregnancy which is restored after the delivery. This may be due to placenta clearance of OC or increased renal clearance during pregnancy. BTM exhibits a significant decrease postpartum with a mean reduction in BMD of 3.5% in 9 months. Bone markers show a slower decline in lactating mothers [95].

But, the most important biologic factor is circadian rhythm in bone forming and bone-resorbing activities, which is usually more pronounced in bone desorption markers. Levels of BTM are highest in the early morning and lowest in the afternoon and evening [96]. The higher circadian variation has been shown for urinary collagen crosslinks and cross-linked telopeptides, and it seems to levels of urinary biomarkers can, accordingly vary up to 30% during the day [97]. The excretion rate of the urine marker may be quantified in a 24 hour urine collection but these kinds of samples are subject to collection errors due to a lack of compliance of patients and reliability of collection. To avoid these inaccuracies markers values must be corrected by urinary creatinine levels because creatinine output has been reported to be fairly constant with time and to correlate with body mass [98]. But mostly an additional blood sample must be demanded for the analyses of formation marker and other laboratory parameters [99].

Other biological factors can influenced in serum BTM levels such as gender, ethnicity, disease, geographical location and medication. Glucocorticoid treatment is the most common cause of drug-associated osteoporosis [100]. Consistent with these outcomes *Dovio et al.* demonstrated that glucocorticoid therapy is associated with decreased levels of bone formation markers (OC, PINP, BALP) and the effect is dose-dependent [101].

Therefore, is very important that clinicians are familiar with situations when BTM can expected to be high such as in children, pregnancy and after recent fracture, because all of these biological factor cannot be controlled.

Analytical Variability

During the past decade considerable progress has been made in the development of fully automated platforms for the study of BTM, improving greatly the analytical variability of these markers. However, can be still large variations inter-laboratory despite using the same method. The recorded inter-laboratory CVs are 39% for NTx, 30% for PYDs and 68–71% for OC [102]. Obviously, several factors contribute to the overall inter-laboratory variation of assay results: availability of multiple immunoassays using different epitopes and reagents, instability of reagents and calibrators and others. Ideally, a laboratory should describe its own reference ranges for each specific group of patients to acommodate age, gender and ethnicity variations and also, reference methods and international standards need to be developed [93].

CLINICAL UTILITY OF BTMS

BTM reflect the rate of bone while DXA scans provide information on BMD; therefore the use of BTM is recommended mainly for monitoring treatment efficacy of osteoporosis as the changes that occur are more rapid and larger when compared to BMD changes but not recommended for the diagnosis of osteoporosis.

Prediction of Bone Loss: Several studies have shown that increased levels of BTM in postmenopausal women are related with loss bone mass and consequently with higher risk of fracture bone. Overall, desorption markers were shown to be more strongly predictive of the rate of bone loss than formation markers [103].

The relationship between rate of bone loss and BTM have been studied in different groups of patients. *Bauer et al.* demonstrated that men >65years of age in the highest quartile of PINP showed greater hip bone loss, but no significant association with others BTM like CTx and TRACP5b was established [25]. Also, longitudinal studies have shown two characteristic groups of untreated postmenopausal women: women with high bone turnover lose significantly more BMD than women with normal or low bone turnover [104]. Women with high bone turnover present elevated concentrations of BTM in their blood and urine compared with low bone turnover women [105]. In prospective studies on postmenopausal women elevated levels of desorption markers at the beginning were associated with a significant loss of bone mass of the lumbar spine after 1 year compared with low initial marker levels [106]. In a long-term study the decrease of BMD was measured in the fore arm. Women with high bone turnover presents 2–6 fold higher loss in bone mass compared with women with low bone turnover [107]. Therefore it seems that bone markers are suitable to identify patients with fast bone loss.

Prediction of Fractures: Large population studies have demonstrated that high bone turnover is associated with increased risk of osteoporotic fractures and that bone markers, as measured by a single or combination of BTM, may be useful in prediction of fractures independent of BMD. There is a greater association between bone desorption markers and the risk of fracture than that for bone formation markers. This statement have been described in three prospective studies namely Rotterdam, EPIDOS and OFELY [103, 104, 108, 109], setting that high levels of desorption markers were associated with increased risk of hip, vertebral and non-vertebral fractures in healthy postmenopausal women [104] and therefore it could predict an elevated risk of osteoporotic bone fracture. For example, 7,598 healthy women aged >75 years were evaluated in EPIDOS study demonstrating that biomarkers of bone desorption such CTx or free DPD were predictive of hip fracture, but no significant association was found between bone formation markers like OC, BALP and urine NTx and bone fracture [104].

Some studies have evaluated the relationship between rates of bone turnover and fracture risk [110, 111]. Various prospective studies of BTM for fracture prediction in untreated men and women were reviewed by *Vasikaran et al.* in 2011 and they concluded that BTM may offer clinical potential to assess fracture risk independent of BMD and which might therefore complement conventional fracture risk assessment [56]. Only a few studies about the association between bone markers and fracture risk in men have been done; one of these is the Dubbo Osteoporosis Epidemiology Study in Australia [112 - 114] which concluded that major bone desorption was associated with increased of bone fractures, independent of BMD.

In conclusion, findings from in the population studies suggest that quantify of BTM may improve fracture risk prediction when used in conjunction with BMD in both postmenopausal women and men [93].

Monitoring of Treatment Efficacy: Measurements of BMD by DEXA is used to diagnose osteoporosis but unfortunately it has several limitations in the monitoring of treatment efficacy because bone changes are slow and only apparent after twenty months, approximately. Nevertheless, BTM reflect the rate of bone turnover, being suitable to monitoring therapeutic response [114]. Nevertheless, during antiresorptive therapy resorption biomarkers decline about 30–70% compared with initial values within 3–6 months, reaching a plateau thereafter [115]. Similarly, the formation markers decrease, but the decrease is less pronounced and it occurs somewhat later. Yearly controls of BTM levels are adequate and an insufficient drop in marker concentrations indicates failure in therapy or flawed compliance of the patient [116, 117].

The primary application of the BTM is monitoring efficacy of therapies intended for treatment of osteoporosis. Bisphosphonates, such as alendronate and zoledronic acid, and the monoclonal antibody denosumab are widely used for the prevention of osteoporotic fracture. Significant decrease of BTM with this treatment has been demonstrated accompanied with decrease the number of osteoclasts [17, 118].

Treatment with alendronate decreases bone desorption around 1 month after treatment initiation. Decreases in bone desorption precede decreases in bone formation that occurs around 3-6 months after treatment initiation. After approximately 6 months of continuous treatment, a steady state of bone turnover was attained, leading to the increase in bone density [119]. Alendronate treatment did not impair bone mineralization, induce the formation of woven bone or have any other adverse effects on bone quality [120]. In treatment with denosumab bone desorption falls rapidly and remains suppressed and bone formation decreases more slowly [121].

Currently, there are available anabolic agents which stimulate osteoblastic activity, such as Teriparatide. Treatment with teriparatide is associated with an increase in bone formation biomarkers and also this correlated with an improvement of bone structures as confirmed by biopsies [122]. This change is accompanied with a delayed, but significant, increase in desorption markers [123, 124]. PINP, show a rapid increase in response to teriparatide, which is noticeable during the first week of therapy. Levels of PINP at baseline and three months post treatment can be used to monitoring patients treated with teriparatide. A change of greater than 10μg /L reflects a good response [125].

Therefore, BTM monitoring may provide information supplemental to BMD measured by DEXA and be a useful aid in managing patients receiving osteoporosis treatment.

A significant response in either BMD or bone turnover can only be assumed, when within a single individual the change in signal is greater than the imprecision of the measurement. Estimating the significance of marker changes by therapy is essential for an assessment of surveillance data. This can by calculated by the "least significant change" (LSC):

LSC= 1.96 X √2 x √[CVa2 + CVi2]

= 2.77 x √[CVi2 + CVa2]

LSC: least significant change

√: square root

CVa: analytic coefficient of variation

CVi: individual

If change of marker level (actual—previous), which is expressed as an absolute percentage from the mean of the actual and previous levels exceeds the least significant change, a significant difference can be assumed. Least significant change is about 20–50% depending on the marker, meaning that changes according to this magnitude prove to have a therapeutic effect. A change >25% should be considered significant when we evaluated formation markers while for most bone desorption markers (serum and urine) LSC is around 60-80% [39].

LIMITATIONS

Despite BTM have shown clinically interesting the current available biomarkers have some limitations. These include: 1) Someanalytes are not bone-specific, as type I collagen is widely distributed in different organs. 2) They reflect quantitative changes but do not provide information on structural abnormalities of bone or remodeling rate of different bone compartments. 3) BTM reflects mainly the function of osteoblast or osteoclast but currently the activity of osteocytes is less known although these cells play a pivotal role in the maintenance of skeletal integrity [23]. Further studies are needed to confirm the potential clinical value of novel biomarkers of osteocytes activity. 4) BTM have an important variable nature. Nowadays, the most relevant sources for variation in the BTMs are the ones that cannot be controlled such as age and gender, menopause, diseases... all of which alter the BTM levels, and, additionally, two easily controllable parameters, circadian rhythm and food intakc [56, 91].

Recent studies in the identification of important pathways in bone physiology have led to the development of potential new biomarkers. These include osteoclastic enzymes, regulators of bone cell activity, non-collagenous matrix proteins, or their fragments, markers of bone matrix properties, and circulating mRNA that could also be of value as early biomarkers.

CONCLUSIONS AND PERSPECTIVES DISCUSION

The ability of bone markers to reflect bone remodeling in physiological and pathological states has been demonstrated, both molecular level and the clinical trials. BTM have the potential to serve as useful tools in diagnostic of osteoporosis, predictive of bone loss and in the monitoring of osteoporosis therapy to identify non-responders patients. Nevertheless their use is greatly hindered by biological and analytical variability and it could be postulated that no perfect BTM (or gold standard) exists. To address this, the International Osteoporosis Foundation (IOF) and International Federation of Clinical Chemistry and Laboratory Medicine (IFCC) formed a joint working group (IOF–IFCC Bone Marker Standards Working Group) to standardize bone markers. The aim of the group was to produce recommendations on the applications of BTM in fracture risk prediction and monitoring of osteoporosis treatment based on current available evidence. The IOF-IFCC Group recommend to choose serum CTx and PINP as the analytes of choice for bone desorption and formation, respectively. Such molecules should be used in future clinical trials and observational studies in osteoporosis, in order to enhance laboratory consistency, enlarge the international experience of the clinical application of BTM to osteoporosis and facilitate their inclusion in routine clinical practice [126]. Serum CTx and PINP had been chosen as the reference standard for bone desorption and bone formation, respectively because: 1) The intra-individual CV is low for both biomarkers: < 10% for CTx and < 9.1% for PINP [127]; 2) Compared to other biomarkers, CTx and PINP are considered to reflect, to an acceptable degree, the metabolic process they represent. CTx is likely to be derived from osteoclast driven bone desorption, although none of the degradation products of type I collagen are specific for bone. It is believed, however, that most PINP is produced during bone formation; 3) From an analytical point of view, they are measurable indifferently in either serum or plasma (EDTA); 4) All currently available automated or manual assays show good precision and a low analytical CV; 5) Unlike other BTM, they promptly respond to antiosteoporotic treatments [128].

Monitoring drug treatment of osteoporosis is an established application of BTM, because biomarkers allow an early judgement of therapeutic success or failure within months compared to late decisions obtained from BMD reports. This has a very important economic aspect. Nevertheless by a reasonable application of BTM in the monitoring of bone therapy, BMD will not lose relevance in the surveillance of bone disease, although its importance can be restricted to controls in larger time intervals, which may also help to save costs [37]. BTM are increasingly becoming good bone turnover biomarkers and their association with BMD measurement can provide a better view of bone status. The within-person variation represents a large part of total variation which limits the utility of the BTMs for monitoring antiosteoporotic treatment in clinical practice. This variability depends both on analytical variability and on various factors of preanalytical variability. Thus, is very important that biological samples be obtained in a standardized way in order to limit preanalytical variability [129].

In the ongoing search for novel ways to improve the diagnosis of bone diseases and the development of treatments for osteoporosis several regulators of bone turnover have been considered as potential BTM. Among such regulators is OPG, a protein that inhibits osteoclastogenesis and decreases bone desorption. Although more studies are needed about this, it's known that levels of the protein are higher in patients with osteoporosis than in controls [72]. Similarly, levels of SCL, which is secreted by osteocytes and inhibits osteoblastogenesis resulting in a decrease in bone formation, are higher in postmenopausal than premenopausal women. Preliminary data on the potential use of SCL in managing osteoporosis are inconclusive and this issue is an area for future development [85, 130].

An important step in the identification of the reference BTM is to standardize the measurement of each marker to obtain comparable values, irrespective of the laboratory in which the measurement is made, as well of the method utilized [131].

In conclusion, BTM have potential for clinical use in many applications related to skeletal disorders. Nevertheless, more specific biomarkers as well as improvement and standardization of measurement techniques are needed to enhance reliability and facilitate the use of BTM in clinical practice. Thus, research efforts should be extended to all BTMs to explore their physiology and usefulness in monitoring metabolic bone disease.

CONFLICT OF INTEREST

The author (editor) declares no conflict of interest, financial or otherwise.

ACKNOWLEDGEMENTS

Declared none.

REFERENCES

[1] Taichman RS. Blood and bone: two tissues whose fates are intertwined to create the hematopoietic stem-cell niche. Blood 2005; 105: 2631-9. [http://dx.doi.org/10.1182/blood-2004-06-2480]

[2] Wojnar R. Bone and cartilage – its structure and physical properties. In: Öchsner A, Ahmed W, Eds. Biomechanics of Hard Tissues: Modeling, Testing, and Materials 10 AD. 1-75. [http://dx.doi.org/10.1002/9783527632732.ch1]

[3] Kini U, Nandeesh BN. Physiology of Bone Formation, Remodeling, and Metabolism; in Springer (ed): Radionuclide and Hybrid Bone Imaging. 2012; pp. 29-57.

[4] Gupta HS, Schratter S, Tesch W, *et al.* Two different correlations between nanoindentation modulus and mineral content in the bone-cartilage interface. J Struct Biol 2005; 149: 138-48. [http://dx.doi.org/10.1016/j.jsb.2004.10.010]

[5] Alford AI, Kozloff KM, Hankenson KD. Extracellular matrix networks in bone remodeling. Int J Biochem Cell Biol 2015; 65: 20-31. [http://dx.doi.org/10.1016/j.biocel.2015.05.008]

[6] Richard Bringhurst F, Marie B. Demay, Stephen M. Krane, Henry M. Kronenberg: Bone and mineral metabolism in health and disease. In: Longo DL, Fauci AS, Kasper DL, Hauser SL, Jameson JL, Loscalzo J, Eds. Harrison's Principles of Internal Medicine, 18e. 2012.

[7] Alliston T. Biological regulation of bone quality. Curr Osteoporos Rep 2014; 12: 366-75. [http://dx.doi.org/10.1007/s11914-014-0213-4]

[8] Bosman FT, Stamenkovic I. Functional structure and composition of the extracellular matrix. J Pathol 2003; 200: 423-8. [http://dx.doi.org/10.1002/path.1437]

[9] Ishikawa Y, Bachinger HP. A molecular ensamble in the rER for procollagen maturation. Biochim Biophys Acta 2013 Oct; 1833(11): 2479-91.

[10] Forlino A, Marini JC. Osteogenesis imperfecta. Lancet 2016 Apr 16; 387(10028): 1657-71.

[11] Malfait F, Symoens S, Goemans N, *et al.* Helical mutations in type I collagen that affect the processing of the amino-propeptide result in an Osteogenesis Imperfecta/Ehlers-Danlos Syndrome overlap syndrome. Orphanet J Rare Dis 2013; 8: 78. [http://dx.doi.org/10.1186/1750-1172-8-78]

[12] Jobling R, D'Souza R, Baker N, *et al.* The collagenopathies: review of clinical phenotypes and molecular correlations. Curr Rheumatol Rep 2014; 16: 394. [http://dx.doi.org/10.1007/s11926-013-0394-3]

[13] Hirvonen MJ, Fagerlund K, Lakkakorpi P, Vaananen HK, Mulari MT. Novel perspectives on the transcytotic route in osteoclasts. Bonekey Rep 2013; 2: 306. [http://dx.doi.org/10.1038/bonekey.2013.40]

[14] Hernandez CJ, Keaveny TM. A biomechanical perspective on bone quality. Bone 2006; 39: 1173-81. [http://dx.doi.org/10.1016/j.bone.2006.06.001]

[15] Seeman E. Osteocytes--martyrs for integrity of bone strength. Osteoporos Int 2006; 17: 1443-8. [http://dx.doi.org/10.1007/s00198-006-0220-0]

[16] Vaananen HK, Laitala-Leinonen T. Osteoclast lineage and function. Arch Biochem Biophys 2008; 473: 132-8.

[http://dx.doi.org/10.1016/j.abb.2008.03.037]

[17] Naylor K, Eastell R. Bone turnover markers: use in osteoporosis. Nat Rev Rheumatol 2012; 8: 379-89. [http://dx.doi.org/10.1038/nrrheum.2012.86]

[18] Khosla S, Melton LJ III, Riggs BL. The unitary model for estrogen deficiency and the pathogenesis of osteoporosis: is a revision needed? J Bone Miner Res 2011; 26: 441-51. [http://dx.doi.org/10.1002/jbmr.262]

[19] Arceo-Mendoza RM, Camacho P. Prediction of fracture risk in patients with osteoporosis: a brief review. Womens Health (Lond Engl) 2015; 11: 477-82. [http://dx.doi.org/10.2217/whe.15.14]

[20] Johnell O, Kanis J. Epidemiology of osteoporotic fractures. Osteoporos Int 2005; 16 (Suppl. 2): S3-7. [http://dx.doi.org/10.1007/s00198-004-1702-6]

[21] Rachner TD, Khosla S, Hofbauer LC. Osteoporosis: now and the future. Lancet 2011; 377: 1276-87. [http://dx.doi.org/10.1016/S0140-6736(10)62349-5]

[22] D'Elia G, Caracchini G, Cavalli L, Innocenti P. Bone fragility and imaging techniques. Clin Cases Miner Bone Metab 2009; 6: 234-46.

[23] Lenora J, Akesson K, Gerdhem P. Effect of precision on longitudinal follow-up of bone mineral density measurements in elderly women and men. J Clin Densitom 2010; 13: 407-12. [http://dx.doi.org/10.1016/j.jocd.2010.04.004]

[24] Bouxsein ML. Determinants of skeletal fragility. Best Pract Res Clin Rheumatol 2005; 19: 897-911. [http://dx.doi.org/10.1016/j.berh.2005.07.004]

[25] Bauer DC, Garnero P, Harrison SL, *et al.* Biochemical markers of bone turnover, hip bone loss, and fracture in older men: the MrOS study. J Bone Miner Res 2009; 24: 2032-8. [http://dx.doi.org/10.1359/jbmr.090526]

[26] Terreni A, Pezzati P. Biochemical markers in the follow-up of the osteoporotic patients. Clin Cases Miner Bone Metab 2012; 9: 80-4.

[27] Brown JP, Albert C, Nassar BA, *et al.* Bone turnover markers in the management of postmenopausal osteoporosis. Clin Biochem 2009; 42: 929-42. [http://dx.doi.org/10.1016/j.clinbiochem.2009.04.001]

[28] Lenora J, Ivaska KK, Gerdhem P. Use of bone turnover markers in osteoporosis. Clin Rev Bone Miner Metab 2010; 8: 1-14. [http://dx.doi.org/10.1007/s12018-009-9042-x]

[29] Eastell R, Hannon RA. Biomarkers of bone health and osteoporosis risk. Proc Nutr Soc 2008; 67: 157-62. [http://dx.doi.org/10.1017/S002966510800699X]

[30] Burch J, Rice S, Yang H, *et al.* Systematic review of the use of bone turnover markers for monitoring the response to osteoporosis treatment: the secondary prevention of fractures, and primary prevention of fractures in high-risk groups. Health Technol Assess 2014; 18: 1-180. [http://dx.doi.org/10.3310/hta18110]

[31] Jung K, Lein M. Bone turnover markers in serum and urine as diagnostic, prognostic and monitoring biomarkers of bone metastasis. Biochim Biophys Acta 2014; 1846: 425-38.

[32] Bauer D, Krege J, Lane N, *et al.* National Bone Health Alliance Bone Turnover Marker Project: current practices and the need for US harmonization, standardization, and common reference ranges. Osteoporos Int 2012; 23: 2425-33. [http://dx.doi.org/10.1007/s00198-012-2049-z]

[33] Iki M, Tamaki J, Fujita Y, *et al.* Serum undercarboxylated osteocalcin levels are inversely associated with glycemic status and insulin resistance in an elderly Japanese male population: Fujiwara-kyo Osteoporosis Risk in Men (FORMEN) Study. Osteoporos Int 2012; 23: 761-70.

[http://dx.doi.org/10.1007/s00198-011-1600-7]

[34] Cepelak I, Cvoriščec D. Biochemical markers of bone remodeling – review. Biochem Med 2009; 19: 17-35.
[http://dx.doi.org/10.11613/BM.2009.003]

[35] Ivaska KK, Kakonen SM, Gerdhem P, Obrant KJ, Pettersson K, Vaananen HK. Urinary osteocalcin as a marker of bone metabolism. Clin Chem 2005; 51: 618-28.
[http://dx.doi.org/10.1373/clinchem.2004.043901]

[36] Ivaska KK, Hentunen TA, Vaaraniemi J, Ylipahkala H, Pettersson K, Vaananen HK. Release of intact and fragmented osteocalcin molecules from bone matrix during bone resorption *in vitro*. J Biol Chem 2004; 279: 18361-9.
[http://dx.doi.org/10.1074/jbc.M314324200]

[37] Bieglmayer C, Dimai HP, Gasser RW, *et al.* Biomarkers of bone turnover in diagnosis and therapy of osteoporosis: a consensus advice from an Austrian working group. Wien Med Wochenschr 2012; 162: 464-77.
[http://dx.doi.org/10.1007/s10354-012-0133-9]

[38] Zhang Q, Riddle RC, Clemens TL. Bone and the regulation of global energy balance. J Intern Med 2015; 277: 681-9.
[http://dx.doi.org/10.1111/joim.12348]

[39] Seibel MJ. Biochemical markers of bone turnover: part I: biochemistry and variability. Clin Biochem Rev 2005; 26: 97-122.

[40] Lee AJ, Hodges S, Eastell R. Measurement of osteocalcin. Ann Clin Biochem 2000; 37(Pt 4): 432-46.
[http://dx.doi.org/10.1177/000456320003700402]

[41] Bernardi D, Zaninotto M, Plebani M. Requirements for improving quality in the measurement of bone markers. Clin Chim Acta 2004; 346: 79-86.
[http://dx.doi.org/10.1016/j.cccn.2004.02.036]

[42] Mahakala A, Thoutreddy S, Kleerekoper M. Prevention and treatment of postmenopausal osteoporosis. Treat Endocrinol 2003; 2: 331-45.
[http://dx.doi.org/10.2165/00024677-200302050-00005]

[43] Samoszuk M, Leuther M, Hoyle N. Role of serum P1NP measurement for monitoring treatment response in osteoporosis. Biomarkers Med 2008; 2: 495-508.
[http://dx.doi.org/10.2217/17520363.2.5.495]

[44] Chen P, Satterwhite JH, Licata AA, *et al.* Early changes in biochemical markers of bone formation predict BMD response to teriparatide in postmenopausal women with osteoporosis. J Bone Miner Res 2005; 20: 962-70.
[http://dx.doi.org/10.1359/JBMR.050105]

[45] Garnero P, Vergnaud P, Hoyle N. Evaluation of a fully automated serum assay for total N-terminal propeptide of type I collagen in postmenopausal osteoporosis. Clin Chem 2008; 54: 188-96.
[http://dx.doi.org/10.1373/clinchem.2007.094953]

[46] Garnero P, Borel O, Delmas PD. Evaluation of a fully automated serum assay for C-terminal cross-linking telopeptide of type I collagen in osteoporosis. Clin Chem 2001; 47: 694-702.

[47] Delmas PD, Eastell R, Garnero P, Seibel MJ, Stepan J. The use of biochemical markers of bone turnover in osteoporosis. Committee of Scientific Advisors of the International Osteoporosis Foundation. Osteoporos Int 2000; 11 (Suppl. 6): S2-S17.
[http://dx.doi.org/10.1007/s001980070002]

[48] Simsek B, Karacaer O, Karaca I. Urine products of bone breakdown as markers of bone resorption and clinical usefulness of urinary hydroxyproline: an overview. Chin Med J (Engl) 2004; 117: 291-5.

[49] Fujimoto D, Suzuki M, Uchiyama A, Miyamoto S, Inoue T. Analysis of pyridinoline, a cross-linking

compound of collagen fibers, in human urine. J Biochem 1983; 94: 1133-6. [http://dx.doi.org/10.1093/oxfordjournals.jbchem.a134457]

[50] Colwell A, Russell RG, Eastell R. Factors affecting the assay of urinary 3-hydroxy pyridinium crosslinks of collagen as markers of bone resorption. Eur J Clin Invest 1993; 23: 341-9. [http://dx.doi.org/10.1111/j.1365-2362.1993.tb02034.x]

[51] Halleen JM, Tiitinen SL, Ylipahkala H, Fagerlund KM, Vaananen HK. Tartrate-resistant acid phosphatase 5b (TRACP 5b) as a marker of bone resorption. Clin Lab 2006; 52: 499-509.

[52] Nenonen A, Cheng S, Ivaska KK, *et al.* Serum TRACP 5b is a useful marker for monitoring alendronate treatment: comparison with other markers of bone turnover. J Bone Miner Res 2005; 20: 1804-12. [http://dx.doi.org/10.1359/JBMR.050403]

[53] Halleen JM, Alatalo SL, Suominen H, Cheng S, Janckila AJ, Vaananen HK. Tartrate-resistant acid phosphatase 5b: a novel serum marker of bone resorption. J Bone Miner Res 2000; 15: 1337-45. [http://dx.doi.org/10.1359/jbmr.2000.15.7.1337]

[54] Rissanen JP, Suominen MI, Peng Z, Halleen JM. Secreted tartrate-resistant acid phosphatase 5b is a Marker of osteoclast number in human osteoclast cultures and the rat ovariectomy model. Calcif Tissue Int 2008; 82: 108-15. [http://dx.doi.org/10.1007/s00223-007-9091-4]

[55] Ivaska KK, Lenora J, Gerdhem P, Akesson K, Vaananen HK, Obrant KJ. Serial assessment of serum bone metabolism markers identifies women with the highest rate of bone loss and osteoporosis risk. J Clin Endocrinol Metab 2008; 93: 2622-32. [http://dx.doi.org/10.1210/jc.2007-1508]

[56] Vasikaran S, Eastell R, Bruyere O, *et al.* Markers of bone turnover for the prediction of fracture risk and monitoring of osteoporosis treatment: a need for international reference standards. Osteoporos Int 2011; 22: 391-420. [http://dx.doi.org/10.1007/s00198-010-1501-1]

[57] Garnero P, Cloos P, Sornay-Rendu E, Qvist P, Delmas PD. Type I collagen racemization and isomerization and the risk of fracture in postmenopausal women: the OFELY prospective study. J Bone Miner Res 2002; 17: 826-33. [http://dx.doi.org/10.1359/jbmr.2002.17.5.826]

[58] Garnero P, Bauer DC, Mareau E, *et al.* Effects of PTH and alendronate on type I collagen isomerization in postmenopausal women with osteoporosis: the PaTH study. J Bone Miner Res 2008; 23: 1442-8. [http://dx.doi.org/10.1359/jbmr.080413]

[59] Delmas PD, Adami S, Strugala C, *et al.* Intravenous ibandronate injections in postmenopausal women with osteoporosis: one-year results from the dosing intravenous administration study. Arthritis Rheum 2006; 54: 1838-46. [http://dx.doi.org/10.1002/art.21918]

[60] Wheater G, Elshahaly M, Tuck SP, Datta HK, van Laar JM. The clinical utility of bone marker measurements in osteoporosis. J Transl Med 2013; 11: 201. [http://dx.doi.org/10.1186/1479-5876-11-201]

[61] Stork S, Stork C, Angerer P, *et al.* von SC, Rambeck W: Bone sialoprotein is a specific biochemical marker of bone metabolism in postmenopausal women: a randomized 1-year study. Osteoporos Int 2000; 11: 790-6. [http://dx.doi.org/10.1007/s001980070058]

[62] Kogianni G, Noble BS. The biology of osteocytes. Curr Osteoporos Rep 2007; 5: 81-6. [http://dx.doi.org/10.1007/s11914-007-0007-z]

[63] Niedzwiedzki T, Filipowska J. Bone remodeling in the context of cellular and systemic regulation: the

role of osteocytes and the nervous system. J Mol Endocrinol 2015; 55: R23-36.
[http://dx.doi.org/10.1530/JME-15-0067]

[64] Kearns AE, Khosla S, Kostenuik PJ. Receptor activator of nuclear factor kappaB ligand and osteoprotegerin regulation of bone remodeling in health and disease. Endocr Rev 2008; 29: 155-92.
[http://dx.doi.org/10.1210/er.2007-0014]

[65] Boyce BF, Yao Z, Zhang Q, *et al.* New roles for osteoclasts in bone. Ann N Y Acad Sci 2007; 1116: 245-54.
[http://dx.doi.org/10.1196/annals.1402.084]

[66] Nakashima T, Takayanagi H. New regulation mechanisms of osteoclast differentiation. Ann N Y Acad Sci 2011; 1240: E13-8.
[http://dx.doi.org/10.1111/j.1749-6632.2011.06373.x]

[67] Yan T, Riggs BL, Boyle WJ, Khosla S. Regulation of osteoclastogenesis and RANK expression by TGF-beta1. J Cell Biochem 2001; 83: 320-5.
[http://dx.doi.org/10.1002/jcb.1200]

[68] Malyankar UM, Scatena M, Suchland KL, Yun TJ, Clark EA, Giachelli CM. Osteoprotegerin is an alpha vbeta 3-induced, NF-kappa B-dependent survival factor for endothelial cells. J Biol Chem 2000; 275: 20959-62.
[http://dx.doi.org/10.1074/jbc.C000290200]

[69] Boyce BF, Xing L. The RANKL/RANK/OPG pathway. Curr Osteoporos Rep 2007; 5: 98-104.
[http://dx.doi.org/10.1007/s11914-007-0024-y]

[70] Khosla S. Minireview: the OPG/RANKL/RANK system. Endocrinology 2001; 142: 5050-5.
[http://dx.doi.org/10.1210/endo.142.12.8536]

[71] Khosla S, Arrighi HM, Melton LJ III, *et al.* Correlates of osteoprotegerin levels in women and men. Osteoporos Int 2002; 13: 394-9.
[http://dx.doi.org/10.1007/s001980200045]

[72] Jabbar S, Drury J, Fordham JN, Datta HK, Francis RM, Tuck SP. Osteoprotegerin, RANKL and bone turnover in postmenopausal osteoporosis. J Clin Pathol 2011; 64: 354-7.
[http://dx.doi.org/10.1136/jcp.2010.086595]

[73] Eghbali-Fatourechi G, Khosla S, Sanyal A, Boyle WJ, Lacey DL, Riggs BL. Role of RANK ligand in mediating increased bone resorption in early postmenopausal women. J Clin Invest 2003; 111: 1221-30.
[http://dx.doi.org/10.1172/JCI200317215]

[74] Mezquita-Raya P, de la HM, Garcia DF, *et al.* The contribution of serum osteoprotegerin to bone mass and vertebral fractures in postmenopausal women. Osteoporos Int 2005; 16: 1368-74.
[http://dx.doi.org/10.1007/s00198-005-1844-1]

[75] Liu JM, Zhao HY, Ning G, *et al.* Relationships between the changes of serum levels of OPG and RANKL with age, menopause, bone biochemical markers and bone mineral density in Chinese women aged 20-75. Calcif Tissue Int 2005; 76: 1-6.
[http://dx.doi.org/10.1007/s00223-004-0007-2]

[76] Chapurlat RD, Confavreux CB. Novel biological markers of bone: from bone metabolism to bone physiology. Rheumatology (Oxford) 2016 Oct; 55(10): 1714-25.
[http://dx.doi.org/10.1093/rheumatology/kev410]

[77] Bekker PJ, Holloway DL, Rasmussen AS, *et al.* A single-dose placebo-controlled study of AMG 162, a fully human monoclonal antibody to RANKL, in postmenopausal women. J Bone Miner Res 2004; 19: 1059-66.
[http://dx.doi.org/10.1359/JBMR.040305]

[78] Anastasilakis AD, Toulis KA, Polyzos SA, Anastasilakis CD, Makras P. Long-term treatment of osteoporosis: safety and efficacy appraisal of denosumab. Ther Clin Risk Manag 2012; 8: 295-306.

[http://dx.doi.org/10.2147/TCRM.S24239]

[79] Cummings SR, San MJ, McClung MR, *et al.* Denosumab for prevention of fractures in postmenopausal women with osteoporosis. N Engl J Med 2009; 361: 756-65. [http://dx.doi.org/10.1056/NEJMoa0809493]

[80] Bonewald LF. Osteocytes: a proposed multifunctional bone cell. J Musculoskelet Neuronal Interact 2002; 2: 239-41.

[81] Noble BS. The osteocyte lineage. Arch Biochem Biophys 2008; 473: 106-11. [http://dx.doi.org/10.1016/j.abb.2008.04.009]

[82] You L, Temiyasathit S, Lee P, *et al.* Osteocytes as mechanosensors in the inhibition of bone resorption due to mechanical loading. Bone 2008; 42: 172-9. [http://dx.doi.org/10.1016/j.bone.2007.09.047]

[83] Li X, Zhang Y, Kang H, *et al.* Sclerostin binds to LRP5/6 and antagonizes canonical Wnt signaling. J Biol Chem 2005; 280: 19883-7. [http://dx.doi.org/10.1074/jbc.M413274200]

[84] Arasu A, Cawthon PM, Lui LY, *et al.* Serum sclerostin and risk of hip fracture in older Caucasian women. J Clin Endocrinol Metab 2012; 97: 2027-32. [http://dx.doi.org/10.1210/jc.2011-3419]

[85] Ardawi MS, Al-Kadi HA, Rouzi AA, Qari MH. Determinants of serum sclerostin in healthy pre- and postmenopausal women. J Bone Miner Res 2011; 26: 2812-22. [http://dx.doi.org/10.1002/jbmr.479]

[86] Brown EM. The calcium-sensing receptor: physiology, pathophysiology and CaR-based therapeutics. Subcell Biochem 2007; 45: 139-67. [http://dx.doi.org/10.1007/978-1-4020-6191-2_6]

[87] Peacock M. Calcium metabolism in health and disease. Clin J Am Soc Nephrol 2010; 5 (Suppl. 1): S23-30. [http://dx.doi.org/10.2215/CJN.05910809]

[88] Lomeo A, Bolner A. Stability of several biochemical markers of bone metabolism. Clin Chem 2000; 46: 1200-2.

[89] Seibel MJ. Molecular markers of bone turnover: biochemical, technical and analytical aspects. Osteoporos Int 2000; 11 (Suppl. 6): S18-29. [http://dx.doi.org/10.1007/s001980070003]

[90] Meddah B, Kamel S, Giroud C, Brazier M. Effects of ultraviolet light on free and peptide-bound pyridinoline and deoxypyridinoline cross-links. Protective effect of acid pH against photolytic degradation. J Photochem Photobiol B 2000; 54: 168-74. [http://dx.doi.org/10.1016/S1011-1344(00)00012-9]

[91] Hannon R, Eastell R. Preanalytical variability of biochemical markers of bone turnover. Osteoporos Int 2000; 11 (Suppl. 6): S30-44. [http://dx.doi.org/10.1007/s001980070004]

[92] Sato J, Hasegawa K, Tanaka H, Morishima T. Urinary N-telopeptides of type I collagen in healthy children. Pediatr Int 2010; 52: 398-401. [http://dx.doi.org/10.1111/j.1442-200X.2010.03086.x]

[93] Hlaing TT, Compston JE. Biochemical markers of bone turnover - uses and limitations. Ann Clin Biochem 2014; 51: 189-202. [http://dx.doi.org/10.1177/0004563213515190]

[94] Ho AY, Kung AW. Determinants of peak bone mineral density and bone area in young women. J Bone Miner Metab 2005; 23: 470-5. [http://dx.doi.org/10.1007/s00774-005-0630-7]

[95] Naylor KE, Iqbal P, Fledelius C, Fraser RB, Eastell R. The effect of pregnancy on bone density and bone turnover. J Bone Miner Res 2000; 15: 129-37. [http://dx.doi.org/10.1359/jbmr.2000.15.1.129]

[96] Qvist P, Christgau S, Pedersen BJ, Schlemmer A, Christiansen C. Circadian variation in the serum concentration of C-terminal telopeptide of type I collagen (serum CTx): effects of gender, age, menopausal status, posture, daylight, serum cortisol, and fasting. Bone 2002; 31: 57-61. [http://dx.doi.org/10.1016/S8756-3282(02)00791-3]

[97] Ju HS, Leung S, Brown B, *et al.* Comparison of analytical performance and biological variability of three bone resorption assays. Clin Chem 1997; 43: 1570-6.

[98] Heymsfield SB, Arteaga C, McManus C, Smith J, Moffitt S. Measurement of muscle mass in humans: validity of the 24-hour urinary creatinine method. Am J Clin Nutr 1983; 37: 478-94.

[99] Huber F, Traber L, Roth HJ, Heckel V, Schmidt-Gayk H. Markers of bone resorption--measurement in serum, plasma or urine? Clin Lab 2003; 49: 203-7.

[100] van Staa TP, Laan RF, Barton IP, Cohen S, Reid DM, Cooper C. Bone density threshold and other predictors of vertebral fracture in patients receiving oral glucocorticoid therapy. Arthritis Rheum 2003; 48: 3224-9. [http://dx.doi.org/10.1002/art.11283]

[101] Dovio A, Perazzolo L, Osella G, *et al.* Immediate fall of bone formation and transient increase of bone resorption in the course of high-dose, short-term glucocorticoid therapy in young patients with multiple sclerosis. J Clin Endocrinol Metab 2004; 89: 4923-8. [http://dx.doi.org/10.1210/jc.2004-0164]

[102] Seibel MJ, Lang M, Geilenkeuser WJ. Interlaboratory variation of biochemical markers of bone turnover. Clin Chem 2001; 47: 1443-50.

[103] Garnero P, Sornay-Rendu E, Claustrat B, Delmas PD. Biochemical markers of bone turnover, endogenous hormones and the risk of fractures in postmenopausal women: the OFELY study. J Bone Miner Res 2000; 15: 1526-36. [http://dx.doi.org/10.1359/jbmr.2000.15.8.1526]

[104] Garnero P, Hausherr E, Chapuy MC, *et al.* Markers of bone resorption predict hip fracture in elderly women: the EPIDOS Prospective Study. J Bone Miner Res 1996; 11: 1531-8. [http://dx.doi.org/10.1002/jbmr.5650111021]

[105] Srivastava AK, Vliet EL, Lewiecki EM, *et al.* Clinical use of serum and urine bone markers in the management of osteoporosis. Curr Med Res Opin 2005; 21: 1015-26. [http://dx.doi.org/10.1185/030079905X49635]

[106] Chesnut CH III, Bell NH, Clark GS, *et al.* Hormone replacement therapy in postmenopausal women: urinary N-telopeptide of type I collagen monitors therapeutic effect and predicts response of bone mineral density. Am J Med 1997; 102: 29-37. [http://dx.doi.org/10.1016/S0002-9343(96)00387-7]

[107] Garnero P, Sornay-Rendu E, Duboeuf F, Delmas PD. Markers of bone turnover predict postmenopausal forearm bone loss over 4 years: the OFELY study. J Bone Miner Res 1999; 14: 1614-21. [http://dx.doi.org/10.1359/jbmr.1999.14.9.1614]

[108] Sornay-Rendu E, Munoz F, Garnero P, Duboeuf F, Delmas PD. Identification of osteopenic women at high risk of fracture: the OFELY study. J Bone Miner Res 2005; 20: 1813-9. [http://dx.doi.org/10.1359/JBMR.050609]

[109] Van Daele PL, Seibel MJ, Burger H, *et al.* Case-control analysis of bone resorption markers, disability, and hip fracture risk: the Rotterdam study. BMJ 1996; 312: 482-3. [http://dx.doi.org/10.1136/bmj.312.7029.482]

[110] Gerdhem P, Ivaska KK, Alatalo SL, *et al.* Biochemical markers of bone metabolism and prediction of fracture in elderly women. J Bone Miner Res 2004; 19: 386-93. [http://dx.doi.org/10.1359/JBMR.0301244]

[111] Ivaska KK, Gerdhem P, Vaananen HK, Akesson K, Obrant KJ. Bone turnover markers and prediction of fracture: a prospective follow-up study of 1040 elderly women for a mean of 9 years. J Bone Miner Res 2010; 25: 393-403. [http://dx.doi.org/10.1359/jbmr.091006]

[112] Stathopoulos IS, Ballas EG, Lampropoulou-Adamidou K, Trovas G. A review on osteoporosis in men. Hormones (Athens) 2014; 13: 441-57.

[113] Meier C, Nguyen TV, Center JR, Seibel MJ, Eisman JA. Bone resorption and osteoporotic fractures in elderly men: the dubbo osteoporosis epidemiology study. J Bone Miner Res 2005; 20: 579-87. [http://dx.doi.org/10.1359/JBMR.041207]

[114] Blank RD, Malone DG, Christian RC, *et al.* Patient variables impact lumbar spine dual energy X-ray absorptiometry precision. Osteoporos Int 2006; 17: 768-74. [http://dx.doi.org/10.1007/s00198-005-0050-5]

[115] Tonino RP, Meunier PJ, Emkey R, *et al.* Skeletal benefits of alendronate: 7-year treatment of postmenopausal osteoporotic women. Phase III Osteoporosis Treatment Study Group. J Clin Endocrinol Metab 2000; 85: 3109-15.

[116] Seibel MJ, Naganathan V, Barton I, Grauer A. Relationship between pretreatment bone resorption and vertebral fracture incidence in postmenopausal osteoporotic women treated with risedronate. J Bone Miner Res 2004; 19: 323-9. [http://dx.doi.org/10.1359/JBMR.0301231]

[117] Hochberg MC, Greenspan S, Wasnich RD, Miller P, Thompson DE, Ross PD. Changes in bone density and turnover explain the reductions in incidence of nonvertebral fractures that occur during treatment with antiresorptive agents. J Clin Endocrinol Metab 2002; 87: 1586-92. [http://dx.doi.org/10.1210/jcem.87.4.8415]

[118] Naylor K, Eastell R. Bone turnover markers: use in osteoporosis. Nat Rev Rheumatol 2012; 8: 379-89. [http://dx.doi.org/10.1038/nrrheum.2012.86]

[119] Garnero P, Ferreras M, Karsdal MA, *et al.* The type I collagen fragments ICTP and CTX reveal distinct enzymatic pathways of bone collagen degradation. J Bone Miner Res 2003; 18: 859-67. [http://dx.doi.org/10.1359/jbmr.2003.18.5.859]

[120] Delmas PD, Munoz F, Black DM, *et al.* Effects of yearly zoledronic acid 5 mg on bone turnover markers and relation of PINP with fracture reduction in postmenopausal women with osteoporosis. J Bone Miner Res 2009; 24: 1544-51. [http://dx.doi.org/10.1359/jbmr.090310]

[121] Eastell R, Christiansen C, Grauer A, *et al.* Effects of denosumab on bone turnover markers in postmenopausal osteoporosis. J Bone Miner Res 2011; 26: 530-7. [http://dx.doi.org/10.1002/jbmr.251]

[122] Dobnig H, Sipos A, Jiang Y, *et al.* Early changes in biochemical markers of bone formation correlate with improvements in bone structure during teriparatide therapy. J Clin Endocrinol Metab 2005; 90: 3970-7. [http://dx.doi.org/10.1210/jc.2003-1703]

[123] Yu EW, Neer RM, Lee H, *et al.* Time-dependent changes in skeletal response to teriparatide: escalating *vs.* constant dose teriparatide (PTH 1-34) in osteoporotic women. Bone 2011; 48: 713-9. [http://dx.doi.org/10.1016/j.bone.2010.11.012]

[124] Finkelstein JS, Wyland JJ, Lee H, Neer RM. Effects of teriparatide, alendronate, or both in women with postmenopausal osteoporosis. J Clin Endocrinol Metab 2010; 95: 1838-45. [http://dx.doi.org/10.1210/jc.2009-1703]

[125] Meier C, Brauchli YB, Jick SS, Kraenzlin ME, Meier CR. Use of depot medroxyprogesterone acetate and fracture risk. J Clin Endocrinol Metab 2010; 95: 4909-16.
[http://dx.doi.org/10.1210/jc.2010-0032]

[126] Vasikaran S, Cooper C, Eastell R, *et al.* International Osteoporosis Foundation and International Federation of Clinical Chemistry and Laboratory Medicine position on bone marker standards in osteoporosis. Clin Chem Lab Med 2011; 49: 1271-4.
[http://dx.doi.org/10.1515/CCLM.2011.602]

[127] Vasikaran SD, Morris HA, Cooper C, Kanis JA. Standardising biochemical assessment of bone turnover in osteoporosis. Clin Biochem 2011; 44: 1033-4.
[http://dx.doi.org/10.1016/j.clinbiochem.2011.07.001]

[128] Lombardi G, Lanteri P, Colombini A, Banfi G. Blood biochemical markers of bone turnover: pre-analytical and technical aspects of sample collection and handling. Clin Chem Lab Med 2012; 50: 771-89.
[http://dx.doi.org/10.1515/cclm-2011-0614]

[129] Szulc P. The role of bone turnover markers in monitoring treatment in postmenopausal osteoporosis. Clin Biochem 2012; 45: 907-19.
[http://dx.doi.org/10.1016/j.clinbiochem.2012.01.022]

[130] Mirza FS, Padhi ID, Raisz LG, Lorenzo JA. Serum sclerostin levels negatively correlate with parathyroid hormone levels and free estrogen index in postmenopausal women. J Clin Endocrinol Metab 2010; 95: 1991-7.
[http://dx.doi.org/10.1210/jc.2009-2283]

[131] Panteghini M. Traceability, reference systems and result comparability. Clin Biochem Rev 2007; 28: 97-104.

CHAPTER 7

Vitamin D: from Bone Metabolism to New Applications

Marta M. Castañeda San Cirilo*

Clinical Analysis Department, Santa Lucia General University Hospital, Cartagena, Spain

Abstract: Vitamin D is a fat-soluble vitamin, which has been classically associated with rickets and osteomalacia. The main function of Vitamin D is the regulation of calcium and phosphorus homeostasis. Plasma levels of 25-hydroxyvitamin D or calcidiol are considered the status indicator of Vitamin D in the human body, although it is biologically inert. 25-hydroxyvitamin D is transformed by 1α-hydroxylase to produce 1,25-dihydroxyvitamin D or calcitriol, the active metabolite of the vitamin. Vitamin D performs its actions through the Vitamin D receptor (VDR) that is present in more than 40 different tissues, as well as the enzyme 1α-hydroxylase, so it is now thought that Vitamin D may play a role in other physiological mechanisms. In the recent years, great interest has been a roused in Vitamin D and its involvement in different pathologies. However, the results of the studies are unclear, due to the involvement of mixed patients of different races, diagnostic criteria, latitude and season blood draw and the confounding variables such as age, BMI or sex not taken into account. The purpose of this chapter is to review the most recent publications of Vitamin D, in relation to new applications such as: diabetes, cancer, pregnancy and cardiovascular disease.

Keywords: 1,25-dihydroxyvitamin D, 25-hydroxyvitamin D, Bone metabolism, Breast cancer, Calcidiol, Calcitriol, Cardiovascular diseases, Colorectal cancer, Deficiency, Diabetes mellitus, Gestational diabetes, Myocardial infarction, Pre-eclampsia, Pregnancy, Pregnancy outcomes, Prostate cancer, VDR, VDR polymorphisms, Vitamin D.

INTRODUCTION

Classically, Vitamin D has been associated with rickets and osteomalacia. The main function of Vitamin D is the regulation of calcium and phosphorus metabolism, through its action in three organs: intestine, kidney and bone.

Vitamin D is a fat-soluble vitamin found in very few foods; only 10% of Vitamin

* **Corresponding author Marta M. Castañeda San Cirilo:** Clinical Analysis Department, Santa Lucia General University Hospital, Cartagena, Spain; Tel/Fax: (+34) 968128600; E-mail: martacastasan@yahoo.es

Juan Antonio Vílchez (Ed.)

D in the body is obtained from food intake. Vitamin can be ingested as Vitamin D2 (ergocalciferol) from vegetable ergosterol and as Vitamin D3 (cholecalciferol) from animal cholesterol. However, the main source of Vitamin D is through skin synthesis by exposure to sunlight, which transforms 7-dehydrocholesterol into previtamin D_3. Both Vitamin D from the diet and previtamin D3 formed in the skin, bind to the "Vitamin D Binding Protein" (VDBP) and are transported to the liver where they are hydroxylated at carbon 25, giving 25-hydroxyvitamin D_3 [25(OH)D] or calcidiol. Plasma levels of 25(OH)D are considered the indicator of Vitamin D status in the human body and, although biologically inert, 25(OH)D is the most stable and most abundant circulating form. Later in the kidney, 25(OH)D is transformed by 1α-hydroxylase into 1,25-dihydroxyvitamin D [1,25$(OH)_2$D] or calcitriol, the active metabolite of Vitamin D. This phase of activation is closely regulated by serum levels of PTH, calcium and phosphorus. When low levels of calcium and phosphorus are detected, PTH is secreted, which stimulates the reabsorption of calcium in the kidneys and promotes its elimination from the skeleton and an increase in the renal production of 1,25$(OH)_2$D.

When 1,25$(OH)_2$D enters cells, it binds to the Vitamin D receptor (VDR). This binding promotes increased RNA transcription of the proteins implicated in the activity of Vitamin D. VDR type receptors and 1α-hydroxylase receptors can be found in numerous tissue types such as hematopoietic cells, epidermal cells, pancreatic cells, neurons, placenta, epithelial small intestine and renal tubular cells, chest and prostate. This explains the many non-calcemic actions that Vitamin D can produce in different tissues and the numerous implications of their deficiency for health.

Several factors may influence the prevalence of Vitamin D deficiency such as low sun exposure coupled with the use of sunscreen, obesity, wearing clothes that cover most of the body, malnutrition, low latitude and winter (Table **1**). Moreover, people with dark skin may have reduced cutaneous synthesis of Vitamin D from 50 to 90% by the action of melanin. It is considered that sun exposure at 15 minutes per day is sufficient for adequate Vitamin D synthesis.

Table 1. Causes of Vitamin D deficiency. Modified from Rendina *et al.* [1].

Increased catabolism	Decreased liver synthesis
Low exposure to UV light	Nutritional deficits
Decreased bioavailability	Reduced skin synthesis
Increased catabolism	Hereditable and acquired disorders
Increased urinary loss	

The definition of Vitamin D deficiency is a topic of discussion. In 2011 the Institute of Medicine (IOM) defined Vitamin D deficiency [2] as a concentration of 25-hydroxyvitamin D <12 ng/mL and insufficiency for concentrations between 12–19 ng/mL. However, the Endocrine Society believes that with this definition [3], between 20 to 100% of the population of the United States, Canada and Europe would be deficient in Vitamin D. For this reason, the Endocrine Society has proposed that levels of 25(OH)D below 50 nmol/L (20 ng/mL) be classified as deficient and above 75 nmol/L (30 ng/mL) as sufficient (Table **2**).

Table 2. Cut-off for 25(OH)D proposed by the Endocrine Society and IOM.

Endocrine Society criteria			**IOM criteria**		
Vitamin D status	**25(OH)D (nmol/L)**	**25(OH)D (ng/mL)**	**Vitamin D status**	**25(OH)D (nmol/L)**	**25(OH)D (ng/mL)**
Deficiency	< 50	< 20	Deficiency	< 30	< 12
Insufficiency	75 - 50	30 - 20	Insufficiency	50 - 30	19 - 12
Sufficiency	> 75	> 30	Sufficiency	> 50	≥ 20

The determination of Vitamin D has also created problems due of its hydrophobic nature, its binding to the VDBP and the existence of two forms, $25(OH)D_2$ and $25(OH)D_3$, obtaining different results depending on the method of determination used. Several analytical methods have been developed for the measurement of 25(OH)D, the most important being enzyme immunoassay (ELISA), chemiluminescence assay, high performance liquid chromatography (HPLC) and liquid chromatography tandem mass spectrometry (LC-MS/MS). Currently, LC-MS/MS is considered the gold standard for 25(OH)D quantification.

VITAMIN D AND DIABETES MELLITUS

There is a strong evidence to suggest that Vitamin D activity is essential for pancreatic β-cell function. However, the biological mechanism by which Vitamin D is related to diabetes mellitus is unknown. Some of the ways proposed to justify the role of Vitamin D in the development of diabetes are:

1. Influence of Vitamin D on the action of beta cells through Vitamin D receptors present in pancreatic cells.
2. Vitamin D can change the sensitivity to insulin by stimulating the expression of the insulin receptor.
3. Vitamin D interferes with systemic inflammation.

Type 1 Diabetes

Different authors have suggested that Vitamin D deficiency may play a role in type 1 diabetes mellitus. Experiments with animals have proven that Vitamin D deficiency can inhibit pancreatic insulin secretion and that 1,25-dihydroxyvitamin D may protect β cells against cytokine-induced death.

Several published studies showed a high incidence of Vitamin D deficiency in children and adolescents with type 1 diabetes mellitus [4, 5]. In a Swiss study, Janner *et al.* [6] measured serum 25(OH)D levels in diabetic children and found that the prevalence of Vitamin D deficiency was 60.5%. Similar results were found in an Australian study by Greer *et al.* [7] performed with children and adolescents with type 1 diabetes, in which the prevalence of 25(OH)D deficiency was 43%. Bener *et al.* [8] showed that the prevalence of Vitamin D deficiency was significantly higher in children with type 1 diabetes mellitus (90.6%) compared to non-diabetic children (85.3%). The study by Svoren *et al.* [9] conducted in North American, found that there was greater Vitamin D deficiency among older diabetic children than among younger and that the deficiency was also greater among those with longer duration of disease. However, other studies found different results and for example, Bierschenk *et al.* [10] did not found significant differences when they measured 25(OH)D levels in type 1 diabetics and healthy subjects.

Others authors suggest that Vitamin D deficiency may predict the risk of developing type 1 diabetes. Munger *et al.* [11] designed a prospective case-control among 310 patients diagnosed with type 1 diabetes mellitus and 613 controls and measured 25(OH)D in samples collected before the diagnosis of the disease. Non-Hispanic whites with Vitamin D levels above 100 nmol/L had a 44% lower risk of developing type 1 diabetes than those with Vitamin D levels < 75 nmol/L. As in other studies, there was no association between 25(OH)D levels and the risk of type 1 diabetes when studied in non-Hispanic blacks and the Hispanic population. Littorin *et al.* [12] measured 25(OH)D in 459 patients with type 1 diabetes at diagnosis and eight years later. At the time of diagnosis, levels of 25(OH)D were lower in type 1 diabetic patients than in control subjects; eight years later, the concentration of 25(OH)D had decreased in diabetics. So they suggested that 25(OH)D may play a role in the development of type 1 diabetes mellitus but no correlation was found between the levels of 25(OH)D and HbA1c. Therefore, it was concluded that the diabetic state is a reason for low concentrations of 25 (OH)D and is not secondary to any hyperglycemic or insulin-resistant state.

In addition, some studies have suggested that Vitamin D supplementation in the first year of life may reduce the risk of developing type 1 diabetes. Hyppönen *et al.* [13] came to this conclusion in their study conducted in a cohort of 10821 children. They related the children who took vitamin supplements during the first year of life with the diagnosis of rickets and type 1 diabetes. Those who took the recommended dose of Vitamin D were associated with a decreased incidence of type 1 diabetes.

The effect of Vitamin D on metabolic control and the relationship with the need for insulin therapy in type 1 diabetics have also been widely discussed. The *et al.* [14] determined in 1426 participants, 25(OH)D and insulin resistance that was calculated using a validated equation. 45% of patients had 25(OH)D insufficiency (<50 nmol/L) but the association between individuals with higher levels of Vitamin D and insulin resistance was not significant.

Vitamin D performs its actions through the Vitamin D receptor (VDR) that is present in more than 40 different tissues including islet cells of the pancreas, this indicates its potential role in the regulation of the numerous metabolic processes. As a result, the Vitamin D receptor has become a candidate in studies of type 1 diabetes complications.

The VDR gene is located on chromosome 12q12-q14. Polymorphisms in the VDR sequence may lead to a change in gene function or its expression form, which could mean an ineffective action of Vitamin D. *FokI* T>C (in introm 2), *BsmI* A>G (exon 8), *ApaI* G>T (in introm 8) and *TaqI* C>T (in exon 9) are the four polymorphisms in the VDR gene that have been investigated more frequently. However, the results of investigations of VDR polymorphisms and their association with type 1 diabetes are discrepant. The association has been reported in countries such as India, Germany, Japan, the Netherlands, Spain and Brazil [15 - 20], but in other populations of Finland, Romania, Norway, the United States and the United Kingdom [21, 22], no association was observed.

The discrepancies between studies could be due to genetic and environmental factors related to the development of type 1 diabetes. It is believed that VDR polymorphisms could be an important factor in personalizing the therapeutic intake of Vitamin D, however, evidence of a relationship between VDR polymorphisms and the risk of developing type 1 diabetes mellitus is currently limited.

Type 2 Diabetes

Low levels of Vitamin D have been found frequently in patients with type 2 diabetes and impaired glucose tolerance compared to people with normal glucose tolerance. Vitamin D deficiency appears to inhibit insulin release and increase insulin resistance among control subjects. However, it is unclear whether low concentrations of 25(OH)D are correlated with an increased risk of developing type 2 diabetes in the general population. The main studies and their characteristics are summarized in Table **3**. The results were adjusted for potential confounding factors such as age, sex, smoking, season of blood collection, physical activity or body mass index (BMI). In some studies, the association persisted after adjustment for risk factors, whereas in other studies, the association disappeared after adjustment [23].

Table 3. Main studies of the association of plasma levels of 25(OH) D with risk of type 2 diabetes. Modified from Afzal *et al.* [24].

			Mean	Mean					
		Women	Age	BMI	White	Adjustement		Population	
Reference	Year	%	Years	Kg/m^2	%	(0-6)[b]	Design	setting	Diagnosis
Thorand [25]	2011	47	52	27,1	ND	6	Case-cohort	General	Validated questionnaire
Husemoen [26]	2012	52	46	26	100	6	Cohort	General	OGGT, fasting glucose, HbA1c, diagnosis
Gonzalez-Molero [27]	2012	57	50	ND	ND	6	Cohort	General	OGTT, HbA1c
Gagnon [28]	2011	55	51	26,6	92	5	Cohort	General	OGGT, fasting glucose, medication treated
Pittas [29]	2010	100	56	27,8	98	6	Nested Case-Control	US female nurses	Validated questionnaire
Afzal [24]	2012	56	56	25,3	100	6	Cohort	General	Self report, medication treated, nonfasting glucose, registry-based
Knekt [30]	2008	54	ND	ND	100	6	Nested Case-Control	General	Medication treated, registry-based

Afzal *et al.* [24] found a connection between low concentrations of 25(OH)D and increased risk of type 2 diabetes in their study of 9841 Danish participants without diabetes at baseline who were followed for 29 years. The same results were obtained by Gagnon *et al.* [28], noting that for each increase of 25 nmol/L 25(OH)D in serum, it was correlated with a reduction in the risk of type 2 diabetes from 22 to 29%. González-Molero *et al.* [27] studied a Spanish population for 4 years and found the frequency of diabetes in individuals with levels of 25(OH)D $\leq$ 18.5 ng/mL was 12.4% *versus* 4.7% in those with levels > 18.5 ng/mL. They also observed that participants with levels greater than 30 ng/mL did not progress to diabetes.

Recent studies have proposed that people with low levels of 25(OH)D may have an increased risk of progression to impaired fasting glucose [31] or impaired glucose tolerance and that high concentrations of 25(OH)D could be related to a lower risk of developing type 2 diabetes in prediabetic subjects [32 - 34]. People with prediabetes are associated with defects in insulin action and insulin secretion. Vitamin D can influence insulin secretion and modulation of insulin action [35, 36], but the exact mechanism that explains the progression from prediabetes to type 2 diabetes is not yet known.

The results obtained in prediabetic individuals have proposed that Vitamin D supplements should be taken into account to prevent type 2 diabetes in patients with prediabetes [37, 38].

VITAMIN D AND CANCER

Cancer is a public health problem with a growing incidence and is expected to be the most lethal disease from 2020. Vitamin D deficiency is usual in cancer patients. People with cancer may have additional risk factors for the deficiency because of reduced exposure to sunlight, low oral intake, increased age, frequent hospitalizations and medications.

There is little information on the clinical significance of the deficiency of 25 (OH)D in cancer. Some studies have explored the correlation between levels of 25(OH)D and the incidence of cancer and others have examined its correlation with the prognosis [39]. Biologically, it is possible that Vitamin D plays a role in the genesis and progression of cancer. Anticancer qualities have been attributed to its metabolite active, 1,25-dihydroxyvitamin D. *In vivo* studies have demonstrated multiple antiproliferative actions of Vitamin D and VDR polymorphisms have been correlated with cancer prognosis [40].

Colorectal Cancer

Colorectal cancer (CCR) is one of the most commonly diagnosed cancers in men and women. Sporadic colorectal neoplasia has been linked to different lifestyle factors, including Vitamin D status.

Several mechanisms have been proposed to explain an anti-cancer benefit of

Vitamin D: 1,25$(OH)_2$D, the active form of Vitamin D, links with the Vitamin D receptor (VDR), then VDR activated binds with the retinoid X receptor alpha (RXRA) [41, 42]. This complex translocates to the nucleus of the cell and connects to the Vitamin D response element to control the transcription of genes involved in the proliferation, differentiation, inflammation, and apoptosis [43]. In addition, Vitamin D has an anti-inflammatory effect, the expression of *PTGS-2 (COX-2)* and decreasing levels of the inflammatory marker C-reactive protein [44, 45].

The first human evidence suggesting a relationship between Vitamin D and CCR was the study by Garland and Garland which detected a strong inverse correlation between cancer mortality and exposure to ultraviolet solar radiation [46]. Since then, several studies have found a correlation between low concentration of 25(OH)D and the increased risk of colorectal adenomas and cancer [47 - 52]. Thus, in a meta-analysis of nine prospective studies, Ma *et al.* found that an elevation of 10 ng/mL in the plasma concentration of 25(OH)D was related with a 26% lower risk of CCR [53]. However, others researchers found no connection between 25(OH)D levels and the risk of CCR or the association disappeared after adjusting for potential confounding factors [54, 55]. Ogino *et al.* [56] justify the various findings of the study because CCR is a heterogeneous disease. Each tumor develops through different routes, therefore the inhibitory capacity of Vitamin D in colorectal carcinogenesis could change according to the characteristics of each colon tumor [57]. Another explanation for the differences found in the results of the studies could be the variation in sex of the study participants, colorectal subsite investigated (distal *versus* proximal), and endpoint of the study (adenoma incident compared with recurrent); and that some reports have incorporated data of these characteristics within the same study population. Fedirko *et al.* [58] studied the correlation between 25(OH)D levels and the characteristics of the incident adenoma, including multiplicity, size location, histology and atypia; but no significant differences were observed in the association between the concentration of 25(OH)D and the specific characteristics of the adenoma. In contrast, Jacobs *et al.* [59] observed a higher risk of recurrent adenomas among those with concentrations greater than 25(OH)D.

Known genetic mutations represent approximately 30-50% of familial CCR [60]; much of the remaining familial aggregation is unexplained. Two studies have found that SNPs were significantly associated with lower circulating 25(OH)D levels in four gene regions that appear to have functional relevance (Table **4**):

DHCR7: with roles in the synthetic route of Vitamin D [61, 62];

CYP2R1: encodes the C-25 hydroxylase that transforms Vitamin D into the active ligand for the Vitamin D receptor;

CYP24A1: encodes the 24-hydroxylase involved in the degradation of 25-hydroxyvitamin D and 1,25-dihydroxyvitamin D [61, 62];

GC: Specific component of VDBP.

Table 4. Vitamin D metabolic pathway. Modified from Theodoratou *et al.* [63].

SNPs	Substrate	Product
DHCR7	7-dehydro-Cholesterol	Previtamin D
CYP2R1	Previtamin D	25(OH)D
CYP27B1	25(OH)D	1,25$(OH)_2$D
CYP24A1	25(OH)D 1,25$(OH)_2$D	24,25$(OH)_2$D

Some studies have evaluated the association between SNPs and the risk of CCR, as the study published by Dong *et al.* [64], which investigated the correlation between genetic markers CYP24A1 and distal colon cancer, but the relationship they found was very mild. In a meta-analysis, the *BsmI* polymorphisms of the VDR gene were related with an increased risk of colon cancer in Asian population [65, 66]. However, studies of additional genes including Vitamin D receptor (VDR), SNPs *FokI*, *ApaI* and *TaqI*, have yielded inconsistent results [67 - 69].

The different results obtained may be due to the fact that these SNPs represent only a small part of the observed variation in circulating 25(OH)D levels and that those alleles linked to low plasma levels of 25(OH) may not affect CRC risk in the same direction [70]. Currently, intervention studies are being performed with Vitamin D supplements to prevent cancer, but the results are confusing [71 - 73].

Breast Cancer

Breast cancer is one of the most common cancers among women worldwide. Risk factors for the development of breast cancer are: family history, use of hormone replacement therapy (HRT), low physical activity, body and abdominal fatness and reproductive factors such as early menarche, late menopause, nulliparity, late age at first birth or never having breast-fed. Several of these factors have also been linked to plasma levels of 25(OH)D, suggesting that there is a relationship between breast cancer and Vitamin D levels.

In vitro studies have shown that 1,25$(OH)_2$D has anticarcinogenic activity in normal and malignant mammary cells [74 - 76]. Vitamin D appears to act directly on the mammary gland, which expresses the Vitamin D receptor (VDR), so it has been proposed that there is local production of 1,25$(OH)_2$D. Some of the pathways that have been suggested to justify the action of Vitamin D in breast cancer include inhibition of cell proliferation, stimulation of cell differentiation, development of apoptosis, reorganization of extracellular matrix and control of estrogen and androgen receptors [45]. Experimental studies indicate that Vitamin D may reduce the risk of breast cancer [77, 78]. Epidemiological and observational studies suggest an inverse correlation between Vitamin D and breast cancer but the results of the investigations are discrepant.

The time from the measurement of the concentration of 25(OH)D and diagnosis of breast cancer varies considerably between the different investigations. Studies with blood samples collected after the diagnosis of breast cancer have established an inverse association between levels of 25 (OH) D and risk of breast cancer [79 - 86], while most studies that did not detect a significant relationship, levels of 25(OH)D had been measured years before diagnosis [87 - 90].

Some researchers believe that changes in way of life and physical activity after the diagnosis of breast cancer may affect 25(OH)D levels, so the associations detected in these studies, would not be representative [91]. However, others think that almost 80% of incident cases of breast cancer are discovered because of self-examination; therefore, these incident cases lack the severe symptoms of a stage of advanced disease, which could produce a drastic change in lifestyle. Another possible explanation is that 25(OH)D levels in the three months prior to diagnosis are physiologically essential for tumor growth and therefore previous concentrations of 25(OH)D would not be associated with the risk of breast cancer [92].

Another cause of the different results obtained may be the population studied. Some studies were conducted in premenopausal women and others in post-menopausal women, including a high proportion of women with osteoporosis and osteopenia. This would explain that in most studies in premenopausal women no correlation was found between 25(OH)D levels and the risk of breast cancer.

It has been proposed that Vitamin D may be related to the prognosis of cancer. Data from animal experiments suggest that Vitamin D deficiency may influence cancer progression by promoting bone metastases [93]. And thus, Tretli *et al.* [94] reported that women with breast cancer and levels of 25(OH)D > 35 ng/mL had a lower risk of breast cancer mortality compared to women with 25(OH)D < 20 ng/mL. Also Goodwin *et al.* [95] found that women with levels of 25(OH)D < 20 ng/mL had a worse prognosis than those with 25(OH)D ≥ 30 ng/mL. However, a study of 500 breast cancer survivors, found no relationship between concentrations of 25(OH)D and cancer prognosis [96].

Recent studies have deduced that there is not enough evidence to recommend Vitamin D supplements for breast cancer prevention [97].

Prostate Cancer

Prostate cancer is the most commonly diagnosed cancer among the male population worldwide but the knowledge of modifiable risk factors is limited. Advanced age, race, family history, obesity, physical activity, smoking, antioxidants, Vitamin D and calcium have been associated with the incidence, recurrence and survival of prostate cancer [98].

The pathway by which Vitamin D may influence prostate cancer is still unclear. *In vitro* and *in vivo* studies have indicated that 1,25$(OH)_2$D affects tumorigenesis of prostate. In prostate epithelial cells, calcitriol has shown antiproliferative properties and pro-differentiation and pro-apoptotic functions [99 - 101]. Furthermore, in mouse models, it has produced a reduction in tumour size and volume and has inhibited invasion, metastasis and angiogenesis [74, 102].

However, published studies have found an inconsistent relationship between Vitamin D and the risk of prostate cancer. Ahonen *et al.* [103] reported that men with concentrations of 25(OH)D above 40 nmol/L had a 41% lower risk of developing prostate cancer. However, in the studies by Albanes *et al.* [104] and Meyer *et al.* [105], the concentrations of 25(OH)D were related to a linear increase in prostate cancer risk. In contrast, many studies have found no correlation between levels of 25(OH)D and the risk of prostate cancer [106 - 109].

Other researchers have studied whether plasma 25(OH)D levels are related to the survival and progression of disease in men diagnosed with prostate cancer. *In vivo* and epidemiological studies indicate that higher levels of 25(OH)D protect against more aggressive types of prostate cancer. The study by Tretli *et al.* [110] showed a lower risk of mortality in men with higher levels of 25(OH)D in samples collected after diagnosis. Similar results were reported by Fang *et al.* [111] and Shui *et al.* [112]. However, Mikhak *et al.* [113] observed that patients with concentrations of 25(OH)D $>$ 37.5nmol/L had a higher risk of high-grade cancer and Albanes *et al.* [104] reported that men with the highest concentration of 25(OH)D, at the time of diagnosis, had a 1.7 times greater risk of aggressive disease.

The different results can be explained by the time in which the blood was collected, if Vitamin D levels were measured after diagnosis or after treatment. Vitamin D levels may have been modified by treatment [114] and this would explain the absence of correlation of Vitamin D with the risk of the disease. However, tumor progression would be affected by the recent deficiency of

Vitamin D, which would explain the constant association with the aggressive disease [115].

VITAMIN D AND PREGNANCY

Maternal Vitamin D deficiency is common during pregnancy with prevalence rates that vary by ethnicity and sun exposure. Among non-modifiable risk factors associated with Vitamin D deficiency are the winter season and darker skin pigmentation.

It has been observed that fetal 25(OH)D levels (umbilical cord blood) are directly proportional to circulating levels in the mother [116]. An adequate level of Vitamin D is very important for fetal bone health, growth and fetal development. Breast milk contains very little Vitamin D, so maternal levels of 25(OH)D during pregnancy are very important for the Vitamin D status of children during the first months of life.

It has been hypothesized that Vitamin D controls the secretion of hormones associated with pregnancy and the expression of genes related to implantation and placental development. In addition, the Vitamin D receptor (VDR) and the 1α-hydroxylase enzyme are expressed in the placenta suggesting a possible action of Vitamin D in the maternal-fetal environment [117]. During pregnancy levels of $1,25(OH)_2D$ increase due to placental production.

This increment is most significant during the second and third trimester, however, low levels of 25(OH)D have been detected in pregnant women worldwide. In the last years, Vitamin D deficiency during pregnancy has been linked to an increased risk of adverse obstetric outcomes, such as insulin resistance, gestational diabetes, preeclampsia, bacterial vaginosis, preterm delivery, cesarean section and fetal growth restriction. The problem with most published studies is that the results have not been adjusted with dietary factors or sun exposure.

Many supplementation studies are currently underway, because many professionals argue that Vitamin D supplementation should be routinely administered to all first-trimester women to prevent obstetric complications.

Gestational Diabetes

The incidence of Gestational Diabetes (GDM) varies from 2 to 20%, depending on the characteristics of populations and the diagnostic criteria used. GDM is characterized by maternal hyperglycemia and glucose intolerance due to increased insulin resistance and impaired insulin secretion. Similarly, Vitamin D deficiency has been linked to insulin resistance and altered insulin secretion in type 2 diabetes mellitus. However, the connection between levels of 25(OH)D and GDM is unclear.

Some researchers have found that 25(OH)D deficiency in the first trimester of pregnancy is associated with a subsequent diagnosis of GDM. Maghbooli *et al.* [118] observed that levels of 25(OH)D < 12.5 nmol/L were associated with GDM and three meta-analyse proposed that the risk of GDM increased by 40-60% in women with 25(OH)D deficiency [119 - 121]. But others studies found no association between the diagnosis of GDM and 25(OH)D deficiency in the first months of pregnancy [122].

Senti *et al.* [123] reported that maternal deficiency and insufficiency of 25(OH)D are related to markers of impaired glucose metabolism. Therefore, Clifton-Bligh *et al.* [124] observed an inverse relationship between 25(OH)D and fasting glucose, however, Perez-Ferre *et al.* [125] found no variation in the incidence of GDM in women with 25(OH)D above and below 50 nmol/L, but women with levels of 25(OH)D < 50 nmol/L had higher fasting glucose, insulin resistance and HbA1c. Some of the causes that explain the different results obtained between studies are the criteria used for the diagnosis of GDM and that the data were not adjusted with sun exposure or dietary factors.

Preeclampsia

Preeclampsia (PE) is a hypertensive disease related to pregnancy. It is detected after the 20th week of gestation, due to the finding of hypertension (140/90 mmHg) and proteinuria (300 mg/24h). The prevalence is around 8% of all pregnancies and is associated with high morbidity and mortality that affects both the mother and fetus.

All published studies agree that women with preeclampsia have a lower concentration of 25(OH)D compared to normotensive pregnant women [126 - 129]. Robinson *et al.* [130] observed that a 10 ng/mL increase in the concentration of 25(OH)D produced a 63% decrease in the risk of severe PE. And Bodnar *et al.* [131] found that maternal levels of 25(OH)D ≥ 50 nmol/L could reduce the risk of developing severe PE by 40% compared to women with levels of 25(OH)D < 30 nmol/L. The relationship between PE and 25(OH)D deficiency has been found in early [129 - 131], middle [130] and late [126, 129] pregnancy.

In addition, data from studies suggest that Vitamin D may protect against PE, because high levels of 25(OH)D have been linked with a lower incidence of PE. However, other authors found no correlation between maternal 25(OH)D levels and the risk of developing PE [132, 133]. The explanation of the different results observed may be the absence of statistical power, high level of Vitamin D in pregnant women studied and groups with different races [134].

Bacterial Vaginosis

Bacterial vaginosis (BV) is an infectious disease in women that affects women of childbearing age. Its importance during pregnancy is that it can cause miscarriage or premature birth.

The National Health and Nutrition Examination Survey (NHANES) informed that black women were twice as many cases of BV compared to white women [135]. One possible explanation for this is that people with dark skin have reduced formation of Vitamin D cutaneous and therefore the antimicrobial activity of endogenous calcitriol is decreased. Thus Bodnar *et al.* [136] observed that pregnant women with levels of 25(OH)D < 20 nmol/L had a 65% higher risk of developing BV in contrast to women with 25(OH)D > 80 nmol/L: subsequent studies have confirmed these results [137, 138]. Hensel *et al.* [139] performed a second analysis of NHANES data and reported that low concentrations of 25(OH)D had a significant correlation with BV only among pregnant women.

Cesarean Section

Vitamin D deficiency is related to decreased strength and muscle performance, which can lead to dystocia and poor progress during childbirth [134].

Again, the results of the studies are confusing, Merewood *et al.* [140] described a strong association between cesarean section and levels of 25(OH)D < 37.5 nmol/L. Two other studies of pregnant women in the first trimester observed an increased risk of cesarean delivery [125, 141]; but other studies found no correlation between maternal levels of Vitamin D and the type of delivery [142, 143].

A cause that explains the different results observed may be the different reasons for the intervention. Some researchers in their studies distinguished between women who required emergency cesarean section due to lack of progress in labor and those with fetal distress, however, in other studies, the two causes were mixed.

Preterm Delivery

Due to known the potential activity of Vitamin D in immunomodulation and inflammatory response, it has been hypothesized about the possible effect of Vitamin D in the prevention of preterm birth.

Many studies have reported that maternal 25(OH)D levels are not associated with preterm delivery [142 - 145]. However, two studies suggested that 25(OH)D deficiency increased the risk of preterm birth when it was used as a definition of preterm birth, one with less than 35 weeks of gestation [146, 147]; while other studies reported no association between Vitamin D deficiency and premature delivery, when it was considered to be before 37 weeks of gestation. Thus, this might suggest that 25(OH)D deficiency is specially related to an increased risk of very premature delivery [148].

Low Birth Weight and Small For Gestational Age

Newborns small for gestational age (SGA) are defined as those with a birth weight for gestational age below the 10th percentile in the standard growth curves and low birth weight is defined as newborn weight of less than 2500 g independently of gestational age.

Based on the effects of Vitamin D on bone metabolism, some authors have proposed that there is an association between Vitamin D deficiency in pregnant women in the first trimester and low birth weight and SGA. There are few published studies on the relationship between maternal 25(OH)D levels and low birth weight or SGA and the results are discordant.

Leffelaar *et al.* [149] found that risk of giving birth to SGA babies was higher among women with Vitamin D < 12 ng/mL. Identical results were reported by Ertl *et al.* [150]. Bodnar *et al.* [151] found that pregnant women with levels of 25(OH)D <37.5 nmol/L had an increased risk of giving birth to SGA babies; this fact was more evident in white women compared to black women. Surprisingly, women with 25(OH)D levels in the insufficiency range had a lower risk of having SGA babies than women with sufficiency; this result was the same for white and black women. But again, other studies found no correlation between levels of 25(OH)D and birth weight or SGA [152, 153]. Morley *et al.* [154] attempted to explain the differences found between the studies and proposed that studies on maternal Vitamin D levels and birth weight should examine neonatal VDR polymorphisms.

VITAMIN D AND CARDIOVASCULAR DISEASE

Cardiovascular disease can take many forms: atherosclerosis, hypertension, coronary artery disease, heart failure and stroke. According to WHO, is responsible for half of all deaths in the United States. Several studies have proposed that Vitamin D deficiency may be associated with the pathogenesis of cardiometabolic disorders and with their clinical severity, however, the mechanisms that justify the relationship between low concentrations of 25(OH)D and increased risk of cardiovascular disease are not yet known.

Some studies have shown that calcitriol, *via* VDR receptors, influences the morphology, proliferation and growth of cardiac cells. Some of the hypotheses that have been proposed to justify the association between Vitamin D and cardiovascular disease control are that $1,25(OH)_2D$ and VDR can adjust the renin-angiotensin-aldosterone system and both can control the secretion of insulin and homeostasis of extracellular calcium. Furthermore, $1,25(OH)_2D$ regulates the expression of factors that induce the differentiation of osteoblasts in the vessel wall, which may influence the development of vascular calcification. $1,25(OH)_2D$ is also a strong modulator of the immune response and cytokine biosynthesis (Table **5**).

In the Third National Health and Nutrition Examination Survey (NHANES III), with 3,408 participants followed for seven years, levels of 25(OH)D at the start of the study were inversely correlated with the risk of death from all causes, with a more potent relationship with cardiovascular mortality [155]. Similar results were found in the Framingham Offspring Study, which included 1,739 subjects followed for five years, noting that patients with levels of 25(OH)D < 15 ng/mL had an increased risk of cardiovascular events compared to those with ≥ 15 ng/mL [156].

Table 5. Pathophysiological mechanisms that associate Vitamin D deficiency and cardiometabolic disorders. Modified from Rendina *et al.* [1].

↑Proinflamatory cytokines
↑RAAS
↑Insulin resistance
↑PTH
↑Vascular calcification

Hypertension

Low levels of 25(OH)D have been linked with a higher incidence of hypertension (HTN) or increased diastolic blood pressure [157]. HTN is related with poor stimulation of the renin-angiotensin-aldosterone (RAA) system. RAA regulates blood pressure and extracellular body volume through the synthesis of angiotensin II and release of aldosterone. Studies have shown that Vitamin D reduces RAA activity by eliminating renin gene expression.

Among participants in the Health Professionals' Follow-Up Study, the risk of developing hypertension in men with Vitamin D deficiency was found to be 6.13, in contrast to those with concentrations ≥ 30 ng/mL [158]. An inverse correlation between 25(OH)D levels and blood pressure was also observed in the Third

National Health and Nutrition Examination Survey (NHANES III) [159]. Hypertensive patients with Vitamin D deficiency also had an increased risk of cardiovascular events, including myocardial infarction, stroke and heart failure [156]. Wang *et al.* found an association between the risk of hypertension, concentrations of 25(OH)D and Vitamin D receptor polymorphisms [160].

Usual exposure to solar radiation has been demonstrated to increase the concentration of 25(OH)D from 26 nmol/L to 100 nmol/L over a six-week intervention period and blood pressure is reduced by about 6 mmHg in hypertensive patients [161]. But not all studies showed a decreased in blood pressure with Vitamin D, and this was the case in a study in patients with resistant hypertension who received intermittent supplements at high doses of Vitamin D every 2 months for 6 months without lowering blood pressure [162].

In conclusion, observational studies showed that low levels of 25(OH)D are related with high blood pressure, but randomized clinical trials were inconclusive [163].

Atherosclerosis

Several studies have demonstrated a relationship between 25(OH)D levels and markers of inflammation and subclinical atherosclerosis. Vitamin D deficiency stimulates systemic vascular inflammation, and allows atherogenesis and cardiovascular events. Vitamin D represses inflammation through various path-ways including inhibition of the prostaglandin and cyclooxygenase, stimulation of anti-inflammatory cytokines, suppression of adhesion molecule expression and decrease of matrix metalloproteinase 9 [164].

Bonakdaran and Varasteh observed a significant relationship between Vitamin D deficiency and elevated levels of ultrasensitive C-reactive protein (hs-CRP) and the existence of microalbuminuria in type 2 diabetic patients [165]. Amer *et al.* [166] also found an inverse association between 25(OH)D and hs-CRP levels in asymptomatic patients. However, the connection was found only among participants with levels of 25(OH) D ≤ 21 ng/mL, whereas among those who had levels of 25 (OH) D > 21 ng/mL, a positive association was observed between the two variables.

Clinical studies have demonstrated that high concentrations of 25(OH)D are related to lower levels of cholesterol, HDL-C, LDL-C and TG, suggesting that elevated levels of 25(OH)D may improve lipid profile and have a positive influence on cardiovascular health [167, 168]. However, other researchers do not support this theory, since in their studies, no association was found between the levels of HDL-C and 25(OH)D [169]. In addition, in clinical studies with Vitamin D supplements, little evidence of improvement in the lipid profile was found [170].

Vitamin D deficiency has also been linked to vascular stiffness, which is considered a marker of subclinical atherosclerosis [171]. Studies in healthy and diabetic type 2 subjects found an inverse association between 25(OH)D levels and the intima-media thickness (IMT) of the internal carotid artery [172]. Yadav *et al.* [173] found similar results in chronic renal patients in stages 4 and 5 but the relationship was not observed in a study of patients with diabetic nephropathy [174].

In the study of Wakasugi *et al.* [175], it was observed that the synthesis of PGI2 increased significantly in the presence of $1,25(OH)_2D$. PGI2 is involved in the reduction of thrombogenicity, cell adhesion and proliferation of smooth muscle cells, so it was concluded that Vitamin D may act as a protector against the development of atherosclerosis due to its action as a vasoactive agent.

Coronary Heart Disease and Heart Failure

Treatment with $1,25(OH)_2D$ increases the expression of Myotrophin and VDR in cardiac cells and reduces the expression of atrial natriuretic peptide. Therefore, Vitamin D deficiency could explain the development of myocardial hypertrophy and heart failure.

Ameri *et al.* [176] found concentrations of 25(OH)D < 30 ng/mL in almost 100% of patients with heart failure. However, this deficiency can be caused by less exposure to sunlight, malabsorption of Vitamin D due to intestinal edema that occurs in severe heart failure and other comorbidities such as obesity and renal and hepatic insufficiency. Different studies have linked the level of 25(OH)D with parameters of heart failure. In the Ludwigshafen Risk and Cardiovascular Health (LURIC) study, in patients cited for coronary angiography, an inverse association was found between the levels of NT-proBNP and those of 25(OH)D and calcitriol. Levels of 25(OH)D and calcitriol also correlated inversely with impaired left ventricle function [177]. In contrast, in a cohort of patients without heart failure who were followed for 13 years, the risk of developing heart failure was found to be associated with parathyroid hormone levels but not with 25(OH)D [178].

A high incidence of Vitamin D deficiency has been detected among patients with myocardial infarction [179, 180], so Vitamin D deficiency has also been linked to an increased risk of myocardial infarction and coronary heart disease [181, 182]. However, in another study [183], the association between 25(OH)D deficiency and an increased risk of coronary disease was found only among white and Chinese subjects, but not in the Hispanic or black race [184].

Several studies have suggested that the Vitamin D status of people who have suffered a myocardial infarction may influence their prognosis, so those with low Vitamin D levels will have more repetitions of myocardial infarction, worsening of heart failure and restenosis [185 - 187].

Stroke

Vitamin D deficiency has been linked to risk factors for ischemic stroke such as diabetes, HTN and hyperlipidemia; however, the association between the levels of 25(OH)D and stroke is not clear.

In LURIC study conducted with 3316 patients referred for a coronary angiogram, lower levels of 25(OH)D and 1,25$(OH)_2$D were found to be associated with an increased likelihood of fatal stroke [188]. The Copenhagen City Heart Study conducted with 10170 healthy individuals also showed an inverse correlation between 25(OH)D levels and the likelihood of ischemic stroke [189]. A study of patients who had suffered an acute ischemic stroke or transient ischemic attack showed that levels of 25(OH)D lower than 25 nmol/L were linked with lacunes, severe white matter hyper intensity and deep cerebral micro bleeds [190]. However, other studies found no relationship between 25(OH)D levels and the risk of stroke [191, 192].

Some studies have reported that Vitamin D may decrease cognitive impairment in patients who have had a stroke and improved mobility. But all researchers agree that Vitamin D supplements for the prevention and treatment of stroke should not be recommended at this time [193] because there is no clear link between high levels of Vitamin D and low risk of stroke [194].

Peripheral Arterial Disease

25(OH)D deficiency has also been associated with a higher incidence of peripheral arterial disease [195] and with a more rapid decrease in functional performance. Lower levels of Vitamin D have been found in type 2 diabetics with lower extremity arterial disease (LEAD) compared to type 2 diabetics without LEAD [196].

Vitamin D receptors may be located in the vessel wall, so it has been suggested that Vitamin D may be involved in the development of arterial disease. However, peripheral arterial disease is associated with decreased mobility, which can decrease the hours of sunlight exposure and therefore the synthesis of Vitamin D. Therefore, in this situation, Vitamin D deficiency could be considered a consequence rather than a cause [197].

CONCLUDING REMARKS

Biologically it seems possible that Vitamin D plays a role in various physiological mechanisms. However, the exact mechanism by which it can act on different diseases is unknown.

The relationship between low levels of Vitamin D and various pathologies is inconsistent and ambiguous. The results of Vitamin D studies may be affected by confounding factors and also by the effects of other risk factors, in addition to those studied, so it is possible that the associations found are occasional.

It is well known that the autumn and winter months, high latitudes and geographical regions with little exposure to sunlight, are related to a higher proportion of patients with Vitamin D deficiency. However, there are more reasons that may justify Vitamin D deficiency such as age, sex, body mass index, medication, diet, physical activity, genetic differences in the Vitamin D receptor and variations in absorption and vitamin metabolism among individuals. In addition to these factors, others such as Vitamin D supplementation, different diagnostic criteria or definitions of Vitamin D deficiency and various ethnic populations may have influenced the results obtained in the studies.

The method of determination used in the studies is another important problem, since the results between different methods are usually not interchangeable. The HPLC-MS/MS method is currently considered the most accurate, although the most commonly used method is the immunoassay, with many problems of reproducibility and precision. In 2010, the first global standard for quantification of 25(OH)D was manufactured by the NIST (National Institute of Standards and Technology), therefore the results of previous studies are not very useful.

The design of the studies that have been published have also been very different, including transverse, longitudinal observational studies, systematic reviews and randomized controlled trials of Vitamin D supplements. In addition, many of the studies were formed by small samples of patients and with different inclusion criteria for the participants, obtaining results with a low level of significance.

Most published studies are observational and in order to understand the true implication of Vitamin D in different diseases, large randomized controlled trials of Vitamin D supplements are needed. If these studies confirm the connection between Vitamin D deficiency and various diseases, treatment of Vitamin D deficiency would be of great public health benefit.

CONFLICT OF INTEREST

The author (editor) declares no conflict of interest, financial or otherwise.

ACKNOWLEDGEMENTS

Declared none.

REFERENCES

[1] Rendina D, De Filippo G, Muscariello R, *et al.* Vitamin D and cardiometabolic disorders. High Blood Press Cardiovasc Prev 2014; 21(4): 251-6. [http://dx.doi.org/10.1007/s40292-014-0060-5] [PMID: 25200767]

[2] Ross AC, Manson JE, Abrams SA, *et al.* The 2011 report on dietary reference intakes for calcium and vitamin D from the Institute of Medicine: what clinicians need to know. J Clin Endocrinol Metab 2011; 96(1): 53-8. [http://dx.doi.org/10.1210/jc.2010-2704] [PMID: 21118827]

[3] Holick MF, Binkley NC, Bischoff-Ferrari HA, *et al.* Evaluation, treatment, and prevention of vitamin D deficiency: an Endocrine Society clinical practice guideline. J Clin Endocrinol Metab 2011; 96(7): 1911-30. [http://dx.doi.org/10.1210/jc.2011-0385] [PMID: 21646368]

[4] Mohr SB, Garland CF, Gorham ED, *et al.* The association between ultraviolet B irradiance, vitamin D status and incidence rates of type 1 diabetes in 51 regions worldwide. Diabetologia 2008; 51(8): 1391-8. [http://dx.doi.org/10.1007/s00125-008-1061-5] [PMID: 18548227]

[5] Al-Daghri NM, Al-Attas OS, Alokail MS, *et al.* Lower vitamin D status is more common among Saudi adults with diabetes mellitus type 1 than in non-diabetics. BMC Public Health 2014; 14: 153. [http://dx.doi.org/10.1186/1471-2458-14-153] [PMID: 24517121]

[6] Janner M, Ballinari P, Mullis PE, *et al.* High prevalence of vitamin D deficiency in children and adolescents with type 1 diabetes. Swiss Med Wkly 2010; 140: w13091. [PMID: 20853194]

[7] Greer RM, Rogers MA, Bowling FG, *et al.* Australian children and adolescents with type 1 diabetes have low vitamin D levels. Med J Aust 2007; 187(1): 59-60. [PMID: 17605716]

[8] Bener A, Alsaied A, Al-Ali M, *et al.* High prevalence of vitamin D deficiency in type 1 diabetes mellitus and healthy children. Acta Diabetol 2009; 46(3): 183-9. [http://dx.doi.org/10.1007/s00592-008-0071-6] [PMID: 18846317]

[9] Svoren BM, Volkening LK, Wood JR, *et al.* Significant vitamin D deficiency in youth with type 1 diabetes mellitus. J Pediatr 2009; 154(1): 132-4. [http://dx.doi.org/10.1016/j.jpeds.2008.07.015] [PMID: 19187735]

[10] Bierschenk L, Alexander J, Wasserfall C, Haller M, Schatz D, Atkinson M. Vitamin D levels in subjects with and without type 1 diabetes residing in a solar rich environment. Diabetes Care 2009; 32(11): 1977-9. [http://dx.doi.org/10.2337/dc09-1089] [PMID: 19675194]

[11] Munger KL, Levin LI, Massa J, Horst R, Orban T, Ascherio A. Preclinical serum 25-hydroxyvitamin D levels and risk of type 1 diabetes in a cohort of US military personnel. Am J Epidemiol 2013; 177(5): 411-9. [http://dx.doi.org/10.1093/aje/kws243] [PMID: 23380046]

[12] Littorin B, Blom P, Schölin A, *et al.* Lower levels of plasma 25-hydroxyvitamin D among young adults at diagnosis of autoimmune type 1 diabetes compared with control subjects: results from the nationwide Diabetes Incidence Study in Sweden (DISS). Diabetologia 2006; 49(12): 2847-52. [http://dx.doi.org/10.1007/s00125-006-0426-x] [PMID: 17072585]

[13] Hyppönen E, Läärä E, Reunanen A, *et al.* Intake of vitamin D and risk of type 1 diabetes: a birth-cohort study. Lancet 2001; 358(9292): 1500-3. [http://dx.doi.org/10.1016/S0140-6736(01)06580-1] [PMID: 11705562]

[14] The NS, Crandell JL, Lawrence JM, *et al.* Vitamin D in youth with Type 1 diabetes: prevalence of insufficiency and association with insulin resistance in the SEARCH Nutrition Ancillary Study. Diabet Med 2013; 30(11): 1324-32. [http://dx.doi.org/10.1111/dme.12297] [PMID: 23909945]

[15] McDermott MF, Ramachandran A, Ogunkolade BW, *et al.* Allelic variation in the vitamin D receptor influences susceptibility to IDDM in Indian Asians. Diabetologia 1997; 40(8): 971-5. [http://dx.doi.org/10.1007/s001250050776] [PMID: 9267994]

[16] Pani MA, Knapp M, Donner H, *et al.* Vitamin D receptor allele combinations influence genetic susceptibility to type 1 diabetes in Germans. Diabetes 2000; 49(3): 504-7. [http://dx.doi.org/10.2337/diabetes.49.3.504] [PMID: 10868975]

[17] Yokota I, Satomura S, Kitamura S, *et al.* Association between vitamin D receptor genotype and age of onset in juvenile Japanese patients with type 1 diabetes. Diabetes Care 2002; 25(7): 1244. [http://dx.doi.org/10.2337/diacare.25.7.1244] [PMID: 12087029]

[18] Koeleman BP, Valdigem G, Eerligh P, Giphart MJ, Roep BO. Seasonality of birth in patients with type 1 diabetes. Lancet 2002; 359(9313): 1246-7. [http://dx.doi.org/10.1016/S0140-6736(02)08228-4] [PMID: 11955562]

[19] Audí L, Martí G, Esteban C, *et al.* VDR gene polymorphism at exon 2 start codon (FokI) may have influenced Type 1 diabetes mellitus susceptibility in two Spanish populations. Diabet Med 2004; 21(4): 393-4. [http://dx.doi.org/10.1111/j.1464-5491.2004.01126.x] [PMID: 15049946]

[20] De Azevêdo Silva J, Guimarães RL, Brandão LA, *et al.* Vitamin D receptor (VDR) gene polymorphisms and age onset in type 1 diabetes mellitus. Autoimmunity 2013; 46(6): 382-7. [http://dx.doi.org/10.3109/08916934.2013.795952] [PMID: 23721405]

[21] Turpeinen H, Hermann R, Vaara S, *et al.* Vitamin D receptor polymorphisms: no association with type 1 diabetes in the Finnish population. Eur J Endocrinol 2003; 149(6): 591-6. [http://dx.doi.org/10.1530/eje.0.1490591] [PMID: 14641002]

[22] Nejentsev S, Cooper JD, Godfrey L, *et al.* Analysis of the vitamin D receptor gene sequence variants in type 1 diabetes. Diabetes 2004; 53(10): 2709-12. [http://dx.doi.org/10.2337/diabetes.53.10.2709] [PMID: 15448105]

[23] Lim S, Kim MJ, Choi SH, *et al.* Association of vitamin D deficiency with incidence of type 2 diabetes in high-risk Asian subjects. Am J Clin Nutr 2013; 97(3): 524-30. [http://dx.doi.org/10.3945/ajcn.112.048496] [PMID: 23364011]

[24] Afzal S, Bojesen SE, Nordestgaard BG. Low 25-hydroxyvitamin D and risk of type 2 diabetes: a prospective cohort study and metaanalysis. Clin Chem 2013; 59(2): 381-91. [http://dx.doi.org/10.1373/clinchem.2012.193003] [PMID: 23232064]

[25] Thorand B, Zierer A, Huth C, *et al.* Effect of serum 25-hydroxyvitamin D on risk for type 2 diabetes may be partially mediated by subclinical inflammation: results from the MONICA/KORA Augsburg study. Diabetes Care 2011; 34(10): 2320-2. [http://dx.doi.org/10.2337/dc11-0775] [PMID: 21873558]

[26] Husemoen LL, Thuesen BH, Fenger M, *et al.* Serum 25(OH)D and type 2 diabetes association in a general population: a prospective study. Diabetes Care 2012; 35(8): 1695-700.

[http://dx.doi.org/10.2337/dc11-1309] [PMID: 22688545]

[27] González-Molero I, Rojo-Martínez G, Morcillo S, *et al.* Vitamin D and incidence of diabetes: a prospective cohort study. Clin Nutr 2012; 31(4): 571-3. [http://dx.doi.org/10.1016/j.clnu.2011.12.001] [PMID: 22204964]

[28] Gagnon C, Lu ZX, Magliano DJ, *et al.* Serum 25-hydroxyvitamin D, calcium intake, and risk of type 2 diabetes after 5 years: results from a national, population-based prospective study (the Australian Diabetes, Obesity and Lifestyle study). Diabetes Care 2011; 34(5): 1133-8. [http://dx.doi.org/10.2337/dc10-2167] [PMID: 21430082]

[29] Pittas AG, Sun Q, Manson JE, Dawson-Hughes B, Hu FB. Plasma 25-hydroxyvitamin D concentration and risk of incident type 2 diabetes in women. Diabetes Care 2010; 33(9): 2021-3. [http://dx.doi.org/10.2337/dc10-0790] [PMID: 20805275]

[30] Knekt P, Laaksonen M, Mattila C, *et al.* Serum vitamin D and subsequent occurrence of type 2 diabetes. Epidemiology 2008; 19(5): 666-71. [http://dx.doi.org/10.1097/EDE.0b013e318176b8ad] [PMID: 18496468]

[31] Tsur A, Feldman BS, Feldhammer I, Hoshen MB, Leibowitz G, Balicer RD. Decreased serum concentrations of 25-hydroxycholecalciferol are associated with increased risk of progression to impaired fasting glucose and diabetes. Diabetes Care 2013; 36(5): 1361-7. [http://dx.doi.org/10.2337/dc12-1050] [PMID: 23393216]

[32] Deleskog A, Hilding A, Brismar K, Hamsten A, Efendic S, Östenson C-G. Low serum 25-hydroxyvitamin D level predicts progression to type 2 diabetes in individuals with prediabetes but not with normal glucose tolerance. Diabetologia 2012; 55(6): 1668-78. [http://dx.doi.org/10.1007/s00125-012-2529-x] [PMID: 22426800]

[33] Shankar A, Sabanayagam C, Kalidindi S. Serum 25-hydroxyvitamin d levels and prediabetes among subjects free of diabetes. Diabetes Care 2011; 34(5): 1114-9. [http://dx.doi.org/10.2337/dc10-1203] [PMID: 21430085]

[34] Abbasi F, Blasey C, Feldman D, Caulfield MP, Hantash FM, Reaven GM. Low circulating 25-hydroxyvitamin D concentrations are associated with defects in insulin action and insulin secretion in persons with prediabetes. J Nutr 2015; 145(4): 714-9. [http://dx.doi.org/10.3945/jn.114.209171] [PMID: 25740907]

[35] Kayaniyil S, Vieth R, Retnakaran R, *et al.* Association of vitamin D with insulin resistance and beta-cell dysfunction in subjects at risk for type 2 diabetes. Diabetes Care 2010; 33(6): 1379-81. [http://dx.doi.org/10.2337/dc09-2321] [PMID: 20215450]

[36] Mitri J, Dawson-Hughes B, Hu FB, Pittas AG. Effects of vitamin D and calcium supplementation on pancreatic β cell function, insulin sensitivity, and glycemia in adults at high risk of diabetes: the Calcium and Vitamin D for Diabetes Mellitus (CaDDM) randomized controlled trial. Am J Clin Nutr 2011; 94(2): 486-94. [http://dx.doi.org/10.3945/ajcn.111.011684] [PMID: 21715514]

[37] Tabesh M, Azadbakht L, Faghihimani E, Tabesh M, Esmaillzadeh A. Effects of calcium-vitamin D co-supplementation on metabolic profiles in vitamin D insufficient people with type 2 diabetes: a randomised controlled clinical trial. Diabetologia 2014; 57(10): 2038-47. [http://dx.doi.org/10.1007/s00125-014-3313-x] [PMID: 25005333]

[38] Elkassaby S, Harrison LC, Mazzitelli N, *et al.* A randomised controlled trial of high dose vitamin D in recent-onset type 2 diabetes. Diabetes Res Clin Pract 2014; 106(3): 576-82. [http://dx.doi.org/10.1016/j.diabres.2014.08.030] [PMID: 25438937]

[39] Sümbül AT, Sezer A, Kavvasoğlu G, *et al.* Low serum levels of vitamin D in metastatic cancer patients: a case-control study. Med Oncol 2014; 31(3): 861. [http://dx.doi.org/10.1007/s12032-014-0861-3] [PMID: 24493144]

[40] Hauser K, Walsh D, Shrotriya S, Karafa M. Low 25-hydroxyvitamin D levels in people with a solid

tumor cancer diagnosis: the tip of the iceberg? Support Care Cancer 2014; 22(7): 1931-9. [http://dx.doi.org/10.1007/s00520-014-2154-y] [PMID: 24682592]

[41] Lamprecht SA, Lipkin M. Chemoprevention of colon cancer by calcium, vitamin D and folate: molecular mechanisms. Nat Rev Cancer 2003; 3(8): 601-14. [http://dx.doi.org/10.1038/nrc1144] [PMID: 12894248]

[42] Jung S, Qian ZR, Yamauchi M, *et al.* Predicted 25(OH)D score and colorectal cancer risk according to vitamin D receptor expression. Cancer Epidemiol Biomarkers Prev 2014; 23(8): 1628-37. [http://dx.doi.org/10.1158/1055-9965.EPI-14-0229] [PMID: 24920642]

[43] Deeb KK, Trump DL, Johnson CS. Vitamin D signalling pathways in cancer: potential for anticancer therapeutics. Nat Rev Cancer 2007; 7(9): 684-700. [http://dx.doi.org/10.1038/nrc2196] [PMID: 17721433]

[44] Fichera A, Little N, Dougherty U, *et al.* A vitamin D analogue inhibits colonic carcinogenesis in the AOM/DSS model. J Surg Res 2007; 142(2): 239-45. [http://dx.doi.org/10.1016/j.jss.2007.02.038] [PMID: 17574271]

[45] Krishnan AV, Feldman D. Mechanisms of the anti-cancer and anti-inflammatory actions of vitamin D. Annu Rev Pharmacol Toxicol 2011; 51: 311-36. [http://dx.doi.org/10.1146/annurev-pharmtox-010510-100611] [PMID: 20936945]

[46] Garland CF, Garland FC. Do sunlight and vitamin D reduce the likelihood of colon cancer? Int J Epidemiol 1980; 9(3): 227-31. [http://dx.doi.org/10.1093/ije/9.3.227] [PMID: 7440046]

[47] Garland C, Shekelle RB, Barrett-Connor E, Criqui MH, Rossof AH, Paul O. Dietary vitamin D and calcium and risk of colorectal cancer: a 19-year prospective study in men. Lancet 1985; 1(8424): 307-9. [http://dx.doi.org/10.1016/S0140-6736(85)91082-7] [PMID: 2857364]

[48] Gorham ED, Garland CF, Garland FC, *et al.* Optimal vitamin D status for colorectal cancer prevention: a quantitative meta analysis. Am J Prev Med 2007; 32(3): 210-6. [http://dx.doi.org/10.1016/j.amepre.2006.11.004] [PMID: 17296473]

[49] Wei MY, Garland CF, Gorham ED, Mohr SB, Giovannucci E. Vitamin D and prevention of colorectal adenoma: a meta-analysis. Cancer Epidemiol Biomarkers Prev 2008; 17(11): 2958-69. [http://dx.doi.org/10.1158/1055-9965.EPI-08-0402] [PMID: 18990737]

[50] Zheng XE, Lipka S, Li T, *et al.* The relationship of vitamin D status, smoking, and colorectal adenoma: a retrospective study in an ethnically diverse community. J Steroid Biochem Mol Biol 2013; 136: 280-3. [http://dx.doi.org/10.1016/j.jsbmb.2012.09.018] [PMID: 23000288]

[51] Yin L, Grandi N, Raum E, Haug U, Arndt V, Brenner H. Meta-analysis: Serum vitamin D and colorectal adenoma risk. Prev Med 2011; 53(1-2): 10-6. [http://dx.doi.org/10.1016/j.ypmed.2011.05.013] [PMID: 21672549]

[52] Yamaji T, Iwasaki M, Sasazuki S, Sakamoto H, Yoshida T, Tsugane S. Association between plasma 25-hydroxyvitamin D and colorectal adenoma according to dietary calcium intake and vitamin D receptor polymorphism. Am J Epidemiol 2012; 175(3): 236-44. [http://dx.doi.org/10.1093/aje/kwr295] [PMID: 22193171]

[53] Ma Y, Zhang P, Wang F, Yang J, Liu Z, Qin H. Association between vitamin D and risk of colorectal cancer: a systematic review of prospective studies. J Clin Oncol 2011; 29(28): 3775-82. [http://dx.doi.org/10.1200/JCO.2011.35.7566] [PMID: 21876081]

[54] Adams SV, Newcomb PA, Burnett-Hartman AN, White E, Mandelson MT, Potter JD. Circulating 25-hydroxyvitamin-D and risk of colorectal adenomas and hyperplastic polyps. Nutr Cancer 2011; 63(3): 319-26. [http://dx.doi.org/10.1080/01635581.2011.535960] [PMID: 21432725]

[55] LePane CA, Singh G, Spanier-Stiasny JA, Svinarich DM, Rasansky RJ, Hoffman SM. Implications of serum 25-Hydroxyvitamin D on the prevalence of neoplastic polyps: A cross-sectional study. Gastroenterol Res 2011; 4(2): 43-50.
[PMID: 27942313]

[56] Ogino S, Chan AT, Fuchs CS, Giovannucci E. Molecular pathological epidemiology of colorectal neoplasia: an emerging transdisciplinary and interdisciplinary field. Gut 2011; 60(3): 397-411.
[http://dx.doi.org/10.1136/gut.2010.217182] [PMID: 21036793]

[57] Colussi D, Brandi G, Bazzoli F, Ricciardiello L. Molecular pathways involved in colorectal cancer: implications for disease behavior and prevention. Int J Mol Sci 2013; 14(8): 16365-85.
[http://dx.doi.org/10.3390/ijms140816365] [PMID: 23965959]

[58] Fedirko V, Bostick RM, Goodman M, Flanders WD, Gross MD. Blood 25-hydroxyvitamin D3 concentrations and incident sporadic colorectal adenoma risk: a pooled case-control study. Am J Epidemiol 2010; 172(5): 489-500.
[http://dx.doi.org/10.1093/aje/kwq157] [PMID: 20650953]

[59] Jacobs ET, Hibler EA, Lance P, Sardo CL, Jurutka PW. Association between circulating concentrations of 25(OH)D and colorectal adenoma: a pooled analysis. Int J Cancer 2013; 133(12): 2980-8.
[PMID: 23754630]

[60] Aaltonen L, Johns L, Järvinen H, Mecklin JP, Houlston R. Explaining the familial colorectal cancer risk associated with mismatch repair (MMR)-deficient and MMR-stable tumors. Clin Cancer Res 2007; 13(1): 356-61.
[http://dx.doi.org/10.1158/1078-0432.CCR-06-1256] [PMID: 17200375]

[61] Ahn J, Yu K, Stolzenberg-Solomon R, *et al.* Genome-wide association study of circulating vitamin D levels. Hum Mol Genet 2010; 19(13): 2739-45.
[http://dx.doi.org/10.1093/hmg/ddq155] [PMID: 20418485]

[62] Wang TJ, Zhang F, Richards JB, *et al.* Common genetic determinants of vitamin D insufficiency: a genome-wide association study. Lancet 2010; 376(9736): 180-8.
[http://dx.doi.org/10.1016/S0140-6736(10)60588-0] [PMID: 20541252]

[63] Theodoratou E, Palmer T, Zgaga L, *et al.* Instrumental variable estimation of the causal effect of plasma 25-hydroxy-vitamin D on colorectal cancer risk: a mendelian randomization analysis. PLoS One 2012; 7(6): e37662.
[http://dx.doi.org/10.1371/journal.pone.0037662] [PMID: 22701574]

[64] Dong LM, Ulrich CM, Hsu L, *et al.* Vitamin D related genes, CYP24A1 and CYP27B1, and colon cancer risk. Cancer Epidemiol Biomarkers Prev 2009; 18(9): 2540-8.
[http://dx.doi.org/10.1158/1055-9965.EPI-09-0228] [PMID: 19706847]

[65] Touvier M, Chan DS, Lau R, *et al.* Meta-analyses of vitamin D intake, 25-hydroxyvitamin D status, vitamin D receptor polymorphisms, and colorectal cancer risk. Cancer Epidemiol Biomarkers Prev 2011; 20(5): 1003-16.
[http://dx.doi.org/10.1158/1055-9965.EPI-10-1141] [PMID: 21378269]

[66] Bai YH, Lu H, Hong D, Lin CC, Yu Z, Chen BC. Vitamin D receptor gene polymorphisms and colorectal cancer risk: a systematic meta-analysis. World J Gastroenterol 2012; 18(14): 1672-9.
[http://dx.doi.org/10.3748/wjg.v18.i14.1672] [PMID: 22529698]

[67] Ochs-Balcom HM, Cicek MS, Thompson CL, *et al.* Association of vitamin D receptor gene variants, adiposity and colon cancer. Carcinogenesis 2008; 29(9): 1788-93.
[http://dx.doi.org/10.1093/carcin/bgn166] [PMID: 18628249]

[68] Park K, Woo M, Nam J, Kim JC. Start codon polymorphisms in the vitamin D receptor and colorectal cancer risk. Cancer Lett 2006; 237(2): 199-206.
[http://dx.doi.org/10.1016/j.canlet.2005.05.048] [PMID: 16019132]

[69] Wong HL, Seow A, Arakawa K, Lee HP, Yu MC, Ingles SA. Vitamin D receptor start codon polymorphism and colorectal cancer risk: effect modification by dietary calcium and fat in Singapore Chinese. Carcinogenesis 2003; 24(6): 1091-5.
[http://dx.doi.org/10.1093/carcin/bgg059] [PMID: 12807755]

[70] Hiraki LT, Qu C, Hutter CM, *et al.* Genetic predictors of circulating 25-hydroxyvitamin d and risk of colorectal cancer. Cancer Epidemiol Biomarkers Prev 2013; 22(11): 2037-46.
[http://dx.doi.org/10.1158/1055-9965.EPI-13-0209] [PMID: 23983240]

[71] Morton A, Hardy J, Morton A, *et al.* Vitamin D deficiency in patients with malignancy in Brisbane. Support Care Cancer 2014; 22(8): 2223-7.
[http://dx.doi.org/10.1007/s00520-014-2218-z] [PMID: 24664359]

[72] Martinez-Alonso. Montserrat, Dusso A, Ariza G, Nabal M: The effect on quality of life of vitamin D administration for advanced cancer treatment (VIDAFACT. study): protocol of a randomised controlled trial. BMJ Open 2014; 1-7.

[73] Lazzeroni M, Serrano D, Pilz S, Gandini S. Vitamin D supplementation and cancer: review of randomized controlled trials. Anticancer Agents Med Chem 2013; 13(1): 118-25.
[http://dx.doi.org/10.2174/187152013804487281] [PMID: 23094929]

[74] Deeb KK, Trump DL, Johnson CS. Vitamin D signalling pathways in cancer: potential for anticancer therapeutics. Nat Rev Cancer 2007; 7(9): 684-700.
[http://dx.doi.org/10.1038/nrc2196] [PMID: 17721433]

[75] James SY, Mackay AG, Colston KW. Effects of 1,25 dihydroxyvitamin D3 and its analogues on induction of apoptosis in breast cancer cells. J Steroid Biochem Mol Biol 1996; 58(4): 395-401.
[http://dx.doi.org/10.1016/0960-0760(96)00048-9] [PMID: 8903423]

[76] Matthews D, LaPorta E, Zinser GM, Narvaez CJ, Welsh J. Genomic vitamin D signaling in breast cancer: Insights from animal models and human cells. J Steroid Biochem Mol Biol 2010; 121(1-2): 362-7.
[http://dx.doi.org/10.1016/j.jsbmb.2010.03.061] [PMID: 20412854]

[77] Welsh J. Vitamin D and breast cancer: insights from animal models. Am J Clin Nutr 2004; 80(6) (Suppl.): 1721S-4S.
[PMID: 15585794]

[78] Bertone-Johnson ER. Vitamin D and breast cancer. Ann Epidemiol 2009; 19(7): 462-7.
[http://dx.doi.org/10.1016/j.annepidem.2009.01.003] [PMID: 19230714]

[79] Abbas S, Linseisen J, Slanger T, *et al.* Serum 25-hydroxyvitamin D and risk of post-menopausal breast cancer-results of a large case-control study. Carcinogenesis 2008; 29(1): 93-9.
[http://dx.doi.org/10.1093/carcin/bgm240] [PMID: 17974532]

[80] Abbas S, Chang-Claude J, Linseisen J. Plasma 25-hydroxyvitamin D and premenopausal breast cancer risk in a German case-control study. Int J Cancer 2009; 124(1): 250-5.
[http://dx.doi.org/10.1002/ijc.23904] [PMID: 18839430]

[81] Crew KD, Gammon MD, Steck SE, *et al.* Association between plasma 25-hydroxyvitamin D and breast cancer risk. Cancer Prev Res (Phila) 2009; 2(6): 598-604.
[http://dx.doi.org/10.1158/1940-6207.CAPR-08-0138] [PMID: 19470790]

[82] Lowe LC, Guy M, Mansi JL, *et al.* Plasma 25-hydroxy vitamin D concentrations, vitamin D receptor genotype and breast cancer risk in a UK Caucasian population. Eur J Cancer 2005; 41(8): 1164-9.
[http://dx.doi.org/10.1016/j.ejca.2005.01.017] [PMID: 15911240]

[83] Dorgan JF, Stanczyk FZ, Kahle LL, Brinton LA. Prospective case-control study of premenopausal serum estradiol and testosterone levels and breast cancer risk. Breast Cancer Res 2010; 12(6): R98.
[http://dx.doi.org/10.1186/bcr2779] [PMID: 21087481]

[84] Green AK, Hankinson SE, Bertone-Johnson ER, Tamimi RM. Mammographic density, plasma

vitamin D levels and risk of breast cancer in postmenopausal women. Int J Cancer 2010; 127(3): 667-74.
[http://dx.doi.org/10.1002/ijc.25075] [PMID: 19960434]

[85] Rejnmark L, Tietze A, Vestergaard P, *et al.* Reduced prediagnostic 25-hydroxyvitamin D levels in women with breast cancer: a nested case-control study. Cancer Epidemiol Biomarkers Prev 2009; 18(10): 2655-60.
[http://dx.doi.org/10.1158/1055-9965.EPI-09-0531] [PMID: 19789365]

[86] Fedirko V, Torres-Mejía G, Ortega-Olvera C, *et al.* Serum 25-hydroxyvitamin D and risk of breast cancer: results of a large population-based case-control study in Mexican women. Cancer Causes Control 2012; 23(7): 1149-62.
[http://dx.doi.org/10.1007/s10552-012-9984-z] [PMID: 22622862]

[87] Freedman DM, Chang SC, Falk RT, *et al.* Serum levels of vitamin D metabolites and breast cancer risk in the prostate, lung, colorectal, and ovarian cancer screening trial. Cancer Epidemiol Biomarkers Prev 2008; 17(4): 889-94.
[http://dx.doi.org/10.1158/1055-9965.EPI-07-2594] [PMID: 18381472]

[88] McCullough ML, Stevens VL, Patel R, *et al.* Serum 25-hydroxyvitamin D concentrations and postmenopausal breast cancer risk: a nested case control study in the Cancer Prevention Study-II Nutrition Cohort. Breast Cancer Res 2009; 11(4): R64.
[http://dx.doi.org/10.1186/bcr2356] [PMID: 19715600]

[89] Almquist M, Bondeson A, Bondeson L, Malm J, Manjer J. Serum levels of vitamin D, PTH and calcium and breast cancer risk a prospective nested case-control study. Int J Cancer J Int du cancer 2010; 127: 2159-68.

[90] Eliassen AH, Spiegelman D, Hollis BW, Horst RL, Willett WC, Hankinson SE. Plasma 25-hydroxyvitamin D and risk of breast cancer in the Nurses' Health Study II. Breast Cancer Res 2011; 13(3): R50.
[http://dx.doi.org/10.1186/bcr2880] [PMID: 21569367]

[91] Chlebowski RT. Vitamin D and breast cancer: interpreting current evidence. Breast Cancer Res 2011; 13(4): 217.
[http://dx.doi.org/10.1186/bcr2846] [PMID: 21884640]

[92] Bluff JE, Brown NJ, Reed MW, Staton CA. Tissue factor, angiogenesis and tumour progression. Breast Cancer Res 2008; 10(2): 204.
[http://dx.doi.org/10.1186/bcr1871] [PMID: 18373885]

[93] Ooi LL, Zheng Y, Zhou H, *et al.* Vitamin D deficiency promotes growth of MCF-7 human breast cancer in a rodent model of osteosclerotic bone metastasis. Bone 2010; 47(4): 795-803.
[http://dx.doi.org/10.1016/j.bone.2010.07.012] [PMID: 20638491]

[94] Tretli S, Schwartz GG, Torjesen PA, Robsahm TE. Serum levels of 25-hydroxyvitamin D and survival in Norwegian patients with cancer of breast, colon, lung, and lymphoma: a population-based study. Cancer Causes Control 2012; 23(2): 363-70.
[http://dx.doi.org/10.1007/s10552-011-9885-6] [PMID: 22193397]

[95] Goodwin PJ, Ennis M, Pritchard KI, Koo J, Hood N. Prognostic effects of 25-hydroxyvitamin D levels in early breast cancer. J Clin Oncol 2009; 27(23): 3757-63.
[http://dx.doi.org/10.1200/JCO.2008.20.0725] [PMID: 19451439]

[96] Jacobs ET, Thomson CA, Flatt SW, *et al.* Vitamin D and breast cancer recurrence in the Women's Healthy Eating and Living (WHEL) Study. Am J Clin Nutr 2011; 93(1): 108-17.
[http://dx.doi.org/10.3945/ajcn.2010.30009] [PMID: 20980485]

[97] Chung M, Lee J, Terasawa T, Lau J, Trikalinos TA. Vitamin D with or without calcium supplementation for prevention of cancer and fractures: an updated meta-analysis for the U.S. Preventive Services Task Force. Ann Intern Med 2011; 155(12): 827-38.
[http://dx.doi.org/10.7326/0003-4819-155-12-201112200-00005] [PMID: 22184690]

[98] Wilson KM, Giovannucci EL, Mucci LA. Lifestyle and dietary factors in the prevention of lethal prostate cancer. Asian J Androl 2012; 14(3): 365-74. [http://dx.doi.org/10.1038/aja.2011.142] [PMID: 22504869]

[99] Holick MF. Vitamin D: its role in cancer prevention and treatment. Prog Biophys Mol Biol 2006; 92(1): 49-59. [http://dx.doi.org/10.1016/j.pbiomolbio.2006.02.014] [PMID: 16566961]

[100] Krishnan AV, Feldman D. Mechanisms of the anti-cancer and anti-inflammatory actions of vitamin D. Annu Rev Pharmacol Toxicol 2011; 51: 311-36. [http://dx.doi.org/10.1146/annurev-pharmtox-010510-100611] [PMID: 20936945]

[101] Welsh J. Cellular and molecular effects of vitamin D on carcinogenesis. Arch Biochem Biophys 2012; 523(1): 107-14. [http://dx.doi.org/10.1016/j.abb.2011.10.019] [PMID: 22085499]

[102] Kovalenko PL, Zhang Z, Yu JG, Li Y, Clinton SK, Fleet JC. Dietary vitamin D and vitamin D receptor level modulate epithelial cell proliferation and apoptosis in the prostate. Cancer Prev Res (Phila) 2011; 4(10): 1617-25. [http://dx.doi.org/10.1158/1940-6207.CAPR-11-0035] [PMID: 21836023]

[103] Ahonen MH, Tenkanen L, Teppo L, Hakama M, Tuohimaa P. Prostate cancer risk and prediagnostic serum 25-hydroxyvitamin D levels (Finland). Cancer Causes Control 2000; 11(9): 847-52. [http://dx.doi.org/10.1023/A:1008923802001] [PMID: 11075874]

[104] Albanes D, Mondul AM, Yu K, *et al.* Serum 25-hydroxy vitamin D and prostate cancer risk in a large nested case-control study. Cancer Epidemiol Biomarkers Prev 2011; 20(9): 1850-60. [http://dx.doi.org/10.1158/1055-9965.EPI-11-0403] [PMID: 21784952]

[105] Meyer HE, Robsahm TE, Bjørge T, Brustad M, Blomhoff R. Vitamin D, season, and risk of prostate cancer: a nested case-control study within Norwegian health studies. Am J Clin Nutr 2013; 97(1): 147-54. [http://dx.doi.org/10.3945/ajcn.112.039222] [PMID: 23193007]

[106] Platz EA, Leitzmann MF, Hollis BW, Willett WC, Giovannucci E. Plasma 1,25-dihydroxy- and 25-hydroxyvitamin D and subsequent risk of prostate cancer. Cancer Causes Control 2004; 15(3): 255-65. [http://dx.doi.org/10.1023/B:CACO.0000024245.24880.8a] [PMID: 15090720]

[107] Ahn J, Albanes D, Berndt SI, *et al.* Vitamin D-related genes, serum vitamin D concentrations and prostate cancer risk. Carcinogenesis 2009; 30(5): 769-76. [http://dx.doi.org/10.1093/carcin/bgp055] [PMID: 19255064]

[108] Faupel-Badger JM, Diaw L, Albanes D, Virtamo J, Woodson K, Tangrea JA. Lack of association between serum levels of 25-hydroxyvitamin D and the subsequent risk of prostate cancer in Finnish men. Cancer Epidemiol Biomarkers Prev 2007; 16(12): 2784-6. [http://dx.doi.org/10.1158/1055-9965.EPI-07-0672] [PMID: 18086789]

[109] Barnett CM, Nielson CM, Shannon J, *et al.* Serum 25-OH vitamin D levels and risk of developing prostate cancer in older men. Cancer Causes Control 2010; 21(8): 1297-303. [http://dx.doi.org/10.1007/s10552-010-9557-y] [PMID: 20383574]

[110] Tretli S, Hernes E, Berg JP, Hestvik UE, Robsahm TE. Association between serum 25(OH)D and death from prostate cancer. Br J Cancer 2009; 100(3): 450-4. [http://dx.doi.org/10.1038/sj.bjc.6604865] [PMID: 19156140]

[111] Fang F, Kasperzyk JL, Shui I, *et al.* Prediagnostic plasma vitamin D metabolites and mortality among patients with prostate cancer. PLoS One 2011; 6(4): e18625. [http://dx.doi.org/10.1371/journal.pone.0018625] [PMID: 21494639]

[112] Shui IM, Mucci LA, Kraft P, *et al.* Vitamin D-related genetic variation, plasma vitamin D, and risk of lethal prostate cancer: a prospective nested case-control study. J Natl Cancer Inst 2012; 104(9): 690-9. [http://dx.doi.org/10.1093/jnci/djs189] [PMID: 22499501]

[113] Mikhak B, Hunter DJ, Spiegelman D, Platz EA, Hollis BW, Giovannucci E. Vitamin D receptor (VDR) gene polymorphisms and haplotypes, interactions with plasma 25-hydroxyvitamin D and 1,25-dihydroxyvitamin D, and prostate cancer risk. Prostate 2007; 67(9): 911-23. [http://dx.doi.org/10.1002/pros.20570] [PMID: 17440943]

[114] Holt SK, Kolb S, Fu R, Horst R, Feng Z, Stanford JL. Circulating levels of 25-hydroxyvitamin D and prostate cancer prognosis. Cancer Epidemiol 2013; 37(5): 666-70. [http://dx.doi.org/10.1016/j.canep.2013.07.005] [PMID: 23972671]

[115] Murphy AB, Nyame Y, Martin IK, *et al.* Vitamin D deficiency predicts prostate biopsy outcomes. Clin Cancer Res 2014; 20(9): 2289-99. [http://dx.doi.org/10.1158/1078-0432.CCR-13-3085] [PMID: 24789033]

[116] Bouillon R, Van Baelen H, De Moor P. 25-hydroxyvitamin D and its binding protein in maternal and cord serum. J Clin Endocrinol Metab 1977; 45(4): 679-84. [http://dx.doi.org/10.1210/jcem-45-4-679] [PMID: 914972]

[117] Álvarez-Fernández I, Prieto B, Rodríguez V, Ruano Y, Escudero AI, Álvarez FV. Role of vitamin D and sFlt-1/PlGF ratio in the development of early- and late-onset preeclampsia. Clin Chem Lab Med 2015; 53(7): 1033-40. [http://dx.doi.org/10.1515/cclm-2014-1039] [PMID: 25544745]

[118] Maghbooli Z, Hossein-Nezhad A, Karimi F, Shafaei AR, Larijani B. Correlation between vitamin D3 deficiency and insulin resistance in pregnancy. Diabetes Metab Res Rev 2008; 24(1): 27-32. [http://dx.doi.org/10.1002/dmrr.737] [PMID: 17607661]

[119] Aghajafari F, Nagulesapillai T, Ronksley PE, Tough SC, O'Beirne M, Rabi DM. Association between maternal serum 25-hydroxyvitamin D level and pregnancy and neonatal outcomes: systematic review and meta-analysis of observational studies. BMJ 2013; 346: f1169. [http://dx.doi.org/10.1136/bmj.f1169] [PMID: 23533188]

[120] Wei S-Q, Qi H-P, Luo Z-C, Fraser WD. Maternal vitamin D status and adverse pregnancy outcomes: a systematic review and meta-analysis. J Matern Fetal Neonatal Med 2013; 26(9): 889-99. [http://dx.doi.org/10.3109/14767058.2013.765849] [PMID: 23311886]

[121] Poel YH, Hummel P, Lips P, Stam F, van der Ploeg T, Simsek S. Vitamin D and gestational diabetes: a systematic review and meta-analysis. Eur J Intern Med 2012; 23(5): 465-9. [http://dx.doi.org/10.1016/j.ejim.2012.01.007] [PMID: 22726378]

[122] Baker AM, Haeri S, Camargo CA Jr, Stuebe AM, Boggess KA. First-trimester maternal vitamin D status and risk for gestational diabetes (GDM) a nested case-control study. Diabetes Metab Res Rev 2012; 28(2): 164-8. [http://dx.doi.org/10.1002/dmrr.1282] [PMID: 21818838]

[123] Senti J, Thiele DK, Anderson CM. Maternal vitamin D status as a critical determinant in gestational diabetes. J Obstet Gynecol Neonatal Nurs 2012; 41(3): 328-38. [http://dx.doi.org/10.1111/j.1552-6909.2012.01366.x] [PMID: 22834880]

[124] Clifton-Bligh RJ, McElduff P, McElduff A. Maternal vitamin D deficiency, ethnicity and gestational diabetes. Diabet Med 2008; 25(6): 678-84. [http://dx.doi.org/10.1111/j.1464-5491.2008.02422.x] [PMID: 18544105]

[125] Perez-Ferre N, Torrejon MJ, Fuentes M, *et al.* Association of low serum 25-hydroxyvitamin D levels in pregnancy with glucose homeostasis and obstetric and newborn outcomes. Endocr Pract 2012; 18(5): 676-84. [http://dx.doi.org/10.4158/EP12025.OR] [PMID: 22548949]

[126] Xu L, Lee M, Jeyabalan A, Roberts JM. The relationship of hypovitaminosis D and IL-6 in preeclampsia. Am J Obstet Gynecol 2014; 210(2): 149.e1-7. [http://dx.doi.org/10.1016/j.ajog.2013.09.037] [PMID: 24080305]

[127] Wei SQ, Audibert F, Luo ZC, *et al.* Maternal plasma 25-hydroxyvitamin D levels, angiogenic factors,

and preeclampsia. Am J Obstet Gynecol 2013; 208(5): 390.e1-6.
[http://dx.doi.org/10.1016/j.ajog.2013.03.025] [PMID: 23618499]

[128] Baker AM, Haeri S, Camargo CA Jr, Espinola JA, Stuebe AM. A nested case-control study of midgestation vitamin D deficiency and risk of severe preeclampsia. J Clin Endocrinol Metab 2010; 95(11): 5105-9.
[http://dx.doi.org/10.1210/jc.2010-0996] [PMID: 20719829]

[129] Bodnar LM, Catov JM, Simhan HN, Holick MF, Powers RW, Roberts JM. Maternal vitamin D deficiency increases the risk of preeclampsia. J Clin Endocrinol Metab 2007; 92(9): 3517-22.
[http://dx.doi.org/10.1210/jc.2007-0718] [PMID: 17535985]

[130] Robinson CJ, Alanis MC, Wagner CL, Hollis BW, Johnson DD. Plasma 25-hydroxyvitamin D levels in early-onset severe preeclampsia. Am J Obstet Gynecol 2010; 203(4): 366.e1-6.
[http://dx.doi.org/10.1016/j.ajog.2010.06.036] [PMID: 20692641]

[131] Bodnar LM, Simhan HN, Catov JM, *et al.* Maternal vitamin D status and the risk of mild and severe preeclampsia. Epidemiology 2014; 25(2): 207-14.
[http://dx.doi.org/10.1097/EDE.0000000000000039] [PMID: 24457526]

[132] Shand AW, Nassar N, Von Dadelszen P, Innis SM, Green TJ. Maternal vitamin D status in pregnancy and adverse pregnancy outcomes in a group at high risk for pre-eclampsia. BJOG 2010; 117(13): 1593-8.
[http://dx.doi.org/10.1111/j.1471-0528.2010.02742.x] [PMID: 21040394]

[133] Powe CE, Seely EW, Rana S, *et al.* First trimester vitamin D, vitamin D binding protein, and subsequent preeclampsia. Hypertension 2010; 56(4): 758-63.
[http://dx.doi.org/10.1161/HYPERTENSIONAHA.110.158238] [PMID: 20733087]

[134] Christesen HT, Falkenberg T, Lamont RF, Jørgensen JS. The impact of vitamin D on pregnancy: a systematic review. Acta Obstet Gynecol Scand 2012; 91(12): 1357-67.
[http://dx.doi.org/10.1111/aogs.12000] [PMID: 22974137]

[135] Allsworth JE, Peipert JF. Prevalence of bacterial vaginosis: 2001-2004 National Health and Nutrition Examination Survey data. Obstet Gynecol 2007; 109(1): 114-20.
[http://dx.doi.org/10.1097/01.AOG.0000247627.84791.91] [PMID: 17197596]

[136] Bodnar LM, Krohn MA, Simhan HN. Maternal vitamin D deficiency is associated with bacterial vaginosis in the first trimester of pregnancy. J Nutr 2009; 139(6): 1157-61.
[http://dx.doi.org/10.3945/jn.108.103168] [PMID: 19357214]

[137] Davis LM, Chang SC, Mancini J, Nathanson MS, Witter FR, O'Brien KO. Vitamin D insufficiency is prevalent among pregnant African American adolescents. J Pediatr Adolesc Gynecol 2010; 23(1): 45-52.
[http://dx.doi.org/10.1016/j.jpag.2009.05.005] [PMID: 19643639]

[138] Dunlop A, Taylor R, Tangpricha V, Fortunato S, Menon R. Maternal vitamin D, folate, and polyunsaturated fatty acid status and bacterial vaginosis during pregnancy. Infect Dis Obs Gynecol 2011; 2011: 216217.

[139] Hensel KJ, Randis TM, Gelber SE, Ratner AJ. Pregnancy-specific association of vitamin D deficiency and bacterial vaginosis. Am J Obstet Gynecol 2011; 204(1): 41.e1-9.
[http://dx.doi.org/10.1016/j.ajog.2010.08.013] [PMID: 20887971]

[140] Merewood A, Mehta SD, Chen TC, Bauchner H, Holick MF. Association between vitamin D deficiency and primary cesarean section. J Clin Endocrinol Metab 2009; 94(3): 940-5.
[http://dx.doi.org/10.1210/jc.2008-1217] [PMID: 19106272]

[141] Scholl TO, Chen X, Stein P. Maternal vitamin D status and delivery by cesarean. Nutrients 2012; 4(4): 319-30.
[http://dx.doi.org/10.3390/nu4040319] [PMID: 22606373]

[142] Fernández-Alonso AM, Dionis-Sánchez EC, Chedraui P, González-Salmerón MD, Pérez-López FR.

First-trimester maternal serum 25-hydroxyvitamin D□ status and pregnancy outcome. Int J Gynaecol Obstet 2012; 116(1): 6-9.
[http://dx.doi.org/10.1016/j.ijgo.2011.07.029] [PMID: 21959069]

[143] Zhou J, Su L, Liu M, *et al.* Associations between 25-hydroxyvitamin D levels and pregnancy outcomes: a prospective observational study in southern China. Eur J Clin Nutr 2014; 68(8): 925-30.
[http://dx.doi.org/10.1038/ejcn.2014.99] [PMID: 24865483]

[144] Schneuer FJ, Roberts CL, Guilbert C, *et al.* Effects of maternal serum 25-hydroxyvitamin D concentrations in the first trimester on subsequent pregnancy outcomes in an Australian population. Am J Clin Nutr 2014; 99(2): 287-95.
[http://dx.doi.org/10.3945/ajcn.113.065672] [PMID: 24257720]

[145] Baker AM, Haeri S, Camargo CA Jr, Stuebe AM, Boggess KA. A nested case-control study of first-trimester maternal vitamin D status and risk for spontaneous preterm birth. Am J Perinatol 2011; 28(9): 667-72.
[http://dx.doi.org/10.1055/s-0031-1276731] [PMID: 21500145]

[146] Bodnar LM, Rouse DJ, Momirova V, *et al.* Maternal 25-hydroxyvitamin d and preterm birth in twin gestations. Obstet Gynecol 2013; 122(1): 91-8.
[http://dx.doi.org/10.1097/AOG.0b013e3182941d9a] [PMID: 23743453]

[147] Bodnar LM, Klebanoff MA, Gernand AD, *et al.* Maternal vitamin D status and spontaneous preterm birth by placental histology in the US Collaborative Perinatal Project. Am J Epidemiol 2014; 179(2): 168-76.
[http://dx.doi.org/10.1093/aje/kwt237] [PMID: 24124195]

[148] Moon RJ, Harvey NC, Cooper C. Influence of maternal vitamin D status on obstetric outcomes and the foetal skeleton. Eur J Endocrinol 2015; 173: 1-33.
[http://dx.doi.org/10.1530/EJE-14-0826] [PMID: 25850828]

[149] Leffelaar ER, Vrijkotte TG, van Eijsden M. Maternal early pregnancy vitamin D status in relation to fetal and neonatal growth: results of the multi-ethnic Amsterdam Born Children and their Development cohort. Br J Nutr 2010; 104(1): 108-17.
[http://dx.doi.org/10.1017/S000711451000022X] [PMID: 20193097]

[150] Ertl R, Yu CK, Samaha R, Akolekar R, Nicolaides KH. Maternal serum vitamin D at 11-13 weeks in pregnancies delivering small for gestational age neonates. Fetal Diagn Ther 2012; 31(2): 103-8.
[http://dx.doi.org/10.1159/000333810] [PMID: 22261570]

[151] Bodnar LM, Catov JM, Zmuda JM, *et al.* Maternal serum 25-hydroxyvitamin D concentrations are associated with small-for-gestational age births in white women. J Nutr 2010; 140(5): 999-1006.
[http://dx.doi.org/10.3945/jn.109.119636] [PMID: 20200114]

[152] Mehta S, Hunter DJ, Mugusi FM, *et al.* Perinatal outcomes, including mother-to-child transmission of HIV, and child mortality and their association with maternal vitamin D status in Tanzania. J Infect Dis 2009; 200(7): 1022-30.
[http://dx.doi.org/10.1086/605699] [PMID: 19673647]

[153] Baker PN, Wheeler SJ, Sanders TA, *et al.* A prospective study of micronutrient status in adolescent pregnancy. Am J Clin Nutr 2009; 89(4): 1114-24.
[http://dx.doi.org/10.3945/ajcn.2008.27097] [PMID: 19244368]

[154] Morley R, Carlin JB, Pasco JA, Wark JD, Ponsonby AL. Maternal 25-hydroxyvitamin D concentration and offspring birth size: effect modification by infant VDR genotype. Eur J Clin Nutr 2009; 63(6): 802-4.
[http://dx.doi.org/10.1038/ejcn.2008.55] [PMID: 19018272]

[155] Ginde AA, Scragg R, Schwartz RS, Camargo CA Jr. Prospective study of serum 25-hydroxyvitamin D level, cardiovascular disease mortality, and all-cause mortality in older U.S. adults. J Am Geriatr Soc 2009; 57(9): 1595-603.
[http://dx.doi.org/10.1111/j.1532-5415.2009.02359.x] [PMID: 19549021]

[156] Wang TJ, Pencina MJ, Booth SL, *et al.* Vitamin D deficiency and risk of cardiovascular disease. Circulation 2008; 117(4): 503-11. [http://dx.doi.org/10.1161/CIRCULATIONAHA.107.706127] [PMID: 18180395]

[157] Pilz S, Tomaschitz A, Ritz E, Pieber TR. Vitamin D status and arterial hypertension: a systematic review. Nat Rev Cardiol 2009; 6(10): 621-30. [http://dx.doi.org/10.1038/nrcardio.2009.135] [PMID: 19687790]

[158] Forman JP, Giovannucci E, Holmes MD, *et al.* Plasma 25-hydroxyvitamin D levels and risk of incident hypertension. Hypertension 2007; 49(5): 1063-9. [http://dx.doi.org/10.1161/HYPERTENSIONAHA.107.087288] [PMID: 17372031]

[159] Ullah M, Uwaifo G, Nicholas W, Koch C. Does vitamin D deficiency cause hypertension? Current evidence from clinical studies and potential mechanisms. Int J Endocrinol 2010; 2010: 579-640. [http://dx.doi.org/10.1155/2010/579640]

[160] Wang L, Ma J, Manson JE, Buring JE, Gaziano JM, Sesso HD. A prospective study of plasma vitamin D metabolites, vitamin D receptor gene polymorphisms, and risk of hypertension in men. Eur J Nutr 2013; 52(7): 1771-9. [http://dx.doi.org/10.1007/s00394-012-0480-8] [PMID: 23262750]

[161] Krause R, Bühring M, Hopfenmüller W, Holick MF, Sharma AM. Ultraviolet B and blood pressure. Lancet 1998; 352(9129): 709-10. [http://dx.doi.org/10.1016/S0140-6736(05)60827-6] [PMID: 9728997]

[162] Witham MD, Ireland S, Houston JG, *et al.* Vitamin D therapy to reduce blood pressure and left ventricular hypertrophy in resistant hypertension: randomized, controlled trial. Hypertension 2014; 63(4): 706-12. [http://dx.doi.org/10.1161/HYPERTENSIONAHA.113.02177] [PMID: 24420547]

[163] Tamez H, Kalim S, Thadhani RI. Does vitamin D modulate blood pressure? Curr Opin Nephrol Hypertens 2013; 22(2): 204-9. [http://dx.doi.org/10.1097/MNH.0b013e32835d919b] [PMID: 23299053]

[164] Mozos I, Marginean O. Links between vitamin D deficiency and cardiovascular diseases. Biomed Res Int 2015; 2015: 1-12.

[165] Bonakdaran S, Varasteh AR. Correlation between serum 25 hydroxy vitamin D3 and laboratory risk markers of cardiovascular diseases in type 2 diabetic patients. Saudi Med J 2009; 30(4): 509-14. [PMID: 19370277]

[166] Amer M, Qayyum R. Relation between serum 25-hydroxyvitamin D and C-reactive protein in asymptomatic adults (from the continuous National Health and Nutrition Examination Survey 2001 to 2006). Am J Cardiol 2012; 109(2): 226-30. [http://dx.doi.org/10.1016/j.amjcard.2011.08.032] [PMID: 21996139]

[167] Ponda MP, Huang X, Odeh MA, Breslow JL, Kaufman HW. Vitamin D may not improve lipid levels: a serial clinical laboratory data study. Circulation 2012; 126(3): 270-7. [http://dx.doi.org/10.1161/CIRCULATIONAHA.111.077875] [PMID: 22718799]

[168] Jorde R, Figenschau Y, Hutchinson M, Emaus N, Grimnes G. High serum 25-hydroxyvitamin D concentrations are associated with a favorable serum lipid profile. Eur J Clin Nutr 2010; 64(12): 1457-64. [http://dx.doi.org/10.1038/ejcn.2010.176] [PMID: 20823896]

[169] Karhapää P, Pihlajamäki J, Pörsti I, *et al.* Diverse associations of 25-hydroxyvitamin D and 1,25-dihydroxy-vitamin D with dyslipidaemias. J Intern Med 2010; 268(6): 604-10. [http://dx.doi.org/10.1111/j.1365-2796.2010.02279.x] [PMID: 20831628]

[170] Wang H, Xia N, Yang Y, Peng DQ. Influence of vitamin D supplementation on plasma lipid profiles: a meta-analysis of randomized controlled trials. Lipids Health Dis 2012; 11: 42. [http://dx.doi.org/10.1186/1476-511X-11-42] [PMID: 22433171]

[171] Reis JP, von Mühlen D, Michos ED, *et al.* Serum vitamin D, parathyroid hormone levels, and carotid atherosclerosis. Atherosclerosis 2009; 207(2): 585-90.
[http://dx.doi.org/10.1016/j.atherosclerosis.2009.05.030] [PMID: 19539290]

[172] Targher G, Bertolini L, Padovani R, *et al.* Serum 25-hydroxyvitamin D3 concentrations and carotid artery intima-media thickness among type 2 diabetic patients. Clin Endocrinol (Oxf) 2006; 65(5): 593-7.
[http://dx.doi.org/10.1111/j.1365-2265.2006.02633.x] [PMID: 17054459]

[173] Yadav AK, Banerjee D, Lal A, Jha V. Vitamin D deficiency, CD4+CD28null cells and accelerated atherosclerosis in chronic kidney disease. Nephrology (Carlton) 2012; 17(6): 575-81.
[http://dx.doi.org/10.1111/j.1440-1797.2012.01611.x] [PMID: 22489804]

[174] Zang L, Fu P, Huang YQ, *et al.* [Vitamin D deficiency and carotid artery intima-media thickness and coronary calcification in patients with diabetic nephropathy]. Sichuan Da Xue Xue Bao Yi Xue Ban 2012; 43(3): 420-424, 450.
[PMID: 22812250]

[175] Wakasugi M, Noguchi T, Inoue M, *et al.* Vitamin D3 stimulates the production of prostacyclin by vascular smooth muscle cells. Prostaglandins 1991; 42(2): 127-36.
[http://dx.doi.org/10.1016/0090-6980(91)90072-N] [PMID: 1775635]

[176] Ameri P, Ronco D, Casu M, *et al.* High prevalence of vitamin D deficiency and its association with left ventricular dilation: an echocardiography study in elderly patients with chronic heart failure. Nutr Metab Cardiovasc Dis 2010; 20(9): 633-40.
[http://dx.doi.org/10.1016/j.numecd.2010.01.002] [PMID: 20399085]

[177] Pilz S, März W, Wellnitz B, *et al.* Association of vitamin D deficiency with heart failure and sudden cardiac death in a large cross-sectional study of patients referred for coronary angiography. J Clin Endocrinol Metab 2008; 93(10): 3927-35.
[http://dx.doi.org/10.1210/jc.2008-0784] [PMID: 18682515]

[178] Wannamethee SG, Welsh P, Papacosta O, Lennon L, Whincup PH, Sattar N. Elevated parathyroid hormone, but not vitamin D deficiency, is associated with increased risk of heart failure in older men with and without cardiovascular disease. Circ Heart Fail 2014; 7(5): 732-9.
[http://dx.doi.org/10.1161/CIRCHEARTFAILURE.114.001272] [PMID: 25104043]

[179] Goleniewska B, Kacprzak M, Zielińska M. Vitamin D level and extent of coronary stenotic lesions in patients with first acute myocardial infarction. Cardiol J 2014; 21(1): 18-23.
[PMID: 23677723]

[180] Lee JH, Gadi R, Spertus JA, Tang F, O'Keefe JH. Prevalence of vitamin D deficiency in patients with acute myocardial infarction. Am J Cardiol 2011; 107(11): 1636-8.
[http://dx.doi.org/10.1016/j.amjcard.2011.01.048] [PMID: 21439530]

[181] Giovannucci E, Liu Y, Hollis BW, Rimm EB. 25-hydroxyvitamin D and risk of myocardial infarction in men: a prospective study. Arch Intern Med 2008; 168(11): 1174-80.
[http://dx.doi.org/10.1001/archinte.168.11.1174] [PMID: 18541825]

[182] Hlaing SM, Garcia LA, Contreras JR, Norris KC, Ferrini MG, Artaza JN. 1,25-Vitamin D3 promotes cardiac differentiation through modulation of the WNT signaling pathway. J Mol Endocrinol 2014; 53(3): 303-17.
[http://dx.doi.org/10.1530/JME-14-0168] [PMID: 25139490]

[183] Robinson-Cohen C, Hoofnagle AN, Ix JH, *et al.* Racial differences in the association of serum 25-hydroxyvitamin D concentration with coronary heart disease events. JAMA 2013; 310(2): 179-88.
[http://dx.doi.org/10.1001/jama.2013.7228] [PMID: 23839752]

[184] Deleskog A, Piksasova O, Silveira A, *et al.* Serum 25-hydroxyvitamin D concentration, established and emerging cardiovascular risk factors and risk of myocardial infarction before the age of 60 years. Atherosclerosis 2012; 223(1): 223-9.

[http://dx.doi.org/10.1016/j.atherosclerosis.2012.04.014] [PMID: 22652526]

[185] Ng LL, Sandhu JK, Squire IB, Davies JE, Jones DJ. Vitamin D and prognosis in acute myocardial infarction. Int J Cardiol 2013; 168(3): 2341-6. [http://dx.doi.org/10.1016/j.ijcard.2013.01.030] [PMID: 23415169]

[186] Correia LC, Sodré F, Garcia G, *et al.* Relation of severe deficiency of vitamin D to cardiovascular mortality during acute coronary syndromes. Am J Cardiol 2013; 111(3): 324-7. [http://dx.doi.org/10.1016/j.amjcard.2012.10.006] [PMID: 23174181]

[187] Kunadian V, Ford GA, Bawamia B, Qiu W, Manson JE. Vitamin D deficiency and coronary artery disease: a review of the evidence. Am Heart J 2014; 167(3): 283-91. [http://dx.doi.org/10.1016/j.ahj.2013.11.012] [PMID: 24576510]

[188] Pilz S, Dobnig H, Fischer JE, *et al.* Low vitamin d levels predict stroke in patients referred to coronary angiography. Stroke 2008; 39(9): 2611-3. [http://dx.doi.org/10.1161/STROKEAHA.107.513655] [PMID: 18635847]

[189] Brøndum-Jacobsen P, Nordestgaard BG, Schnohr P, Benn M. 25-hydroxyvitamin D and symptomatic ischemic stroke: an original study and meta-analysis. Ann Neurol 2013; 73(1): 38-47. [http://dx.doi.org/10.1002/ana.23738] [PMID: 23225498]

[190] Chung PW, Park KY, Kim JM, *et al.* 25-hydroxyvitamin D status is associated with chronic cerebral small vessel disease. Stroke 2015; 46(1): 248-51. [http://dx.doi.org/10.1161/STROKEAHA.114.007706] [PMID: 25424481]

[191] Skaaby T, Husemoen LL, Pisinger C, *et al.* Vitamin D status and incident cardiovascular disease and all-cause mortality: a general population study. Endocrine 2013; 43(3): 618-25. [http://dx.doi.org/10.1007/s12020-012-9805-x] [PMID: 23015273]

[192] Gupta A, Prabhakar S, Modi M, Bhadada SK, Lal V, Khurana D. Vitamin D status and risk of ischemic stroke in North Indian patients. Indian J Endocrinol Metab 2014; 18(5): 721-5. [PMID: 25285293]

[193] Pilz S, Tomaschitz A, Drechsler C, Zittermann A, Dekker JM, März W. Vitamin D supplementation: a promising approach for the prevention and treatment of strokes. Curr Drug Targets 2011; 12(1): 88-96. [http://dx.doi.org/10.2174/138945011793591563] [PMID: 20795935]

[194] Theodoratou E, Tzoulaki I, Zgaga L, Ioannidis JP. Vitamin D and multiple health outcomes: umbrella review of systematic reviews and meta-analyses of observational studies and randomised trials. BMJ 2014; 348: g2035. [http://dx.doi.org/10.1136/bmj.g2035] [PMID: 24690624]

[195] Melamed ML, Muntner P, Michos ED, *et al.* Serum 25-hydroxyvitamin D levels and the prevalence of peripheral arterial disease: results from NHANES 2001 to 2004. Arterioscler Thromb Vasc Biol 2008; 28(6): 1179-85. [http://dx.doi.org/10.1161/ATVBAHA.108.165886] [PMID: 18417640]

[196] Zhou W, Ye S. Relationship between serum 25-hydroxyvitamin D and lower extremity arterial disease in type 2 diabetes mellitus patients and the analysis of the intervention of vitamin D. J Diabetes Res 2015; 2015 6 pages

[197] Norman PE, Powell JT. Vitamin D and cardiovascular disease. Circ Res 2014; 114(2): 379-93. [http://dx.doi.org/10.1161/CIRCRESAHA.113.301241] [PMID: 24436433]

CHAPTER 8

New Research About Biomarkers and Atrial Fibrillation

Diana Hernández-Romero[1,*], Vanessa Roldán[2], Mariano Valdés[1] and Francisco Marín[1]

[1] *Department of Cardiology, Virgen de la Arrixaca Clinic University Hospital, Murcia, Spain; Instituto de Investigación Biomédica-Virgen de la Arrixaca, IMIB-Arrixaca, Murcia*

[2] *Department of Hematology, Morales Meseguer University Hospital, Murcia, Spain; Instituto de Investigación Biomédica-Virgen de la Arrixaca, IMIB-Arrixaca*

Abstract: Atrial fibrillation (AF) is the most common sustained chronic cardiac arrhythmia in clinical practice, increasing the risk of stroke, thromboembolism and mortality. The pathophysiology of AF is complex, including inflammation, oxidative stress, structural remodeling with apoptosis or fibrosis, leading to remodeling in the atria. However, the underlying mechanisms involved in the development of AF are not fully understood. Biomarkers, mainly determined in peripheral blood may increase our knowledge of the pathophysiology of AF with important implications in the assessment of AF diagnosis, prognosis, or therapy decision-making. The aim of this chapter is to provide an exhaustive overview of the knowledge about classical and novel biomarkers in AF and their implications, diagnostic and therapeutic potential.

Keywords: Adiponectin, Atrial fibrillation, Biomarkers, BTP, CRP, Cystatin D-dimer, E-sel, GDF-15, IL-6,NT-proBNP, Micro particles, MiRNAs, Mortality, P-sel, rTPA, sTM, Stroke, Thromboembolism, Troponin, VWF.

INTRODUCTION

Atrial fibrillation (AF) is the most common sustained cardiac arrhythmia associated with pronounced morbidity, high risk of stroke, thromboembolism and mortality [1, 2]. In addition to these complications, many patients with AF have impaired quality of life, with poor cognitive function, thus increasing health care costs. Approximately 1–2% of the general population is affected by AF, with a rising prevalence up to 8–10% for population aged >80 years [3 - 6]. The pathophysiology of AF is complex and the precise mechanisms for the deve-

* **Corresponding author Diana Hernández-Romero:** Department of Cardiology, Virgen de la Arrixaca Clinic University Hospital, Murcia, Spain; Instituto de Investigación Biomédica-Virgen de la Arrixaca, IMIB-Arrixaca, Murcia; Tel/Fax: 868888151; E-mail: dianahr@um.es

lopment of AF are not completely understood, but are thought to be multifactorial. Initial stages of AF may be caused by the effect of different pathophysiological processes, including inflammation, oxidative stress and fibrosis. The process includes a structural remodeling as well as affected atrial electrophysiological properties promoting the initiation and perpetuation of AF [7, 8]. There are different clinical presentations of AF: AF usually presents in paroxysmal form and is defined by self-termination within 7 days. In persistent AF, termination only occurs by pharmacological or direct-current electric cardio version, whereas in permanent AF, no restoration to sinus rhythm is possible or advisable [9]. Currently, management of AF focuses on the reduction of symptoms and the prevention of future complications. Many AF patients remain asymptomatic with the occurrence of fatal or severe complications as first manifestation of this arrhythmia. In consequence, improved diagnostic techniques have identified various biomarkers that may play an important role in the prediction of AF and derived outcomes.

Food and Drug Administration defines biomarker as any measurable indicator that is potentially useful along the whole spectrum of the disease process; research and development of new therapies, diagnosis, prognosis and monitoring progression of a disease or response to treatment [10]. The principal characteristics for an ideal biomarker are; easy to obtain with minimum risk or discomfort for the patient, it may also increase or decrease its concentration over the course of disease progression and thus be useful for the disease prognosis within an individual. Finally, a reliable biomarker has a detection method that could be both sensitive and specific and strongly reproducible among clinical laboratories [10].

The major reason for the current limited effectiveness of the therapies in AF may be the fact that they do not address the underlying pathophysiology of AF, thus, only target a single mechanism, whereas AF is a multifactorial, progressive process [11]. Hence, the understanding of the processes underlying AF is crucial for the development of novel therapeutic approaches [12].

PATHOPHYSIOLOGICAL MECHANISMS OF AF DEVELOPMENT

It has been proposed that the natural history of AF progresses from paroxysmal to persistent to permanent AF due to structural and electrical remodeling and to the disease progression [9]. Focal ectopic firing usually initiates AF, which can then be maintained by either rapid focal ectopic firing or re-entry in a vulnerable substrate [13 - 15]. The underlying causes of this focal ectopic firing or re-entry substrates are determined by atrial remodeling, involving structural and electrical remodeling as well as calcium homeostasis and neurohormonal disequilibria [16].

Electrical remodeling in AF is the consequence of shortening of the atrial effective refractory period (ERP), caused by shortening of the action potential duration (APD) [17, 18]. Subsequently, ADP is decreased, due to the enhancement of inward rectifier K^+ current (I_{K1}) or acetylcholine-dependent K^+ current, (I_{KACh}), or decrease in inward L-type Ca^{2+} current (ICaL) [13, 14, 19, 20]. Decreased ADP increases AF vulnerability and maintenance. Changes in Ca^{2+} handling promote ectopic activity. Electric remodeling participates of different clinically relevant processes such early AF recurrence or progression from paroxysmal to persistent forms [9].

Cardiac structural remodeling, mainly atrial fibrosis, is slower to develop than electrical remodeling [9, 16]. Increased atrial dimensions predict the occurrence of AF [21], enhancing complex re-entrant arrhythmia mechanisms [22]. In addition, AF itself promotes atrial fibrosis [23]. Renin–angiotensin–aldosterone system, transforming growth factor-β1 (TGF-β1) signaling, and the oxidative stress pathway have been identified in fibrotic remodeling, including the [9, 14, 24, 25]. Despite precise mechanisms underlying atrial fibrosis are still not completely understood, fibrosis promotes AF progression to permanent forms, indicating that its development is a possible therapeutic target (Fig. **1**) [9].

The autonomic nervous system is important in AF [26]. Vagal activation enhances I_{KACh}, which results in shortening APD and favours re-entry. Adrenergic stimulation increases Ca^{2+} diastolic leak, promoting delayed after-depolarizations. Atrial autonomic hyper-innervation enhances spontaneous AF in experimental models and, perhaps, in patients with AF [26].

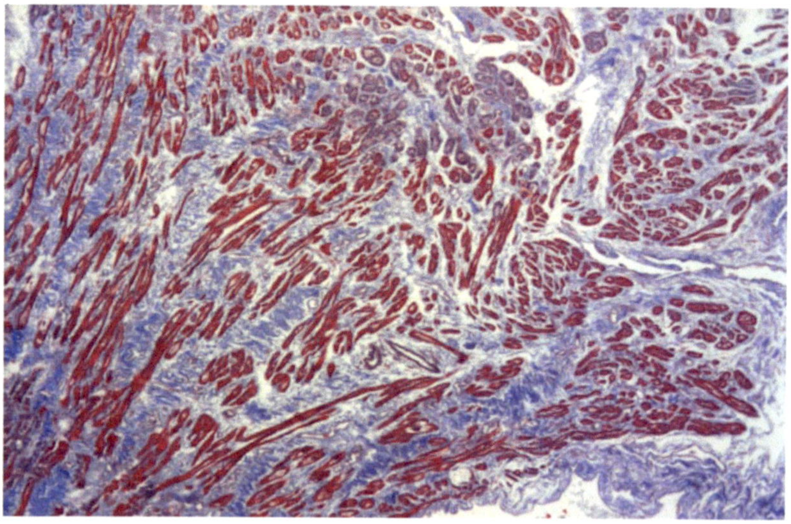

Fig. (1). Connective tissue deposition and myocardial fibrosis in atrial tissue assessed by Masson's Trichrome staining. Magnification x100.

ESTABLISHED BIOMARKERS OF CARDIOVASCULAR PATHOLOGY

Various biomarkers determined in blood have been studied to improve the prediction or identification of the high risk patients (Fig. (**2**), Table **1**).

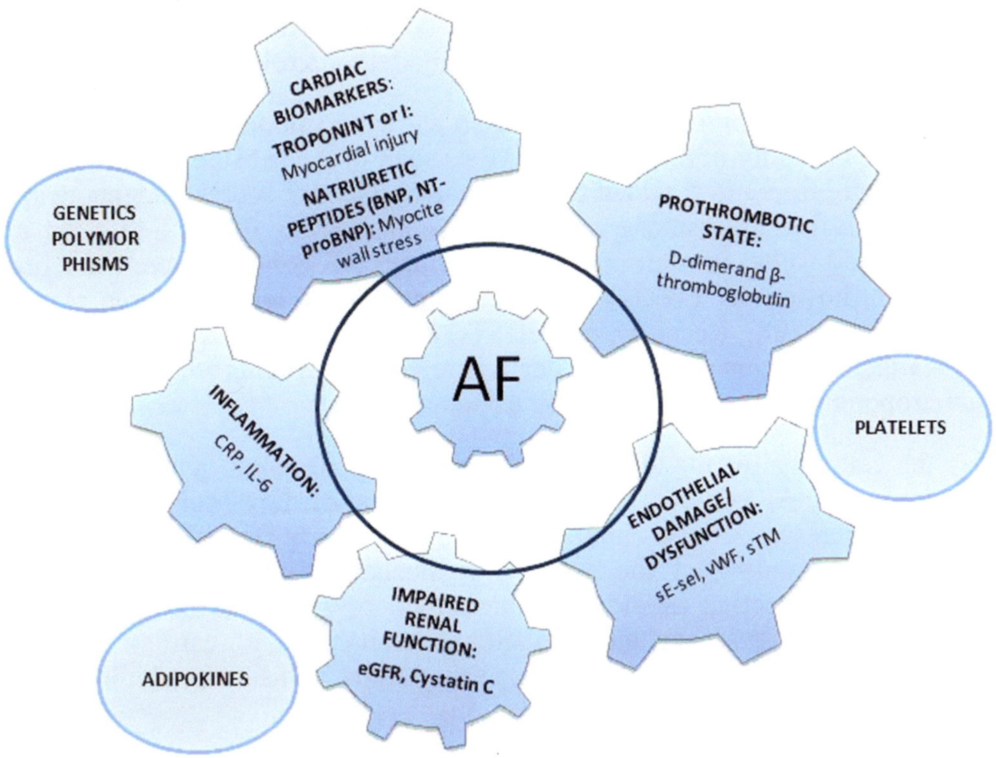

Fig. (2). Different pathways involved in the AF pathophysiology related to various biomarkers (Obtained from Vílchez *et al.*, 2014).

Biomarkers of Myocardial Injury: Cardiac Troponins

Troponin together with tropomyosin forms a regulatory protein complex involved in muscular contraction [27]. Cardiac troponins are known as sensitive and specific biomarkers of myocardial injury [28]. They have been proposed as biomarkers of cardiac remodeling that promotes AF occurrence [29], and found associated with different events within this pathologic context. Troponin levels have been shown to be elevated in patients with different cardiopathies such as stable coronary artery disease, heart failure, and elderly apparently healthy population, being associated with poor outcomes and mortality [30 - 32]. In a substudy of the Randomized Evaluation of Long Term Anticoagulant Therapy (RE-LY) trial in 6189 patients with AF and treated with either warfarin or dabigatran, they evaluated cardiac troponins as markers of high risk of stroke or systemic embolism [33]. A subestudy of the Apixaban for the Prevention of Stroke in Subjects with Atrial Fibrillation (ARISTOTLE) corroborated that troponin levels were related to the risk of stroke, with independency of the population base-line characteristics and other biomarkers [34]. The troponin I level provided significant incremental prognostic information in comparison with $CHADS_2$ and CHA_2DS_2-VASc to a predictive model for stroke outcomes.

These biomarkers of myocardial injury also show an important contribution in prediction of mortality in patients with AF as seen in the substudy from the RE-LY trial, where troponin I, added to thromboembolic risk scores, independently improved the risk assessment for cardiovascular death [33]. Roldán *et al.* confirmed the results, in a chronic anticoagulated AF cohort, where poor prognosis as cardiovascular events and mortality associated with increased plasma troponin T levels [28]. Recently, an ARISTOTLE substudy showed that the risk of myocardial infarction and cardiac death is highest in patients with raised levels of both troponins, being even higher for troponin I [34].

Today the causality for the association between high troponin and AF remains unknown. In fact, AF could be caused by concomitant cardiovascular risk factors, or troponin increase may be simply due to a sick heart. Despite the lack of a complete understanding of the underlying mechanism, cardiac troponin has become a very attractive candidate as biomarker of prognosis in AF patients. Hence, the strong evidence and the global availability of cardiac troponin determination in most hospitals worldwide abrogate its inclusion in addition to the recommended clinical stroke risk stratification.

Biomarkers of Heart Wall Stress: Natriuretic Peptides

Pre–proBNP is synthesized in wall of the ventricular myocardium in response to volume expansion or pressure overload. Subsequently, the peptide is cleaved first to proBNP and then to the biologically active BNP and the inactive amino-terminal fragment (NTproBNP). The release of BNP results in improved myocardial relaxation and serves an important regulatory role in response to acute increases in ventricular volume by opposing the vasoconstriction, sodium retention, and antidiuretic effects of the activated renin-angiotensin-aldosterone system [35]. Natriuretic peptides are extraordinary markers of LV function being a simple and effective tool for evaluation and diagnosis of heart failure or LV dysfunction [36].

Initial studies observed higher levels of BNP in patients with AF in comparison with matched controls in sinus rhythm [37, 38]. Different authors have studied the role of NT-proBNP levels in AF. In the RE-LY substudy, the levels of NT-proBNP correlated with the risk of thromboembolic events and cardiovascular mortality even after adjusting by confounding factors. Patients within the highest quartiles of NT-proBNP resulted in a doubled risk of stroke or systemic embolism and fivefold higher risk for cardiovascular mortality [33]. In the larger ARISTOTLE study, a relevant strong association between increased risk of ischemic stroke and elevated NT-proBNP levels was observed, as well as doubled risk of death [39]. Hence, the contribution of NT-proBNP to the risk stratification profiles significantly improved the prediction of mortality and cardiovascular events. Recently Roldán *et al.*, showed in a real-world cohort of stable AF patients in anticoagulant treatment that NT-proBNP complemented information of prognostic risk prediction for stroke/systemic embolism when added to a previous established clinical risk score (CHA_2DS_2-VASc) [40].

A plausible underlying pathophysiologic mechanism suggests that atrial dysfunction is an established risk factor of thrombus formation in AF and thereby explains the relation between natriuretic peptides and thromboembolic events in AF [41]. In fact, elevated BNP levels were strongly associated with new onset AF in patients with acute ischemic stroke. Including determinations of NT-proBNP values could help to implement better stratification scales for risk evaluation in AF patients in routine clinical practice.

Biomarkers of Renal Function

The prevalence of AF is higher in impaired renal function populations compared with the general population. In addition, AF prevalence increases in patients with general chronic kidney disease presenting low glomerular filtration rates (GFR) [42, 43]. GFR is estimated form serum levels of endogenous filtration markers such as creatinine and is accepted as useful index of renal function. Data from the ATRIA study concluded that chronic kidney disease increased the risk of thromboembolism and stroke in AF independently of other risk factors. Decrease in GFR and presence of proteinuria associated with increased risk [44]. Similar findings were published based on ARISTOTLE trial population where it was observed higher rates of stroke as renal function was impaired [45]. Roldán *et al* reported that impaired renal function associated with cardiovascular events, mortality and bleeding [46]. However reported data based on c-statistics and the integrated discrimination improvement, showing that adding CKD to the stroke risk scores did not independently improve the predictive value of current clinical scores [47].

BTP and cystatin C have been proposed as more reliable markers of renal function than serum creatinine, particularly when of low reductions in GFR occurred [48, 49]. Beta-Trace protein (BTP) is a lipocalin glycoprotein localized in myocardial cells, atrial and ventricular endocardial cells [50]. Vílchez *et al.*, demonstrated that the addition of BTP improves both the predictive value for thrombotic events of the clinical CHA_2DS_2-VASc score as well as for major bleeding prediction in the HAS-BLED score [51]. Cystatin C is a proposed to reflect micro vascular renal dysfunction and significantly improves risk stratification in both elderly and in coronary artery disease populations [52]. In ARISTOTLE substudy high cystatin C levels were independently associated with increased rates of systemic embolism, stroke and major bleedings [45].

Biomarkers of Inflammation

Interleukin-6 (IL-6), a circulating cytokine produced by macrophages, T lymphocytes fibroblasts and endothelial cells, is the inflammatory marker that best correlates to AF [53]. C-reactive protein (CRP) is an established inflammatory biomarker synthesized in hepatocytes as an acute-phase reactant. Preliminary results from RE-LY biomarker substudy [54] showed that top quartile levels of IL-6 doubled the stroke risk and higher quartiles of CRP independently associated with cardiovascular death in adjusted analysis. Conway *et al.,* [55] observed the association between both CRP and IL-6 and a composite outcome of stroke and death in AF patients. A larger subestudy of the Stroke Prevention in Atrial Fibrillation III trial showed the prognostic value of CRP to a composite of ischaemic stroke, myocardial infarction or vascular death and to all cause mortality [56]. A substudy of ARIC cohort, confirmed the results on hsCRP improving the CHA_2DS_2-VASc prediction of mortality [57].

High IL6 and TnT have been reported to significantly associate with stroke/TIA, systemic embolism, long-term cardiovascular events and death, even after adjusting for $CHADS_2$ score [28]. Despite this could suggest an inflammatory state in AF patients, it seems more plausible that clinical variables and comorbidities explain the increase in pro-inflammatory biomarkers.

Biomarkers of Prothrombotic State

AF promotes changes in blood flow with effect in vessel walls and atrial dilatation, reflected in blood constituents fulfilling Virchow`s triad for thrombogenesis [58]. Plasma D-dimer is a fibrin degradation product and is a marker of intravascular thrombogenesis. D-dimer levels have been associated with thromboembolic events in non-valvular AF patients, even under anticoagulation treatment [59, 60]. However, a study by Roldán *et al.*, did not find that D-dimer levels were related to the prognosis in an anticoagulated AF cohort [61]. Data from trials as RE-LY or ARISTOTLE patients showed a significant association between D-dimer levels and the risk of stroke, cardiovascular death and major bleeding events independent of established risk factors including

CHA_2DS_2-VASc variables [62, 63]. These results suggest that D-dimer may also contribute as clinically relevant risk biomarker in AF.

Abnormal concentrations of prothrombin fragment 1 +2 (F1+2) reflecting *in vivo* thrombin generation, is reported to be elevated in AF and is suppressed by anticoagulation in a dose-dependent manner. In the Third Stroke Prevention in Atrial Fibrillation (SPAF III) study, elevated F1+2 levels, as index of thrombogenesis, were associated with a clinical risk factor for stroke in AF [64]. Furthermore, F1+2 levels resulted increased in participants who subsequently suffered thromboembolic events, but with only marginal statistical differences.

A dysfunction in the fibrinolytic system dysfunction has been also related to increased risk of thrombosis. It has been observed that, in AF patients, elevated levels of plasminogen activator inhibitor, PAI-1, caused hypofibrinolytic state, with no increase in plasmin-antiplasmin, PAP, complex [65]. Moreover, increased levels of t-PA antigen levels significantly associated with composite card-iovascular endpoints' in AF patients [66]. In consequence, high levels of D-dimer and t-PA antigen could be useful biomarkers of cardiovascular risk in AF patients.

Biomarkers of Endothelial Dysfunction

Atrial fibrillation provides hypercoagulable state with altered hemostasis and coagulation [58]. Despite it has been proven the immediate improvement of endothelial function after restoration of sinus rhythm by catheter ablation or electrical cardio version, more sustained injury and shedding of endothelial cells could be relevant mechanisms involved in longer-term thromboembolic complications. These changes in the hypercoagulable state appear to be persistent after sinus rhythm restoration contributing to thromboembolic complications [67].

vWF, which is synthesized by vascular endothelial cells and promotes platelet adhesion and aggregation, has been proposed as an established biomarker of endothelial dysfunction [61]. High plasma vWF levels associated with predictors of stroke and stroke risk stratification schemes, as well as with major bleeding even in anticoagulated permanent AF patients [66 - 68]. Roldan *et al.* confirmed these results whereby increased plasma vWF levels resulted associated with increased thrombotic and major bleeding risks, showing an additive effect on the HAS-BLED score [61].

Endothelial dysfunction, as demonstrated by impaired flow-mediated dilatation FMD and raised vWF and E-selectin, is present in AF [67]. Krishnamoorthy *et al.* reported that high plasma vWf and sE-sel levels associated with an increased risk of ischaemic stroke in 'real-world' patients with AF [69]. Plasma levels of sTM, a soluble form of thrombomodulin, may contribute to the hypercoagulable state in

AF and is a recognized marker of endothelial dysfunction, since sTM levels have been found to be lower in patients with persistent AF [70].

Platelets

It is under discussion whether platelet activation in AF might be the result of the associated comorbidities such as hypertension or vascular disease, rather than a real enhancement of the prothrombotic state in AF per se [58]. Hence, it has been proposed that platelet activation might even play a role in clinical progression of AF, since increased high soluble CD40L level was found to be predictive for vascular events in patients with AF [56, 71]. On the other hand, Choudhry *et al.*, showed higher levels of platelet micro particles and soluble P-selectin in AF patients in comparison to healthy controls, however, no difference was shown when comparing with disease patients in sinus rhythm [72].

P-selectin levels can predispose to thrombosis and vascular events [73]. In the Rotterdam Study, plasma P-selectin levels associated to clinical worse outcomes in AF, suggesting a role of platelets in the prothrombotic state [74]. Moreover, Hayashi *et al.* published an increase in P-selectin expression and micro particles in platelets in AF induction models [75]. Increased beta-thromboglobulin (BTG), a specific marker of platelet activation, raised in patients with AF [58, 76]. However, in the SPAF III study, BTG levels were not predictive of thromboembolic events [64].

Table 1. Biomarkers related to AF in clinical studies.

Biomarker	Principal findings	Study	References
Troponin	Indicator of high risk of stroke, systemic embolism or mortality	RE-LY	[33]
	Independently associated with risk of stroke, CV death and MI	ARISTOTLE	[34]
	Prognostic value of hsTnT associated with CV events	Prospective real-world cohort	[28]
NT-proBNP	Correlation with risk of thromboembolic events and CV mortality Association with risk of stroke and death Prognostic added information to clinical risk scores	RE-LY ARISTOTLE Prospective real-world cohort	[33] [39] [40]
GFR	Inversely associated with risk of thromboembolism and stroke	ATRIA	[44]
	Association with risk of stroke	ARISTOTLE	[45]
	Association with cardiovascular events, mortality and bleeding	Prospective real-world cohort	[46]

(Table 1) contd.....

Biomarker	Principal findings	Study	References
BTP	Added information to clinical risk score for thrombotic and bleeding risks	Consecutive AF patients	[51]
Cystatin C	increased rates of stroke, systemic embolism and major bleedings	ARISTOTLE	[45]
IL-6	Associated with stroke risk IL-6 and CRP associated with a composite outcome of stroke and death in AF Assessed clinical usefulness for CV events	RE-LY Prospective chronic AF cohort Prospective real-world cohort	[54] [55] [28]
CRP	Associated with CV death IL-6 and CRP associated with a composite outcome of stroke and death in AF Associated to all cause mortality and a composite outcome of ischaemic stroke, MI or CV death Association of CRP with all cause and CV death	RE-LY Prospective chronic AF cohort SPAF III ARIC	[54] [55] [56] [57]
D-dimer	Predictor of CV events Associated with thromboembolic events	Prospective chronic AF cohort RE-LY	[59,60]
F1+2	Relation with stroke	SPAF III	[64]
t-PA-PAI	Predictive value for CV events	Prospective real-world cohort	[65]
	Association with combined cardiovascular events	Prospective chronic AF cohort	[66]
vWF	Association with stroke and with major bleeding Association with thrombotic events and major bleeding	SPAF III Prospective real-world cohort	[68] [61]
E-selectin	Relationship with CV events	Real-world AF community patients	[69]
CD40L	Predictive value for stroke and myocardial infarction	Prospective non-valvular AF cohort	[56,71]
P-selectin	Prediction of clinical adverse outcomes in AF	Rotterdam Study	[74]

GFR: glomerular filtration rate; (F1+2): prothrombin fragment 1 +2; BTP: beta trace protein; IL-6: interleukin-6; CRP: C-reactive protein; t-PA-PAI: tissue plasminogen activator-inhibitor plasminogen activator; vWF: Von Willebrand Factor.

NEW PROPOSED BIOMARKERS

MicroRNAs

Cardiac remodeling has been defined as an adaptive, regulatory process of cardiac myocytes that occurs over time in order to maintain homeostasis against external stresses [77]. There are two main forms of cardiac remodeling: a structural remodeling, affecting cardiac tissue organization and an electrical remodeling with alteration in the electrical properties of the myocardial cell [78].

MicroRNAs (miRNAs) are small noncoding RNAs (~22 nucleotides) first discovered in Caenorhabditis elegans in 1993 by Lee *et al.* [79]. The suppressive mechanism includes mRNA degradation or blockage of mRNA translation of different target genes [80]. The main function of miRNAs is to repress expression of target gene(s) at the post-transcriptional level. In the human genome ~1,200 miRNAs have been identified to date, regulating more than 60% of protein-coding genes [81]. Each miRNA may regulates several target genes [82] by annealing to complementary sequences in the 3'-untranslated regions (3'UTR) of their target mRNAs, causing mRNA to be cleaved or repressing the translational machinery [83].

The pathophysiological role of miRNAs has been investigated in different diseases, including cardiac diseases [84, 85]. The most recent research on the roles of miRNAs in AF initiation and perpetuation proves that changes in gene expression induced by miRNA could participate in the structural and electrical remodeling responsible for AF maintenance [86].

miRNAs and Cardiac Structural Changes

Structural remodeling, leaded by atrial fibrosis, is directly related with the pathophysiological substrate of the onset of AF [87]. Several miRNAs have been investigated into this concept and related to AF initiation and maintenance (Table **2**).

miR-1: is the second more abundant miRNA in human heart after miR133 [88]. miR-1 have been shown to inhibit cardiac hypertrophy by targeting the cardiac transcription factor Hand2 [89] and perform pro-apoptotic roles [90]. Data indicating its involvement in ventricular myocytes failure and myocardial arrhythmias [91] have been reported, suggesting miR-1 as a trigger of the promotion of the arrhythmogenic substrate.

miR-21: has been proved to participate in the AF development in humans by its target Sprouty. miR-21 up regulation is consistently associated with cardiac remodeling related to AF. Blocking of its endogenous synthesis reduces extracellular matrix formation related to cardiac fibrosis and remodeling [92]. Roles preventing myocyte damage under oxidative stress or as oncomir with anti-apoptotic functions have also been reported [93, 94].

miR-26: regulates the expression of transient receptor potential channel (TRPC3), which regulates Ca^{2+} entry into the cell. This entry was reported increased in dogs with AF [95], enhancing atrial fibrosis, since it has been demonstrated a down regulation of miR-26 in these atrial fibroblasts.

miR-29: targets different genes from the extracellular matrix such as collagens, elastin or fibrillin [96], and its expression inversely correlates with protein levels in their development of AF [97]. Its use as biomarker is supported by reports indicating decreased miR-29b expression in atrial tissue and blood in AF patients [97].

miR-30: its expression is regulated in a similar way than miR-133, thus reduced in dogs with AF [98], and contributes to ventricular fibrosis [99].

miR-133: is the most common miRNA in human heart [88] and its down regulation contributes to ventricular fibrosis during cardiac hypertrophy by connective tissue growth factor (CTGF) depression [99] as described for miR-30. Levels of miR-133 where down regulated in AF, when compared with sham controls in dogs [98]. In addition, different roles have been proposed, such as hypertrophy prevention, suppression of embryonic cardiomyocyte proliferation or apoptosis prevention [100, 101].

miR-590: it is plausible that down regulation of miR-590 and miR-133 are responsible for the fibrogenesis in dog models, since their expression have been reported to be reduced whereas TGF-β1 and its type 2 receptor resulted up regulated in nicotine-treated dogs, causing atrial fibrosis and AF [102].

miRNAs and Electrical Remodeling

A shortening in the action potential duration, mainly due to changes in ionic channels and connexins expression in atrial myocytes, has been found to underlie atrial electrical remodeling in AF [103]. Different reports have studied the contribution of different miRNAs in the atrial electrical changes and their contribution to AF progression.

miR-1: its involvement in electrical remodeling was proposed in coronary patients [104]. Several studies have investigated the involvement of miR-1 in atrial electrical components [103, 105, 106] and its participation in AF development is clear, but further experimentation is needed to elucidate the exact underlying mechanism.

miR-26: regulates a subunit of the ion channel I_{K1} increasing the inward-rectifier current, and its down regulation has been found to enhance vulnerability to AF. Accordingly, miR-26 overexpression reduces this vulnerability [107].

miR-155: has been reported to be increased in AF patients compared with controls. There are data supporting the role of miR-155 in the arrhythmia outcomes where it might target expression of critical ion channel genes, including CACNA1C and KCNA4, encoding the cardiac L-type Ca^{2+} channel (I_{CaL}) α1c and the initial component of the transient outward potassium current (I_{to1}), respectively [108]. Thus, mRNA-155 might be utilized as a biomarker for AF detection and prognosis.

miR-208a: has been published that its lack results in deficiencies in Cx40 transcript expression, an atrial molecule expressed in cardiomyocytes of the His bundle and Purkinje fibers [109]. Hence, miR-208 presents a role in electric remodeling perpetuating AF [86].

miR-328: is up regulated in patients with AF. Its overexpression increase vulnerability to AF. miR-328 has been reported to affect calcium ion channels, resulting in proarrhythmic effect mediated by I_{CaL} down regulation in animal models [110].

miR-499: up regulates in atrial tissue from AF patients [111]. It down regulates KCNN3, the small conductance calcium-activated potassium channel protein 3, associated with AF development [112].

Table 2. Principal miRNAs involved in AF occurrence.

Pathophysiological pathway	miRNA	Target gene(s)	Biological effect related to AF	References
Structural remodeling	miR-1	*HAND2*	Decrease in cardiomyocyte proliferation. Pro-apoptotic	[88-91]
	miR-21	*SPRY1*	Myocardial fibrosis. Role in oxidative stress and apoptosis	[92-94]
	miR-26	*TRPC3*	Calcium homeostasis. Atrial fibrosis	[90]
	miR-29	*COL1A1/COL1A2/ELN/FBRN1*	Decrease in collagen and other ECM proteins	[96, 97]
	miR-30 miR-133	*CTGF* *CTGF*	Inhibition of ventricular fibrosis Inhibition of cardiac fibrosis. Anti-apoptotic role	[88, 98, 99]
	miR-590	*TGFB1/TGFB2*	Atrial fibrosis	[102]

(Table 2) contd.....

Pathophysiological pathway	miRNA	Target gene(s)	Biological effect related to AF	References
Electrical remodeling	miR-1	*KCNJ2/KCNE1/KCNB2*	Shortening in atrial effective refractory period	[103, 105, 106]
	miR-26	*KCNJ2*	Increase in the inward-rectifier current	[107]
Electrical remodeling	miR-155 miR-208a miR-328 miR-499	*CACNA1C/ KCNA4* *GJA5* *CACNA1C/CACNB1* *KCNN3*	Regulation of calcium channels Remodeling in His bundle and Purkinje fibers Shortening in atrial potential duration Affects calcium channels	[108, 109, 110, 112]

Microparticles

Microparticles (MPs) are microvesicles between (0.1-1.0 μm), characterized by series of typical morphological features such as cytoskeleton reorganization and exposure of phosphatidylserine (PS) [113]. These MPs are produced by shedding of vascular related cells as a consequence of stress conditions, including cellular activation and apoptosis but can even occur in resting no-activated platelets [114 - 116]. Circulating MPs have been found increased in patients with several cardiovascular risk factors, coronary artery disease and ACS [117], correlating to parameters of platelet activation and endothelial dysfunction [118].

It has been suggested enhanced thrombogenicity could be due to a sustained and enhanced generation of MPs during AF [119]. It has been reported that procoagulant MPs increase in patients with AF compared to patients with cardiovascular risk factors in sinus rhythm and in them when compared with patients in sinus rhythm without cardiovascular risk factors [119]. Furthermore patients with permanent or persistent AF increased by 2-fold procoagulant MPs in comparison to age-matched controls [120].

High levels of platelet-derived MPs (PMPs) have been associated with AF [121, 122]. Choudhry *et al.* reported increased levels of PMPs and P-selectin in AF patients when compared with healthy controls, but no difference was found between AF patients and disease-matched controls [123], suggesting the underlying comorbidities as causal reasons for PMPs increase. Similar explanation has been suggested for the shedding of endothelial-derived MPs (EMPs) during AF, indicating that its increase is due to cardiovascular risk factors rather than AF itself [119]. On the contrary, data indicating differences in EMPs levels persisting after adjustment for potential confounding factors have also been published [124].

Several roles have been proposed for MPs during AF: i) it has been observed that MPs harbor PS, providing additional phospholipid surface for thrombin generation assembling complexes, as well as induce TF activity both increasing thrombogenicity [119]; ii) MPs seem to interact with extracellular matrix components, supporting the potential relevance in matrix and electrophysiological remodeling in AF settings [125, 126]; iii) MPs can diminish NO production, contributing to endothelial dysfunction in apoptotic smooth muscle cells [127]; iv) MPs participate in inflammation promoting prothrombotic pathways and stimulating the release of proinflammatory endothelial cytokines [128]; v) several studies have suggested a direct contribution to MPs to stroke [129, 130], and their monitoring during thrombotic risk assessment in AF seems to be useful [119].

In spite of there are increasing data suggesting MPs relevance in different pathophysiological processes liked to AF, however, strong experimentalevidence indicating a direct link between thromboembolic risk and enhanced shedding of MPs is still lacking. Their interest in the clinical practice with additive value to current clinical risk scores guarantees further investigation.

Adiponectin

Adiponectin presents anti-inflammatory, atherogenic and antihypertrophic functions associated with different established risk factors for AF, such as inflammation, diabetes, obesity, myocardial infarction and heart failure [131]. Hernández- Romero *et al.* found that low levels of adiponectin independently associated with adverse cardiovascular events but only in female AF patients. The apparent lack of association in men could be due to testosterone decreasing adiponectin production [132].

These dates confirmed that adiponectin could exert a protective role against cardiovascular diseases.

Growth Differentiation Factor 15

Growth Differentiation Factor 15 (GDF-15) is a divergent member of the transforming growth factor-β family secreted from a different types of cells as adipocytes and myocytes. GDF-15 is produced in situations of stress such as cellular ischemia and mechanical and oxidative stress. Wallentin *et al.,* reported how GDF-15 presented additive prognostic value for major bleeding in AF patients under oral anticoagulation treatment, even after adjusting for clinical and risk factors as well as for CHA_2DS_2-VASc score, and biomarkers [133]. However, at the moment, there are no further studies corroborating the mechanistic role of GDF-15 in AF.

USE OF BIOMARKERS FOR RISK STRATIFICATION IN AF

AF increases the risk of stroke and death. Assessment of stroke risk in AF patients is mainly based on clinical risk scores such as $CHADS_2$ and CHA_2DS_2-VASc. The current guidelines on AF recommend the use of CHA_2DS_2–VASc score (1 point each for presence of congestive heart failure, hypertension, diabetes mellitus, vascular disease, age >65 and sex category (female gender) and 2 points to age > 75 years and prior stroke/AIT [134]). to assess thromboembolic risk [134, 135]. Oral anticoagulation (OAC) has been demonstrated to effectively reduce stroke risk and mortality rates in patients with AF, but also increases bleeding risk [136]. AF patients with CHA_2DS_2–VASc ≥ 2 should be considered for OAC treatment and patients with CHA_2DS_2–VASc = 1 would have considered indication to initiate OAC for preventing stroke. Subjects categorized to be OAC eligible will be exposed to an increased risk of major bleeding [137], but not to treat some patients could expose them to the risk of fatal and devastating strokes. Stroke is intimately close related to bleeding risk; hence, evaluation of OAC initiation needs to weigh the benefit from stroke prevention against assumption of the bleeding risk. Risk factors described for thromboembolic risk factors also appear as bleeding risk factors (*e.g.* advanced age or uncontrolled hypertension) [138]. The HAS-BLED (Hypertension, Abnormal renal/liver function, Stroke, Bleeding history or predisposition, Labile International Normalized Ratio, Elderly, and Drugs/alcohol concomitantly) has been establish to assess bleeding risk in AF patients [139]. A score of > 3 indicates "high risk"; however, this does not contraindicate OAC therapy [135] but close monitoring is required.

As detailed before in this chapter, the potential utility of biomarkers in refining risk stratification and improving the predictive power of clinical risk scores have been reported in several studies. Many new candidate biomarkers have been proposed with advances in genomics, proteomics and molecular pathology suggesting important roles in prediction of related outcomes in AF [63, 140]. The interest in developing new biomarkers that could provide additional refinement is highly justified. In spite of that, to the moment, none of them has already been included as part of clinical risks scores.

NEW CHALLENGES FOR BIOMARKERS IN AF

Clinical AF is a complex condition, where different pathways participate in diverse pathological entities with distinct mechanisms. Currently, there are increasing data supporting use of new biomarkers in the diagnosis and prognosis and treatment response in AF. However several important limitations must be solved before their extended and accepted use including decision on normal and abnormal ranges (especially relevant in miRNAs with high variably expression profiles, variations due to clinical and demographical variables, the underlying pathology…), development of improved high performed, more quantitative methods with available and cost-effective techniques or knowledge of potential interactions. All this features must be taking into account to understand these different pathophysiological roles and develop appropriate therapeutic targets [141].

Biomarkers could be useful in refine clinical risk stratification scores and may also increase our knowledge on AF pathogenesis. The design of a multimarker strategy leading to a better overall risk stratification, as with coronary artery disease [142] or acute coronary syndrome [143] has also been reported. Also, these biomarkers could be used as indicators of ongoing processes such as thrombogenesis, to test antithrombotic therapies and help decision-making on dose selection [144].

It has been proposed that tailoring different antithrombotic options to individual patients based on biomarker profiles has not been explored in patients with AF, and this could support interesting hypotheses for future trials [145]. In addition the implementation of future therapies is also sprinkled with technological difficulties and handicaps to be confronted; for instance, interfering with a specific miRNA expression could influence different pathways since each miRNA can regulate several targets [86]. Moreover, miR-mimics, synthesized nucleic acids binding mRNAs similarly as the natural miRNAs have been used to recover lowered miRNA levels in cardiovascular disease [146]. On the contrary, blocking miRNA expression when its overexpression is causing a disease (*e.g.* antisense inhibitor oligonucleotides, AMOs) is also possible. Both strategies have been published to be successful [146, 147]. However new available miRNA-based therapeutic strategies in human are still far away. Directed delivery of miRNAs to their specific targets is a real major challenge, although in theory, miRNAs supposes a powerful tool to subjugate AF [86].

Biomarkers might provide useful information for patients' selection as well as for decisions about the benefit from reducing the combination of stroke, systemic or venous embolism, myocardial infarction, cardiovascular death, or major bleeding. Hence, biomarkers are useful to stratify the risk of death in patients with AF, and can provide relevant information to aid in the selection of appropriate antithrombotic therapy or in the follow-up of patient's response to different therapies for the treatment of AF [145].

In summary, different biomarkers have been studied within the context of the AF.

Some of them have demonstrated to add valuable information for risk stratification and prognosis, whereas more newly studied biomarkers are still developing their potential. Future investigations will shed light about pathophysiological implications and therapeutic options related to the use of biomarkers in AF.

CONFLICT OF INTEREST

None declared in relation to this manuscript for all authors. VR has received funding for consultancy and lecturing from Bristol-Myers-Squibb, Bayer and Boehringer Ingelheim. FM has received funding for research, consultancy and lecturing from Abbott, Boston Scientifics, Bayer, Astra Zeneca, Daiichi-Sankyo, BMS/Pfizer and Boehringer Ingelheim.

ACKNOWLEDGEMENTS

D Hernandez-Romero holds a research contract by the Instituto de Salud Carlos III, Madrid, Spain RD12/0042/0049.

REFERENCES

[1] Wolf PA, Abbott RD, Kannel WB. Atrial fibrillation as an independent risk factor for stroke: the Framingham Study. Stroke 1991; 22: 983-8. [http://dx.doi.org/10.1161/01.STR.22.8.983]

[2] Benjamin EJ, Wolf PA, D′ Agostino RB, Silbershatz H, Kannel WB, Levy D. Impact of atrial fibrillation on the risk of death. The Framingham Heart Study. Circulation 1998; 98: 946-52. [http://dx.doi.org/10.1161/01.CIR.98.10.946]

[3] Ahmad Y. Lip, Lane, DA. Recent developments in understanding epidemiology and risk determinants of atrial fibrillation as a cause of stroke. Can J Cardiol 2013; 29 (Suppl. 7): S4-S13. [http://dx.doi.org/10.1016/j.cjca.2013.03.009]

[4] Andrade J, Khairy P, Dobrev D, Nattel S. The clinical profile and pathophysiology of atrial fibrillation: relationships among clinical features, epidemiology, and mechanisms. Circ Res 2014; 114: 1453-68. [http://dx.doi.org/10.1161/CIRCRESAHA.114.303211]

[5] Naccarelli GV, Varker H, Lin J, Schulman KL. Increasing prevalence of atrial fibrillation and flutter in the United States. Am J Cardiol 2009; 104: 1534-9. [http://dx.doi.org/10.1016/j.amjcard.2009.07.022]

[6] Murphy NF, Simpson CR, Jhund PS, *et al.* A national survey of the prevalence, incidence, primary care burden and treatment of atrial fibrillation in Scotland. Heart 2007; 93: 606-12. [http://dx.doi.org/10.1136/hrt.2006.107573]

[7] Daoud EG, Bogun F, Goyal R, *et al.* Effect of atrial fibrillation on atrial refractoriness in humans. Circulation 1996; 94(7): 1600-6. [http://dx.doi.org/10.1161/01.CIR.94.7.1600]

[8] Frustaci A, Chimenti C, Belocci F, Morgante E, Russo MA, Maseri A. Histological substrate of atrial biopsies in patients with lone atrial fibrillation. Circulation 1997; 96: 1180-4. [http://dx.doi.org/10.1161/01.CIR.96.4.1180]

[9] Iwasaki Y, Nishida K, Kato T, Nattel S. Atrial fibrillation pathophysiology implications for

management. Circulation 2011; 124: 2264-74.
[http://dx.doi.org/10.1161/CIRCULATIONAHA.111.019893]

[10] Felix. Biomarker Qualification Pilot Process at the US Food and Drug Administration. AAPS J 2007; 9(1): E105-8.
[http://dx.doi.org/10.1208/aapsj0901010]

[11] Sovari AA, Dudley SC. Antioxidant therapy for atrial fibrillation: lost in translation. Heart 2012; 98: 1615-6.
[http://dx.doi.org/10.1136/heartjnl-2012-302328]

[12] Nattel S. From guidelines to bench: implications of unresolved clinical issues for basic investigations of atrial fibrillation mechanisms. Can J Cardiol 2011; 27: 19-26.
[http://dx.doi.org/10.1016/j.cjca.2010.11.004]

[13] Nattel S. New ideas about atrial fibrillation 50 years on. Nature 2002; 415: 219-26.
[http://dx.doi.org/10.1038/415219a]

[14] Wakili R, Voigt N, Kaab S, Dobrev D, Nattel S. Recent advances in the molecular pathophysiology of atrial fibrillation. J Clin Invest 2011; 121: 2955-68.
[http://dx.doi.org/10.1172/JCI46315]

[15] Schotten U, Verheule S, Kirchhof P, Goette A. Pathophysiological mechanisms of atrial fibrillation: a translational appraisal. Physiol Rev 2011; 91: 265-325.
[http://dx.doi.org/10.1152/physrev.00031.2009]

[16] Nattel S, Burstein B, Dobrev D. Atrial remodeling and atrial fibrillation: mechanisms and implications. Circ Arrhythm Electrophysiol 2008; 1: 62-73.
[http://dx.doi.org/10.1161/CIRCEP.107.754564]

[17] Wijffels MC, Kirchhof CJ, Dorland R, Allessie MA. Atrial fibrillation begets atrial fibrillation. A study in awake chronically instrumented goats. Circulation 1995; 92: 1954-68.
[http://dx.doi.org/10.1161/01.CIR.92.7.1954]

[18] Yue L, Feng J, Gaspo R, Li GR, Wang Z, Nattel S. Ionic remodeling underlying action potential changes in a canine model of atrial fibrillation. Circ Res 1997; 81: 512-25.
[http://dx.doi.org/10.1161/01.RES.81.4.512]

[19] Makary S, Voigt N, Maguy A, *et al.* Differential protein kinase C isoform regulation and increased constitutive activity of acetylcholine-regulated potassium channels in atrial remodeling. Circ Res 2011; 109: 1031-43.
[http://dx.doi.org/10.1161/CIRCRESAHA.111.253120]

[20] Brundel BJ, Van Gelder IC, Henning RH, *et al.* Ion channel remodeling is related to intraoperative atrial effective refractory periods in patients with paroxysmal and persistent atrial fibrillation. Circulation 2001; 103: 684-90.
[http://dx.doi.org/10.1161/01.CIR.103.5.684]

[21] Parkash R, Green MS, Kerr CR, *et al.* The association of left atrial size and occurrence of atrial fibrillation: a prospective cohort study from the Canadian Registry of Atrial Fibrillation. Am Heart J 2004; 148: 649-54.
[http://dx.doi.org/10.1016/j.ahj.2004.04.029]

[22] Zou R, Kneller J, Leon LJ, Nattel S. Substrate size as a determinant of fibrillatory activity maintenance in a mathematical model of canine atrium. Am J Physiol Heart Circ Physiol 2005; 289: H1002-12.
[http://dx.doi.org/10.1152/ajpheart.00252.2005]

[23] Burstein B, Qi XY, Yeh YH, Calderone A, Nattel S. Atrial cardiomyocyte tachycardia alters cardiac fibroblast function: a novel consideration in atrial remodeling. Cardiovasc Res 2007; 76: 442-52.
[http://dx.doi.org/10.1016/j.cardiores.2007.07.013]

[24] Burstein B, Libby E, Calderone A, Nattel S. Differential behaviors of atrial *versus* ventricular fibroblasts: a potential role for platelet-derived growth factor in atrial-ventricular remodeling

differences. Circulation 2008; 117: 1630-41.
[http://dx.doi.org/10.1161/CIRCULATIONAHA.107.748053]

[25] Rahmutula D, Marcus GM, Wilson EE, Din C, Xiao Y, Paquet A. Molecular basis of selective atrial fibrosis due to overexpression of transforming growth factor-β1. Cardiovasc Res 2003; 99: 769-79.
[http://dx.doi.org/10.1093/cvr/cvt074]

[26] Chen PS, Chen LS, Fishbein MC, Lin SF, Nattel S. Role of the autonomic nervous system in atrial fibrillation: pathophysiology and therapy. Circ Res 2014; 114: 1500-15.
[http://dx.doi.org/10.1161/CIRCRESAHA.114.303772]

[27] Collinson PO, Boa FG, Gaze DC. Measurement of cardiac troponins. Ann Clin Biochem 2001; 38: 423-49.
[http://dx.doi.org/10.1177/000456320103800501]

[28] Roldán V, Marín F, Diaz J, *et al.* High sensitivity cardiac troponin T and interleukin-6 predict adverse cardiovascular events and mortality in anticoagulated patients with atrial fibrillation. J Thromb Haemost 2012; 10(8): 1500-7.
[http://dx.doi.org/10.1111/j.1538-7836.2012.04812.x]

[29] Hernández-Romero D, Vílchez JA, Lahoz Á, *et al.* High-sensitivity troponin T as a biomarker for the development of atrial fibrillation after cardiac surgery. Eur J Cardiothorac Surg 2014; 45(4): 733-8.
[http://dx.doi.org/10.1093/ejcts/ezt488]

[30] Zethelius B, Johnston N, Venge P. Troponin I as a predictor of coronary heart disease and mortality in 70-year-old men: a community-based cohort study. Circulation 2006; 113(8): 1071-8.
[http://dx.doi.org/10.1161/CIRCULATIONAHA.105.570762]

[31] Omland T, de Lemos JA, Sabatine MS, *et al.* A sensitive cardiac troponin T assay in stable coronary artery disease. N Engl J Med 2009; 361(26): 2538-47.
[http://dx.doi.org/10.1056/NEJMoa0805299]

[32] Horwich TB, Patel J, MacLellan WR, Fonarow GC. Cardiac troponin I is associated with impaired hemodynamics, progressive left ventricular dysfunction, and increased mortality rates in advanced heart failure. Circulation 2003; 108(7): 833-8.
[http://dx.doi.org/10.1161/01.CIR.0000084543.79097.34]

[33] Hijazi Z, Oldgren J, Andersson U, *et al.* Cardiac biomarkers are associated with an increased risk of stroke and death in patients with atrial fibrillation: a randomized evaluation of long-term anticoagulation therapy (RE-LY) substudy. Circulation 2012; 125: 1605-16.
[http://dx.doi.org/10.1161/CIRCULATIONAHA.111.038729]

[34] Hijazi Z, Walletin L, Siegbahn A, *et al.* High-sensitivity troponin T and risk stratification in patients with atrial fibrillation during treatment with apixaban or warfarin (ARISTOTLE). J Am Coll Cardiol 2014; 63(1): 52-61.
[http://dx.doi.org/10.1016/j.jacc.2013.07.093]

[35] Nakagawa O, Ogawa Y, Itoh H, *et al.* Rapid transcriptional activation and early mRNA turnover of brain natriuretic peptide in cardiocyte hypertrophy: Evidence for brain natriuretic peptide as an "emergency" cardiac hormone against ventricular overload. J Clin Invest 1995; 96: 1280-7.
[http://dx.doi.org/10.1172/JCI118162]

[36] Daniels LB, Maisel AS. Natriuretic peptides. J Am Coll Cardiol 2007; 50(25): 2357-68.
[http://dx.doi.org/10.1016/j.jacc.2007.09.021]

[37] Ellinor PT, Low AF, Patton KK, Shea MA, Macrae CA. Discordant atrial natriuretic peptide and brain natriuretic peptide levels in lone atrial fibrillation. J Am Coll Cardiol 2005; 45: 82-6.
[http://dx.doi.org/10.1016/j.jacc.2004.09.045]

[38] Shelton RJ, Clark AL, Goode K, Rigby AS, Cleland JG. The diagnostic utility of N-terminal pro--type natriuretic peptide for the detection of major structural heart disease in patients with atrial fibrillation. Eur Heart J 2006; 27: 2353-61.

[http://dx.doi.org/10.1093/eurheartj/ehl233]

[39] Hijazi Z, Wallentin L, Siegbahn A, *et al.* N-terminal pro-B-type natriuretic peptide for risk assessment in patients with atrial fibrillation: insights from the ARISTOTLE Trial (Apixaban for the Prevention of Stroke in Subjects with Atrial Fibrillation). J Am Coll Cardiol 2013; 61(22): 2274-84. [http://dx.doi.org/10.1016/j.jacc.2012.11.082]

[40] Roldán V, Vílchez JA, Manzano-Fernández S, *et al.* Usefulness of N-terminal pro–B-type natriuretic peptide levels for stroke risk prediction in anticoagulated patients with atrial fibrillation. Stroke 2014; 45: 696-701. [http://dx.doi.org/10.1161/STROKEAHA.113.003338]

[41] Okada Y, Shibazaki K, Kimura K, *et al.* Brain natriuretic peptide is a marker associated with thrombus in stroke patients with atrial fibrillation. J Neurol Sci 2011; 301: 86-9. [http://dx.doi.org/10.1016/j.jns.2010.10.017]

[42] Ananthapanyasut W, Napan S, Rudolph EH, *et al.* Prevalence of atrial fibrillation and its predictors in non-dialysis patients with chronic kidney disease. Clin J Am Soc Nephrol 2010; 5: 173-81. [http://dx.doi.org/10.2215/CJN.03170509]

[43] Deo R, Katz R, Kestenbaum B, *et al.* Impaired kidney function and atrial fibrillation in elderly subjects. J Card Fail 2010; 16: 55-60. [http://dx.doi.org/10.1016/j.cardfail.2009.07.002]

[44] Go AS, Fang MC, Udaltsova N, *et al.* Impact of proteinuria and glomerular filtration rate on risk of thromboembolism in atrial fibrillation (ATRIA) study. Circulation 2009; 119: 1363-9. [http://dx.doi.org/10.1161/CIRCULATIONAHA.108.816082]

[45] Hohnloser SH, Hijazi Z, Thomas L, *et al.* Efficacy of apixaban when compared with warfarin in relation to renal function in patients with atrial fibrillation: insights from the ARISTOTLE trial. Eur Heart J 2012; 33: 2821-22830. [http://dx.doi.org/10.1093/eurheartj/ehs274]

[46] Roldan V, Marin F, Fernandez H, *et al.* Renal impairment in a "real-life" cohort of anticoagulated patients with atrial fibrillation (implications for thromboembolism and bleeding). Am J Cardiol 2013; 111(8): 1159-64. [http://dx.doi.org/10.1016/j.amjcard.2012.12.045]

[47] Roldán V, Marín F, Manzano-Fernández S, *et al.* Does chronic kidney disease improve the predictive value of CHADS2 and CHA2DS2-VASc stroke stratification risk scores for atrial fibrillation? Thromb Haemost 2013; 109(5): 956-60. [http://dx.doi.org/10.1160/TH13-01-0054]

[48] Laterza OF, Price CP, Scott MG. Cystatin C: an improved estimator of glomerular filtration rate? Clin Chem 2002; 48: 699-707.

[49] Newman DJ, Thakkar H, Edwards RG, *et al.* Serum Cystatin C measured by automated immunoassay: a more sensitive marker of changes in GFR than serum creatinine. Kidney Int 1995; 47: 312-8. [http://dx.doi.org/10.1038/ki.1995.40]

[50] Hoffmann A, Conradt HS, Gross G, Nimtz M, Lottspeich F, Wurster U. Purification and chemical characterization of beta-trace protein from human cerebrospinal fluid: its identification as prostaglandin D synthase. J Neurochem 1993; 61(2): 451-6. [http://dx.doi.org/10.1111/j.1471-4159.1993.tb02145.x]

[51] Vílchez JA, Roldán V, Manzano-Fernández S, *et al.* β-Trace protein and prognosis in patients with atrial fibrillation receiving anticoagulation treatment. Chest 2013; 144(5): 1564-70. [http://dx.doi.org/10.1378/chest.13-0922]

[52] Ix JH, Shlipak MG, Chertow GM, Whooley MA. Association of cystatin C with mortality, cardiovascular events, and incident heart failure among persons with coronary heart disease: data from the Heart and Soul Study. Circulation 2007; 115: 173-9.

[http://dx.doi.org/10.1161/CIRCULATIONAHA.106.644286]

[53] Kerr R, Stirling D, Ludlam CA. Interleukin 6 and haemostasis. Br J Haematol 2001; 115(1): 3-12. [http://dx.doi.org/10.1046/j.1365-2141.2001.03061.x]

[54] Aulin JK, Andersson U, Connolly SJ, *et al.* Interleukin-6 and C-reactive protein and risk for death and cardiovascular events in patients with atrial fibrillation. J Am Coll Cardiol 2011; 57: E91. [http://dx.doi.org/10.1016/S0735-1097(11)60091-0]

[55] Conway DS, Buggins P, Hughes E, Lip GY. Prognostic significance of raised plasma levels of interleukin-6 and C-reactive protein in atrial fibrillation. Am Heart J 2004; 148(3): 462-6. [http://dx.doi.org/10.1016/j.ahj.2004.01.026]

[56] Lip GY, Patel JV, Hughes E, Hart RG. High-sensitivity C-reactive protein and soluble CD40 ligand as indices of inflammation and platelet activation in 880 patients with nonvalvular atrial fibrillation: relationship to stroke risk factors, stroke risk stratification schema, and prognosis. Stroke 2007; 38(4): 1229-37. [http://dx.doi.org/10.1161/01.STR.0000260090.90508.3e]

[57] Hermida J, Lopez FL, Montes R, Matsushita K, Astor BC, Alonso A. Usefulness of high sensitivity C reactive protein to predict mortality in patients with atrial fibrillation (from the Atherosclerosis Risk in Communities, ARIC, Study). Am J Cardiol 2012; 109(1): 95-9. [http://dx.doi.org/10.1016/j.amjcard.2011.08.010]

[58] Watson T, Shantsila E, Lip GY. Mechanisms of thrombogenesis in atrial fibrillation: Virchow's triad revisited. Lancet 2009; 373(9658): 155-66. [http://dx.doi.org/10.1016/S0140-6736(09)60040-4]

[59] Asakura H, Hifumi S, Jokaji H, *et al.* Prothrombin F1+2 and thrombin-antihrombin III complex are useful markers of the hypercoagulable stat in atrial fibrillation. Blood Coagul Fibrinolysis 1992; 3: 469-73. [http://dx.doi.org/10.1097/00001721-199203040-00015]

[60] Nozawa T, Inoue H, Hirai T, *et al.* D-dimer level influences thrombo-embolic events in patients with atrial fibrillation. Int J Cardiol 2006; 109(1): 59-65. [http://dx.doi.org/10.1016/j.ijcard.2005.05.049]

[61] Roldán V, Marín F, Muiña B, *et al.* Plasma von willebrand factor levels are an independent risk factor for adverse events including mortality and major bleeding in anticoagulated atrial fibrillation patients. J Am Coll Cardiol 2011; 57(25): 2496-504. [http://dx.doi.org/10.1016/j.jacc.2010.12.033]

[62] Eikelboom J, Hijazi Z, Oldgren J, *et al.* D-dimer is prognostic for stroke, major bleeding and death during anticoagulation of atrial fibrillation-a RELY substudy. Circulation 2010; 122: A18321.

[63] Hijazi Z1. Oldgren J, Siegbahn A, Granger CB, Wallentin L. Biomarkers in atrial fibrillation: a clinical review. Eur Heart J 2013; 34(20): 1475-80. [http://dx.doi.org/10.1093/eurheartj/eht024]

[64] Feinberg WM, Pearce LA, Hart RG, *et al.* Markers of thrombin and platelet activity in patients with atrial fibrillation. Stroke 1999; 30: 2547-53. [http://dx.doi.org/10.1161/01.STR.30.12.2547]

[65] Roldán V, Marín F, Marco P, Martínez JG, Calatayud R, Sogorb F. Hypofibrinolysis in atrial fibrillation. Am Heart J 1998; 136: 956-60. [http://dx.doi.org/10.1016/S0002-8703(98)70149-8]

[66] Vene N, Mavri A, Kosmelj K, Stegnar M. High D-dimer levels predict cardiovascular events in patients with chronic atrial fibrillation during oral anticoagulant therapy. Thromb Haemost 2003; 90(6): 1163-72.

[67] Freestone B, Chong AY, Blann AD, Lip GY. The effects of direct current cardioversion for persistent atrial fibrillation on indices of endothelial damage/dysfunction. Thromb Res 2006; 118: 479-85.

[http://dx.doi.org/10.1016/j.thromres.2005.10.004]

[68] Lip GY, Lane D, Van WC, Hart RG. Additive role of plasma von Willebrand factor levels to clinical factors risk stratification of patients with atrial fibrillation. Stroke 2006; 37(9): 2294-300. [http://dx.doi.org/10.1161/01.STR.0000236840.00467.84]

[69] Krishnamoorthy S, Khoo CW, Lim HS, *et al.* Prognostic role of plasma von Willebrand factor and soluble E-selectin levels for future cardiovascular events in a 'real-world' community cohort of patients with atrial fibrillation. Eur J Clin Invest 2013; 43(10): 1032-8. [http://dx.doi.org/10.1111/eci.12140]

[70] Freestone B, Chong AY, Nuttall S, Blann AD, Lip GY. Soluble E-selectin, von Willebrand factor, soluble thrombomodulin, and total body nitrate/nitrite product as indices of endothelial damage/dysfunction in paroxysmal, persistent, and permanent atrial fibrillation. Chest 2007; 132(4): 1253-8. [http://dx.doi.org/10.1378/chest.07-1185]

[71] Ferro D, Loffredo L, Polimeni L, *et al.* Soluble CD40 ligand predicts ischemic stroke and myocardial infarction in patients with nonvalvular atrial fibrillation. Arterioscler Thromb Vasc Biol 2007; 27: 2763-8. [http://dx.doi.org/10.1161/ATVBAHA.107.152777]

[72] Choudhury A, Chung I, Blann AD, Lip GY. Elevated platelet microparticle levels in nonvalvular atrial fibrillation: relationship to P-selectin and antithrombotic therapy. Chest 2007; 131(3): 809-15. [http://dx.doi.org/10.1378/chest.06-2039]

[73] Chirinos JA, Castrellon A, Zambrano JP, *et al.* Digoxin use is associated with increased platelet and endothelial cell activation in patients with nonvalvular atrial fibrillation. Heart Rhythm 2005; 2(5): 525-9. [http://dx.doi.org/10.1016/j.hrthm.2005.01.016]

[74] Heeringa J, Conway DS, van der Kuip DA, *et al.* A longitudinal population-based study of prothrombotic factors in elderly subjects with atrial fibrillation: the Rotterdam study 1990-1999. J Thromb Haemost 2006; 4: 1944-9. [http://dx.doi.org/10.1111/j.1538-7836.2006.02115.x]

[75] Hayashi M, Takeshita K, Inden Y, *et al.* Platelet activation and induction of tissue factor in acute and chronic atrial fibrillation: involvement of mononuclear cell-platelet interaction. Thromb Res 2011; 128(6): e113-8. [http://dx.doi.org/10.1016/j.thromres.2011.07.013]

[76] Kamath S, Chin BS, Blann AD, Lip GY. A study of platelet activation in paroxysmal, persistent and permanent atrial fibrillation. Blood Coagul Fibrinolysis 2002; 13(7): 627-36. [http://dx.doi.org/10.1097/00001721-200210000-00008]

[77] Nattel S. Electrophysiologic remodeling: are ion channels static players or dynamic movers. J Cardiovasc Electrophysiol 1999; 10: 1553-6. [http://dx.doi.org/10.1111/j.1540-8167.1999.tb00214.x]

[78] Schoonderwoerd BA, Van Gelder IC, Van Veldhuisen DJ, Van den Berg MP, Crijns HJ. Electrical and structural remodeling: role in the genesis and maintenance of atrial fibrillation. Prog Cardiovasc Dis 2005; 48: 153-68. [http://dx.doi.org/10.1016/j.pcad.2005.06.014]

[79] Lee RC, Feinhaum RL, Ambros V. The C. elegans heterochronic gene *lin-4* encodes small RNAs with antisense complementarity to *lin-14*. Cell 1993; 75: 843-54. [http://dx.doi.org/10.1016/0092-8674(93)90529-Y]

[80] Shi KH, Tao H, Yang JJ, Wu JX, Xu SS, Zhan HY. Role of microRNAs in atrial fibrillation: new insights and perspectives. Cell Signal 25: 2079-84. [http://dx.doi.org/10.1016/j.cellsig.2013.06.009]

[81] Friedman RC, Farh JJ, Burge CB, Bartel DP. Most mammalian mRNAs are conserved targets of microRNAs. Genome Res 2009; 19: 92-105.
[http://dx.doi.org/10.1101/gr.082701.108]

[82] Lewis BP, Burge CB, Bartel DP. Conserved seed pairing, often flanked by adenosines, indicates that thousands of human genes are microRNA targets. Cell 2005; 120: 15-20.
[http://dx.doi.org/10.1016/j.cell.2004.12.035]

[83] Lai EC. Micro RNAs are complementary to 3_ UTR sequence motifs that mediate negative post-transcriptional regulation. Nat Genet 2002; 30: 363-4.
[http://dx.doi.org/10.1038/ng865]

[84] van Rooij E, Olson EN. MicroRNA therapeutics for cardiovascular disease: opportunities and obstacles. Nat Rev Drug Discov 2012; 11: 860-72.
[http://dx.doi.org/10.1038/nrd3864]

[85] van Rooij E, Purcell AL, Levin AA. Developing microRNA therapeutics. Circ Res 2012; 110: 496-507.
[http://dx.doi.org/10.1161/CIRCRESAHA.111.247916]

[86] Orenes-Piñero E, Quintana-Giner M, Romero-Aniorte AI, Valdés M, Marín F. Novel biomarkers in cardiology: MicroRNAs in atrial fibrillation. Arch Cardiol Mex 2015; 85(3): 225-9.
[http://dx.doi.org/10.1016/j.acmx.2015.01.005]

[87] Luo X, Yang B, Nattel S. MicroRNAs and atrial fibrillation: mechanisms and translational potential. Nat Rev Cardiol 2015; 12: 80-90.
[http://dx.doi.org/10.1038/nrcardio.2014.178]

[88] Luo X, Zhang H, Xiao J, Wang Z. Regulation of human cardiac ion channel genes by microRNAs: theoretical perspective and pathophysiological implications. Cell Physiol Biochem 2010; 25: 571-86.
[http://dx.doi.org/10.1159/000315076]

[89] Srivastava D, Thomas T, Lin Q, Kirby ML, Brown D, Olson EN. Regulation of cardiac mesodermal and neural crest development by the βHLH transcription factor, δHAND. Nat Genet 1997; 16: 154-60.
[http://dx.doi.org/10.1038/ng0697-154]

[90] Xu C, Lu Y, Pan Z, Chu W, Luo X, Lin H, *et al.* The muscle-specific microRNAs miR-1 and miR-133 produce opposing effects on apoptosis by targeting HSP60, HSP70 and caspase-9 in cardiomyocytes. J Cell Sci 2007; 120: 3045-52.
[http://dx.doi.org/10.1242/jcs.010728]

[91] Yang B, Lin H, Xiao J, *et al.* The muscle-specific microRNA miR-1 regulates cardiac arrhythmogenic potential by targeting GJA1 and KCNJ2. Nat Med 2007; 13: 486-91.
[http://dx.doi.org/10.1038/nm1569]

[92] Roy S, Khanna S, Hussain SR, Biswas S, Azad A, Rink C. MicroRNA expression in response to murine myocardial infarction: miR-21 regulates fibroblast metalloprotease-2 *via* phosphatase and tensin homologue. Cardiovasc Res 2009; 82: 21-9.
[http://dx.doi.org/10.1093/cvr/cvp015]

[93] Cheng Y, Liu X, Zhang S, Lin Y, Yang J, Zhang C. MicroRNA-21 protects against the H2O2-induced injury on cardiac myocytes via its target gene PDCD4. Mol Cell Cardiol 2009; 47: 5-14.

[94] Esquela-Kerscher A, Oncomirs Slack FJ. Oncomirs - microRNAs with a role in cancer. Nat Rev Cancer 2006; 6: 259-69.
[http://dx.doi.org/10.1038/nrc1840]

[95] Harada M, Luo X, Qi XY, *et al.* Transient receptor potential canonical-3 channel-dependent fibroblast regulation in atrial fibrillation. Circulation 2012; 126: 2051-64.
[http://dx.doi.org/10.1161/CIRCULATIONAHA.112.121830]

[96] van Rooij E, Sutherland LB, Thatcher JE, DiMaio JM, Naseem RH, Marshall WS. Dysregulation of

microRNAs after myocardial infarction reveals a role of miR-29 in cardiac fibrosis. Proc Natl Acad Sci USA 2008; 105: 13027-32.
[http://dx.doi.org/10.1073/pnas.0805038105]

[97] Dawson K, Wakili R, Ordög B, Clauss S, Chen Y, Iwasaki Y. MicroRNA29: a mechanistic contributor and potential biomarker in atrial fibrillation. Circulation 2013; 127: 1466-75.
[http://dx.doi.org/10.1161/CIRCULATIONAHA.112.001207]

[98] Li H, Li S, Yu B, Liu S. Expression of miR-133 and miR-30 in chronic atrial fibrillation in canines. Mol Med Rep 2012; 5: 1457-60.

[99] Duisters RF, Tijsen AJ, Schroen B, *et al.* miR-133 and miR-30 regulate connective tissue growth factor: implications for a role of microRNAs in myocardial matrix remodeling. Circ Res 2009; 104: 170-8.
[http://dx.doi.org/10.1161/CIRCRESAHA.108.182535]

[100] Care A, Catalucci D, Felicetti F, *et al.* MicroRNA-133 controls cardiac hypertrophy. Nat Med 2007; 13: 613-8.
[http://dx.doi.org/10.1038/nm1582]

[101] Liu N, Bezprozvannaya S, Williams AH, *et al.* microRNA- 133a regulates cardiomyocyte proliferation and suppresses smooth muscle gene expression in the heart. Genes Dev 2008; 22: 3242-54.
[http://dx.doi.org/10.1101/gad.1738708]

[102] Shan H, Zhang Y, Lu Y, *et al.* Downregulation of miR-133 and miR-590 contributes to nicotine-induced atrial remodelling in canines. Cardiovasc Res 2009; 83: 465-72.
[http://dx.doi.org/10.1093/cvr/cvp130]

[103] Girmatsion Z, Biliczki P, Bonauer A, Wimmer-Greinecker G, Scherer M, Moritz A. Changes in microRNA-1 expression and IK1 up-regulation in human atrial fibrillation. Heart Rhythm 2009; 6(12): 1802-9.
[http://dx.doi.org/10.1016/j.hrthm.2009.08.035]

[104] Yang B, Lin H, Xiao J, *et al.* The muscle-specific microRNA miR-1 regulates cardiac arrhythmogenic potential by targeting GJA1 and KCNJ2. Nat Med 2007; 13: 486-91.
[http://dx.doi.org/10.1038/nm1569]

[105] Zicha S, Moss I, Allen B, *et al.* Molecular basis of species-specific expression of repolarizing K+ currents in the heart. Am J Physiol Heart Circ Physiol 2003; 285: H1641-9.
[http://dx.doi.org/10.1152/ajpheart.00346.2003]

[106] Harada M, Luo X, Murohara T, Yang B, Dobrev D, Nattel S. MicroRNA regulation and cardiac calcium signaling: role in cardiac disease and therapeutic potential. Circ Res 2014; 114: 689-705.
[http://dx.doi.org/10.1161/CIRCRESAHA.114.301798]

[107] Luo X, Pan Z, Shan H, *et al.* MicroRNA-26 governs profibrillatory inward-rectifier potassium current changes in atrial fibrillation. J Clin Invest 2013; 123: 1939-51.
[http://dx.doi.org/10.1172/JCI62185]

[108] Wang J, Song S, Xie C, *et al.* MicroRNA profiling in the left atrium in patients with non-valvular paroxysmal atrial fibrillation. BMC Cardiovasc Disord 2015; 29: 15-97.

[109] Lo CW. Role of gap junctions in cardiac conduction and development: insights from the connexin knockout mice. Circ Res 2000; 87: 346-8.
[http://dx.doi.org/10.1161/01.RES.87.5.346]

[110] Lu Y, Zhang Y, Wang N, *et al.* MicroRNA-328 contributes to adverse electrical remodeling in atrial fibrillation. Circulation 2010; 122: 2378-87.
[http://dx.doi.org/10.1161/CIRCULATIONAHA.110.958967]

[111] Ling TY, Wang XL, Chai Q, *et al.* Regulation of the SK3 channel by microRNA-499-potential role in atrial fibrillation. Heart Rhythm 2013; 10: 1001-9.
[http://dx.doi.org/10.1016/j.hrthm.2013.03.005]

[112] Ellinor PT, Lunetta KL, Glazer NL, *et al.* Common variants in KCNN3 are associated with lone atrial fibrillation. Nat Genet 2010; 42: 240-4.
[http://dx.doi.org/10.1038/ng.537]

[113] Lacroix R, Robert S, Poncelet P, Kasthuri RS, Key NS, Dignat-George F. Standardization of platelet-derived microparticle enumeration by flow cytometry with calibrated beads: results of the International Society on Thrombosis and Haemostasis SSC Collaborative workshop. J Thromb Haemost 2010; 8: 2571-4.
[http://dx.doi.org/10.1111/j.1538-7836.2010.04047.x]

[114] George JN, Pickett EB, Sauceman S, *et al.* Platelet surface glycoproteins: studies on resting and activated platelets and platelet membrane microparticles in normal subjects, and observations in patients during adult respiratory distress syndrome and cardiac surgery. J Clin Invest 1986; 78: 340-8.
[http://dx.doi.org/10.1172/JCI112582]

[115] Cauwenberghs S, Feijge MA, Harper AG, Sage SO, Curvers J, Heemskerk JW. Shedding of procoagulant microparticles from unstimulated platelets by integrin-mediated destabilization of actin cytoskeleton. FEBS Lett 2006; 580(22): 5313-20.
[http://dx.doi.org/10.1016/j.febslet.2006.08.082]

[116] Montoro-García S, Shantsila E, Hernández-Romero D, *et al.* Small-size platelet microparticles trigger platelet and monocyte functionality and modulate thrombogenesis *via* P-selectin. Br J Haematol 2014; 166(4): 571-80.
[http://dx.doi.org/10.1111/bjh.12913]

[117] Rautou PE, Vion AC, Amabile N, *et al.* Microparticles, vascular function, and atherothrombosis. Circ Res 2011; 109: 593-606.
[http://dx.doi.org/10.1161/CIRCRESAHA.110.233163]

[118] Morel O, Pereira B, Averous G, *et al.* Increased levels of procoagulant tissue factor-bearing microparticles within the occluded coronary artery of patients with ST-segment elevation myocardial infarction: role of endothelial damage and leukocyte activation. Atherosclerosis 2009; 204: 636-41.
[http://dx.doi.org/10.1016/j.atherosclerosis.2008.10.039]

[119] Jesel L, Abbas M, Toti F, Cohen A, Arentz T, Morel O. Microparticles in atrial fibrillation: a link between cell activation or apoptosis, tissue remodelling and thrombogenicity. Int J Cardiol 2013; 168(2): 660-9.
[http://dx.doi.org/10.1016/j.ijcard.2013.03.031]

[120] Ederhy S, Di Angelantonio E, Mallat Z, *et al.* Levels of circulating procoagulant microparticles in nonvalvular atrial fibrillation. Am J Cardiol 2007; 100: 989-94.
[http://dx.doi.org/10.1016/j.amjcard.2007.04.040]

[121] Akar JG, Jeske W, Wilber DJ. Acute onset human atrial fibrillation is associated with local cardiac platelet activation and endothelial dysfunction. J Am Coll Cardiol 2008; 51: 1790-3.
[http://dx.doi.org/10.1016/j.jacc.2007.11.083]

[122] Wang H, Yan HM, Tang MX, *et al.* Increased serum levels of microvesicles in nonvalvular atrial fibrillation determinated by ELISA using a specific monoclonal antibody AD-1. Clin Chim Acta 2010; 411: 1700-4.
[http://dx.doi.org/10.1016/j.cca.2010.07.005]

[123] Choudhury A, Chung I, Blann AD, Lip GY. Elevated platelet microparticle levels in nonvalvular atrial fibrillation: relationship to p-selectin and antithrombotic therapy. Chest 2007; 131: 809-15.
[http://dx.doi.org/10.1378/chest.06-2039]

[124] Chirinos JA, Castrellon A, Zambrano JP, *et al.* Digoxin use is associated with increased platelet and endothelial cell activation in patients with nonvalvular atrial fibrillation. Heart Rhythm 2005; 2: 525-9.
[http://dx.doi.org/10.1016/j.hrthm.2005.01.016]

[125] Morel O, Toti F, Hugel B, *et al.* Procoagulant microparticles: disrupting the vascular homeostasis

equation? Arterioscler Thromb Vasc Biol 2006; 26: 2594-604. [http://dx.doi.org/10.1161/01.ATV.0000246775.14471.26]

[126] Lozito TP, Tuan RS. Endothelial cell microparticles act as centers of matrix metalloproteinsase-2 (MMP-2) activation and vascular matrix remodeling. J Cell Physiol 2012; 227: 534-49. [http://dx.doi.org/10.1002/jcp.22744]

[127] Essayagh S, Brisset AC, Terrisse AD, *et al.* Microparticles from apoptotic vascular smooth muscle cells induce endothelial dysfunction, a phenomenon prevented by beta3-integrin antagonists. Thromb Haemost 2005; 94: 853-8.

[128] Mesri M, Altieri DC. Leukocyte microparticles stimulate endothelial cell cytokine release and tissue factor induction in a JNK1 signaling pathway. J Biol Chem 1999; 274: 23111-8. [http://dx.doi.org/10.1074/jbc.274.33.23111]

[129] Lee YJ. Elevated platelet microparticles in transient ischemic attacks, lacunar infarcts, and multiinfarct dementias. Thromb Res 1993; 72: 295-304. [http://dx.doi.org/10.1016/0049-3848(93)90138-E]

[130] Lukasik M, Michalak S, Dworacki G, *et al.* Reactive leptin resistance and the profile of platelet activation in acute ischaemic stroke patients. Thromb Haemost 2012; 108: 107-18. [http://dx.doi.org/10.1160/TH11-12-0860]

[131] Rienstra M, Sun JX, Lubitz SA, *et al.* Plasma resistin, adiponectin, and risk of incident atrial fibrillation: the Framingham Offspring Study. Am Heart J 2012; 163(1): 119-24. [http://dx.doi.org/10.1016/j.ahj.2011.09.029]

[132] Hernandez-Romero D, Jover E, Marin F, *et al.* The prognostic role of the adiponectin levels in atrial fibrillation. Eur J Clin Invest 2013; 43(2): 168-73. [http://dx.doi.org/10.1111/eci.12028]

[133] Wallentin L, Hijazi Z, Andersson U, *et al.* Growth differentiation factor 15, a marker of oxidative stress and inflammation, for risk assessment in patients with atrial fibrillation: insights from ARISTOTLE trial. Circulation 2014; 130(21): 1847-58. [http://dx.doi.org/10.1161/CIRCULATIONAHA.114.011204]

[134] Craig T, January L, Wann S, *et al.* 2014 AHA/ACC/HRS Guideline for the Management of Patients with Atrial Fibrillation. J Am Coll Cardiol 2014; 64(21): 2246-80. [http://dx.doi.org/10.1016/j.jacc.2014.03.021]

[135] Camm JA, Lip GY, De Caterina R, *et al.* 2012 focused update of the ESC Guidelines for the management of atrial fibrillation: an update of the 2010 ESC Guidelines for the management of atrial fibrillation--developed with the special contribution of the European Heart Rhythm Association. Europace 2012; 14(10): 1385-413. [http://dx.doi.org/10.1093/europace/eus305]

[136] Singer DE, Chang Y, Fang MC, *et al.* Should patient characteristics influence target anticoagulation intensity for stroke prevention in nonvalvular atrial fibrillation? The ATRIA study. Circ Cardiovasc Qual Outcomes 2009; 2(4): 297-304. [http://dx.doi.org/10.1161/CIRCOUTCOMES.108.830232]

[137] Providencia R, Paiva L, Barra S. Risk stratification of patients with atrial fibrillation biomarkers and other future perspectives. World J Cardiol 2012; 4(6): 195-200. [http://dx.doi.org/10.4330/wjc.v4.i6.195]

[138] Lip GY, Andreotti F, Fauchier L, *et al.* Bleeding risk assessment and management in atrial fibrillation patients. Executive summary of a position document from the European Heart Rhythm Association (EHRA), endorsed by the European Society of Cardiology (ESC) working group on thrombosis. Thromb Haemost 2011; 106(6): 997-1011. [http://dx.doi.org/10.1160/TH11-10-0690]

[139] Pisters R, Lane DA, Nieuwlaat de Vos CB, Crijns HJ, Lip GY. A novel user-friendly score (HAS-

BLED) to assess 1-year risk of major bleeding in patients with atrial fibrillation: the Euro Heart Survey. Chest 2010; 138(5): 1093-100.
[http://dx.doi.org/10.1378/chest.10-0134]

[140] López-Cuenca A, Marín F, Roldán V, *et al.* Hernández- Romero D, Valdés M, Lip GY. Genetic polymorphisms and atrial fibrillation: Insights into the prothrombotic state and thromboembolic risk. Ann Med 2010; 42(8): 562-75.
[http://dx.doi.org/10.3109/07853890.2010.507601]

[141] Rodríguez-Serrano AI, Esteve-Pastor MA, Hernández-Romero D, Valdés M, Roldán V, Marín F. The new role of biomarkers in atrial fibrillation. In: Patel VB, Preedy VR, Eds. Biomarkers in Cardiovascular Disease Springer Science+Business Media Dordrecht. 2015.
[http://dx.doi.org/10.1007/978-94-007-7741-5_36-1]

[142] Lindahl B, Toss H, Siegbahn A, Venge P, Wallentin L. Markers of myocardial damage and inflammation in relation to long-term mortality in unstable coronary artery disease. FRISC Study Group. Fragmin during Instability in Coronary Artery Disease. N Engl J Med 2000; 343(16): 1139-47.
[http://dx.doi.org/10.1056/NEJM200010193431602]

[143] Tello-Montoliu A, Marin F, Roldan V, *et al.* A multimarker risk stratification approach to non-ST elevation acute coronary syndrome: implications of troponin T, CRP, NT pro-BNP and fibrin D-dimer levels. J Intern Med 2007; 262(6): 651-8.
[http://dx.doi.org/10.1111/j.1365-2796.2007.01871.x]

[144] Vílchez JA, Roldán V, Hernández-Romero D, Valdés M, Lip GY, Marín F. Biomarkers in atrial fibrillation: an overview. Int J Clin Pract 2014; 68(4): 434-43.
[http://dx.doi.org/10.1111/ijcp.12304]

[145] Marín F, Roldán V. GDF-15 and risk stratification in atrial fibrillation. Nat Rev Cardiol 2015; 12: 8-9.
[http://dx.doi.org/10.1038/nrcardio.2014.190]

[146] Xiao J, Yang B, Lin H, Lu Y, Luo X, Wang Z. Novel approaches for gene-specific interference *via* manipulating actions of microRNAs: examination on the pacemaker channel genes HCN2 and HCN4. J Cell Physiol 2007; 212: 285-92.
[http://dx.doi.org/10.1002/jcp.21062]

[147] Thum T, Galuppo P, Wolf C, Fiedler J, Kneitz S, van Laake LW. MicroRNAs in the human heart: a clue to fetal gene reprogramming in heart failure. Circulation 2007; 116: 258-67.
[http://dx.doi.org/10.1161/CIRCULATIONAHA.107.687947]

CHAPTER 9

Current Trends in Biomarkers of Acute Coronary Syndrome

Carmen María Puche-Morenilla[1,*], Luis García de Guadiana Romualdo[2] and Juan Antonio Vílchez[2]

[1] *Clinical Analysis Department, Clinical University Hospital Virgen de la Arrixaca, Murcia, Spain*

[2] *Clinical Analysis Department, Santa Lucia General University Hospital, Cartagena, Spain*

Abstract: Evaluation of patients presented to the emergency department with a complaint of chest pain or other signs and/or symptoms suggestive of acute coronary syndrome (ACS) is expensive and time-consuming. Nowadays, cardiac troponins (cardiac troponins I and T) are established as the standard biomarkers for prognostic evaluation and the detection of myocardial lesion of patients with ACS. Several studies have demonstrated that increases in biomarkers upstream from biomarkers of myocardial injury such as acute phase reactants, markers of inflammation, plaque destabilization and rupture biomarkers, cellular adhesion molecules and biomarkers of ischemia may identify patients with higher risk of having a cardiac event. The focus of this review is to provide information on biomarkers, specifically cardiac troponin, interleukin-6 and copeptin, which has become very important to improve the diagnosis of acute coronary syndrome and to predict prognosis following an actual event.

Keywords: Acute coronary syndrome, Acute chest pain, Acute-phase reactants, Angina, Atherosclerosis, Cardiac biomarkers, Cardiac troponin, Copeptin, Delta, Diagnosis, Electrocardiograh, Emergency departments, High sensitivity assays, Inflammation, Interleukin, Ischemia, Myocardial infarction, Myocardial necrosis, Plaque vulnerability, Prognosis.

INTRODUCTION

Incidence of Acute Coronary Syndrome

In the United States, more than 8 million people visit hospital for symptoms consistent with myocardial ischemia annually, which makes this the second most frequent cause of emergency departments (ED) encounters in adults [1]. Hence,

* **Corresponding author Carmen María Puche-Morenilla:** Clinical Analysis Department, Clinical University Hospital Virgen de la Arrixaca, Murcia, Spain; Tel/Fax: 968128600; E-mail: carmenma100@gmail.com

the challenge for clinicians is rapid identification of those with a benign cause who can be discharged from the ED and those who require ingress for immediate management. Error to detect myocardial syndrome of such patients from the ED may exceed 2%, with a high risk adjusted mortality ratio that is nearly 2-fold that of patients hospitalized for ACS, and it is also associated with substantial liability [2].

Among patients presented with chest pain to the ED, expected disease prevalence was 5–10% ST-segment-elevation myocardial infarction (STEMI) and up to 25% non-ST-segment-elevation MI (NSTE-MI). However, more than 50% exhibited other cardiac and non-cardiac conditions, some mimicking NSTE-MI. Thus, rapid, adequate management for patients with acute coronary syndrome (ACS) must be adjusted against appreciation of patients with non-critical syndromes for whom hospitalization are unnecessary, expensive, and an ineffective use of limited resources.

ACSs present a complex and heterogeneous pathophysiology with high morbidity and mortality mainly due to new cardiac ischaemic events [3]. To solve this fact, an increasing array of diagnostic strategies have been used, including chest pain units (CPUs), new cardiac biomarkers and risk scores [4]. Nevertheless, despite of these possibilities, clinical judgment is essential for optimal interpretation and evaluation of patients.

Definition and Classification of Acute Coronary Syndromes

ACS is a complex syndrome with a heterogeneous etiology caused by acute myocardial ischemia due to sudden and reduced blood flow to leading the heart, caused by atherosclerotic plaque erosion or rupture, with subsequent formation of thrombus. The most frequent clinical presentation is intermittent chest pain secondary to coronary artery sub-occlusion [5].

According to 12-lead electrocardiogram (ECG), patients with ACS are classified in two categories [6].

1) Patients with acute chest pain and persistent ST-segment elevation. This condition is named as ST-elevation ACS and reflects an acute total coronary occlusion. These patients will develop an STEMI. The aim of treatment in most patients is urgent reperfusion by primary angioplasty or fibrinolytic management.

2) Patients with acute chest pain but no persistent ST-segment elevation [non-ST elevation ACS (NSTEACS)]. ECG changes may include transitory ST-segment elevation, persistent or transitory ST-segment depression, T-wave inversion, flat T waves or pseudo-normalization of T waves or the ECG may be normal. The

pathological correlate at the myocardial level is cardiomyocyte necrosis [NSTE-myocardial infarction (NSTEMI)] or myocardial ischemia [unstable angina (UA)], recognized by cardiac biomarkers levels.

Definition of Acute Myocardial Infarction

According to the last universal definition of myocardial infarction, acute myocardial infarction considers cardiomyocyte necrosis in a clinical setting consistent with acute myocardial ischemia. Several criteria are part of the diagnosis of acute MI, that is the detection of an increase and/or decrease of a cardiac biomarker, if possible high-sensitivity cardiac troponin, with at least one value above the 99th percentile of the upper reference limit and at least one of the following criteria:

1. Symptoms of ischemia
2. New or apparent new significant ST-T wave changes or left bundle branch block on 12-lead ECG
3. pathological Q waves on ECG
4. Imaging evidence of new loss of viable myocardium or regional wall motion abnormality and
5. Intracoronary thrombus identified on angiography or autopsy.

Universal definition of myocardial infarction involves type 1 myocardial infarction, characterized by atherosclerotic plaque erosion or rupture with resulting intraluminal thrombus in one or more coronary arteries leading to decreased myocardial blood flow and/or distal embolization and subsequent myocardial necrosis, and type 2 myocardial infarction, in which a condition other than coronary plaque instability contributes to an imbalance between myocardial oxygen supply and demand. Mechanisms involve endothelial dysfunction, coronary artery spasm, tachyarrhythmias, bradyarrhythmias, anemia, respiratory disease, hypotension and severe hypertension.

The universal definition of myocardial infarction also includes type 3 myocardial infarction, myocardial infarction resulting in death if biomarkers are not available and type 4 and 5 myocardial infarction, due to percutaneous coronary intervention and coronary artery bypass grafting, respectively.

Moreover, unstable angina is defined as myocardial ischaemia in a situation of rest or of minimal efforts in default of cardiomyocyte necrosis.

Pathogenesis of Acute Coronary Syndrome

The pathogenesis of ACSs syndromes includes from the appearance of atherosclerotic plaque, to plaque progression and vascular remodeling, to plaque destabilization, to finally plaque rupture or erosion and thrombus formation.

ACS is a life-threatening manifestation of atherosclerosis normally accelerate by acute thrombosis, induced by a ruptured or eroded atherosclerotic plaque, with or without concomitant vasoconstriction, inducing a rapid and critical reduction in blood flow. In the complex process of plaque disruption, inflammation was revealed as a key pathophysiological element. On rare occasions, ACS may have a non-atherosclerotic etiology such as arteritis, trauma, dissection, thromboembolism, congenital anomalies, cocaine abuse, and complications of cardiac catheterization [7].

Atherosclerosis

Atherosclerosis is a chronic disease that can remain asymptomatic for decades. It is triggered by multiple factors and consists of intra intimal accumulation of intra cellular and extracellular oxidized LDL, macrophages, T cells, smooth muscle cells, proteoglycan, collagen, calcium, and necrotic debris. Low endothelial shear stress (ESS) can contribute to atherosclerotic plaque formation, vulnerability, and rupture. Intimal accumulation of oxidized LDL-C, called fatty streaks, are the earliest histopathologic stage of atherosclerosis [8].

Adhesion molecules expressed by endothelial cells mediate the rolling and adhesion of circulating leukocytes on the endothelial surface. Chemoattractant chemokines initiate transmigration of leukocytes into the intima. Monocytes infiltrate beneath the endothelium, differentiate to macrophages, phagocytose the oxidized LDL-C and transform into foam cells. Foam cells generate cytokines, growth factors, reactive oxygen species (ROS) and matrix degrading enzymes, sustaining atherosclerosis progression. The intensity of oxidized LDL-C accumulation in the subendothelial space is a major stimulus for the starting inflammatory process. The accumulation of lipid-laden foam cells constitute the intermediate lesions or pathologic intimal thickening, which evolve through several stages of progression.

Arterial Remodeling

Arterial remodeling includes a cascade of structural and morphological changes of a vessel wall in response to several stimuli including changes in blood flow and pressure, and acute injury; all are common findings in atherosclerotic plaques. The pathogenesis of this disease is not completely understood and remains debated. Many hypotheses have been discussed. Arterial wall neovascularisation of atherosclerotic plaques has a potential role in modulating lesion formation and structural changes of the arterial wall, by nourishing the growing plaque. Several angiogenic growth factors and receptors are implicated in coronary wall angiogenesis such as vascular endothelial growth factor (VEGF), estrogens and interleukin-8. Angiotensin II *via* Angiotensin type 1 (AT1) receptors is another trigger of plaque neovascularisation and remodeling. Activation of NADPH oxidase by various triggers such as Angiotensin II and mechanical stretch promotes ROS production and ROS-mediated pathways leading to vascular remodelling [9].

Plaque Destabilisation

Typically a vulnerable plaque is described as having a thin fibrous cap and a rich superficial lipid core [10]. Weakening of the fibrous cap is due mainly to accelerated degradation of collagen and other matrix components. Many factors contribute to plaque vulnerability.

Inflammation

Unstable or ruptured atherosclerotic plaques are based by the existence of foam cells, macrophages, lymphocytes, and mast cells as it shows histological analysis of atherosclerotic coronary arteries taken from patients who died of ACS. Macrophages and to a lesser extent T lymphocytes were the dominant cell type at the site of plaque rupture or erosion. These inflammatory cells are activated, showing ongoing inflammation at the site of plaque disruption [11]. Also, the shoulder regions of eccentric plaques are sites of predilection for active inflammation as is shown by the high number of activated mast cells, an inflammatory cell type that can induce matrix degeneration by release of both tryptase and chymase and can actively contribute to plaque destabilization preceding rupture [12].

Renin-Angiotensin System

There is evidence of angiotensin converting enzyme (ACE), angiotensin II, and AT1 receptors are related to the plaque. An increased activity of ACE was found within culprits lesions in the setting of ACS, likely related to a local secretion. Angiotensin II increases the likelihood of plaque progression and rupture *via* AT1 receptors by regulating the gene expression of various bioactive substances mainly interleukin-6 (IL-6), metalloproteinases, and other growth factors and cytokines. Angiotensin II also activates multiple intracellular signalling cascades in coronary endothelial and smooth muscle cells. Furthermore, Angiotensin II enhances plaque neovascularisation [13].

Plaque Rupture

The fibrous cap rupture, is due to an imbalance between synthesis and breakdown of extracellular matrix collagen and other matrix components leading to thinning of the cap. This fact, predispose the cap to spontaneous rupture or rupture in response to a variety of triggers. Plaque rupture primarily occurs in yellowish plaques with an increased lipid core and thin fibrous cap. Rupture of the thin fibrous cap exposes blood flow to the lipid core. The lipid core is believed to be highly thrombogenic when exposed to circulating blood. The enhanced thrombogenicity of the lipid core has been attributed to the high levels of functionally active tissue factor most likely derived from the death of macrophages inside the plaque. Furthermore, oxidized lipids in the lipid core may also directly promote platelet aggregation [14].

The thrombus is usually occlusive in acute ST-elevation myocardial infarction (STEMI) and non-occlusive in non-ST-elevation myocardial infarction (NSTE MI). Episodes of plaque disruption and thrombosis may be subclinical and do not always result in ACS. Healing process may play an integral key in the progression of atherosclerosis, having the potential to cause sudden plaque growth.

Plaque Erosion

Younger patients, women and smokers are the most likely to suffer from plaque erosion. Culprit lesion do not have a large lipid core but instead present a proteoglycan-rich matrix, a deep lipid core, the prevalence of inflammation is lower with less macrophages and T cells and more smooth muscle cells compared to culprit lesions in plaque ruptures. Plaque erosion is defined as acute thrombus in direct contact with the intimal plaque without rupture of a lipid pool. The plaque luminal surface is irregular and eroded [15].

Enhanced platelet aggregability, increased circulating tissue factor levels, and depressed fibrinolytic state can trigger thrombosis. Also, activated circulating leucocytes may transfer active tissue factor by shedding microparticles and transferring them onto adherent platelets. Virmani *et al.* [15] have also showed yet another pathological variant where a calcified nodule within the plaque erodes through the surface of the plaque leading to thrombosis.

Early Evaluation of Cardiac Chest Pain–Beyond History and Electrocardiograph

The critical decision of the clinicians for the diagnosis and management of ACS is based on the patient history, physical examination, 12-lead ECG, cardiac biomarkers results, and non-invasive risk stratification.

The resting 12-lead ECG is the first-line diagnostic tool in the assessment of patients with suspected ACS. It is the most readily available tool for identifying patients with STEMI who are likely to have MI. Establishing the correct diagnosis in patients without ST-segment elevation, however, can be more challenging. The ECG is often not diagnostic for acute chest pain, and in fact, the sensitivity of the baseline ECG for detecting acute MI is only 60% [16].The physical examination can also be unsuitable; specifically, atypical chest pain is often difficult to discriminate from chest pain of cardiac origin, and up to 33% of patients with ACS have no chest pain [17].

These imperfect strategies generate inappropriate and costly management decisions. Of patients with acute chest pain who have initial diagnostic findings of ACS and are admitted to the hospital approximately half are later found not to suffer from ACS. Misdiagnosis has been notified to be the main reason of treatment delays [18].

Besides, undetected infarctions remain a serious public health issue and represent the leading cause of malpractice cases in the emergency department. Therefore, cardiac biomarkers are a fundamental part in the management of ACS patients, mainly for differential diagnosis of UA and NSTEMI. In patients with suspected ACS, biomarkers help clinical assessment and 12-lead ECG in the diagnosis, treatment and risk stratification.

Biomarkers for Myocardial Infarction

Progress in the understanding of the pathophysiology of ACS has permitted to the marked increase in development of biomarkers for management of patients with ACS. Overall, a biomarker has been defined as "a characteristic that is reliably and accurately measured and evaluated as an indicator of normal biological processes, pathogenic processes, or pharmacologic responses to a therapeutic intervention" [19]. The ideal characteristics of a cardiac marker are listed in Table **1**.

Table 1. Characteristics of an ideal cardiac biomarker.

High sensitivity	**Abundant in myocardium**
High specificity	**Absent in non-myocardial tissue** **Not detectable in blood from non-diseased subjects**
Release	**Rapid release following myocardial tissue injury for an early diagnosis and in quantities that are in proportion to the size of myocardial injury** **Long half-life in blood for late diagnosis, yet not so long as to mask a recurrent injury**

(Table 1) contd.....

Analytical	**Cost effective** **Short turnaround time (< 60 minutes for central laboratory according to NACB guidelines [31])** **Precise and accurate**
Clinical	**Ability to influence therapy and so improve patient outcome** **Validated by clinical studies**

Biochemical markers for myocardial infarction emerged in 1950s. The concept that tissue injury resulted in enzyme release that could subsequently be measured was the innovation that began the era of diagnostic enzymology.

The measurement of enzymatic biomarkers (Aspartate transaminase (AST), Lactate dehydrogenase (LDH) and creatine kinase (CK) as part of the diagnostic strategies for patients with suspected acute myocardial infarction is included in the World Health Organization (WHO) definition. These biomarkers had a low specificity due to their wide distribution in other tissues different to cardiac tissue. In 2007, NACB recommended that total CK, CK (MB isoenzyme), measured by activity assays, AST and LDH should not be used as biomarkers for the diagnosis of myocardial infarction [20]. Similarly, according to the universal definition of myocardial infarction, published in 2007, measurement of total CK is not recommended for the diagnosis of myocardial infarction, because of the great skeletal muscle distribution and the lack of specificity of this enzyme.

The recognition of the clinical value of CK isoenzymes was followed by the development of improved assays for the MB isoenzyme (CK-MB), implying and increase of specificity regarding to total CK. The inmunoinhibition methods for measurement of CK-MB activity were widely adopted due to low cost and availability on routine chemistry analyzer. Later, the development of inmunoassays for CK-MB in the 1980s was the beginning of a new era in the measurement of cardiac biomarkers, in which the measurement of catalytic activity was replaced by assays for measurement of "protein mass". However, due to the clear diagnostic superiority of measurement of cardiac troponin, measurement of CK-MB, by mass assays, is only recommended when cardiac troponin assays are not available [21].

In 1999, NACB Guidelines [22] recommended the inclusion for diagnosis of acute myocardial infarction of an early biomarker, reliably increased in blood within 6 h after onset of symptoms, such as myoglobin, with a high sensitivity but low specificity. However, nowadays its measurement is not recommended for diagnosis of myocardial infarction [23, 24]. In 1990s the cardiac biomarkers field emerged into what may be termed the 'era of cardiac troponin'. Currently, cardiac troponin (cTn) is the favourite marker for diagnosis and management of suspected MI patients [25, 26].

For predicting ischemic events, especially mortality, in patients with ACS, a variety of other biomarkers have been found to have independent value [20, 27]; however, few of these newer biomarkers are commercially available or have been validated in an undifferentiated population, such as patients presenting to the ED with acute chest pain. Currently, only 2 of these biomarkers, B-type natriuretic peptide and high-sensitivity C-reactive protein (CRP), are available for routine use, and only B-type natriuretic peptide is used in the ED. The utility of B-type natriuretic peptide has been showed in many publications, and elevations of this marker provide powerful risk stratification in ACS patients [28, 29]. However, B-type natriuretic peptide should not be considered a specific biomarker of ischemia, because abnormal values frequently occur in ED patients and are more likely to identify those who have systolic dysfunction related to heart failure rather than ACS [30]. High sensitivity CRP has showed value in predicting long-term cardiac events, but its role in the acute setting of ED patients presenting with chest pain is less well defined, and currently available data do not suggest it is importance in this context [20, 31].

TROPONIN

Cardiac troponins (cTn) were generally recommended as the biochemical gold standard for the investigation of patients with ACS in 1999 by the National Academy of Clinical Biochemistry (NACB) [22], and in 2000 by the European Society of Cardiology and the American College of Cardiology (ESC/ACC) [32]. Both clinical and analytical aspects of these guidelines were revised in 2007 [26] and 2012, in order to establish a more universal definition of acute myocardial infarction. Nowadays, cTn (cTnI and T) are the standard biomarkers for the detection of myocardial injury and prognostic evaluation of patients with ACS.

Cardiac troponins are components of the contractile apparatus of cardiomyocytes and are released during myocardial necrosis in patients with ACS [33]. They are proteins that are part of the contractile apparatus of skeletal and cardiac muscle tissue but they are not present in smooth muscle tissue. The troponin complex consists of 3 interacting and functionally distinct proteins (troponin I, T, and C) [34], which are the products of different genes. Troponin T and I are ideally suitable for the detection of myocardial damage as they are expressed as cardio-specific isoforms.

Most of cardiac troponin is structurally bound within the thin filaments of the contractile apparatus. A small percentage of protein remains free in the cytosol. This percentage is approximately 2–4% for cTnI and 6–8% for cTnT [35]. Damage to cardiac myocytes resulting in loss of membrane integrity causes the release of cTn into the circulation. In patients with MI, levels of cardiac troponin rise rapidly; cTn begins to rise within 3– 4 h after the onset of myocardial injury and remain increased for up to 4 –7 days for cTnI and 10 –14 days for cTnT when contemporary assays are used [20] and usually within 1 h, if using high-sensitivity assays, after symptom onset and remain elevated for a variable period of time (usually several days).

Classification of Troponin Assays

Currently, cTn assays are classified according two criteria (Table **2**) [36]:

1. The total imprecision, expressed as coefficient of variation (% CV), of each assay at the 99th percentile upper reference limit concentration of a healthy population. It is been further recommended that only cTn assays with optimal total imprecision (CV ≤10%) at the 99th percentile value, termed "guideline acceptable") be used. Although a better imprecision at low cTn concentrations improves the value of cTn as both diagnostic and risk indicator [20] use of cTn assays with intermediate imprecision (10% to 20% CV) at the 99th percentile does not lead to significant patient misclassification when interpreting serial cTn results [37]. These assays are termed as "clinically usable". Assays with a total imprecision of > 20% CV at the 99th percentile according to these requirements are "not acceptable".

2. The number of specimens from normal individuals having cTn concentrations that are measurable below the 99th percentile.

Table 2. Designation of cTn assays.

Acceptance designation	Total imprecisión at the 99th percentile, CV%
Guideline acceptable Clinically usable Not acceptable	≤ 10 > 10 to ≤ 20 > 20
Assay designation	**Measurable normal values below the 99th percentile, %**
Level 4 (third generation, hs) Level 3 (second generation, hs) Level 2 (first generation, hs) Level 1 (contemporary)	≥ 95 75 to < 95 50 to < 75 < 50
CV: Coefficient of variation; hs: high sensitivity	

According to these criteria the ultimate goal is to have all assays be "third generation (level 4), guideline acceptable."

More recently, IFCC Task Force on Clinical Applications of Cardiac Bio-Markers establishes that an assay should measure cTn above the limit of detection in ≥

50% of healthy subjects to be classified as high sensitivity assay [38].

Clinical Implications of hs cTn Assays

Compared with contemporary assays, hs cTn assays have a higher negative predictive value and reduce the "troponin-blind" interval leading to earlier detection of acute MI and consequently a faster treatment. Due to the higher sensitivity, the time interval to the second cTn determination can be shortened with the use of these assays, allowing more rapid approaches to rule in or rule out ACSs.

2007 NACB guidelines about utilization of biomarkers in ACSs recommend that blood should be obtained for testing at hospital presentation followed by serial sampling with timing of sampling based on the clinical, although for most patients, blood should be obtained for testing at hospital presentation and at 6-9 h [20]; in the last definition of myocardial infarction, collection of blood samples on first assessment and 3-6 h later, and later if further ischaemic episodes occur, or when the timing of the initial symptoms is unclear, is recommended. A similar timing of sampling is recommended by American Heart Association (AHA) when contemporary assays for measurement of cTn are used [24].

Recently, the new European Society of Cardiology guidelines for NSTEMI management recommend a rapid rule-out protocol at 0 and 3 hours if hs cTn assays are available [39]. Given the number of ED patients who undergo serial marker testing, this shortened evaluation period has the potential to affect ED crowding [40]. Indeed, according to results of High Sensitivity Cardiac Troponin T Assay for Rapid Rule-out of Acute Myocardial Infarction (TRAPIDAMI) [41] and Advantageous Predictors of Acute Coronary Syndrome Evaluation (APACE) [42] trials, a rapid rule-out and rule-in protocol at 0 and 1 hour is recommended if a hs cTn assay with a validated 0 h/1 h algorithm is available.

Therefore, these new assays allow more rapid approaches to rule in or rule out ACSs than with previous cTn assay generations which is one of the most important benefits. The problem is that hs cTn is also more sensitive for the detection of myocardial damage unrelated to acute myocardial ischemia. So that, the increase in early diagnostic sensitivity of hs cTn assays for ACS comes at the cost of a reduced ACS specificity, because more patients with other causes of acute or chronic myocardial injury without overt myocardial ischemia are identified than with previous cTn assays [43]. Other causes of elevations of hs cTn in absence of significant coronary artery disease are listed in Table **3**.

Specificity of hs cTn could be further improved by serial sampling [43, 44]. Serial measurement to determine changing troponin concentration over time (delta cTn) supplies an important way of distinguishing between acute and chronic causes of cTn rise (*e.g.* acute MI *versus* chronic heart failure), in patients with only minor elevations of cTn just above the 99th percentile, which significantly increase with the introduction of hs-cTn assays [45 - 47].

Table 3. Elevations of hs cTn in absence of significant coronary artery disease (CAD).

Acute myocardial damage related to secondary myocardial ischemia (MI type 2)	**Tachycardia or bradycardia Aortic dissection with involvement of coronary ostia Severe aortic valve stenosis Hypertrophic cardiomyopathy Hypo- or hyper-tension (*e.g.*, hemorrhagic shock, hypertensive emergency) Acute heart failure without significant concomitant CAD Severe pulmonary embolism or pulmonary hypertension Coronary vasculitis, *e.g.*, systemic lupus erythematosus Coronary endothelial dysfunction (spasm) without significant CAD (*e.g.*, cocaine abuse) Coronary embolism**
Acute myocardial damage not related to myocardial ischemia	**Cardiac contusion or cardiac procedures Radiofrequency or cryoablation therapy for arrhythmias Rhabdomyolysis with cardiac involvement Myocarditis Myocardial drug toxicity or poisoning, *e.g.*, anthracyclines or herceptin**
Other causes	**Apical ballooning syndrome Renal failure Severe acute neurological diseases, *e.g.*, stroke, trauma Infiltrative diseases, *e.g.*, amyloidosis, sarcoidosis, scleroderma, haemocromathosis Extreme exertion Sepsis Acute respiratory failure Frequent defibrillator shocks Hypo- and hyperthyroidism Ischemic stroke**

Emphasis on the dynamic change in cTn concentrations from baseline to follow-up ("delta") has intensified as more patients with non-ischemic conditions have elevations when high-sensitivity assays are used [48]. However, the degree to which the cTn concentration must change and the optimum timing over which dynamic changes should be calculated has not been specified. Some research have used various delta criteria including ≥10-30% deltas [43, 49], receiver operating characteristics (ROC) curves to derive optimum percentage changes [43, 44] and standard deviations [49]. The NACB laboratory medicine practice guidelines recommends the use of ≥ 20% delta (the level exceeding that due to analytical variation at the 99th percentile) for cTn concentrations from elevated baseline values [22], but this has yet to be confirmed. Many clinicians have worked with a change in the cTn concentration of ≥ 20% as a practical cut-off. Recent research has suggested that this cut-off value regarding significant rise of troponin level needs to be higher in the lower cTn range. An absolute kinetic change of 9.2ng/L hs-TnT has been proposed to rule in or rule out NSTEMI. Despite this [50], guidance on the precise utility of this method remains poorly defined [51, 52]. And nowadays [53], there is no accord about which delta value should be used for distinguishing between acute and chronic causes of cTn rise [54].

High Sensitivity Troponin Assays for Prognostic Use

The identification of patients at risk for acute cardiovascular events requires a change.

Cardiac troponin level giving clinicians an idea of the prognosis following an infarct as this level is dependent on infarct size [55].

It has also been demonstrated that the magnitude of troponin level elevation correlates with risk of future cardiac events or death and aids the identification of patients with greater disease severity who may benefit from more aggressive therapy [56, 57]. In point of fact, the size of the infarcted area may be predicted based on peak cTnI levels or cTnT levels at 72 hours [58 - 60].

With the development of more accurate hs-cTn assays, these allows the identification of some high risk patients from within the vast population of acute chest pain who would otherwise go undetected using conventional cTn [61 - 64].

A first prognostic hs cTnT study indeed showed a significant association between increased cTnT concentrations and the incidence of cardiovascular events in patients with stable chronic heart failure [65]. Omland *et al.* [66] showed a significant relation between increased hs cTnT concentrations and the incidence of cardiovascular death and heart failure in patients with stable coronary artery disease (CAD).

Mingels *et al.* [67] demonstrated that hs-cTnT was a significant predictor for the composite endpoint of late revascularizations, ACS and cardiac mortality in patients with symptoms of chest discomfort suspected for CAD. De Filippi *et al.* performed serial measures of hs-cTnT in community-dwelling older adults [64]. They found a significant association between baseline hs-cTnT concentrations, changes in hs-cTnT concentrations and the evolution of heart failure and cardiovascular death. De Lemos *et al.* found an relation between increased hs-cTnT and structural heart disease, especially left ventricular hypertrophy, and subsequent risk for all-cause mortality [63].

Bandstein *et al.* [68] reported that patients with chest pain who have an initial hs-cTnT level of < 5 ng/l and no signs of ischemia on ECG have a minimal risk of MI or death within 30 days and can be safely discharged directly from the ED.

COPEPTIN

Arginine vasopressin (AVP), also named antidiuretic hormone, is a nonapeptide generated in the hypothalamus. AVP is released from the neurohypophysis into the blood to induce water conservation by the kidney, contributing to the regulation of osmotic and cardiovascular homeostasis [69]. AVP is derived from a larger precursor peptide (pre-provasopressin) along with 2 other peptides, neurophysin II and copeptin [70].

Vasopressin is difficult to measure because it is unstable and rapidly cleared [71]. However, copeptin, a 39-amino acid glycopeptides that comprises the C-terminal part of the AVP precursor, was found to be a stable and sensitive surrogate marker of AVP release [72].

There are certain confounding factors for the interpretation of copeptin levels. In healthy volunteers, the median plasma level of copeptin was found to be 4.2 pmol/L. A gender difference in the median plasma levels was also reported previously (male: 5.2 female: 3.7 pmol/L) [72, 73]. Copeptin levels were found to be higher in the male volunteers compared with female. Especially in men, there is a strong relationship between copeptin and decreased glomerular filtration rate, probably due to decreased renal copeptin clearance. However, unlike other biomarkers, plasma levels are generally similar in all age groups [72, 73]. Furthermore, corticosteroids appear to inhibit copeptin release [74].

Copeptin, is a marker for severe, hemodynamic stress and has been studied as a marker for the early rule-out of MI in several observational trials with very promising results [75, 76].

Although the physiopathology of copeptin release in acute myocardial infarction is unclear, plasma levels peak early at patient admission and decrease from the second day [75, 77].

There are various hypotheses to explain the rapid release of AVP/copeptin after AMI. A likely explanation is that AVP responds rapidly as part of the endocrine stress axis, resulting in release of adrenocorticotropic hormone and cortisol. Copeptin is a rapid and immediate biomarker of the individual stress response

[78]. An alternative trigger of AVP/copeptin secretion from the posterior pituitary could be baroreceptor stimulation by the threat of hypotension as a result of the AMI or direct damage to the cardiac baroreceptors. The latter possibility is supported by the fact that the highest copeptin elevation after AMI is seen in patients with STEMI [76].

Gu *et al.* study [79] confirmed the findings of previous studies, which showed that copeptin levels in AMI patients were maximal on admission in contrast to cTnT [75, 76] and dropped to a plateau level after the first day [77].

With increasing time after onset of symptoms, Gu *et al.* observed decreasing levels of copeptin, in contrast to increasing levels of troponin T. The fall in copeptin levels may reflect a mechanism of adaptation by the endogenous stress system facing a continuous stress such as AMI or may be the consequence of the resolution or at least reduction of chest pain after the onset of AMI, or both. This extends and corroborates findings in 132 patients with AMI and blood sampling for 5 days after diagnosis of AMI, showing a copeptin peak with similar levels on day 1 and falling levels thereafter, until reaching a plateau by day 3 to 5 [77].

Consistent with the other studies [20, 80], Gu *et al.* noticed that CK-MB and cTnT begin to rise within several hours after the onset of myocardial injury and reach their peak value in the first 24 h. Concordantly, hs-cTnT follows a similar release pattern as cTnT, despite a higher sensitivity and improved precision at the lower limit of detection [81, 82]. The only conventional cardiac biomarker that could approach the early presence of marker levels above normal ranges is myoglobin, whose levels begin to rise 1 h after myocardial injury [20]. Despite its quick release into the circulation, however, myoglobin is not of additional value in a diagnostic strategy including cardiac troponin and is not recommended as a diagnostic biomarker in the routine clinical practice [83].

The release of necrosis markers from cardiomyocytes is believed to be delayed after cell damage, and this might explain the weakness in diagnostic performance of conventional Tn assays early after chest pain onset [72]. Because troponin assays lack sensitivity in the early hours after onset of AMI, adding a marker that is sensitive in these early hours and acts independently from cardiomyocyte injury to this conventional biomarker of myocardial necrosis may complement and improve the diagnostic accuracy of a measurement on admission.

Patients with UA have lower copeptin concentrations than those with NSTEMI or even have levels that are similar to those in patients with other causes of non-ischemic chest pain [76]. These data suggest that AMI induces a higher level of endogenous stress than unstable angina does, potentially related at least in part to the more prolonged course of chest pain in patients with AMI [84]. Ischemia, as long as not accompanied by necrosis (*i.e.*, unstable angina), does not seem to be a stronger trigger of copeptin release than are other causes of chest pain. Agree with this, a recent study showing comparable increases in copeptin levels during exercise in patients with or without exercise-induced ischemia [85]. Thus, most authors define copeptin as a biomarker to rule out MI and not other types of ACS [86].

Copeptin Assays for Diagnostic Use

To date, conventional cardiac biomarkers are not especially useful in immediately ruling out AMI in patients presenting early with symptoms that are suggestive for an AMI but with non-diagnostic findings on ECG. The findings suggest that copeptin may be a valuable marker on top of cTnT for the detection of AMI. In this case, adding a copeptin to a cTnT measurement on admission improved the discriminative value compared to a cTnT measurement alone [75, 76, 87].

Reichlin *et al.* [76], found that the diagnostic performance of a combination of the cTnT concentration resulted with a conventional fourth-generation assay plus the copeptin concentration at initial presentation was significantly better for ruling out MI than cTnT alone (AUC, 0.97 *vs* 0.86; $P< 0.001$), with an NPV of 99.7%.

Consistent with this fact are those for 1386 patients reported by Keller *et al.* [75], who found that the combination of the copeptin concentration and the cTnT concentration at admission obtained with a fourth-generation assay improved the c-statistic from 0.84 for cTnT alone to 0.93 in the overall population ($P < 0.001$), with an NPV of 94.8%. However, they included high percentages of patients with AMI (21.6%) and patients with STEMI among AMI (31.1%), which might limit the extrapolation of their data to all-comer populations with lower risk (such as those treated in the ED).

Findings of Giannitsis *et al.* [88] add the important information that copeptin improves the rapid rule-out of non-STEMI, even if the hs-cTnT assay is used as the diagnostic standard of MI, as other studies suggest [75, 89].

If the ED physicians follow the current guideline recommendations, all patients with a non diagnostic ECG require an ED stay of at least 3 to 6 h or admission to the hospital to rule out an AMI. The multicenter CHOPIN (Copeptin Helps in the Early Detection of Patients With Acute Myocardial Infarction) trial demonstrated that adding copeptin allowed 58% of all patients with non diagnostic ECG to be ruled out for an AMI, whereas the number of undetected AMI cases only increased by 0.2% (from 32 to 35 of 1,967 patients). Assuming a 3-h delay before the second cTnI is determined, the availability of copeptin at baseline would reduce the time to decision for AMI diagnosis from an average of 3 to 1.8 h (43%

reduction). The benefit is most prominent in those patients with an intermediate risk of AMI (5% to 25%), as assessed by ED physicians. A low copeptin level in these patients (<14 pmol/l) identifies a subset with a risk comparable to those who present with a low pretest probability (<5%) for AMI. Most (72%) of the AMIs detected who had normal copeptin values were in those who were not low risk [90].

These facts have great clinical implications. The rapid and suitable exclusion of AMI in patients presenting with chest pain is one of the large needs in clinical practice. Because of the delayed increase in troponins and normal or unspecific ECG findings, 10 million patients worldwide require prolonged monitoring and serial blood sampling each year before AMI can safely be excluded. The additional costs associated with the remaining diagnostic uncertainty after the first troponin measurement are estimated to exceed several billion U.S. dollars each year [91]. Thus, the improvement in the early rule out of AMI offered by copeptin testing may have the potential to improve allocation of resources in the ED and to markedly reduce total management cost [92]. Acute MI could be rapidly and adequately ruled out at admission in two-thirds of patients, and only the remaining one-third of patients would need monitoring and serial blood sampling, with roughly one-half of them (positive predictive value 46%) finally suffering from AMI.

The results of the COPED-MIRRO study [93] are in agreement with those of various research [75, 76] that have evaluated the role of copeptin in patients with non-traumatic chest pain in the ED, most of which reported an NPV above 90%. Nevertheless, the comparison is difficult due to the essential characteristics of these studies. One of the important points is the cut-off used for copeptin. Previous studies have established different values, although most used the cut-off of 14 pmol/l recommended by the manufacturer. With this cut-off, the NPV of isolated copeptin ranged from 54 to 94%. Some studies using lower cut-offs have reported slightly higher NPVs of between 86 and 98%. Even in the study by Reichlin *et al.*, the cut-off of copeptin was reduced from 14 to 9 pmol/l, increasing the NPV from 99.7 to 100%, to rule out MI. Although in this study the result was obtained on combining patients with and without ST elevation. In the case by Charpentier *et al.* [94] carried out exclusively in patients with NSTEMI, the reduction in the cut-off from 14 to 12 pmol/l showed almost no modification in the NPV (91.9–92.4).

Charpentier *et al.* showed that copeptin added to contemporary cTnI improved the early diagnostic accuracy of NSTEMI in ED chest pain patients but that the sensitivity of the combination (90%) was insufficient to safely rule out NSTEMI at presentation. The use of sensitive cardiac troponin assays has been shown to further improve the precision of the combination for acute myocardial infarction diagnosis [95]; however, this improvement was inconsistent in patients with low or intermediate risk of acute coronary syndrome when patients with STEMI were excluded or not analyzed along with NSTEMI patients [96].

Moreover, according to the COPED MIRRO study, the analysis of copeptin on arrival to the ED in patients with acute chest pain suggestive of ACS, in whom the first troponin determinations were negative with a non-diagnostic ECG, does not avoid the use for a second troponin determination and thus requires patients to prolong their ED stay 6 h to determine the most appropriate therapeutic approach to follow. Similar results were recently reported by Esteban-Torrella *et al.* [97].

In summary, as suggested by some studies, copeptin improves the diagnostic performance but cannot rapidly rule out NSTEMI at presentation to an ED. Recently, the European Society of Cardiology guidelines for NSTEMI management conclude that the incremental value of copeptin over hs cTn assays remains to be fully elucidated.

Recent studies in copeptin assays have led to a newly developed sensitive assay differing from the conventional assay by a lower detection level in a majority of healthy people. The newly developed high sensitivity copeptin (hs-copeptin) assay clearly shows improved analytical performances as compared with the conventional assay, with a lower detection limit (1.9 pmol/L) and 10% functional sensitivity (3 pmol/L) [98].

Study of Sebbane *et al.* assess the newly developed hs-copeptin assay as a diagnostic tool for AMI in ED patients with chest pain. Their results highlighted the added value of hs-copeptin in conjunction with hs cTnT for early rule out of AMI, particularly NSTEMI, at ED admission in a prospective cohort of patients presenting to the ED with chest pain and onset within the previous 12 hours [98].

The aim of Duchenne *et al.* study [99] was to determine whether hs-copeptin could rule out diagnosis of acute MI without prolonged monitoring and serial blood sampling in patients with suspected NSTEMI and hs-cTnT below the 99th centile at presentation to ED. In this study, hs-copeptin does not add a diagnostic value at admission to ED for patients with suspected acute coronary syndrome without ST-segment elevation and with hs-cTnT below the 99th centile.

Copeptin Assays for Prognostic Use

Stress hormones including copeptin have been proposed as prognostic markers in acute illness as magnitude of individual stress is generally well correlated with the severity of the disease. Therefore, copeptin, as a marker of stress, may have the potential to predict clinical outcome, and may help determine the therapeutic strategy in patients with acute illness [100].

Copeptin was shown to be an excellent prognostic marker in the elderly ED patients with non-specific symptoms regardless of the final diagnosis [101]. This suggests that copeptin works well as a marker of non-specific stress response, and hence disease severity in a specific age group in whom atypical presentation, even in the presence of a serious condition, is common possibly due to attenuated physiological response [102, 103].

Several studies have demonstrated that copeptin is a strong predictor of death in patients after AMI with or without signs or symptoms of heart failure [77, 104].

When copeptin was measured in 980 patients who had experienced an acute MI, it was found to be elevated in patients who died or were readmitted with heart failure, compared with survivors. Copeptin and N-terminal pro-BNP were independent predictors of death or heart failure at 60 days, helping to stratify patients into low-, intermediate- or high-risk groups [77]. Nevertheless, in other study, the prognostic value of copeptin was even greater compared with those of BNP and NT-proBNP [104].

Morgenthaler *et al.* have demonstrated that an elevated plasma copeptin following admission with NSTEMI is strongly associated with adverse outcomes and has independent prognostic value when included in a multivariate risk model. This study demonstrated that copeptin could be used as a prognostic biomarker following admission with NSTEMI, which supports previous works of the same author which demonstrated that copeptin was an independent predictor for adverse events following STEMI [105].

In a multicenter study, it was confirmed that copeptin is a strong marker for mortality and morbidity in patients with heart failure after an acute MI [104]. Kelly *et al.* [106] reported an association of copeptin with the degree of left ventricular remodeling after AMI.

Other findings [107] support previous studies investigating the prognostic role of copeptin in different various settings such as AMI, unstable angina and heart failure [108, 109]. Interestingly, an animal study documented the release of vasopressin from cardiomyocytes, indicating a vasopressin system in rat hearts [110]. This experiment demonstrated that vasopressin is synthesized and released on stimulation with cardiac pressure overload or stimulation with nitric oxide, resulting in systemic effects such as osmoregulation, regulation of cardiac function, perfusion and cardiac neurohormone secretion. These findings could explain the higher copeptin concentrations in patients with pre-existing CAD even in the absence of AMI because of increased nitric oxide synthesis [111].

Copeptin may have not only short term prognostic relevance, but also identify patients at higher long-term risk. Increased copeptin concentrations in elderly patients with symptoms of heart failure were associated with an increased risk of all-cause mortality after a lengthy median follow-up of 13 years [112].

A study of patients treated in an outpatient clinic for chronic stable heart failure showed that copeptin and hs-cTnT levels, both single and combined, are powerful predictors of death and hospitalization. This suggests that simultaneous assessment of myocardial damage and the activated vasopressin system, through single measurement of hs-cTnT and copeptin levels, might be of prognostic relevance [113].

According to a study [107], copeptin levels measured at presentation provided powerful information for the risk of death during the subsequent 12 months. Only patients with both raised copeptin and cTn levels were at markedly increased risk, whereas patients with raised hs-cTnT levels and normal copeptin levels had a similar prognosis to those with normal hs-cTnT levels. The clinical use of highly-sensitive cTn assays leads to a marked increase in patients with increased cTn in settings other than AMI [81, 82]. This has caused some confusion among clinicians and makes the interpretation of raised troponin values more challenging. Having a second marker that identifies patients with a good prognosis despite a raised hs-cTn level is therefore of great clinical value. Copeptin might help emergency physicians to tailor treatment and allocate resources accordingly. Tailored treatment in high-risk patients may include consultation of specialists, early invasive strategy or admission to the intensive care unit. Whether this risk stratification-guided strategy might affect outcome needs to be evaluated prospectively.

In the ED, copeptin is in competition with other novel biomarkers that are also investigated for their potential use as prognosticators. The potential advantage of AVP and copeptin over these competing markers in the ED lies in its more central role as one of the key hormones in the body. It is not limited to a single organ system, but is triggered by many disease processes also outside the cardiovascular system. This non-specificity, with respect to a precise diagnostic role, is its strength as a more generalized marker for severe disturbances in patient physiology. In future head-to-head comparisons of biomarkers or biomarker panel strategies copeptin is likely to be one of the strongest candidates.

INTERLEUKIN-6

IL-6 is a ubiquitous cytokine, produced by many cell types including activated monocytes/macrophages and endothelial cells (ECs), as well as by adipose tissue. IL-6 is able to stimulate macrophages to secrete monocyte chemo attractant protein-1 (MCP-1) and participates in the proliferation of smooth muscle cells (SMCs). In addition, ECs stimulated by IL-6, express intercellular adhesion molecule-1 (ICAM-1) [13].

In a murine model of atherosclerosis, injection of excessive amounts of re-combinant IL-6 resulted in enhanced fatty lesion development. Furthermore, IL-6 represents the principal procoagulant cytokine [114], but its most important function is the amplification of the inflammatory cascade through which IL-6 at least in part might exerts its direct proatherogenic effects in the arterial wall. Indeed, large amounts of IL-6 have been found in human atherosclerotic plaque [115], in particular within the shoulder region of stable and unstable plaque, where it localized with the AT-1 receptor [116]. Up regulation of the AT-1 receptor by IL-6 has led to increased angiotensin II–mediated vasoconstriction, enhanced free oxygen radical production and the development of endothelial dysfunction [13].

IL-6 also promotes production of hepatic acute-phase reactants, including C reactive protein (CRP) [117, 118]. Increased concentrations of CRP have been reported in unstable angina and in AMI [119]. Liuzzo *et al.* [120] have shown that concentrations of CRP and serum amyloid A in unstable angina increase independently of myocardial cell injury, as shown by normal concentrations of creatine kinase and troponin T. Therefore, hs-CRP levels can reflect the prognosis of ACS patients [121]; however, hs-CRP is a non-specific inflammatory factor, and therefore has poor specificity.

Numerous studies have demonstrated that the inflammatory level is increased in the ACS patients [122 - 124]. Shantsila *et al.* [125] highlighted that monocytes were actively involved in the pathological processes related to ACS, which promoted the synthesis of pro-inflammatory molecules, such as IL-6, tumor necrosis factor-α (TNF-α) and hs-CRP.

The evidence of elevated levels of acute-phase proteins in unstable angina is in line with a growing body of evidence suggesting that inflammation plays a pivotal role in acute coronary syndromes and is an indirect sign of the increased production of interleukins from monocytes, macrophages, and endothelial cells. IL-6 is involved in a number of other processes, because it has pro-inflammatory and pro-coagulant properties that may have a role in the pathophysiology of ACS [126].

Interleukin-6 Assays for Diagnostic Use

IL-6 and other cytokines were significantly increased in ACS patients [127]. Several studies demonstrates that elevated levels of IL-6 are common in unstable angina [118, 128]. In a study, elevated IL-6 levels on admission were significantly increased in patients with AMI compared with patients with non-ischemic chest pain or UA [129]. Burazor *et al.* found that IL-6 levels in ACS patients were higher than those of healthy controls [130].

In addition, Maier *et al.* have demonstrated in patients with ACS that IL-6 levels were markedly higher at the site of coronary plaque rupture than in the systemic circulation [131].

Thus suggests the importance of IL-6 in the pathophysiology of ACS and confirms the role of IL-6 as a key mediator of inflammatory reactions including hepatic synthesis and release of CRP and fibrinogen.

However, the relationship between IL-6 and ACS is not consistent across other studies [130]. One study showed that IL-6 levels in patients with unstable angina were significantly higher than those in patients with ACS [132]. So, the relationship between IL-6 and ACS needs further clarification.

Myocardial ischaemia biomarkers do not require myocardial cell death for release. These include biomarkers of inflammation and platelet activation. As the prototypical acute phase reactants, C-reactive protein (CRP) and interleukin-6 (IL-6) have been the focus of investigations for the diagnosis of ACS [133].

Interleukin-6 Assays for Prognostic Use

NSTEMI, has been showed to have a wide spectrum of risk for new acute ischemic events and mortality. The only biomarker used in clinical guidelines and risk score scales is troponin, a marker of myocardial necrosis. Nevertheless, patients with troponin levels below the standardized cut-off point, called troponin-negative, are still at risk of presenting new events [3].

The process that leads to eventual plaque erosion or rupture involves a number of inflammatory mechanisms, including endothelial dysfunction, leukocyte migration, extracellular matrix degradation, and platelet activation. The optimal inflammatory biomarker would provide a method for quantitating cardiovascular-specific inflammation, thereby predicting the risk of recurrent atherothrombosis. Because they address a separate aspect of ACS pathophysiology, biomarkers of inflammation may provide unique information to the clinician separate from that provided by biomarkers of myocyte necrosis and hemodynamic stress.

IL-6 is a biomarker of inflammation which has been related to prognosis in NSTEMI [134]. In fact, several studies have shown relationship between the IL-6 level and mortality in patients with NSTE-ACS.

Interleukin-6 serum levels showed to be an independent predictor of adverse events among low–moderate risk patients with troponin-negative NSTE-ACS at 6 months follow up [135].

Among patients with unstable angina, an increase in IL-6 levels that occurred 48 hours after admission compared with the admission value was associated with the combined end point of death, myocardial infarction (MI), or refractory angina [126, 136].

The Fragmin and Fast Revascularization During Instability in Coronary Artery Disease II (FRISC II) [137] study not only confirmed the predictive power of IL-6, but also demonstrated that patients with high IL-6 level might benefit most from an early invasive strategy.

In addition, several prospective studies have consistently shown that baseline levels of IL-6 are a potent predictor of future CV end points in apparently healthy asymptomatic subjects from the general population [138].

Currently, however, data regarding the long-term prognostic value of IL-6 in patients with pre-existent CAD are limited and controversial. In addition, the large circadian variations in IL-6 levels and lack of confirmatory studies limit the applications of IL-6 as a biomarker in ACS.

DISCUSSION

Biomarkers complement clinical assessment and 12-lead ECG in the diagnosis, risk stratification and treatment of patients with suspected NSTE-ACS. Measurement of a biomarker of cardiomyocyte injury, preferably high-sensitivity cardiac troponin, is mandatory in all patients with suspected NSTE-ACS. Cardiac troponins are more sensitive and specific markers of cardiomyocyte injury than creatine kinase (CK), its MB isoenzyme (CK-MB) and myoglobin. If the clinical presentation is compatible with myocardial ischaemia, then a dynamic elevation of cardiac troponin above the 99th percentile of healthy individuals indicates MI. Data from large multicentre studies have consistently shown that sensitive and high-sensitivity cardiac troponin assays increase diagnostic accuracy for MI at the time of presentation as compared with conventional assays, especially in patients presenting early after chest pain onset, and allow for a more rapid 'rule-in' and 'rule-out' of MI. Other life-threatening conditions presenting with chest pain, such as aortic dissection and pulmonary embolism, may also result in elevated troponin levels and should be considered as differential diagnoses.

Among the multitude of additional biomarkers evaluated for the diagnosis of NSTE-ACS, copeptin and interleukin-6 seem to have clinical relevance. Assessment of copeptin, the C-terminal part of the vasopressin prohormone, may quantify the endogenous stress level in multiple medical conditions including MI. As the level of endogenous stress appears to be invariably high at the onset of MI, the added value of copeptin to conventional (less sensitive) cardiac troponin assays is substantial. Copeptin may have some added value even over high-sensitivity cardiac troponin in the early rule out of MI.

The evidence of elevated levels of acute-phase proteins in unstable angina is in line with a growing body of evidence suggesting that inflammation plays a pivotal role in acute coronary syndromes and is an indirect sign of the increased production of interleukins from monocytes, macrophages, and endothelial cells. IL-6 levels were significantly increased in ACS patients, so that interleukin-6 could be useful for diagnosis.

Beyond diagnostic utility, cardiac troponin levels add prognostic information in terms of short- and long-term mortality to clinical and ECG variables. While high-sensitivity cardiac troponin T and I seem to have comparable diagnostic accuracy, high-sensitivity cardiac troponin T has greater prognostic accuracy. The higher the high-sensitivity troponin levels at presentation, the greater the risk of death. Multiple biomarkers have been associated with mortality in NSTE-ACS, several of them conferring additive prognostic value to cardiac troponin. The extensively validated natriuretic peptides (*i.e.* B-type natriuretic peptide, N-terminal pro-B-type natriuretic peptide and midregional pro-A-type natriuretic peptide) provide prognostic information on top of cardiac troponin. To some extent, the same applies to high sensitivity C-reactive protein and novel biomarkers such as interleukin-6 and copeptin. Several studies have demonstrated that copeptin is a strong predictor of death in patients after acute myocardial infarction with or without signs or symptoms of heart failure. The serum IL-6 levels can be used to determine the stability of plaque, and have some relevance in the ACS process, showing great value for judgments in ACS prognosis.

However, clinical judgment is essential for optimal interpretation and assessment of these markers.

CONFLICT OF INTEREST

The author (editor) declares no conflict of interest, financial or otherwise.

ACKNOWLEDGEMENTS

Declared none.

REFERENCES

[1] Pitts SR, Niska RW, Xu J, Burt CW. National Hospital Ambulatory Medical Care Survey: 2006 emergency department summary. Natl Health Stat Rep 2008; 6: 1-38.

[2] Pope JH, Aufderheide TP, Ruthazer R, *et al.* Missed diagnoses of acute cardiac ischemia in the emergency department. N Engl J Med 2000; 342: 1163-70. [http://dx.doi.org/10.1056/NEJM200004203421603]

[3] Antman EM, Cohen M, Bernink PJ, *et al.* The TIMI risk score for unstable angina/non-ST elevation MI: A method for prognostication and therapeutic decision making. JAMA 2000; 284: 835-42. [http://dx.doi.org/10.1001/jama.284.7.835]

[4] Diercks DB, Kirk JD, Amsterdam EA. Chest pain units: management of special populations. Cardiol Clin 2005; 23: 549-57. [viii.]. [http://dx.doi.org/10.1016/j.ccl.2005.08.015]

[5] Thygesen K, Alpert JS, White HD. Universal definition of myocardial infarction. Eur Heart J 2007; 28: 2525-38. [http://dx.doi.org/10.1093/eurheartj/ehm355]

[6] Roffi M, Patrono C, Collet JP, *et al.* 2015 ESC Guidelines for the management of acute coronary syndromes in patients presenting without persistent ST-segment elevation: Task Force for the Management of Acute Coronary Syndromes in Patients Presenting without Persistent ST-Segment Elevation of the European Society of Cardiology (ESC). Eur Heart J 2016; 37: 267-315. [http://dx.doi.org/10.1093/eurheartj/ehv320]

[7] Bassand JP, Hamm CW, Ardissino D, *et al.* Guidelines for the diagnosis and treatment of non-S- -segment elevation acute coronary syndromes. Eur Heart J 2007; 28: 1598-660. [http://dx.doi.org/10.1093/eurheartj/ehm161]

[8] Hamm C. Benefits and risks of the new markers for early diagnosis of myocardial infarction. Internist (Berl) 2006; 47 (Suppl. 1): S60. [http://dx.doi.org/10.1007/s00108-006-1701-0]

[9] Libby P. Molecular bases of the acute coronary syndromes. Circulation 1995; 91: 2844-50. [http://dx.doi.org/10.1161/01.CIR.91.11.2844]

[10] Sukhova GK, Schonbeck U, Rabkin E, *et al.* Evidence for increased collagenolysis by interstitial collagenases-1 and -3 in vulnerable human atheromatous plaques. Circulation 1999; 99: 2503-9. [http://dx.doi.org/10.1161/01.CIR.99.19.2503]

[11] van der Wal AC, Becker AE, van der Loos CM, Das PK. Site of intimal rupture or erosion of thrombosed coronary atherosclerotic plaques is characterized by an inflammatory process irrespective of the dominant plaque morphology. Circulation 1994; 89: 36-44. [http://dx.doi.org/10.1161/01.CIR.89.1.36]

[12] Laine P, Kaartinen M, Penttila A, Panula P, Paavonen T, Kovanen PT. Association between myocardial infarction and the mast cells in the adventitia of the infarct-related coronary artery. Circulation 1999; 99: 361-9. [http://dx.doi.org/10.1161/01.CIR.99.3.361]

[13] Schieffer B, Schieffer E, Hilfiker-Kleiner D, *et al.* Expression of angiotensin II and interleukin 6 in human coronary atherosclerotic plaques: potential implications for inflammation and plaque instability. Circulation 2000; 101: 1372-8. [http://dx.doi.org/10.1161/01.CIR.101.12.1372]

[14] Dirksen MT, van der Wal AC, van den Berg FM, van der Loos CM, Becker AE. Distribution of inflammatory cells in atherosclerotic plaques relates to the direction of flow. Circulation 1998; 98: 2000-3.
[http://dx.doi.org/10.1161/01.CIR.98.19.2000]

[15] Farb A, Burke AP, Tang AL, *et al.* Coronary plaque erosion without rupture into a lipid core. A frequent cause of coronary thrombosis in sudden coronary death. Circulation 1996; 93: 1354-63.
[http://dx.doi.org/10.1161/01.CIR.93.7.1354]

[16] Panteghini M. Acute coronary syndrome: biochemical strategies in the troponin era. Chest 2002; 122: 1428-35.
[http://dx.doi.org/10.1378/chest.122.4.1428]

[17] Canto JG, Shlipak MG, Rogers WJ, *et al.* Prevalence, clinical characteristics, and mortality among patients with myocardial infarction presenting without chest pain. JAMA 2000; 283: 3223-9.
[http://dx.doi.org/10.1001/jama.283.24.3223]

[18] Dadkhah S, Sharain K, Sharain R, *et al.* The value of bedside cardiac multibiomarker assay in rapid and accurate diagnosis of acute coronary syndromes. Crit Pathw Cardiol 2007; 6: 76-84.
[http://dx.doi.org/10.1097/HPC.0b013e318053d1c9]

[19] Biomarkers and surrogate endpoints: preferred definitions and conceptual framework. Clin Pharmacol Ther 2001; 69: 89-95.
[http://dx.doi.org/10.1067/mcp.2001.113989]

[20] Morrow DA, Cannon CP, Jesse RL, *et al.* National Academy of Clinical Biochemistry Laboratory Medicine Practice Guidelines: clinical characteristics and utilization of biochemical markers in acute coronary syndromes. Clin Chem 2007; 53: 552-74.
[http://dx.doi.org/10.1373/clinchem.2006.084194]

[21] Collinson PO, Garrison L, Christenson RH. Cardiac biomarkers - A short biography. Clin Biochem 2015; 48: 197-200.
[http://dx.doi.org/10.1016/j.clinbiochem.2014.11.014]

[22] Wu AH, Apple FS, Gibler WB, Jesse RL, Warshaw MM, Valdes R Jr. National Academy of Clinical Biochemistry Standards of Laboratory Practice: recommendations for the use of cardiac markers in coronary artery diseases. Clin Chem 1999; 45: 1104-21.

[23] Thygesen K, Mair J, Katus H, *et al.* Recommendations for the use of cardiac troponin measurement in acute cardiac care. Eur Heart J 2010; 31: 2197-204.
[http://dx.doi.org/10.1093/eurheartj/ehq251]

[24] Amsterdam EA, Wenger NK, Brindis RG, *et al.* 2014 AHA/ACC Guideline for the Management of Patients with Non-ST-Elevation Acute Coronary Syndromes: a report of the American College of Cardiology/American Heart Association Task Force on Practice Guidelines. J Am Coll Cardiol 2014; 64: e139-228.
[http://dx.doi.org/10.1016/j.jacc.2014.09.017]

[25] Thygesen K, Alpert JS, Jaffe AS, Simoons ML, Chaitman BR, White HD. Third universal definition of myocardial infarction. Eur Heart J 2012; 33: 2551-67.
[http://dx.doi.org/10.1093/eurheartj/ehs184]

[26] Alpert JS, Thygesen K, Antman E, Bassand JP. Myocardial infarction redefined--a consensus document of The Joint European Society of Cardiology/American College of Cardiology Committee for the redefinition of myocardial infarction. J Am Coll Cardiol 2000; 36: 959-69.
[http://dx.doi.org/10.1016/S0735-1097(00)00804-4]

[27] Jaffe AS, Babuin L, Apple FS. Biomarkers in acute cardiac disease: the present and the future. J Am Coll Cardiol 2006; 48: 1-11.
[http://dx.doi.org/10.1016/j.jacc.2006.02.056]

[28] Morrow DA, de Lemos JA, Blazing MA, *et al.* Prognostic value of serial B-type natriuretic peptide

testing during follow-up of patients with unstable coronary artery disease. JAMA 2005; 294: 2866-71. [http://dx.doi.org/10.1001/jama.294.22.2866]

[29] Sabatine MS, Morrow DA, de Lemos JA, *et al.* Multimarker approach to risk stratification in non-ST elevation acute coronary syndromes: simultaneous assessment of troponin I, C-reactive protein, and B-type natriuretic peptide. Circulation 2002; 105: 1760-3. [http://dx.doi.org/10.1161/01.CIR.0000015464.18023.0A]

[30] Vanzetto G, Jacon P, Calizzano A, *et al.* N-terminal pro-brain natriuretic peptide predicts myocardial ischemia and is related to postischemic left-ventricular dysfunction in patients with stable coronary artery disease. J Nucl Cardiol 2007; 14: 835-42. [http://dx.doi.org/10.1016/j.nuclcard.2007.07.011]

[31] Apple FS, Smith SW, Pearce LA, Murakami MM. Assessment of the multiple-biomarker approach for diagnosis of myocardial infarction in patients presenting with symptoms suggestive of acute coronary syndrome. Clin Chem 2009; 55: 93-100. [http://dx.doi.org/10.1373/clinchem.2008.102905]

[32] Alpert JS, Thygesen K, Antman E, Bassand JP. Myocardial infarction redefined--a consensus document of The Joint European Society of Cardiology/American College of Cardiology Committee for the redefinition of myocardial infarction. J Am Coll Cardiol 2000; 36: 959-69. [http://dx.doi.org/10.1016/S0735-1097(00)00804-4]

[33] Gupta S, de Lemos JA. Use and misuse of cardiac troponins in clinical practice. Prog Cardiovasc Dis 2007; 50: 151-65. [http://dx.doi.org/10.1016/j.pcad.2007.01.002]

[34] Filatov VL, Katrukha AG, Bulargina TV, Gusev NB. Troponin: structure, properties, and mechanism of functioning. Biochemistry (Mosc) 1999; 64: 969-85.

[35] Katus HA, Remppis A, Scheffold T, Diederich KW, Kuebler W. Intracellular compartmentation of cardiac troponin T and its release kinetics in patients with reperfused and non reperfused myocardial infarction. Am J Cardiol 1991; 67: 1360-7. [http://dx.doi.org/10.1016/0002-9149(91)90466-X]

[36] Apple FS. A new season for cardiac troponin assays: it's time to keep a scorecard. Clin Chem 2009; 55: 1303-6. [http://dx.doi.org/10.1373/clinchem.2009.128363]

[37] Apple FS, Parvin CA, Buechler KF, Christenson RH, Wu AH, Jaffe AS. Validation of the 99th percentile cutoff independent of assay imprecision (CV) for cardiac troponin monitoring for ruling out myocardial infarction. Clin Chem 2005; 51: 2198-200. [http://dx.doi.org/10.1373/clinchem.2005.052886]

[38] Apple FS, Jaffe AS, Collinson P, *et al.* IFCC educational materials on selected analytical and clinical applications of high sensitivity cardiac troponin assays. Clin Biochem 2015; 48: 201-3. [http://dx.doi.org/10.1016/j.clinbiochem.2014.08.021]

[39] Roffi M, Patrono C, Collet JP, *et al.* 2015 ESC guidelines for the management of acute coronary syndromes in patients presenting without persistent ST-segment elevation. Eur Heart J 2015; 37: ehv320.

[40] Twerenbold R, Jaffe A, Reichlin T, Reiter M, Mueller C. High-sensitive troponin T measurements: what do we gain and what are the challenges? Eur Heart J 2012; 33: 579-86. [http://dx.doi.org/10.1093/eurheartj/ehr492]

[41] Mueller C, Giannitsis E, Christ M, *et al.* Multicenter evaluation of a 0-hour/1-hour algorithm in the diagnosis of myocardial infarction with high-sensitivity cardiac troponin T. Ann Emerg Med 2016; 68: 76-87. [http://dx.doi.org/10.1016/j.annemergmed.2015.11.013]

[42] Rubini GM, Twerenbold R, Jaeger C, *et al.* One-hour rule-in and rule-out of acute myocardial

infarction using high-sensitivity cardiac troponin I. Am J Med 2015; 128: 861-70.
[http://dx.doi.org/10.1016/j.amjmed.2015.01.046]

[43] Giannitsis E, Becker M, Kurz K, Hess G, Zdunek D, Katus HA. High-sensitivity cardiac troponin T for early prediction of evolving non-ST-segment elevation myocardial infarction in patients with suspected acute coronary syndrome and negative troponin results on admission. Clin Chem 2010; 56: 642-50.
[http://dx.doi.org/10.1373/clinchem.2009.134460]

[44] Apple FS, Pearce LA, Smith SW, Kaczmarek JM, Murakami MM. Role of monitoring changes in sensitive cardiac troponin I assay results for early diagnosis of myocardial infarction and prediction of risk of adverse events. Clin Chem 2009; 55: 930-7.
[http://dx.doi.org/10.1373/clinchem.2008.114728]

[45] Agewall S, Giannitsis E, Jernberg T, Katus H. Troponin elevation in coronary *vs.* non-coronary disease. Eur Heart J 2011; 32: 404-11.
[http://dx.doi.org/10.1093/eurheartj/ehq456]

[46] Morrow DA, Antman EM. Evaluation of high-sensitivity assays for cardiac troponin. Clin Chem 2009; 55: 5-8.
[http://dx.doi.org/10.1373/clinchem.2008.117218]

[47] Wu AH, Jaffe AS. The clinical need for high-sensitivity cardiac troponin assays for acute coronary syndromes and the role for serial testing. Am Heart J 2008; 155: 208-14.
[http://dx.doi.org/10.1016/j.ahj.2007.10.016]

[48] White HD. Will new higher-precision troponins lead to clarity or confusion? Curr Opin Cardiol 2008; 23: 292-5.
[http://dx.doi.org/10.1097/HCO.0b013e3283050808]

[49] Kavsak PA, MacRae AR, Yerna MJ, Jaffe AS. Analytic and clinical utility of a next-generation, highly sensitive cardiac troponin I assay for early detection of myocardial injury. Clin Chem 2009; 55: 573-7.
[http://dx.doi.org/10.1373/clinchem.2008.116020]

[50] Anderson JL, Adams CD, Antman EM, *et al.* ACC/AHA 2007 guidelines for the management of patients with unstable angina/non-ST-Elevation myocardial infarction: a report of the American College of Cardiology/American Heart Association Task Force on Practice Guidelines (Writing Committee to Revise the 2002 Guidelines for the Management of Patients With Unstable Angina/Non-ST-Elevation Myocardial Infarction) developed in collaboration with the American College of Emergency Physicians, the Society for Cardiovascular Angiography and Interventions, and the Society of Thoracic Surgeons endorsed by the American Association of Cardiovascular and Pulmonary Rehabilitation and the Society for Academic Emergency Medicine. J Am Coll Cardiol 2007; 50: e1-e157.
[http://dx.doi.org/10.1016/j.jacc.2007.02.013]

[51] Jaffe AS. Chasing troponin: how low can you go if you can see the rise? J Am Coll Cardiol 2006; 48: 1763-4.
[http://dx.doi.org/10.1016/j.jacc.2006.08.006]

[52] Jaffe AS, Apple FS. Refining our criteria: a critical challenge. Am J Clin Pathol 2009; 131: 11-3.
[http://dx.doi.org/10.1309/AJCPFYDY7B2AYDZE]

[53] Mair J. High-sensitivity cardiac troponins in everyday clinical practice. World J Cardiol 2014; 6: 175-82.
[http://dx.doi.org/10.4330/wjc.v6.i4.175]

[54] Thygesen K, Mair J, Giannitsis E, *et al.* How to use high-sensitivity cardiac troponins in acute cardiac care. Eur Heart J 2012; 33: 2252-7.
[http://dx.doi.org/10.1093/eurheartj/ehs154]

[55] Antman EM, Tanasijevic MJ, Thompson B, *et al.* Cardiac-specific troponin I levels to predict the risk of mortality in patients with acute coronary syndromes. N Engl J Med 1996; 335: 1342-9.

[http://dx.doi.org/10.1056/NEJM199610313351802]

[56] Lindahl B, Venge P, Wallentin L. Troponin T identifies patients with unstable coronary artery disease who benefit from long-term antithrombotic protection. Fragmin in Unstable Coronary Artery Disease (FRISC) Study Group. J Am Coll Cardiol 1997; 29: 43-8. [http://dx.doi.org/10.1016/S0735-1097(96)00447-0]

[57] Morrow DA, Antman EM, Tanasijevic M, *et al.* Cardiac troponin I for stratification of early outcomes and the efficacy of enoxaparin in unstable angina: a TIMI-11B substudy. J Am Coll Cardiol 2000; 36: 1812-7. [http://dx.doi.org/10.1016/S0735-1097(00)00942-6]

[58] Licka M, Zimmermann R, Zehelein J, Dengler TJ, Katus HA, Kubler W. Troponin T concentrations 72 hours after myocardial infarction as a serological estimate of infarct size. Heart 2002; 87: 520-4. [http://dx.doi.org/10.1136/heart.87.6.520]

[59] Gallegos RP, Swingen C, Xu XJ, *et al.* Infarct extent by MRI correlates with peak serum troponin level in the canine model. J Surg Res 2004; 120: 266-71. [http://dx.doi.org/10.1016/j.jss.2004.02.025]

[60] Metzler B, Hammerer-Lercher A, Jehle J, *et al.* Plasma cardiac troponin T closely correlates with infarct size in a mouse model of acute myocardial infarction. Clin Chim Acta 2002; 325: 87-90. [http://dx.doi.org/10.1016/S0009-8981(02)00296-6]

[61] Bonaca M, Scirica B, Sabatine M, *et al.* Prospective evaluation of the prognostic implications of improved assay performance with a sensitive assay for cardiac troponin I. J Am Coll Cardiol 2010; 55: 2118-24. [http://dx.doi.org/10.1016/j.jacc.2010.01.044]

[62] Sanchis J, Bardaji A, Bosch X, *et al.* N-terminal pro-brain natriuretic peptide and high-sensitivity troponin in the evaluation of acute chest pain of uncertain etiology. A PITAGORAS substudy. Rev Esp Cardiol (Engl Ed) 2013; 66: 532-8. [http://dx.doi.org/10.1016/j.rec.2012.11.004]

[63] de Lemos JA, Drazner MH, Omland T, *et al.* Association of troponin T detected with a highly sensitive assay and cardiac structure and mortality risk in the general population. JAMA 2010; 304: 2503-12. [http://dx.doi.org/10.1001/jama.2010.1768]

[64] deFilippi CR, de Lemos JA, Christenson RH, *et al.* Association of serial measures of cardiac troponin T using a sensitive assay with incident heart failure and cardiovascular mortality in older adults. JAMA 2010; 304: 2494-502. [http://dx.doi.org/10.1001/jama.2010.1708]

[65] Latini R, Masson S, Anand IS, *et al.* Prognostic value of very low plasma concentrations of troponin T in patients with stable chronic heart failure. Circulation 2007; 116: 1242-9. [http://dx.doi.org/10.1161/CIRCULATIONAHA.106.655076]

[66] Omland T, de Lemos JA, Sabatine MS, *et al.* A sensitive cardiac troponin T assay in stable coronary artery disease. N Engl J Med 2009; 361: 2538-47. [http://dx.doi.org/10.1056/NEJMoa0805299]

[67] Mingels AM, Joosen IA, Versteylen MO, *et al.* High-sensitivity cardiac troponin T: risk stratification tool in patients with symptoms of chest discomfort. PLoS One 2012; 7: e35059. [http://dx.doi.org/10.1371/journal.pone.0035059]

[68] Bandstein N, Ljung R, Johansson M, Holzmann MJ. Undetectable high-sensitivity cardiac troponin T level in the emergency department and risk of myocardial infarction. J Am Coll Cardiol 2014; 63: 2569-78. [http://dx.doi.org/10.1016/j.jacc.2014.03.017]

[69] Singh RG. The physiology and emerging roles of antidiuretic hormone. Int J Clin Pract 2002; 56: 777-

82.

[70] De Bree FM, Burbach JP. Structure-function relationships of the vasopressin prohormone domains. Cell Mol Neurobiol 1998; 18: 173-91. [http://dx.doi.org/10.1023/A:1022564803093]

[71] Kluge M, Riedl S, Erhart-Hofmann B, Hartmann J, Waldhauser F. Improved extraction procedure and RIA for determination of arginine8-vasopressin in plasma: role of premeasurement sample treatment and reference values in children. Clin Chem 1999; 45: 98-103.

[72] Morgenthaler NG, Struck J, Alonso C, Bergmann A. Assay for the measurement of copeptin, a stable peptide derived from the precursor of vasopressin. Clin Chem 2006; 52: 112-9. [http://dx.doi.org/10.1373/clinchem.2005.060038]

[73] Morgenthaler NG, Struck J, Jochberger S, Dunser MW. Copeptin: clinical use of a new biomarker. Trends Endocrinol Metab 2008; 19: 43-9. [http://dx.doi.org/10.1016/j.tem.2007.11.001]

[74] Lemaire LC, Giebelen IA, Struck J, *et al.* The influence of corticosteroids on the release of novel biomarkers in human endotoxemia. Intensive Care Med 2008; 34: 518-22. [http://dx.doi.org/10.1007/s00134-007-0955-x]

[75] Keller T, Tzikas S, Zeller T, *et al.* Copeptin improves early diagnosis of acute myocardial infarction. J Am Coll Cardiol 2010; 55: 2096-106. [http://dx.doi.org/10.1016/j.jacc.2010.01.029]

[76] Reichlin T, Hochholzer W, Stelzig C, *et al.* Incremental value of copeptin for rapid rule out of acute myocardial infarction. J Am Coll Cardiol 2009; 54: 60-8. [http://dx.doi.org/10.1016/j.jacc.2009.01.076]

[77] Khan SQ, Dhillon OS, O'Brien RJ, *et al.* C-terminal provasopressin (copeptin) as a novel and prognostic marker in acute myocardial infarction: Leicester Acute Myocardial Infarction Peptide (LAMP) study. Circulation 2007; 115: 2103-10. [http://dx.doi.org/10.1161/CIRCULATIONAHA.106.685503]

[78] Katan M, Morgenthaler N, Widmer I, *et al.* Copeptin, a stable peptide derived from the vasopressin precursor, correlates with the individual stress level. Neuroendocrinol Lett 2008; 29: 341-6.

[79] Gu YL, Voors AA, Zijlstra F, *et al.* Comparison of the temporal release pattern of copeptin with conventional biomarkers in acute myocardial infarction. Clin Res Cardiol 2011; 100: 1069-76. [http://dx.doi.org/10.1007/s00392-011-0343-y]

[80] Antman EM, Anbe DT, Armstrong PW, *et al.* ACC/AHA guidelines for the management of patients with ST-elevation myocardial infarction; A report of the American College of Cardiology/American Heart Association Task Force on Practice Guidelines (Committee to Revise the 1999 Guidelines for the Management of patients with acute myocardial infarction). J Am Coll Cardiol 2004; 44: E1-E211. [http://dx.doi.org/10.1016/j.jacc.2004.07.014]

[81] Keller T, Zeller T, Peetz D, *et al.* Sensitive troponin I assay in early diagnosis of acute myocardial infarction. N Engl J Med 2009; 361: 868-77. [http://dx.doi.org/10.1056/NEJMoa0903515]

[82] Reichlin T, Hochholzer W, Bassetti S, *et al.* Early diagnosis of myocardial infarction with sensitive cardiac troponin assays. N Engl J Med 2009; 361: 858-67. [http://dx.doi.org/10.1056/NEJMoa0900428]

[83] Eggers KM, Oldgren J, Nordenskjold A, Lindahl B. Diagnostic value of serial measurement of cardiac markers in patients with chest pain: limited value of adding myoglobin to troponin I for exclusion of myocardial infarction. Am Heart J 2004; 148: 574-81. [http://dx.doi.org/10.1016/j.ahj.2004.04.030]

[84] Lotze U, Lemm H, Heyer A, Muller K. Combined determination of highly sensitive troponin T and copeptin for early exclusion of acute myocardial infarction: first experience in an emergency

department of a general hospital. Vasc Health Risk Manag 2011; 7: 509-15. [http://dx.doi.org/10.2147/VHRM.S21753]

[85] Staub D, Morgenthaler NG, Buser C, *et al.* Use of copeptin in the detection of myocardial ischemia. Clin Chim Acta 2009; 399: 69-73. [http://dx.doi.org/10.1016/j.cca.2008.09.016]

[86] von HS. Stojakovic T, Bigalke B: Copeptin--a novel marker in acute myocardial infarction. Clin Chem 2012; 58: 307-9. [http://dx.doi.org/10.1373/clinchem.2011.173658]

[87] Chenevier-Gobeaux C, Freund Y, Claessens YE, *et al.* Copeptin for rapid rule out of acute myocardial infarction in emergency department. Int J Cardiol 2013; 166: 198-204.

[88] Giannitsis E, Kehayova T, Vafaie M, Katus HA. Combined testing of high-sensitivity troponin T and copeptin on presentation at prespecified cutoffs improves rapid rule-out of non-ST-segment elevation myocardial infarction. Clin Chem 2011; 57: 1452-5. [http://dx.doi.org/10.1373/clinchem.2010.161265]

[89] Meune C, Zuily S, Wahbi K, Claessens YE, Weber S, Chenevier-Gobeaux C. Combination of copeptin and high-sensitivity cardiac troponin T assay in unstable angina and non-ST-segment elevation myocardial infarction: a pilot study. Arch Cardiovasc Dis 2011; 104: 4-10. [http://dx.doi.org/10.1016/j.acvd.2010.11.002]

[90] Maisel A, Mueller C, Neath SX, *et al.* Copeptin helps in the early detection of patients with acute myocardial infarction: primary results of the CHOPIN trial (Copeptin Helps in the early detection Of Patients with acute myocardial INfarction). J Am Coll Cardiol 2013; 62: 150-60. [http://dx.doi.org/10.1016/j.jacc.2013.04.011]

[91] Forberg JL, Henriksen LS, Edenbrandt L, Ekelund U. Direct hospital costs of chest pain patients attending the emergency department: a retrospective study. BMC Emerg Med 2006; 6: 6. [http://dx.doi.org/10.1186/1471-227X-6-6]

[92] Mueller C, Laule-Kilian K, Scholer A, *et al.* Use of B-type natriuretic peptide for the management of women with dyspnea. Am J Cardiol 2004; 94: 1510-4. [http://dx.doi.org/10.1016/j.amjcard.2004.08.029]

[93] Llorens P, Sanchez M, Herrero P, Martin-Sanchez FJ, Pinera P, Miro O. The utility of copeptin in the emergency department for non-ST-elevation myocardial infarction rapid rule out: COPED-MIRRO study. Eur J Emerg Med 2014; 21: 220-9. [http://dx.doi.org/10.1097/MEJ.0b013e3283632f8b]

[94] Charpentier S, Maupas-Schwalm F, Cournot M, Elbaz M, Botella JM, Lauque D. Combination of copeptin and troponin assays to rapidly rule out non-ST elevation myocardial infarction in the emergency department. Acad Emerg Med 2012; 19: 517-24. [http://dx.doi.org/10.1111/j.1553-2712.2012.01350.x]

[95] Giavarina D, Carta M, Fortunato A, Wratten ML, Hartmann O, Soffiati G. Copeptin and high sensitive troponin for a rapid rule out of acute myocardial infarction? Clin Lab 2011; 57: 725-30.

[96] Karakas M, Januzzi JL Jr, Meyer J, *et al.* Copeptin does not add diagnostic information to high-sensitivity troponin T in low- to intermediate-risk patients with acute chest pain: results from the rule out myocardial infarction by computed tomography (ROMICAT) study. Clin Chem 2011; 57: 1137-45. [http://dx.doi.org/10.1373/clinchem.2010.160192]

[97] Esteban-Torrella P, Garcia dG-R, Consuegra-Sanchez L, *et al.* Usefulness of copeptin in discarding non-ST elevation acute myocardial infarction in patients with acute chest pain and negative first troponin I. Med Intensiva 2015; 39: 477-82. [http://dx.doi.org/10.1016/j.medin.2014.11.008]

[98] Sebbane M, Lefebvre S, Kuster N, *et al.* Early rule out of acute myocardial infarction in ED patients:

value of combined high-sensitivity cardiac troponin T and ultrasensitive copeptin assays at admission. Am J Emerg Med 2013; 31: 1302-8.
[http://dx.doi.org/10.1016/j.ajem.2013.04.033]

[99] Duchenne J, Mestres S, Dublanchet N, *et al.* Diagnostic accuracy of copeptin sensitivity and specificity in patients with suspected non-ST-elevation myocardial infarction with troponin I below the 99th centile at presentation. BMJ Open 2014; 4: e004449.
[http://dx.doi.org/10.1136/bmjopen-2013-004449]

[100] Katan M, Christ-Crain M. The stress hormone copeptin: a new prognostic biomarker in acute illness. Swiss Med Wkly 2010; 140: w13101.

[101] Nickel CH, Ruedinger J, Misch F, *et al.* Copeptin and peroxiredoxin-4 independently predict mortality in patients with nonspecific complaints presenting to the emergency department. Acad Emerg Med 2011; 18: 851-9.
[http://dx.doi.org/10.1111/j.1553-2712.2011.01126.x]

[102] Nickel CH, Bingisser R, Morgenthaler NG. The role of copeptin as a diagnostic and prognostic biomarker for risk stratification in the emergency department. BMC Med 2012; 10: 7.
[http://dx.doi.org/10.1186/1741-7015-10-7]

[103] Nemec M, Koller MT, Nickel CH, *et al.* Patients presenting to the emergency department with non-specific complaints: the Basel Non-specific Complaints (BANC) study. Acad Emerg Med 2010; 17: 284-92.
[http://dx.doi.org/10.1111/j.1553-2712.2009.00658.x]

[104] Voors AA, Von HS, Anker SD, *et al.* C-terminal provasopressin (copeptin) is a strong prognostic marker in patients with heart failure after an acute myocardial infarction: results from the OPTIMAAL study. Eur Heart J 2009; 30: 1187-94.
[http://dx.doi.org/10.1093/eurheartj/ehp098]

[105] Struck J, Morgenthaler NG, Bergmann A. Copeptin, a stable peptide derived from the vasopressin precursor, is elevated in serum of sepsis patients. Peptides 2005; 26: 2500-4.
[http://dx.doi.org/10.1016/j.peptides.2005.04.019]

[106] Kelly D, Squire IB, Khan SQ, *et al.* C-terminal provasopressin (copeptin) is associated with left ventricular dysfunction, remodeling, and clinical heart failure in survivors of myocardial infarction. J Card Fail 2008; 14: 739-45.
[http://dx.doi.org/10.1016/j.cardfail.2008.07.231]

[107] Potocki M, Reichlin T, Thalmann S, *et al.* Diagnostic and prognostic impact of copeptin and high-sensitivity cardiac troponin T in patients with pre-existing coronary artery disease and suspected acute myocardial infarction. Heart 2012; 98: 558-65.
[http://dx.doi.org/10.1136/heartjnl-2011-301269]

[108] Narayan H, Dhillon OS, Quinn PA, *et al.* C-terminal provasopressin (copeptin) as a prognostic marker after acute non-ST elevation myocardial infarction: Leicester Acute Myocardial Infarction Peptide II (LAMP II) study. Clin Sci (Lond) 2011; 121: 79-89.
[http://dx.doi.org/10.1042/CS20100564]

[109] Potocki M, Breidthardt T, Mueller A, *et al.* Copeptin and risk stratification in patients with acute dyspnea. Crit Care 2010; 14: R213.
[http://dx.doi.org/10.1186/cc9336]

[110] Hupf H, Grimm D, Riegger GA, Schunkert H. Evidence for a vasopressin system in the rat heart. Circ Res 1999; 84: 365-70.
[http://dx.doi.org/10.1161/01.RES.84.3.365]

[111] Yoon Y, Song J, Hong SH, Kim JQ. Plasma nitric oxide concentrations and nitric oxide synthase gene polymorphisms in coronary artery disease. Clin Chem 2000; 46: 1626-30.

[112] Alehagen U, Dahlstrom U, Rehfeld JF, Goetze JP. Association of copeptin and N-terminal proBNP

concentrations with risk of cardiovascular death in older patients with symptoms of heart failure. JAMA 2011; 305: 2088-95.
[http://dx.doi.org/10.1001/jama.2011.666]

[113] Tentzeris I, Jarai R, Farhan S, *et al.* Complementary role of copeptin and high-sensitivity troponin in predicting outcome in patients with stable chronic heart failure. Eur J Heart Fail 2011; 13: 726-33.
[http://dx.doi.org/10.1093/eurjhf/hfr049]

[114] Kerr R, Stirling D, Ludlam CA. Interleukin 6 and haemostasis. Br J Haematol 2001; 115: 3-12.
[http://dx.doi.org/10.1046/j.1365-2141.2001.03061.x]

[115] Rus HG, Vlaicu R, Niculescu F. Interleukin-6 and interleukin-8 protein and gene expression in human arterial atherosclerotic wall. Atherosclerosis 1996; 127: 263-71.
[http://dx.doi.org/10.1016/S0021-9150(96)05968-0]

[116] Schieffer B, Selle T, Hilfiker A, *et al.* Impact of interleukin-6 on plaque development and morphology in experimental atherosclerosis. Circulation 2004; 110: 3493-500.
[http://dx.doi.org/10.1161/01.CIR.0000148135.08582.97]

[117] Holvoet P. Relations between metabolic syndrome, oxidative stress and inflammation and cardiovascular disease. Verh K Acad Geneeskd Belg 2008; 70: 193-219.

[118] Biasucci LM, Liuzzo G, Fantuzzi G, *et al.* Increasing levels of interleukin (IL)-1Ra and IL-6 during the first 2 days of hospitalization in unstable angina are associated with increased risk of in-hospital coronary events. Circulation 1999; 99: 2079-84.
[http://dx.doi.org/10.1161/01.CIR.99.16.2079]

[119] Kashiwagi M, Tanaka A, Kitabata H, *et al.* Relationship between coronary arterial remodeling, fibrous cap thickness and high-sensitivity C-reactive protein levels in patients with acute coronary syndrome. Circ J 2009; 73: 1291-5.
[http://dx.doi.org/10.1253/circj.CJ-08-0968]

[120] Liuzzo G, Biasucci LM, Gallimore JR, *et al.* The prognostic value of C-reactive protein and serum amyloid a protein in severe unstable angina. N Engl J Med 1994; 331: 417-24.
[http://dx.doi.org/10.1056/NEJM199408183310701]

[121] Nakachi T, Kosuge M, Hibi K, *et al.* C-reactive protein elevation and rapid angiographic progression of nonculprit lesion in patients with non-ST-segment elevation acute coronary syndrome. Circ J 2008; 72: 1953-9.
[http://dx.doi.org/10.1253/circj.CJ-08-0185]

[122] Anogeianaki A, Angelucci D, Cianchetti E, *et al.* Atherosclerosis: a classic inflammatory disease. Int J Immunopathol Pharmacol 2011; 24: 817-25.
[http://dx.doi.org/10.1177/039463201102400401]

[123] Tuttolomondo A, Di RD, Pecoraro R, Arnao V, Pinto A, Licata G. Atherosclerosis as an inflammatory disease. Curr Pharm Des 2012; 18: 4266-88.
[http://dx.doi.org/10.2174/138161212802481237]

[124] Mulvihill NT, Foley JB. Inflammation in acute coronary syndromes. Heart 2002; 87: 201-4.
[http://dx.doi.org/10.1136/heart.87.3.201]

[125] Shantsila E, Lip GY. Monocytes in acute coronary syndromes. Arterioscler Thromb Vasc Biol 2009; 29: 1433-8.
[http://dx.doi.org/10.1161/ATVBAHA.108.180513]

[126] Biasucci LM, Vitelli A, Liuzzo G, *et al.* Elevated levels of interleukin-6 in unstable angina. Circulation 1996; 94: 874-7.
[http://dx.doi.org/10.1161/01.CIR.94.5.874]

[127] De GL, Brunetti ND, Montrone D, De RF, Cuculo A, Di BM. Subacute inflammatory activation in subjects with acute coronary syndrome and left ventricular dysfunction. Inflammation 2012; 35: 363-70.

[http://dx.doi.org/10.1007/s10753-011-9326-4]

[128] Armstrong EJ, Morrow DA, Sabatine MS. Inflammatory biomarkers in acute coronary syndromes: part I: introduction and cytokines. Circulation 2006; 113: e72-5. [http://dx.doi.org/10.1161/CIRCULATIONAHA.105.595520]

[129] Ferroni P, Rosa A, Di FM, *et al.* Prognostic significance of interleukin-6 measurement in the diagnosis of acute myocardial infarction in emergency department. Clin Chim Acta 2007; 381: 151-6. [http://dx.doi.org/10.1016/j.cca.2007.03.002]

[130] Wang XH, Liu SQ, Wang YL, Jin Y. Correlation of serum high-sensitivity C-reactive protein and interleukin-6 in patients with acute coronary syndrome. Genet Mol Res 2014; 13: 4260-6. [http://dx.doi.org/10.4238/2014.June.9.11]

[131] Maier W, Altwegg LA, Corti R, *et al.* Inflammatory markers at the site of ruptured plaque in acute myocardial infarction: locally increased interleukin-6 and serum amyloid A but decreased C-reactive protein. Circulation 2005; 111: 1355-61. [http://dx.doi.org/10.1161/01.CIR.0000158479.58589.0A]

[132] Sarrafzadegan N, Sadeghi M, Ghaffarpasand F, *et al.* Interleukin-6 and E-selectin in acute coronary syndromes and stable angina pectoris. A comparative study. Herz 2012; 37: 926-30. [http://dx.doi.org/10.1007/s00059-012-3648-3]

[133] Moe KT, Wong P. Current trends in diagnostic biomarkers of acute coronary syndrome. Ann Acad Med Singapore 2010; 39: 210-5.

[134] Beygui F, Silvain J, Pena A, *et al.* Usefulness of biomarker strategy to improve GRACE score's prediction performance in patients with non-ST-segment elevation acute coronary syndrome and low event rates. Am J Cardiol 2010; 106: 650-8. [http://dx.doi.org/10.1016/j.amjcard.2010.04.019]

[135] Garcia-Salas JM, Tello-Montoliu A, Manzano-Fernandez S, *et al.* Interleukin-6 as a predictor of cardiovascular events in troponin-negative non-ST elevation acute coronary syndrome patients. Int J Clin Pract 2014; 68: 294-303. [http://dx.doi.org/10.1111/ijcp.12245]

[136] Fisman EZ, Benderly M, Esper RJ, *et al.* Interleukin-6 and the risk of future cardiovascular events in patients with angina pectoris and/or healed myocardial infarction. Am J Cardiol 2006; 98: 14-8. [http://dx.doi.org/10.1016/j.amjcard.2006.01.045]

[137] Lindmark E, Diderholm E, Wallentin L, Siegbahn A. Relationship between interleukin 6 and mortality in patients with unstable coronary artery disease: effects of an early invasive or noninvasive strategy. JAMA 2001; 286: 2107-13. [http://dx.doi.org/10.1001/jama.286.17.2107]

[138] Ridker PM, Rifai N, Stampfer MJ, Hennekens CH. Plasma concentration of interleukin-6 and the risk of future myocardial infarction among apparently healthy men. Circulation 2000; 101: 1767-72. [http://dx.doi.org/10.1161/01.CIR.101.15.1767]

CHAPTER 10

Physiopathology of Thrombotic Diseases and Platelet Derived Biomarkers

Silvia Montoro- García* and **Sara María Martínez-Sánchez**

Department of Cardiovascular Risk, Faculty of Health Sciences, Universidad Católica San Antonio de Murcia (UCAM), Campus de los Jerónimos, Guadalupe, Murcia, Spain

Abstract: Thromboembolic disease includes a wide concept enclosing arterial and venous thrombosis. However, their physiopathologies show different effectors. The haemostatic system is a very complex mechanism with several roles such as maintenance of vascular permeability, production of an occlusive thrombus if vascular injury and its lysis. Biomarkers in cardiovascular medicine are useful non-invasive and sensitive tools adding information of the disease state and the possible management of these patients. The current chapter will attempt to provide the reader an insight into the mechanisms that accompany the pathological development of thrombus, particularly platelet-related. The endothelium, platelets and coagulation factors continue to attract widespread attention because of two independent mechanisms involved in atherothrombosis, aspects of both of which will be addressed in the following sections together with novel biomarkers such as microparticles, miRNAs or monocyte-platelet aggregates.

Keywords: Atherothrombosis, Biomarkers, Cytokines, Function test, Glycoprotein IIb/IIIa, Glycoprotein Ia, Glycoprotein VI, Mean platelet volume, Membrane receptor, miRNA, Microparticles, Monocytes, P-selectin, Pathophysiology, Platelet functional assay, Platelet glycoproteins, Platelet-monocyte aggregate, Platelets, Polymorphism, Venous thromboembolism.

INTRODUCTION

Thromboembolism is the leading cause of death worldwide and its prevalence is much greater in developed countries. The incidence increased to between 2 and 7 per 1000 among those aged ≥70 years [1]. Thromboembolic disease includes a wide concept enclosing arterial and venous thrombosis. However, their physiopathologies show different effectors which will be discussed in the current chapter.

* **Corresponding author Silvia Montoro-García:** Department of Cardiovascular Risk, Faculty of Health Sciences, Universidad Católica San Antonio de Murcia (UCAM), Campus de los Jerónimos, Guadalupe, Murcia, Spain; Tel: (+34) 968278638; E-mail: smontoro@ucam.edu

Venous thrombosis is common in inferiors extremities and may be accompanied by a dangerous complication such as pulmonary embolism (PE). In developed countries, deep venous thrombosis (DVT) appears only in 0.1-0.2% of the population, and its survival is relatively high (69% after 90 days) [2]. However, its prevalence increases with age being 1% for individuals above 65 years old and cause cancer [3].

On the other hand, arterial thrombosis encloses occlusive processes occurring in the arterial system such as coronary arteries (ischemia, myocardium infarct), cerebrovascular arteries (transient ischemic attack and stroke) and peripheral vascular disease (PVD). Those cases are more frequent than DVT and lead mortality in developed countries [4].

Although, the resulting event in each case is similar: a thrombotic vascular occlusion, the pathogenetic mechanisms are distinct, resulting in distinct therapeutic approaches and biomarker strategies [5]. Due to that, important efforts have been made on the haemostatic research in order to identify molecular and biochemical changes that could explain a permanent hypercoagulate state.

Platelets, a Reliable Biomarker?

The haemostatic system is a very complex mechanism with several roles such as maintenance of vascular permeability, production of an occlusive thrombus if vascular injury and its lysis. It could be even simplified as two main blocks: cellular enclosing platelets, leukocytes and endothelium *vs.* plasmatic (coagulation factors and fibrinolytic proteins). Blood generally contains 150 000-350 000 platelets/μL, which provide an easily available biomarker source for research and clinical use.

In the DVT, the thrombus is rich in fibrin and plenty of vascular elements entrapped in the fibrin system (red blood cells mainly). Known as "*red thrombus*", is formed in a vascular wall injury-free that could explain the apparition of the thrombus [6]. Therefore, the proposed processes are related to an excessive activation of coagulation factors or due to deficiency in the regulation mechanisms (for instance fibrinolysis). Clinical evidences have justified the efficacy of the anticoagulant and thrombolytic therapy in the prophylaxis of DVT [7, 8]. The place of aspirin in DVT prophylaxis remains controversial [9], however, several features of platelets have been shown related to DVT prevalence, as shown later.

Concerning the arterial thrombosis, pathogenetic mechanisms and risk factors differ from the above for DVT. The resulting occlusive thrombus is rich in platelets and poor in red blood cells, we refer to it as a "*white thrombus*". Besides, there is generally a lesion in the vascular wall resulting from an atherosclerotic process [10]. As such, the pathophysiological mechanism is due to platelet activation in the injured wall, followed by cellular activation leading to final thrombin activation *in situ*. Thus, the therapeutic strategy for arterial thrombosis is based mainly antiaggregant drugs inhibiting platelet activation [11].

Research over the last few years has been focused on obtaining clinical data that are not always extrapolated to the practice or translated in benefits for patients. Biomarkers in cardiovascular medicine are useful non-invasive and sensitive tools adding information of the disease state and the possible dealing of these patients. The approach of the use of biomarkers has shown an exponential increase in basic and clinical analyses, known as translational research. That is why translational research these days, one of the main challenges of the biomedical investigation [12]. Better knowledge of the pathological state of patients with thrombosis would irrefutably allow better stratification risk and monitorization. To date, 10-15% of the diagnosis is based on clinical and electrocardiographic data; however it does not always allow to determine early states of the disease, in this case biomarkers display an emerging importance [13]. In fact, it has been described that molecular biomarkers could represent around 80-90% of the translational success [14]. Nonetheless, before this takes place, researchers have to be able to coordinate basic and clinical science.

The current chapter will attempt to provide the reader with an insight into the mechanisms that accompany pathological development of thrombus, particularly platelet-related. The endothelium, platelets and coagulation factors continue to gain widespread attention because of two independent mechanisms (*e.g.* inflammation and thrombosis) involved in atherothrombosis, aspects of both of which will be addressed in the following sections together with novel biomarkers such as microparticles, miRNAs or monocyte-platelet aggregates.

PHYSIOPATHOLOGY OF VENOUS THROMBOSIS

The pathogenesis of venous thromboembolism (VTE) is multifactorial owing to the interaction of both genetic and environmental factors in combination. On average, identified genetic risk factors for VTE provide explanations of barely 30% thrombophilic disorders (Factor V Leiden, prothrombin gene or Factor II, methylentetrahydrofolate reductase enzyme mutation). Over the time, randomized clinical trials and *in vitro* studies demonstrated that increased platelet counts activate fibrin production and hypercoagulability of the blood [15]. However, there are still a remarkable number of VTE events where none of the known risk factors can be identified. In fact, platelet role in genetic susceptibility to thrombotic disease remains controversial, particularly in venous thrombosis.

Associations between VTE and Polymorphisms of Platelet Glycoproteins

Glycoprotein Ia

The C807 and T807 alleles of the glycoprotein Ia (GPIa) gene, have been showed associated with low or high platelet GPIa-IIa density and therefore, with the rate of platelet adhesion to type I collagen. Hereafter, the polymorphism in such a gene could predispose to thrombotic diseases. Dinauer *et al.* were first to look at allelic frequencies of GPIa in different racial groups and their relationship with VTE [16]. However, they could not find differences in the allelic frequency of two silent polymorphisms GPIa 807C/T and 873G/A in VTE patients and random patients. Similarly, the GPIa 807C/T and 873G/A dimorphisms were not shown as risk factors for VTE in a more recent paper [17]. In contrast to acenocoumarinic drugs, acetylsalicylic acid is the most effective antithrombotic drug in retinal vein occlusion. Hereafter, inherited platelet GP variations might have a major impact on small vessel venous thrombosis. In accordance, retinopathy was showed 3.4 fold higher in diabetic patients with the GPIa/IIa TT/AA subtype compared to patients with the CC/GG subtype [18]. Another study identified the GPIa/IIa subtype CT/GA leading to an intermediate prothrombotic risk in the majority of RVO patients rather than the higher risk of TT/AA genotype [19].

Glycoprotein IIb/IIIa

Platelet aggregation is also mediated by the GPIIb/IIIa (integrin aIIbb3) (around 80,000 per platelet), representing 15% of external membrane receptors [20]. It was suggested that the substitution of Leu22Pro in carriers of PIA2 polymorphism of GPIIIa, induces a conformational shift in the receptor thus influencing its activation and increased platelet aggregability [21]. Very little research has been published investigating the relation between such GPIIb/IIIa, its genetic polymorphisms PIA1/PIA2 and VTE in different populations. Ivanov *et al.* described the PIA2 polymorphism in the 35.3% of patients with pulmonary embolism (PE) and represented only the 14.3% of the control group, which again was comparable to other thrombophilic mutations such the prothrombin gene and the FVL [22]. This could indicate an implication of platelets to the PE however there is a lack of further reports in order to confirm this.

Glycoprotein VI

Two elegant studies have found an association between single nucleotide polymorphisms in the gene GP6 and the risk of VTE [23, 24]. The gene GP6 codifies for the glycoprotein VI (GPVI), a platelet receptor for collagen that develops an essential role during platelet adhesion and aggregation. The resulting consequences of the Ser219Pro change are tough to be related with the receptor expression as carriers of the proline 219 allele (approximately 22% of the population) display 10% lower GPVI expression levels than the serine identical carriers. These results effectively showed that GPVI levels are well-controlled and could have a critical effect in on platelet activation under shear rate such as idiopathic VTE [25].

Although much evidence supports the involvement of platelet GP polymorphisms in major arterial disease, these findings do not provide evidence of a determinant platelet genetic marker in venous vessel thrombosis. In this regard, it remains still inconsistent and further studies are necessary to elucidate the common genetic link in the majority of VTE patients which suggests causality or predispose to a more severe course.

Platelet Function Testing

Platelet Functional Assay (PFA-100®)

The most interesting association when investigating on various platelet function assays was found regarding sticky platelet syndrome (SPS). SPS is a thrombophilic platelet function disorder, inherited possibly in autosomal-dominant fashion, and characterized by an inherent platelet aggregation *in vitro* in response to low concentrations of the epinephrine (Epi) and/or adenosine diphosphate (ADP). Three different types are characterized (type I: Epi+ADP; type II: Epi; type III: ADP). Clinically it is characterized by venous (1/3) and/or arterial (2/3) thrombosis. SPS was found in 18-28% of patients with otherwise unexplained thromboembolic events (TE) [26, 27]. Particularly, one report found an unequal response to low dose Epi in 68% of patients with thrombosis compared to 36% of controls ($P=0.034$) which might be of predictive value [28]. The same study did not find a significant correlation between PFA-Epi ($r=-0.171$) or PFA-ADP ($r=0.096$) closure times. Besides, platelet abnormalities in the presence of low dose Epi has been described in 14-34% of a healthy population [29 - 31].

In contrast, increased platelet aggregation with Epi or ADP could not be found in a study on patients with previous unprovoked VTE; however enhanced maximal platelet aggregation with low concentrations of TRAP-6 ($P=0.006$) and collagen ($P=0.0076$), respectively, was reported, indicating that platelet hyperreagibility might represent an independent risk factor for VTE [32].

Thrombelastography and Rotational Thrombelastometry

Thrombelastography (TEG) and rotational thrombelastometry (ROTEM) are global haemostatic function tests that reflect the sum of platelet function, coagulation proteases and inhibitors as well as the fibrinolytic system. In ROTEM information on platelet function is contained in INTEM and EXTEM. Regarding feasibility of ROTEM in prediction of VTE, one study demonstrated significantly lower clot formation time (CFT) and higher α-angle and larger maximum clot firmness (MCF) in patients with susceptibility for postsurgical VTE [33]. MCF, CFT and α-angle were significantly influenced towards hypercoagulability in patients with acute VTE compared to healthy subjects in INTEM and EXTEM. However, this finding might be confounded by pathophysiologically elevated fibrinogen and factor VIII levels in acute VTE [34]. One study limited by small patient numbers found no significant differences for INTEM and EXTEM in patients with cirrhotic or non-cirrhotic portal vein thrombosis compared to control subjects [35].

Several studies have reported that TEG may predict the occurrence of postoperative VTE [36 - 42]. The sensitivity and specificity ranged from 0% to 100% and 62% to 92%, respectively. The positive and negative likelihood ratios ranged from 1.43 to 7.67 and 0.28 to 0.96, respectively. The diagnostic odds ratio ranged from 1.5 to 27.7 when comparing patients with VTE and patients without VTE [36, 37, 41, 43, 44]. On the other hand, TEG was not predictive for VTE development in high risk for VTE trauma ICU patients [45]. Also other studies reported hypercoagulability but failed to find an association with outcomes [46].

Also, TEG was not conclusive when comparing patients with a history of VTE and healthy subjects [47] but showed a significantly shortened clot formation time and an accelerated maximum velocity of clot propagation in patients with previous DVT compared to a control group [48].

P-selectin

P-selectin (CD62P) is a platelet activation marker stored in platelets' α-granules and endothelial cells' Weibel Palade bodies [49]. It is redistributed to the cell surface upon platelet activation, *e.g.* thrombin and platelet bound-Pselectin is measured through flow cytometry. It acts through the principal counter-receptor P-selectin glycoprotein ligand-1 (PSGL-1) on leukocytes (and platelets). The PSL-PSGL-1 interaction is responsible for triggering the release of microparticles (MP) from most circulating blood cells and endothelial cells, tissue factor (TF) expression on monocytes, as well as recruitment and aggregation of platelets, activation, tethering and adhesion of leukocytes [49 - 52]. Moreover, it has been showed to enhance thrombus formation by affecting fibrin crosslinking in thrombus [53]. Procoagulatory function is also mediated by triggering the formation of leukocyte-derived microparticles and the transfer of TF to platelets [54, 55].

Clinically, several studies on P-selectin or its soluble form (sPSL) have identified their feasibility as a risk factor for VTE and demonstrated its predictive value for confirmation of VTE with a sensitivity of 68-73% and a specificity of 73-81% [56 - 59]. These results were evaluated in patients with cancer; hazard ratio was 2.6 [95% CI 1.4-4.9] [60]. Soluble PSL has been included and validated in a VTE risk prediction score (VIENNA) [61 - 63].

Other Platelet Parameters and Venous Thromboembolism

Platelet Count

In several population-based cohort studies an increased platelet count was not associated with increased risk of VTE [64 - 66]. However, a case control study on medical inpatients identified elevated platelet count > $350x10^9/L$ at admission as an independent risk factor for venous thrombosis with an odds ratio of 2.5 (95% CI 1.3-4.8) [67]. Elevated platelet count is not a risk factor for VTE in the general population, but is associated with increased risk in high-risk populations such as medical inpatients and cancer patients [63, 68, 69]. A platelet count above $350x10^9/L$ has later been validated as a parameter in a risk-assessment model for chemotherapy-associated venous thrombosis [70]. There is no direct evidence of a causal relationship; however, a malignant environment is suggested to convert high platelet count into a risk factor for VTE.

Elevated platelet counts may be due to inflammation, infection and tumour and it is unknown whether the platelet count could implicate causal aspects or simply reflects the underlying disease. In an attempt to demonstrate causal relationship, a recent population-based cohort study showed that pre-cancer platelet counts above the 80th percentile ($295x10^9/L$) was associated with a 2-fold higher risk of symptomatic VTE in subjects later developing a cancer disease, but not in cancer-free subjects. In addition a synergistic effect with high leucocyte counts was identified. This lead the authors to suggest that the platelet count together with interactions of platelets and leukocytes may play a role in the pathogenesis of cancer-related VTE [71].

Mean Platelet Volume

Platelet size as described by mean platelet volume (MPV) is a parameter associated with platelet reactivity. In contrast to the platelet count, an increased MPV has been identified as a predictor of particularly unprovoked VTE in the general population, suggesting platelet reactivity to play a role in the pathogenesis of VTE [64]. An increased MPV was associated with recurrence of venous and arterial thrombosis in patients with antiphospholipid antibody syndrome too [72]. In cancer patients, the opposite trend has been observed as Riedl and colleagues reported that high MPV was associated with a decreased risk of venous thrombosis [73]. In addition, a recent retrospective study in cancer patients found a significant association of low MPV prior to treatment start and the first VTE episode [74]. Immature platelet count is also a parameter associated with platelet reactivity but its association with venous thromboembolism has so far not been explored.

Microparticles

Membrane-shed submicron MPs (0.1-1 μm), released after cell activation, mediate in many biological processes because of various membrane proteins and bioactive lipids from their parental cells (mainly platelets, endothelial cells and monocytes) and upon stimuli [75]. Cellular MPs and other microvesicles types, mainly apoptotic bodies and exosomes, are differentiate in terms of size and sedimentation properties. Although MPs are usually smaller than apoptotic bodies (>1.5 μm), exosomes (released from the cell's endosomal compartment) are smaller than MPs and more homogenous in size (40-100 nm). Few characteristics differentiate these membranous vesicles, whilst exosomes are cup-shaped vesicles of endocytic origin and with a density of 1.10-1.21 g/mL, non-exosomal vesicles can be of irregular shape and have higher densities (>1.23 g/mL) [76].

Long-time regarded to be passive elements, "cellular debris" or "platelet dust" [77], they were first considered to signify non-specific and irrelevant cell debris produced during cell death. However, it is now irrefutable that the release of MPs is a regulated process that occurs in different cells types and that their composition depends on the stimulus applied and different agonists produce distinct populations of MPs with unique protein compositions [78]. The release of a MP is preceded by specific membrane changes in the parental cell, such as the shift of phosphatidylserine, cytoskeleton disruption, and changes in the local concentration of specific intracellular molecules, all of which result in the budding of vesicles that further becomes MPs [79].

Their release might reflect an activate state and be involved in the 'transfer' of signals arising from dysfunctional endothelium and vascular cells [80]. Several *in vitro* and *in vivo* studies have demonstrated that MPs are involved in the regulation of diverse pathobiological processes, such as sepsis, cellular activation and apoptosis [81]. Certainly, MPs have been recognised as true vectors in the transcellular exchange of biologic signals. Depending upon the pathological status, MPs of different origin could contain different lipidic messengers [82], carry surface receptors from parental cells [83], and even transfer mRNA [84] and proteins encapsulated inside them [85] to other cellular targets thus stimulating the coagulation cascade (Fig. **1**). For instance, the presence of the phosphatidylserine on MP surface provides a binding site for the prothrombinase complex and Factor IIa, resulting in a 50- to 100- fold higher procoagulant activity compared to activated platelets [86]. Besides, addition of platelet MP (PMP) directly to platelets resulted in further platelet degranulation and P-selectin redistribution to the membrane (P=0.019) in a dose and time-dependent manner [87].

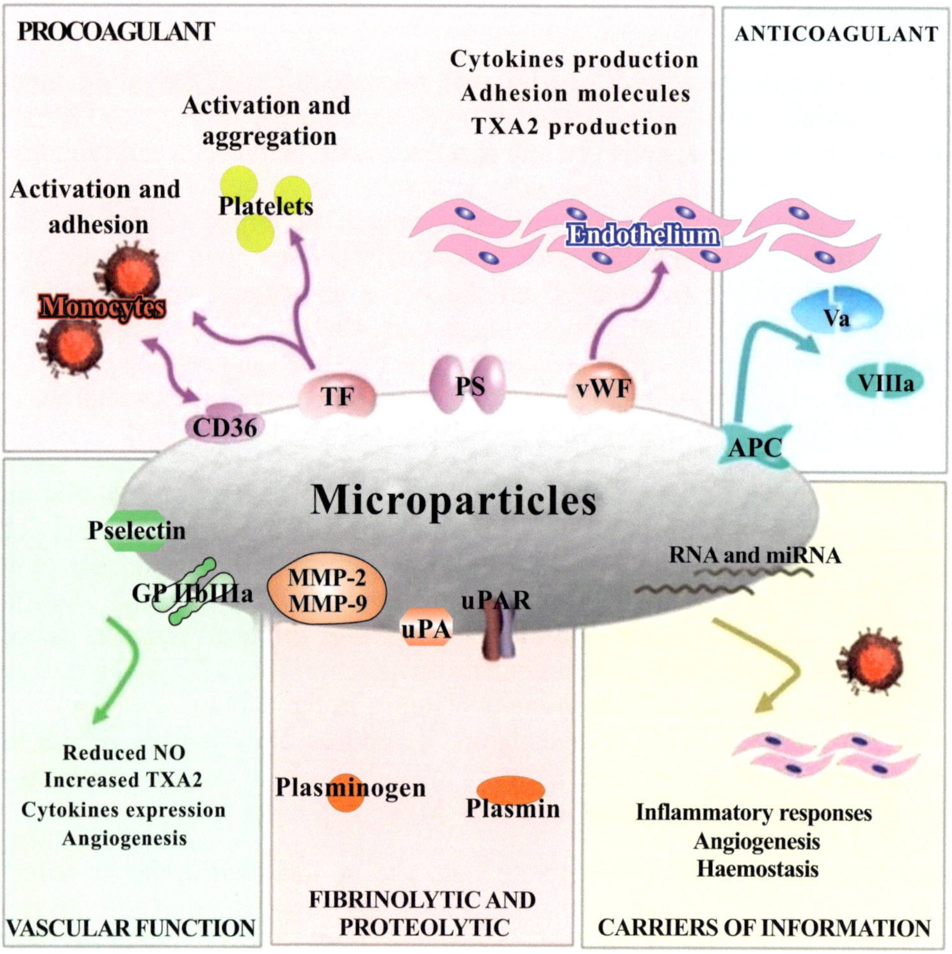

Fig. (1). Biological effects of MPs (adapted from Montoro-García *et al*. 2011).

However, microparticles are also expressed under normal physiological conditions, suggesting a potential role in general homeostasis. More specifically, in regard to the role of platelets in VTE, PMP have been addressed in one study [88]. Statistically significant higher EMP, PMP and TF+MP levels were found when comparing cancer patients with and without VTE to healthy subjects, and comparing cancer patients with VTE and patients with unprovoked VTE ($P<0.001$ respectively). TF+MP levels were also significantly higher than in patients with cancer without VTE. While this observation had been demonstrated in other studies [89, 90], multivariate analysis failed to show a significant association between elevated TF+MPs and VTE in cancer patients [88].

Another study found no significant PMP elevation in patients with unprovoked VTE compared with healthy controls [91]. This was inconsistent with other study results [92 - 94]. The authors hypothesized a rapid PMP level normalization caused by their rapid binding to leukocytes. Consistently, their study yielded a significant number of platelet-leukocyte conjugates (61.7% *vs*. 39.6%; $P=0.01$).

One study comparing patients with initial DVT, recurrent DVT, and no DVT, respectively, found that patients with recurrent DVT had significantly increased TF+PMP while there was no difference between patients with initial DVT and healthy subjects. However, the number of all PMP showed no significant differences between these groups [95].

The surface of PMP is able to bind to coagulation factors, including factors VIIIa, IXa, Va and Xa. This explains why direct and indirect FXa inhibitors are able to inhibit prothrombinase activity of PMPs [96].

PHYSIOPATHOLOGY OF THROMBOSIS IN ATHEROSCLEROTIC LESIONS

Atherosclerosis is defined as the thickening of the artery walls; it results from the accumulation of fat (lipids), and poorly-controlled production and organisation of collagen, a key structural protein, within the inner layer (intima) of arterial blood vessels. The pathological internal thickening of the arterial vessel is known as a plaque. During the plaque surface rupture, collagen and lipids are exposed to the flowing blood. Various inflammatory cell types (*e.g.*, macrophages, neutrophils, and lymphocytes) play essential roles in the destabilization and subsequent rupture or erosion of an atherosclerotic plaque, leading to the clinical manifestation of atherothrombosis [97]. The mechanisms that trigger the transition of the intimal alteration from 'healthy endothelium' into a non-reversible arteriosclerotic plaque are still poorly understood. Endothelial dysfunction encourages platelets and monocytes to initiate and promote atherosclerosis through the production of cytokines, growth factors and reactive oxygen species [98, 99] leading to the development of clinical complications, such as angina and acute coronary syndromes (ACS) [100]. Platelets, small pieces of megakaryocyte, have evolved to become activated and adhere to exposed collagen after injury of healthy vessels. The metabolism of plaque lipids leads to the production of materials that enhance platelet activation and thrombus deposition. Platelets bind to each other, resulting in a thrombus that, in a damaged but healthy vessel, provides a temporary repair to prevent loss of blood, but which, if occurring on a ruptured plaque, may block the artery entirely. Thus, three cellular components, endothelium, monocytes and platelets, have been suggested to play a critical role in modulating both the inflammatory and thrombotic response that occurs in atherosclerosis [101]. It is still unknown how the different cellular components communicate between each other or the potential modulators of plaque phenotype.

The resulting clinical event, arterial thrombosis, leads to sudden heart attack or to stroke, leading causes of premature death in mid-to-high income economies (WHO Factsheet 310, 2011). Ischemic heart disease is the main cause of death and morbidity in most industrialized countries and represents high costs to public health systems. Although the advances in the diagnosis and prevention are notorious, the incidence is still growing and coronary heart disease is yet one of the dominant causes of mortality and complications in developed countries. Ischemic heart disease refers to a spectrum of clinical presentations from coronary artery disease (CAD) to ACS. In terms of pathology, ACS is almost always associated with endothelial activation followed by rupture of an atherosclerotic plaque and the formation of a platelet-rich thrombus in the infarct-related artery.

Novel Effectors in Atherosclerosis

Interactions and exchange of information between cells stimulate production of adhesion factors (*e.g.* E-selectin, Intercellular adhesion marker [ICAM]...), cytokines (*e.g.* Interleukins), and growth factors (*e.g.* vascular endothelial growth factor [VEGF], platelet growth factor), ... in the endothelium, resulting in the subsequent infiltration of leukocytes into the sub intimal space [101]. In addition, an increase of reactive oxygen species synthesis (endothelial nitric oxide synthase [eNOS]...) and degradative enzymes (metalloproteases...) by endothelial cells can weaken the fibrous cap of atherosclerotic lesions and make it vulnerable to rupture. As already mentioned, MPs are elements released by the three components above implicated in atherothrombosis and may also modulate important aspects of endothelial, monocyte and platelets function, thus providing a novel mechanism by which these vesicles interact with such cells in the disease. On the other hand, MPs have been proposed as package elements of microRNAs (miRNA), but how this process is phenotype-specific is still uncertain. Similarly, the physiological role of these transcripts is mostly unclear. Following, we will discuss the physiological role of MPs, miRNA and monocyte-platelet aggregates (MPAs) since their relevance should be understood before proclaiming them novel biomarkers of thrombosis.

Physiological Role of Microparticles in Atherothrombosis

A large body of clinical studies have associated circulating MPs levels with a greater extent or severity of atherosclerosis [102], which could suggest that their surface composition (bioactive lipids or proteins) and nucleic acids could modulate the regulation of pathological processes [103]. Besides, plaque MPs have been found to increase tumour necrosis factor-α (TNF-α) release by endothelial cells *in vitro*, suggesting that MPs act as vectors of signal transduction and can regulate molecular transduction pathways with further involvement in atherosclerosis, myocardial ischemia and heart failure. Thereafter, these small vesicles from different origins might be involved in the 'transfer' of signals arising from tissue injury or inflammation in many cellular processes, such as apoptosis, cell activation, tumorigenesis or thrombosis [104, 105].

Several of these mediators of inflammation and adhesion (*e.g.* MAC-1) have also been found in the surface of MPs as a result of the damaged endothelium and/or activated platelets [106, 107]. For example, endothelial microparticles (EMPs) were significantly lower in patients with ST-Elevation Myocardial Infarction (STEMI) patients after PCI [108] but not in non-STEMI *vs*. CAD, and they have been proposed as potential biomarkers for assessing endothelial injury [109]. Similarly, PMPs express surface constitutive markers of platelet activation such as P-selectin, tissue factor (TF), von Willebrand factor (vWF), glycoproteins, quemokine receptors or enriched phosphatidylserine surface composition with subsequent prothrombotic activity [87, 110].

Thus, these circulating vesicles could serve as useful biomarkers of vascular injury and potential predictors of cardiovascular mortality and major adverse cardiovascular events, leading to a better evaluation and management of patients [108]. Given the obvious involvement of MPs in ACS; to date, only limited data are available on the real effectors regulating this cross-talk. In this sense, the internal and external composition of MPs might be of paramount importance, as well as its patterns with the pathological progression [87].

As a conclusion, the released MPs reflect an activate state and can act as a catalytic centre for further platelet, monocyte or endothelial binding. Although formation of MPs appears to be a passive shedding of membrane microvesicles from dysfunctional endothelium and vascular cells, increasing evidences suggests that they may represent a potential candidate for the regulation of pathological processes.

miRNAs as Potential Biomarkers

On the other hand, miRNAs are non-coding single stranded RNAs of around 22 nucleotides. Their existence in animal was confirmed one decade ago [111]. Since it, they have been irrefutably identified as a new class of negative gene regulators, being estimated that 30% of human genes are targeted by these miRNAs [112]. Processes that occur during atherosclerotic progression–cell proliferation, migration, fibrosis and apoptosis, among others– have been recently demonstrated to be modulated by miRNAs molecules [113]. Besides, while the majority of miRNAs are found intracellularly, a handful of miRNAs have been detected outside cells becoming then a possible exportation system of information. The mechanism of how miRNAs are released into the circulation is unclear. Recent studies suggest that only 10% of the circulating levels might be incorporated in membranous vesicles [114], being this package process very selective [115]. Therefore, both MPs and miRNAs might play key roles in a variety of cardiovascular processes, as it has been recently reviewed [79, 116]. Furthermore, one interesting study found that the addition of RNAse produces the loss of biological effects of MPs, indicating an essential role for MPs-RNA in the mediating of their activity [117]. Taken together these data imply a role of MPs as signal-messengers between different cell types and tissues or even pathologies, perhaps with miRNA as effectors [118].

Additionally, impaired levels of miRNAs have been associated with several pathological states, for example, miRNA-1 has been proposed as a potential molecular marker for patients with myocardial infarction [119], for instance it has been correlated with impairment of cardiac conduction, necrosis and worse outcome [120]. Interestingly, most of studies focus the effort to identify specific miRNAs from necrotic cardiac tissue more than in early phases of atherosclerosis [121]. It is in this context that interest in biomarkers of the evolvement of ACS is on the rise, providing potentially knowledge on the functional changes occurring during the pathology and thus, with risk stratification.

Science Journal has published that platelets, mainly through PMPs, are involved in proinflammatory processes of the rheumatic arthritis [122]. Similarly, miRNAs are small RNA molecules recently discovered and, according to the current publications, represent an incredible modulation power, not only in cardiovascular pathways. The use of genomics and proteomic techniques to quantify these two biomarkers would irrefutably aid in the diagnosis and management of patients with ACS in the future [123]. In summary, understanding the potential contribution of MPs and miRNAs to the pathogenesis of atherosclerosis, as well as assessing the implicated effectors or pathways will help to determine what leads to arterial wall damage.

Monocytes-Platelet Aggregates in Atherothrombosis

A vulnerable plaque includes a thin fibrous cap with important predominance of macrophages and smaller collagen content, and a large, lipid-rich necrotic core. On top of that there is an overlying thrombus and platelet aggregates.

Monocytes have a role in thrombus propagation contributing to the coagulation cascade during the acute event. Monocytes change their phenotype during an ACS to procoagulant and inflammatory characteristics such as exposure of tissue factor [124], adhesion molecules [125] and scavenger receptors [126, 127]. Besides, the monocytes produce platelet-derived growth factor, which induces mitotic activity of the endothelial cells and vascular smooth muscle cells [128]. There are also the main source of tissue factor [129].

Monocyte-platelet aggregate counts and markers of platelet activation involved in regulation of their function are increased in patients ACS, persisting even after one month of the acute event [125]. Also, MPs derived from monocytes are abundant in ACS [108] and support faster fibrin formation [130].

MPAs have been widely evaluated as a sensitive marker of platelet activation even better than platelet surface P-selectin [131]. Platelet aggregation to monocytes needs monocyte activation, resulting in the already mentioned increased cell-adhesion molecules, cytokine and matrix metalloproteinases production.

As well as simply representing a marker of platelet activation, a recent study has shown that the formation of MPAs results in significant activation of monocytes [132] thereby promoting to a potential mechanism for monocyte migration and a greater inflammatory phenotype [133].

Therefore, evidence points that MPAs should not be simplistically understood as a marker of platelet activation or a pathway of their elimination through phagocytosis by monocytes but rather as a mode of regulation of monocyte (and perhaps platelet) function [132, 134]. However, it is important to mention the complex pathways between platelets, the coagulation factors, as well as other components of the immune system, such as the rest of leukocytes during atherosclerosis.

CONCLUSION

Thereafter, although much evidence supports the involvement of platelets in major arterial disease, these findings do not provide evidence of a determinant platelet marker in venous vessel thrombosis. In this regard, it remains still inconsistent and further studies are necessary to elucidate the common link in the majority of VTE patients which suggests causality or predispose to a more sever course.

Other promising platelet polymorphisms and derived parameters associated with higher risk for thromboembolic disease might also benefit from novel strategies for identification [135]. However, identifying isolated biomarkers as a direct consequence of cardiovascular disease would be an oversimplification and future epidemiologic studies of platelet biomarkers will require larger sample sizes and family based approaches to further elucidate clinically important associations with thrombotic disease. The discovery of novel platelet derived-biomarkers of potential clinical use is foreseen.

FUNDING

This work was partially supported by Catholic University of Murcia and the People Programme (Marie Curie Actions) of the European Union's Seventh Framework Programme (FP7/2007-2013) under REA grant agreement nº 608765.

CONFLICT OF INTEREST

None declared in relation to this chapter for both authors. SMG and SMS have received funding for research from Catholic University of Murcia. SMG has received funding for research from BMS/Pfizer.

ACKNOWLEDGEMENTS

Declared none.

REFERENCES

[1] Day ISCfWT Thrombosis: a major contributor to global disease burden. Thromb Res 2014; 134: 931-8.
[http://dx.doi.org/10.1016/j.thromres.2014.08.014]

[2] Silverstein MD, Heit JA, Mohr DN, Petterson TM, O'Fallon WM, Melton LJ. Trends in the incidence of deep vein thrombosis and pulmonary embolism: a 25-year population-based study. Arch Intern Med 1998; 158: 585-93.
[http://dx.doi.org/10.1001/archinte.158.6.585]

[3] Khorana AA. Cancer-associated thrombosis: updates and controversies. Hematology Am Soc Hematol Educ Program 2012; 626-30.

[4] Lowe GD. Venous and arterial thrombosis: epidemiology and risk factors at various ages. Maturitas 2004; 47: 259-63.
[http://dx.doi.org/10.1016/j.maturitas.2003.12.009]

[5] Jerjes-Sanchez C. Venous and arterial thrombosis: a continuous spectrum of the same disease? Eur Heart J 2005; 26: 3-4.
[http://dx.doi.org/10.1093/eurheartj/ehi041]

[6] Esmon CT. Basic mechanisms and pathogenesis of venous thrombosis. Blood Rev 2009; 23: 225-9.

[http://dx.doi.org/10.1016/j.blre.2009.07.002]

[7] Moll S, Mackman N. Venous thromboembolism: a need for more public awareness and research into mechanisms. Arterioscler Thromb Vasc Biol 2008; 28: 367-9. [http://dx.doi.org/10.1161/ATVBAHA.108.163097]

[8] Popuri RK, Vedantham S. The role of thrombolysis in the clinical management of deep vein thrombosis. Arterioscler Thromb Vasc Biol 2011; 31: 479-84. [http://dx.doi.org/10.1161/ATVBAHA.110.213413]

[9] Cayley WE. Preventing deep vein thrombosis in hospital inpatients. BMJ 2007; 335: 147-51. [http://dx.doi.org/10.1136/bmj.39247.542477.AE]

[10] Lippi G, Franchini M, Targher G. Arterial thrombus formation in cardiovascular disease. Nat Rev Cardiol 2011; 8: 502-12. [http://dx.doi.org/10.1038/nrcardio.2011.91]

[11] Phillips DR, Conley PB, Sinha U, Andre P. Therapeutic approaches in arterial thrombosis. J Thromb Haemost 2005; 3: 1577-89. [http://dx.doi.org/10.1111/j.1538-7836.2005.01418.x]

[12] Homer-Vanniasinkam S, Tsui J. The continuing challenges of translational research: clinician-scientists' perspective Cardiol Res Pract 2012. 246710.

[13] Bai JP, Bell R, Buckman S, *et al.* Translational biomarkers: From preclinical to clinical a report of 2009 aaps/accp biomarker workshop. AAPS J 2011; 13: 274-83. [http://dx.doi.org/10.1208/s12248-011-9265-x]

[14] Bermejo J, Heras M, Segovia J, Alfonso F. Translational cardiovascular medicine. Now or never. Rev Esp Cardiol 2009; 62: 66-8. [http://dx.doi.org/10.1016/S0300-8932(09)70022-3]

[15] Buss DH, Stuart JJ, Lipscomb GE. The incidence of thrombotic and hemorrhagic disorders in association with extreme thrombocytosis: An analysis of 129 cases. Am J Hematol 1985; 20: 365-72. [http://dx.doi.org/10.1002/ajh.2830200408]

[16] Dinauer DM, Friedman KD, Hessner MJ. Allelic distribution of the glycoprotein ia (alpha2-integrin) c807t/g873a dimorphisms among caucasian venous thrombosis patients and six racial groups. Br J Haematol 1999; 107: 563-5. [http://dx.doi.org/10.1046/j.1365-2141.1999.01753.x]

[17] Okumus G, Kiyan E, Arseven O, *et al.* Platelet glycoprotein ia 807c/t and 873g/a polymorphisms in patients with venous thromboembolism. Clin Appl Thromb Hemost 2007; 13: 101-3. [http://dx.doi.org/10.1177/1076029606296422]

[18] Matsubara Y, Murata M, Maruyama T, *et al.* Association between diabetic retinopathy and genetic variations in alpha2beta1 integrin, a platelet receptor for collagen. Blood 2000; 95: 1560-4.

[19] Dodson PM, Haynes J, Starczynski J, *et al.* The platelet glycoprotein ia/iia gene polymorphism c807t/g873a: A novel risk factor for retinal vein occlusion. Eye (Lond) 2003; 17: 772-7. [http://dx.doi.org/10.1038/sj.eye.6700452]

[20] Wagner CL, Mascelli MA, Neblock DS, Weisman HF, Coller BS, Jordan RE. Analysis of gpiib/iiia receptor number by quantification of 7e3 binding to human platelets. Blood 1996; 88: 907-14.

[21] Feng D, Lindpaintner K, Larson MG, *et al.* Increased platelet aggregability associated with platelet gpiiia pla2 polymorphism: The framingham offspring study. Arterioscler Thromb Vasc Biol 1999; 19: 1142-7. [http://dx.doi.org/10.1161/01.ATV.19.4.1142]

[22] Ivanov P, Komsa-Penkova R, Kovacheva K, *et al.* Impact of thrombophilic genetic factors on pulmonary embolism: Early onset and recurrent incidences. Lung 2008; 186: 27-36. [http://dx.doi.org/10.1007/s00408-007-9061-7]

[23] El-Galaly TC, Severinsen MT, Overvad K, *et al.* Single nucleotide polymorphisms and the risk of venous thrombosis: Results from a danish case-cohort study. Br J Haematol 2013; 160: 838-41. [http://dx.doi.org/10.1111/bjh.12132]

[24] Bezemer ID, Bare LA, Doggen CJ, *et al.* Gene variants associated with deep vein thrombosis. JAMA 2008; 299: 1306-14. [http://dx.doi.org/10.1001/jama.299.11.1306]

[25] Best D, Senis YA, Jarvis GE, *et al.* Gpvi levels in platelets: relationship to platelet function at high shear. Blood 2003; 102: 2811-8. [http://dx.doi.org/10.1182/blood-2003-01-0231]

[26] Mammen EF, Barnhart MI, Selik NR, Gilroy J, Klepach GL. "Sticky platelet syndrome": a congenital platelet abnormality predisposing to thrombosis? Folia Haematol Int Mag Klin Morphol Blutforsch 1988; 115: 361-5.

[27] Kubisz P, Stasko J, Holly P. Sticky platelet syndrome. Semin Thromb Hemost 2013; 39: 674-83. [http://dx.doi.org/10.1055/s-0033-1353394]

[28] Hayes C, Kitahara S, Tcherniantchouk O. Decreased threshold of aggregation to low-dose epinephrine is evidence of platelet hyperaggregability in patients with thrombosis. Hematol Rep 2014; 6: 5326. [http://dx.doi.org/10.4081/hr.2014.5326]

[29] Yee DL, Sun CW, Bergeron AL, Dong JF, Bray PF. Aggregometry detects platelet hyperreactivity in healthy individuals. Blood 2005; 106: 2723-9. [http://dx.doi.org/10.1182/blood-2005-03-1290]

[30] Yee DL, Bergeron AL, Sun CW, Dong JF, Bray PF. Platelet hyperreactivity generalizes to multiple forms of stimulation. J Thromb Haemost 2006; 4: 2043-50. [http://dx.doi.org/10.1111/j.1538-7836.2006.02089.x]

[31] Berger JS, Becker RC, Kuhn C, Helms MJ, Ortel TL, Williams R. Hyperreactive platelet phenotypes: Relationship to altered serotonin transporter number, transport kinetics and intrinsic response to adrenergic co-stimulation. Thromb Haemost 2013; 109: 85-92. [http://dx.doi.org/10.1160/TH12-03-0202]

[32] Weber M, Gerdsen F, Gutensohn K, Schoder V, Eifrig B, Hossfeld DK. Enhanced platelet aggregation with trap-6 and collagen in platelet aggregometry in patients with venous thromboembolism. Thromb Res 2002; 107: 325-8. [http://dx.doi.org/10.1016/S0049-3848(02)00351-1]

[33] Hincker A, Feit J, Sladen RN, Wagener G. Rotational thromboelastometry predicts thromboembolic complications after major non-cardiac surgery. Crit Care 2014; 18: 549. [http://dx.doi.org/10.1186/s13054-014-0549-2]

[34] Spiezia L, Marchioro P, Radu C, *et al.* Whole blood coagulation assessment using rotation thrombelastogram thromboelastometry in patients with acute deep vein thrombosis. Blood Coagul Fibrinolysis 2008; 19: 355-60. [http://dx.doi.org/10.1097/MBC.0b013e328309044b]

[35] Rossetto V, Spiezia L, Senzolo M, Rodriguez-Castro KI, Maggiolo S, Simioni P. Whole blood rotation thromboelastometry (ROTEM®) profiles in subjects with non-neoplastic portal vein thrombosis. Thromb Res 2013; 132: e131-4. [http://dx.doi.org/10.1016/j.thromres.2013.06.009]

[36] McCrath DJ, Cerboni E, Frumento RJ, Hirsh AL, Bennett-Guerrero E. Thromboelastography maximum amplitude predicts postoperative thrombotic complications including myocardial infarction. Anesth Analg 2005; 100: 1576-83. [http://dx.doi.org/10.1213/01.ANE.0000155290.86795.12]

[37] Arcelus JI, Traverso CI, Caprini JA. Thromboelastography for the assessment of hypercoagulability during general surgery. Semin Thromb Hemost 1995; 21 (Suppl. 4): 21-6.

[38] Wilson D, Cooke EA, McNally MA, Wilson HK, Yeates A, Mollan RA. Changes in coagulability as measured by thrombelastography following surgery for proximal femoral fracture. Injury 2001; 32: 765-70.
[http://dx.doi.org/10.1016/S0020-1383(01)00139-5]

[39] Kashuk JL, Moore EE, Sabel A, *et al.* Rapid thrombelastography (r-teg) identifies hypercoagulability and predicts thromboembolic events in surgical patients.Surgery 2009; 146: 764-72. discussion 772-764.

[40] Jennings S, Cass AJ, Heather BP, Greenhalgh RM. Coagulation changes during major surgery and relationship to post-operative deep vein thrombosis. J Cardiovasc Surg (Torino) 1981; 22: 327-9.

[41] Cerutti E, Stratta C, Romagnoli R, *et al.* Thromboelastogram monitoring in the perioperative period of hepatectomy for adult living liver donation. Liver Transpl 2004; 10: 289-94.
[http://dx.doi.org/10.1002/lt.20078]

[42] Manchio JV, Gu J, Romar L, *et al.* Disruption of graft endothelium correlates with early failure after off-pump coronary artery bypass surgery. Ann Thorac Surg 2005; 79: 1991-8.
[http://dx.doi.org/10.1016/j.athoracsur.2004.12.054]

[43] Abrahams JM, Torchia MB, McGarvey M, Putt M, Baranov D, Sinson GP. Perioperative assessment of coagulability in neurosurgical patients using thromboelastography. Surg Neurol 2002; 58: 5-11.
[http://dx.doi.org/10.1016/S0090-3019(02)00777-2]

[44] Wen YR, Ho WY, Sun WZ, *et al.* Thromboelastographic study of thrombosis in the implantable central venous access device. Acta Anaesthesiol Sin 1997; 35: 223-8.

[45] Van Haren RM, Valle EJ, Thorson CM, *et al.* Hypercoagulability and other risk factors in trauma intensive care unit patients with venous thromboembolism. J Trauma Acute Care Surg 2014; 76: 443-9.
[http://dx.doi.org/10.1097/TA.0b013e3182a9d11d]

[46] Caprini JA, Zuckerman L, Cohen E, Vagher JP, Lipp V. The identification of accelerated coagulability. Thromb Res 1976; 9: 167-80.
[http://dx.doi.org/10.1016/0049-3848(76)90161-4]

[47] Koopman K, Uyttenboogaart M, Hendriks HG, *et al.* Thromboelastography in patients with cerebral venous thrombosis. Thromb Res 2009; 124: 185-8.
[http://dx.doi.org/10.1016/j.thromres.2008.12.032]

[48] Hvitfeldt Poulsen L, Christiansen K, Sørensen B, Ingerslev J. Whole blood thrombelastographic coagulation profiles using minimal tissue factor activation can display hypercoagulation in thrombosis-prone patients. Scand J Clin Lab Invest 2006; 66: 329-36.
[http://dx.doi.org/10.1080/00365510600672783]

[49] Fijnheer R, Frijns CJ, Korteweg J, *et al.* The origin of p-selectin as a circulating plasma protein. Thromb Haemost 1997; 77: 1081-5.

[50] McEver RP. Adhesive interactions of leukocytes, platelets, and the vessel wall during hemostasis and inflammation. Thromb Haemost 2001; 86: 746-56.

[51] Diacovo TG, Roth SJ, Buccola JM, Bainton DF, Springer TA. Neutrophil rolling, arrest, and transmigration across activated, surface-adherent platelets *via* sequential action of p-selectin and the beta 2-integrin cd11b/cd18. Blood 1996; 88: 146-57.

[52] Celi A, Pellegrini G, Lorenzet R, *et al.* P-selectin induces the expression of tissue factor on monocytes. Proc Natl Acad Sci USA 1994; 91: 8767-71.
[http://dx.doi.org/10.1073/pnas.91.19.8767]

[53] Palabrica T, Lobb R, Furie BC, *et al.* Leukocyte accumulation promoting fibrin deposition is mediated *in vivo* by p-selectin on adherent platelets. Nature 1992; 359: 848-51.
[http://dx.doi.org/10.1038/359848a0]

[54] Hrachovinová I, Cambien B, Hafezi-Moghadam A, *et al.* Interaction of p-selectin and psgl-1 generates microparticles that correct hemostasis in a mouse model of hemophilia a. Nat Med 2003; 9: 1020-5. [http://dx.doi.org/10.1038/nm899]

[55] Rauch U, Bonderman D, Bohrmann B, *et al.* Transfer of tissue factor from leukocytes to platelets is mediated by cd15 and tissue factor. Blood 2000; 96: 170-5.

[56] Blann AD, Noteboom WM, Rosendaal FR. Increased soluble p-selectin levels following deep venous thrombosis: Cause or effect? Br J Haematol 2000; 108: 191-3. [http://dx.doi.org/10.1046/j.1365-2141.2000.01813.x]

[57] Gremmel T, Ay C, Seidinger D, Pabinger I, Panzer S, Koppensteiner R. Soluble p-selectin, d-dimer, and high-sensitivity c-reactive protein after acute deep vein thrombosis of the lower limb. J Vasc Surg 2011; 54: 48S-55S. [http://dx.doi.org/10.1016/j.jvs.2011.05.097]

[58] Ay C, Jungbauer LV, Sailer T, *et al.* High concentrations of soluble p-selectin are associated with risk of venous thromboembolism and the p-selectin thr715 variant. Clin Chem 2007; 53: 1235-43. [http://dx.doi.org/10.1373/clinchem.2006.085068]

[59] Ramacciotti E, Blackburn S, Hawley AE, *et al.* Evaluation of soluble p-selectin as a marker for the diagnosis of deep venous thrombosis. Clin Appl Thromb Hemost 2011; 17: 425-31. [http://dx.doi.org/10.1177/1076029611405032]

[60] Ay C, Simanek R, Vormittag R, *et al.* High plasma levels of soluble p-selectin are predictive of venous thromboembolism in cancer patients: Results from the vienna cancer and thrombosis study (cats). Blood 2008; 112: 2703-8. [http://dx.doi.org/10.1182/blood-2008-02-142422]

[61] Ay C, Dunkler D, Marosi C, *et al.* Prediction of venous thromboembolism in cancer patients. Blood 2010; 116: 5377-82. [http://dx.doi.org/10.1182/blood-2010-02-270116]

[62] Ahlbrecht J, Dickmann B, Ay C, *et al.* Tumor grade is associated with venous thromboembolism in patients with cancer: Results from the vienna cancer and thrombosis study. J Clin Oncol 2012; 30: 3870-5. [http://dx.doi.org/10.1200/JCO.2011.40.1810]

[63] Dickmann B, Ahlbrecht J, Ay C, *et al.* Regional lymph node metastases are a strong risk factor for venous thromboembolism: Results from the vienna cancer and thrombosis study. Haematologica 2013; 98: 1309-14. [http://dx.doi.org/10.3324/haematol.2012.073338]

[64] Braekkan SK, Mathiesen EB, Njølstad I, Wilsgaard T, Størmer J, Hansen JB. Mean platelet volume is a risk factor for venous thromboembolism: The tromsø study, tromsø, norway. J Thromb Haemost 2010; 8: 157-62. [http://dx.doi.org/10.1111/j.1538-7836.2009.03498.x]

[65] Tsai AW, Cushman M, Rosamond WD, Heckbert SR, Polak JF, Folsom AR. Cardiovascular risk factors and venous thromboembolism incidence: The longitudinal investigation of thromboembolism etiology. Arch Intern Med 2002; 162: 1182-9. [http://dx.doi.org/10.1001/archinte.162.10.1182]

[66] van der Bom JG, Heckbert SR, Lumley T, *et al.* Platelet count and the risk for thrombosis and death in the elderly. J Thromb Haemost 2009; 7: 399-405. [http://dx.doi.org/10.1111/j.1538-7836.2008.03267.x]

[67] Zakai NA, Wright J, Cushman M. Risk factors for venous thrombosis in medical inpatients: validation of a thrombosis risk score. J Thromb Haemost 2004; 2: 2156-61. [http://dx.doi.org/10.1111/j.1538-7836.2004.00991.x]

[68] Mandalà M, Barni S, Prins M, *et al.* Acquired and inherited risk factors for developing venous

thromboembolism in cancer patients receiving adjuvant chemotherapy: a prospective trial. Ann Oncol 2010; 21: 871-6.
[http://dx.doi.org/10.1093/annonc/mdp354]

[69] Simanek R, Vormittag R, Ay C, *et al.* High platelet count associated with venous thromboembolism in cancer patients: Results from the vienna cancer and thrombosis study (cats). J Thromb Haemost 2010; 8: 114-20.
[http://dx.doi.org/10.1111/j.1538-7836.2009.03680.x]

[70] Khorana AA, Kuderer NM, Culakova E, Lyman GH, Francis CW. Development and validation of a predictive model for chemotherapy-associated thrombosis. Blood 2008; 111: 4902-7.
[http://dx.doi.org/10.1182/blood-2007-10-116327]

[71] Jensvoll H, Blix K, Brækkan SK, Hansen JB. Platelet count measured prior to cancer development is a risk factor for future symptomatic venous thromboembolism: The tromsø study. PLoS One 2014; 9: e92011.
[http://dx.doi.org/10.1371/journal.pone.0092011]

[72] Rupa-Matysek J, Gil L, Wojtasińska E, Ciepłuch K, Lewandowska M, Komarnicki M. The relationship between mean platelet volume and thrombosis recurrence in patients diagnosed with antiphospholipid syndrome. Rheumatol Int 2014; 34: 1599-605.
[http://dx.doi.org/10.1007/s00296-014-2996-0]

[73] Riedl J, Kaider A, Reitter EM, *et al.* Association of mean platelet volume with risk of venous thromboembolism and mortality in patients with cancer. Results from the vienna cancer and thrombosis study (cats). Thromb Haemost 2014; 111: 670-8.
[http://dx.doi.org/10.1160/TH13-07-0603]

[74] Ferroni P, Guadagni F, Riondino S, *et al.* Evaluation of mean platelet volume as a predictive marker for cancer-associated venous thromboembolism during chemotherapy. Haematologica 2014; 99: 1638-44.
[http://dx.doi.org/10.3324/haematol.2014.109470]

[75] Martínez MC, Tesse A, Zobairi F, Andriantsitohaina R. Shed membrane microparticles from circulating and vascular cells in regulating vascular function. Am J Physiol Heart Circ Physiol 2005; 288: H1004-9.
[http://dx.doi.org/10.1152/ajpheart.00842.2004]

[76] Théry C, Boussac M, Véron P, *et al.* Proteomic analysis of dendritic cell-derived exosomes: A secreted subcellular compartment distinct from apoptotic vesicles. J Immunol 2001; 166: 7309-18.
[http://dx.doi.org/10.4049/jimmunol.166.12.7309]

[77] Wolf P. The nature and significance of platelet products in human plasma. Br J Haematol 1967; 13: 269-88.
[http://dx.doi.org/10.1111/j.1365-2141.1967.tb08741.x]

[78] Peterson DB, Sander T, Kaul S, *et al.* Comparative proteomic analysis of pai-1 and tnf-alpha-derived endothelial microparticles. Proteomics 2008; 8: 2430-46.
[http://dx.doi.org/10.1002/pmic.200701029]

[79] Montoro-Garcia S, Shantsila E, Marin F, Blann A, Lip GY. Circulating microparticles: New insights into the biochemical basis of microparticle release and activity. Basic Res Cardiol 2011; 106: 911-23.
[http://dx.doi.org/10.1007/s00395-011-0198-4]

[80] Mause SF, Weber C. Microparticles: Protagonists of a novel communication network for intercellular information exchange. Circ Res 2010; 107: 1047-57.
[http://dx.doi.org/10.1161/CIRCRESAHA.110.226456]

[81] McGregor L, Martin J, McGregor JL. Platelet-leukocyte aggregates and derived microparticles in inflammation, vascular remodelling and thrombosis. Front Biosci 2006; 11: 830-7.
[http://dx.doi.org/10.2741/1840]

[82] Abid Hussein MN, Böing AN, Biró E, *et al.* Phospholipid composition of *in vitro* endothelial microparticles and their *in vivo* thrombogenic properties. Thromb Res 2008; 121: 865-71. [http://dx.doi.org/10.1016/j.thromres.2007.08.005]

[83] Rukoyatkina N, Begonja AJ, Geiger J, Eigenthaler M, Walter U, Gambaryan S. Phosphatidylserine surface expression and integrin alpha iib beta 3 activity on thrombin/convulxin stimulated platelets/particles of different sizes. Br J Haematol 2009; 144: 591-602. [http://dx.doi.org/10.1111/j.1365-2141.2008.07506.x]

[84] Ratajczak J, Miekus K, Kucia M, *et al.* Embryonic stem cell-derived microvesicles reprogram hematopoietic progenitors: Evidence for horizontal transferof mRNA and protein delivery. Leukemia 2006; 20: 847-56. [http://dx.doi.org/10.1038/sj.leu.2404132]

[85] Patchipulusu S, Turturro M, Hall CL. Monocyte-derived macrophage microparticles impart tissue factor activity to biomaterial surfaces. J Biomed Mater Res A 2010; 92: 724-32. [http://dx.doi.org/10.1002/jbm.a.32320]

[86] Sinauridze EI, Kireev DA, Popenko NY, *et al.* Platelet microparticle membranes have 50- to 100-fold higher specific procoagulant activity than activated platelets. Thromb Haemost 2007; 97: 425-34.

[87] Montoro-Garcia S, Shantsila E, Hernandez-Romero D, *et al.* Small-size platelet microparticles trigger platelet and monocyte functionality and modulate thrombogenesis *via* p-selectin. Br J Haematol 2014; 166: 571-80. [http://dx.doi.org/10.1111/bjh.12913]

[88] Campello E, Spiezia L, Radu CM, *et al.* Endothelial, platelet, and tissue factor-bearing microparticles in cancer patients with and without venous thromboembolism. Thromb Res 2011; 127: 473-7. [http://dx.doi.org/10.1016/j.thromres.2011.01.002]

[89] Tesselaar ME, Romijn FP, Van Der Linden IK, Prins FA, Bertina RM, Osanto S. Microparticle-associated tissue factor activity: A link between cancer and thrombosis? J Thromb Haemost 2007; 5: 520-7. [http://dx.doi.org/10.1111/j.1538-7836.2007.02369.x]

[90] Manly DA, Wang J, Glover SL, *et al.* Increased microparticle tissue factor activity in cancer patients with venous thromboembolism. Thromb Res 2010; 125: 511-2. [http://dx.doi.org/10.1016/j.thromres.2009.09.019]

[91] Chirinos JA, Heresi GA, Velasquez H, *et al.* Elevation of endothelial microparticles, platelets, and leukocyte activation in patients with venous thromboembolism. J Am Coll Cardiol 2005; 45: 1467-71. [http://dx.doi.org/10.1016/j.jacc.2004.12.075]

[92] Jy W, Horstman LL, Wang F, Duncan RC, Ahn YS. Platelet factor 3 in plasma fractions: Its relation to microparticle size and thromboses. Thromb Res 1995; 80: 471-82. [http://dx.doi.org/10.1016/0049-3848(95)00202-2]

[93] Barry OP, Pratico D, Lawson JA, FitzGerald GA. Transcellular activation of platelets and endothelial cells by bioactive lipids in platelet microparticles. J Clin Invest 1997; 99: 2118-27. [http://dx.doi.org/10.1172/JCI119385]

[94] Sims PJ, Rollins SA, Wiedmer T. Regulatory control of complement on blood platelets. Modulation of platelet procoagulant responses by a membrane inhibitor of the c5b-9 complex. J Biol Chem 1989; 264: 19228-35.

[95] Ye R, Ye C, Huang Y, Liu L, Wang S. Circulating tissue factor positive microparticles in patients with acute recurrent deep venous thrombosis. Thromb Res 2012; 130: 253-8. [http://dx.doi.org/10.1016/j.thromres.2011.10.014]

[96] Hérault JP, Perrin B, Jongbloet C, Pflieger AM, Bernat A, Herbert JM. Effect of factor xa inhibitors on the platelet-derived microparticles procoagulant activity *in vitro* and *in vivo* in rats. Thromb Haemost 2000; 84: 668-74.

[97] Borissoff JI, Spronk HM, ten Cate H. The hemostatic system as a modulator of atherosclerosis. N Engl J Med 2011; 364: 1746-60. [http://dx.doi.org/10.1056/NEJMra1011670]

[98] Pamukcu B, Lip GY, Devitt A, Griffiths H, Shantsila E. The role of monocytes in atherosclerotic coronary artery disease. Ann Med 2010; 42: 394-403. [http://dx.doi.org/10.3109/07853890.2010.497767]

[99] Davì G, Patrono C. Platelet activation and atherothrombosis. N Engl J Med 2007; 357: 2482-94. [http://dx.doi.org/10.1056/NEJMra071014]

[100] Falk E. Pathogenesis of atherosclerosis. J Am Coll Cardiol 2006; 47: C7-C12. [http://dx.doi.org/10.1016/j.jacc.2005.09.068]

[101] Quehenberger O. Thematic review series: the immune system and atherogenesis. Molecular mechanisms regulating monocyte recruitment in atherosclerosis. J Lipid Res 2005; 46: 1582-90. [http://dx.doi.org/10.1194/jlr.R500008-JLR200]

[102] Shantsila E, Kamphuisen PW, Lip GY. Circulating microparticles in cardiovascular disease: Implications for atherogenesis and atherothrombosis. J Thromb Haemost 2010; 8: 2358-68. [http://dx.doi.org/10.1111/j.1538-7836.2010.04007.x]

[103] Dean WL, Lee MJ, Cummins TD, Schultz DJ, Powell DW. Proteomic and functional characterisation of platelet microparticle size classes. Thromb Haemost 2009; 102: 711-8.

[104] Castellana D, Toti F, Freyssinet JM. Membrane microvesicles: macromessengers in cancer disease and progression. Thromb Res 2010; 125 (Suppl. 2): S84-8. [http://dx.doi.org/10.1016/S0049-3848(10)70021-9]

[105] Al-Nedawi K, Meehan B, Rak J. Microvesicles: messengers and mediators of tumor progression. Cell Cycle 2009; 8: 2014-8. [http://dx.doi.org/10.4161/cc.8.13.8988]

[106] Wang JG, Manly D, Kirchhofer D, Pawlinski R, Mackman N. Levels of microparticle tissue factor activity correlate with coagulation activation in endotoxemic mice. J Thromb Haemost 2009; 7: 1092-8. [http://dx.doi.org/10.1111/j.1538-7836.2009.03448.x]

[107] Aharon A, Tamari T, Brenner B. Monocyte-derived microparticles and exosomes induce procoagulant and apoptotic effects on endothelial cells. Thromb Haemost 2008; 100: 878-85.

[108] Montoro-Garcia S, Shantsila E, Tapp LD, *et al.* Small-size circulating microparticles in acute coronary syndromes: Relevance to fibrinolytic status, reparative markers and outcomes. Atherosclerosis 2013; 227: 313-22. [http://dx.doi.org/10.1016/j.atherosclerosis.2013.01.028]

[109] Bernal-Mizrachi L, Jy W, Fierro C, *et al.* Endothelial microparticles correlate with high-risk angiographic lesions in acute coronary syndromes. Int J Cardiol 2004; 97: 439-46. [http://dx.doi.org/10.1016/j.ijcard.2003.10.029]

[110] Li X, Cong H. Platelet-derived microparticles and the potential of glycoprotein iib/iiia antagonists in treating acute coronary syndrome. Tex Heart Inst J 2009; 36: 134-9.

[111] Lagos-Quintana M, Rauhut R, Lendeckel W, Tuschl T. Identification of novel genes coding for small expressed RNAs. Science 2001; 294: 853-8. [http://dx.doi.org/10.1126/science.1064921]

[112] Lewis BP, Burge CB, Bartel DP. Conserved seed pairing, often flanked by adenosines, indicates that thousands of human genes are microRNA targets. Cell 2005; 120: 15-20. [http://dx.doi.org/10.1016/j.cell.2004.12.035]

[113] Weber C, Schober A, Zernecke A. MicroRNA in arterial remodelling, inflammation and atherosclerosis. Curr Drug Targets 2010; 11: 950-6.

[http://dx.doi.org/10.2174/138945010791591377]

[114] Arroyo JD, Chevillet JR, Kroh EM, *et al.* Argonaute2 complexes carry a population of circulating microRNA independent of vesicles in human plasma. Proc Natl Acad Sci USA 2011; 108: 5003-8. [http://dx.doi.org/10.1073/pnas.1019055108]

[115] Vickers KC, Palmisano BT, Shoucri BM, Shamburek RD, Remaley AT. microRNA are transported in plasma and delivered to recipient cells by high-density lipoproteins. Nat Cell Biol 2011; 13: 423-33. [http://dx.doi.org/10.1038/ncb2210]

[116] Orenes-Pinero E, Montoro-Garcia S, Patel JV, Valdes M, Marin F, Lip GY. Role of microRNA in cardiac remodelling: New insights and future perspectives. Int J Cardiol 2013; 167: 1651-9. [http://dx.doi.org/10.1016/j.ijcard.2012.09.120]

[117] Skog J, Würdinger T, van Rijn S, *et al.* Glioblastoma microvesicles transport RNA and proteins that promote tumour growth and provide diagnostic biomarkers. Nat Cell Biol 2008; 10: 1470-6. [http://dx.doi.org/10.1038/ncb1800]

[118] Yuan A, Farber EL, Rapoport AL, *et al.* Transfer of microRNA by embryonic stem cell microvesicles. PLoS One 2009; 4: e4722. [http://dx.doi.org/10.1371/journal.pone.0004722]

[119] Cheng Y, Tan N, Yang J, *et al.* A translational study of circulating cell-free microrna-1 in acute myocardial infarction. Clin Sci (Lond) 2010; 119: 87-95. [http://dx.doi.org/10.1042/CS20090645]

[120] Ai J, Zhang R, Li Y, *et al.* Circulating microrna-1 as a potential novel biomarker for acute myocardial infarction. Biochem Biophys Res Commun 2010; 391: 73-7. [http://dx.doi.org/10.1016/j.bbrc.2009.11.005]

[121] Dong S, Cheng Y, Yang J, *et al.* Microrna expression signature and the role of microrna-21 in the early phase of acute myocardial infarction. J Biol Chem 2009; 284: 29514-25. [http://dx.doi.org/10.1074/jbc.M109.027896]

[122] Boilard E, Nigrovic PA, Larabee K, *et al.* Platelets amplify inflammation in arthritis *via* collagen-dependent microparticle production. Science 2010; 327: 580-3. [http://dx.doi.org/10.1126/science.1181928]

[123] O'Donnell CJ, Nabel EG. Genomics of cardiovascular disease. N Engl J Med 2011; 365: 2098-109. [http://dx.doi.org/10.1056/NEJMra1105239]

[124] Altieri DC, Morrissey JH, Edgington TS. Adhesive receptor mac-1 coordinates the activation of factor x on stimulated cells of monocytic and myeloid differentiation: An alternative initiation of the coagulation protease cascade. Proc Natl Acad Sci USA 1988; 85: 7462-6. [http://dx.doi.org/10.1073/pnas.85.20.7462]

[125] Tapp LD, Shantsila E, Wrigley BJ, Pamukcu B, Lip GY. The cd14++cd16+ monocyte subset and monocyte-platelet interactions in patients with st-elevation myocardial infarction. J Thromb Haemost 2012; 10: 1231-41. [http://dx.doi.org/10.1111/j.1538-7836.2011.04603.x]

[126] Tapp LD, Shantsila E, Wrigley BJ, Montoro-Garcia S, Lip GY. Tlr4 expression on monocyte subsets in myocardial infarction. J Intern Med 2013; 273: 294-305. [http://dx.doi.org/10.1111/joim.12011]

[127] Shantsila E, Tapp LD, Wrigley BJ, *et al.* Monocyte subsets in coronary artery disease and their associations with markers of inflammation and fibrinolysis. Atherosclerosis 2014; 234: 4-10. [http://dx.doi.org/10.1016/j.atherosclerosis.2014.02.009]

[128] Polverini PJ, Cotran PS, Gimbrone MA, Unanue ER. Activated macrophages induce vascular proliferation. Nature 1977; 269: 804-6. [http://dx.doi.org/10.1038/269804a0]

[129] Cermak J, Key NS, Bach RR, Balla J, Jacob HS, Vercellotti GM. C-reactive protein induces human peripheral blood monocytes to synthesize tissue factor. Blood 1993; 82: 513-20.

[130] Aleman MM, Gardiner C, Harrison P, Wolberg AS. Differential contributions of monocyte- and platelet-derived microparticles towards thrombin generation and fibrin formation and stability. J Thromb Haemost 2011; 9: 2251-61.
[http://dx.doi.org/10.1111/j.1538-7836.2011.04488.x]

[131] Michelson AD, Barnard MR, Krueger LA, Valeri CR, Furman MI. Circulating monocyte-platelet aggregates are a more sensitive marker of *in vivo* platelet activation than platelet surface p-selectin: Studies in baboons, human coronary intervention, and human acute myocardial infarction. Circulation 2001; 104: 1533-7.
[http://dx.doi.org/10.1161/hc3801.095588]

[132] Passacquale G, Vamadevan P, Pereira L, Hamid C, Corrigall V, Ferro A. Monocyte-platelet interaction induces a pro-inflammatory phenotype in circulating monocytes. PLoS One 2011; 6: e25595.
[http://dx.doi.org/10.1371/journal.pone.0025595]

[133] Ammon C, Kreutz M, Rehli M, Krause SW, Andreesen R. Platelets induce monocyte differentiation in serum-free coculture. J Leukoc Biol 1998; 63: 469-76.

[134] Ollivier V, Arnaud B, Jandrot-Perrus M, Ajzenberg N. Monocytes downregulate the early stage of collagen-induced platelet activation by a pecam-1-dependent mechanism. J Thromb Haemost 2009; 7: 143-51.
[http://dx.doi.org/10.1111/j.1538-7836.2008.03170.x]

[135] Morange PE, Tregouet DA. Deciphering the molecular basis of venous thromboembolism: where are we and where should we go? Br J Haematol 2010; 148: 495-506.
[http://dx.doi.org/10.1111/j.1365-2141.2009.07975.x]

CHAPTER 11

Importance of Biomarkers in Heart Failure: Classic and New Aspects

Juan Antonio Vílchez[1,*], **Pablo Perez-Cañadas**[1], **Francisco Marín**[2] and **Jordi Ordóñez-Llanos**[3]

[1] *Clinical Analysis Department, Santa Lucía General University Hospital, Cartagena, Spain*

[2] *Department of Cardiology, Virgen de la Arrixaca Clinic University Hospital, Murcia, Spain, Instituto de Investigación Biomédica-Virgen de la Arrixaca, IMIB-Arrixaca, Murcia, Spain*

[3] *IIB-Sant Pau Biomedical Research Institute, Hospital Santa Creu i Sant Pau and Department of Biochemistry and Molecular Biology, Universitat Autònoma, Barcelona, Spain*

Abstract: Despite wide range of biomarkers studied in heart failure (HF), only natriuretic peptides have become standard-of-care biomarkers in HF and are currently the elected biomarkers that have crossed the field of research commonly used in daily practice. To achieve this knowledge, many biomarkers related to different patho-physiological effects in HF have been studied, being these pathways: myocardial stretch, myocyte injury, extracellular matrix and fibrotic state, inflammation, renal dysfunction, neurohormonal activation, and oxidative stress. The aim of this review was to summarize the published data about biomarkers studied in HF, focusing on data from randomized prospective clinical trials and large community-based cohorts. We also reviewed the new trends in those biomarkers and their application on this pathology.

Keywords: Acute heart failure, Adrenomedullin, Biomarkers, BNP, C-type natriuretic peptide, Chronic heart failure, Collagen peptides, Copeptin, CRP, Diagnosis, Endothelin, Galectin-3, Heart failure, Neprilysin, NT-proBNP, Prognosis, Review, ST2, Troponin, Urate.

INTRODUCTION

Heart failure (HF) is one of the main causes of morbidity and mortality in western world with 26 million people worldwide that have been diagnosed of HF [1]. Moreover, a 5-year mortality remains unacceptably high, despite recently latest improvements in HF therapy [2]. That's why HF appears as an enormous burden

* **Corresponding author Juan Antonio Vílchez:** Clinical Analysis Department, Santa Lucía General University Hospital, Cartagena, Spain; Tel/Fax: 968128600; E-mail: juananv81@hotmail.com

on health care costs, due to the simple reason of a high number of rehospitalizations [3], and it is expected that prevalence of HF will rise in future years, due to an aging population. In fact, it has been appreciated that how there exists a direct relationship between advancing age and incidence of HF. The incidence of HF nearly doubles for patients >85 years compared with those <75 years [4]. So, HF has to be considered as one of the most prevalent and challenging medical conditions [5].

HF results from structural or functional cardiovascular disorders that cause inadequate systemic perfusion [2]. HF can be defined as a complex mechanical and neurohumoral syndrome manifested by breathlessness (dyspnoea) at normal or mild exertion, fatigue, paroxysmal nocturnal dyspnea and peripheral edema. Indeed, it's an usual clinical syndrome caused by diverse diseases such as hypertension, anemia, COPD, cancer, diabetes mellitus, atrial fibrillation (AF), diseases affecting the pericardium, myocardium, endocardium, cardiac valves or cardiac vessels [6].

HF has been considered to be associated with impaired cardiac contractility and cardiac dilation, but it has become evident that a considerable portion of patients presenting with clinical HF have a normal ejection fraction (EF) [7]. Therefore, two pathophysiologic myocardial mechanisms arise causing HF: systolic dysfunction and diastolic dysfunction:

Systolic Dysfunction

Systolic dysfunction is also known as HF with reduced ejection fraction, defined as a left ventricular EF of <40% (HFrEF). In this case the most common causes of systolic dysfunction are idiopathic dilated cardiomyopathy (DCM), the most frequent etiology and also ischemic heart disease, hypertension or valvular disease [8].

Diastolic Dysfunction

Diastolic dysfunction, also known as HF with preserved ejection fraction (HFpEF), is defined as a left ventricular EF >45% [6]. Some similar states that lead to systolic dysfunction have been found to also cause HFpEF. The most common are hypertension, ischemic heart disease, obstructive hypertrophic cardiomyopathy (HCM), and restrictive cardiomyopathy. It's known how patients with HFpEF usually show abnormal LV relaxation, diastolic distensibility, or diastolic stiffness and a lack of valvular abnormalities detected in echocardiograpy, as was established by the American College of Cardiology/American Heart Association. The final effect is the displacement of LV filling from firstly to latter part of diastole [6, 9].

Nowadays, experts define HF as a complex mechanical and neuro hormonal syndrome manifested by hemodynamic congestion [1]. But, the pathophysiology concept of HF is related to the concept of myocardial remodeling. The neuro hormonal syndrome leads to changes in myocardial tissue characterized by progressive and marked increases in the presence of fibrosis. So, increased stress or injury in the myocardium due to hypertension, diabetes mellitus, or ongoing ischemia can contribute to cardiac remodeling [10].

In fact, age contributes to all forms of HF because it is a major determinant of tissue fibrosis and vascular stiffness, and hypertension appears to accelerate this aging process in the myocardium and the vasculature [1]. It's known how around a 70% of HF patients, have underlying coronary artery disease and are considered to have ischemic cardiomyopathy with zones presenting cardiac fibrosis as a result of ischemia and infarction [11]. On other hand, the epidemiology of non-ischemic cardiomyopathy, which includes the main one-third of individuals with reduced left ventricular ejection fraction (LVEF) and clinical HF, has a less striking relationship with age; however, it is associated with history of hypertension in a majority of cases. Approximately 20% of first-degree family members will be found to have depressed LVEF, which is often asymptomatic and amenable to therapy preventing a worsening [1, 12]. Most forms of non-ischemic cardiomyopathy, as the final histopathological process, have excess deposition of collagen in the extracellular matrix (ECM) and tissue fibrosis presented at necropsy [13].

In this scenario, it is critical to identify the main population that could have risk of developing HF and death. There is no specific diagnostic test for HF, initial evaluation for patients with suggestive symptoms of HF includes clinical assessment through history and physical exam, electrocardiogram, blood tests, and chest X-ray, and candidate biomarkers of HF are suggested to be evaluated in patients with suspected HF, in whom the diagnosis is uncertain. However, existing risk factors such as age, body mass index (BMI), blood pressure, abnormal renal function, diabetes, New York Heart Association (NYHA) functional class, LVEF are not powerful predictors of mortality, incident HF, or could be possible quantifiers of the nature of the pathways of fibrosis, inflammation and injury [14]. That's why, biomarkers are being increasingly used as predictors in HF patients. Moreover, the most recent cardiovascular risk models have been developed to assume the use of biomarkers added to classical risk factors, assessing if their inclusion could improve cardiovascular disease prediction scales [15]. Blood-based biomarkers, could improve our knowledge of the pathophysiology of HF.

BIOMARKERS

Biomarkers are molecules that are easily measured by laboratory techniques. This fact can give us useful information about normal biological processes, abnormal pathophysiology, and prognosis, as well could assist in a differential diagnosis. Cardiovascular pathologies is one of the medicine fields in which biomarkers have shown to be really useful in the management of different pathologies, as ischemic heart disease, acute coronary syndrome (ACS) and HF. The study of biomarkers related to the different pathways involved in HF, Table **1** could be related to clinical and echocardiographic risk factors.

Table 1. Candidate biomarkers in HF.

PATHWAYS	BIOMARKERS STUDIED
Myocardial stretch and neurohormonal activation	B-type natriuretic peptide (BNP) Atrial natriuretic peptide (ANP) N-terminal pro b-type natriuretic peptide (NT-proBNP)
Neurohormonal activation	Endothelin Adrenomedullin Copeptin Neprilysin
Renal dysfunction	Urinary C-type natriuretic peptide
Inflammation	Interleukin (IL-6) Tumor necrosis factorα (TNFα) C-reactive protein (CRP)
Inflammation and profibrotic state	Growth differentiation factor-15 (GDF-15) Suppression of tumorigenicity 2 (ST2)
Myocyte injury	Cardiac troponins (TnI and TnT)
Extracellular matrix and fibrotic state	Collagen peptides Galectin-3
Oxidative stress	Urate

The aim of this review was to summarize the published data about biomarkers studied in HF, with focus on data from randomized prospective clinical trials and large community-based cohorts. We also reviewed the new trends in biomarkers studied and their application on this pathology.

Biomarkers Related to Myocardial Stretch

Wall Stress Markers: Natriuretic Peptides

Natriuretic peptides are classified as neuro hormones and have a diuretic and vasodilator effect, that's why it could be related to neuro hormonal activation too.

Atrial (ANP) and B-type natriuretic peptides (BNP) are predominantly synthesized in atria and left ventricle, respectively, due to an elevated wall tension response. BNP or its amino-terminal cleavage equivalent (N-terminal pro b-type natriuretic peptide (NT-proBNP)) proceeds from a usual amino-acid precursor peptide (proBNP). This peptide is formed by cardiomyocytes in the context of numerous triggers, being the most notably a myocardial stretch [16]. It is known how plasma levels of this peptides positively correlate with cardiac filling pressures, and this fact turn them excellent biomarkers for the presence of LV dysfunction and abnormal LV wall stress [17].

Predominantly, the laboratory assays for BNP and NT-proBNP have been used to diagnostic the HF presence and severity status. In general ways, BNP and NT-proBNP levels are tightly correlated, and both can be adopted in patient care processes, but knowing that their respective absolute concentrations and cut points could not be commutable. The use of BNP and NT-proBNP support the clinical judgment for a correct HF diagnosis or exclusion, in the setting of chronic ambulatory HF [9, 18] or acute decompensated HF [9, 19]. Moreover, the value of natriuretic peptide testing is particularly significant when the etiology of dyspnoea is unclear. Although lower levels of BNP or NT-proBNP exclude the presence of HF and higher levels have conclusive positive predictive value to HF diagnose, the clinicians have to be aware of high levels due to, both peptides, have been related with a large variety of cardiac and non-cardiac causes [20]. BNP and NT-proBNP levels decrease with a correct treatment of chronic HF [21], and this fact correlates with improved clinical outcomes [9, 22].

The natriuretic peptides have been also disturbed in underlying HF pathologies. Elevated NT-proBNP levels are usually associated with incipient LV remodeling and fibrosis, fact that could be used to diagnose unfavorable LV remodeling and major risk of sudden death in HCM patients [23]. Moreover, continuous elevations correlates with the classification of the New York Heart Association (NYHA) suggesting a worst functional capacity and ventricular obstruction [24].

As we are describing, the presence of severe cardiac dysfunction and poor functional NYHA class correlates with higher plasma BNP levels in patients with HCM or dilated cardiomyopathy. Similarly, in HCM has been related higher BNP levels and major severity of HF symptoms. These symptoms are usually associated with a prevalent diastolic dysfunction, and also with systolic dysfunction with a low LV, lower wall hypertrophy and a growing of the cavity by 10-15% of the cases. This scenario usually appears in middle age, and added to sudden cardiac death (SCD), represents the leading cause of death in these patients [25, 26].

This behavior of NT-proBNP levels has been also presented in the recent study by Coats CJ *et al.*, showing a significant difference in NT-proBNP across NYHA classes. Patients who presented baseline levels of NT-proBNP ≥1500 pg/mL, showed an increased risk of clinical degradation and major symptoms of HF at first 2 years. They also approximate a relation of NT-proBNP levels to prognosis, showing how NT-proBNP was a significant predictor of HF and deaths caused by cardiac transplants but not in SCD or appropriate shock from an implantable cardioverter defibrillator, in patients with HCM. This behavior suggests that raised NT-proBNP might identify patients who would benefit from pharmacological therapy to prevent progressive left ventricular dysfunction before the onset of apparent HF [27].

Diagnosis and Prognosis of Acute Heart Failure

Natriuretic peptides have been usually established in acute heart failure (AHF) diagnosis. Maybe, due to their high sensitivity and specificity (NT-proBNP: sensitivity 90%, specificity 84%; BNP: sensitivity 90%, specificity 76%) [28]. A stablished age-indiscriminate cutoff of NT-proBNP of <1800 pg/mL can be used to identify patients at low mortality risk [28]. A recent study by Maisel *et al.*, conclude that patients with BNP <200 pg/mL have a very low 30-day mortality risk and presents a proposed algorithm to stratify patients into low, intermediate, and high risk, incorporating natriuretic peptides in baseline triage with other common clinical characteristics (Fig. **1**) [29]. So, in onset of AHF, patients with BNP levels >1000 pg/mL, NT-proBNP >1800 pg/mL, could represent high risk of HF. Patients with a BNP level from 400 pg/mL to 1000 pg/mL, NT-proBNP levels from 900 pg/mL to 1800 pg/mL, could be at intermediate risk and patients with BNP <400 pg/mL or NT-proBNP <900 pg/mL are considered to be at low risk (Fig. **1**) [29].

Concentrations of natriuretic peptides have also proved to be useful in the AHF prognosis. So, results at discharge or the decreasing levels of these biomarkers have been related to patient evolution. Studies have shown how values above 5189 pg/mL increased fivefold the risk of death in a short term (2.5 months) [30].

Diagnosis and Prognosis of Stable Heart Failure

The use of natriuretic peptides in the diagnosis of HF in primary care has also been accepted, but with differences in clinical guidelines regarding to the chosen cutoff. The guidelines of the European Society of Cardiology suggested 35 pg/mL for BNP and 125 pg/mL for NT-proBNP, although different studies also proposed cutoffs for age [31], a recently Spanish study proposed a cutoff point for diagnosis of 280 pg/mL [32].

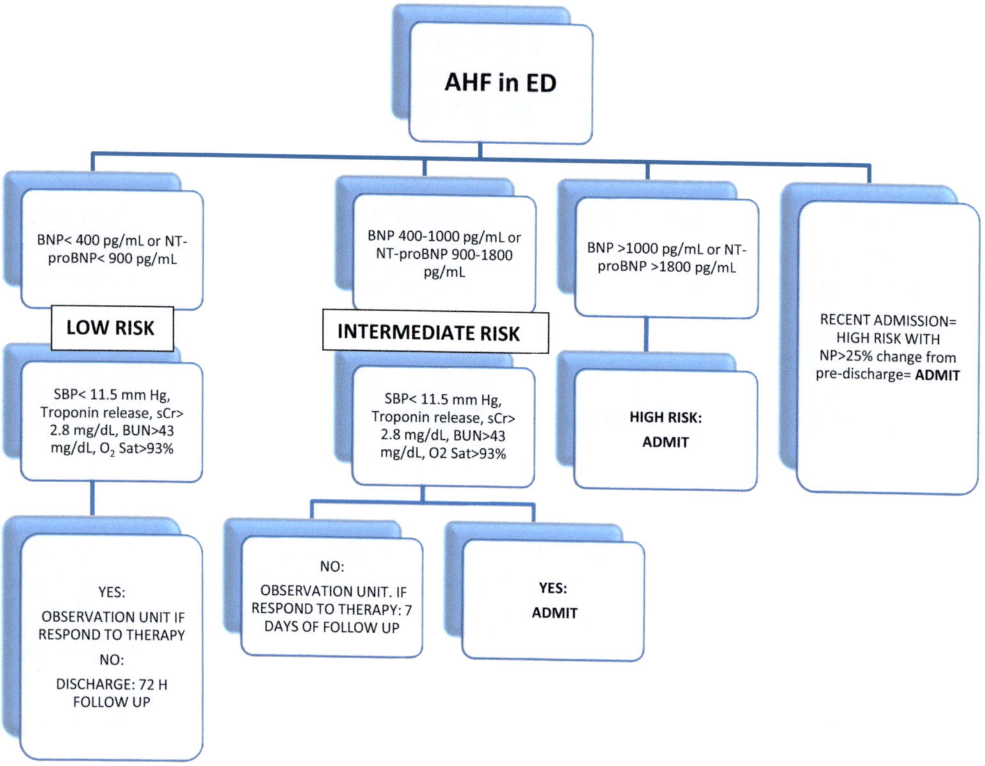

Fig. (1). Algorithm proposed for use of natriuretic peptides in emergency department. Use of natriuretic peptides and subsequent use of common clinical parameters to stratificate. Patients can be divided into low, intermediate, and high risk groups for management. ED, emergency department, AHF, acute heart failure; BNP, B-type natriuretic peptide; NT-proBNP, N-terminal proB-type natriuretic peptide; SBP, systolic blood pressure; SCr, serum creatinine; BUN, blood urea nitrogen; O_2 Sat, O_2 saturation; NP, natriuretic peptides. Adapted from Maisel *et al.*, (2015) [29].

It is also established the prognostic significance of natriuretic peptides levels in patients with chronic HF. Thus, studies range from HF patients with moderate to severe HF. In the Val-HeFT study [33], an increase of 500 pg/mL above baseline NT-proBNP levels, represented a 3.8% increase in mortality and 3% in re-hospitalization in patients with moderate HF. In the multivariate analysis, NT-proBNP showed better prognostic value of variables such as age, functional class, ventricular dilatation or renal dysfunction. Meanwhile, in patients with advanced HF, the COPERNICUS study concluded as NT-proBNP levels above the cutoff point chosen, were a strong predictor of one year mortality (HR: 2.7 CI: 1.7-4.3) [34].

In chronic monitoring of patients with HF it is important to perform serial measurements of natriuretic peptides. Thus, a decrease in levels predicts minor hospitalizations or deaths, while higher levels increase adverse outcomes [35]. It could also help in assessing the response to medical treatment being lower levels of peptides a short term good prognosis. Regarding the monitoring frequency of these biomarkers, a semi-annual or annual monitoring is recommended [36].

In this way, BNP or NT-proBNP "guided" therapy have been deliberated against standard care without BNP or NT-proBNP assessment, to determine whether guided therapy rendered a major and practical achievement of *guideline-directed medical therapy* in patients with HF. However, randomized control trials (RCTs) have brought inconsistent results [37]. Firstly, the positive and negative natriuretic peptide–guided therapy trials are distinct in their study cohorts, with successful trials enrolling younger HF patients. Moreover, a lower level of natriuretics peptides and/or an important level during treatment are consistently presented in the positive "guided" therapy trials [38]. Anyway, although most trials examining the strategy of biomarker "guided" HF management were slight and under-powered, some meta-analyses concluded that BNP-guided therapy reduces all-cause mortality in chronic HF patients compared with usual clinical care [39], especially in patients under 75 years. If the BNP or NT-proBNP value did not fall after aggressive HF therapy, risk for death or hospitalization for HF was significant.

Biomarkers Related to Neurohormonal Activation

Endothelin-1

Endothelin-1 (ET-1) is a potent 21-amino acid vasoconstrictive peptide secreted as a prohormone, and cleaved to proendothelin which is further modified to mature endothelin. It's the most abundant peptide of the endothelin family that also included ET-2 and 3. ET-1 is mainly produced in vascular endothelial cells; its vasoconstrictive action is exerted in autocrine and paracrine way on smooth muscle cells. The half-life of ET-1 in circulation is short (1-2 min), which added to its binding to plasma proteins, difficult an accurately measure. Its precursor, the big-endothelin-1 has a half-life greater but on the other hand is converted to active ET-1 and other peptide of 31 amino acids by the action of ET-1 converting enzyme. Therefore, the best approach to the measurement of this family of vasoconstrictors peptides is the measurement of the carboxy-terminal pre-pro-ET-1 which is more stable and which keeps a concentration correlated to ET-1 secreted [39].

Elevated levels in HF are associated as a secondary release from lungs and myocardial cells in paracrine system and possibly because of a reduced renal clearance, and correlate with both NYHA functional class and LVSD [40]. Increased levels have been associated with poor prognosis in moderate HF [41].

Important evidence of prognostic values of ET-1 is provided in the Val-HeFT study; however, whilst ET-1 was a univariate predictor of outcome, it was outperformed by BNP [33]. Other studies also evaluated the prognostic performances of ET-1 with BNP and NT-proANP in patients with moderate HF being ET-1 the best predictor of cardiac death [42]. But, this is in contrast to the data of other studies that found major superiority of NT-proBNP in patients with advanced HF [43].

In conclusion, the measure of ET-1 is unusual in clinical practice, due to the instability and short half-life of this molecule as previously mentioned, added to the contradictory results found in different HF cohorts.

Adrenomedullin and Midregional Proadrenomedullin

Adrenomedullin (ADM) is a 52-amino acid peptide, originally isolated from cells of human phaeochromocytoma and subsequently detected in heart. The main sites of production of ADM are endothelium and vascular smooth muscle cells [44]. Mechanical stretching and shear stress enhance the ADM production in cardiomyocyte sand vascular smooth muscle cells [45]. ADM perform as an endogenous vasodilator with powerful natriuretic and vasodilatory properties which are increased in hypertension, chronic renal impairment and HF, proportionally to a major disease severity, and correlates with neurohormonal biomarkers including BNP [46]. Elevated plasma levels of ADM have been reported in various diseases, such HF, pulmonary arterial hypertension, myocardial infarction (MI), systemic arterial hypertension, chronic kidney disease, sepsis and pneumonia. The prognostic value of ADM in some of these disorders has been demonstrated [47, 48]. ADM emerged as an independent predictor of 1 year prognosis in a stable systolic HF population [47]. Other studies showed that data above median elevations in ADM confirmed an increased risk of mortality at 18 months independently of age, NYHA class or LVEF [49]. In patients with advanced HF, Gardner *et al.*, demonstrated the superiority of NT-proBNP over ADM [43]. So, we can observe some conflicting results, maybe related to the heterogeneous populations studied that calls for future studies.

Recently, the prognostic value of a new marker, midregional proadrenomedullin (MR-proADM), has been described in risk stratification of patients with HF [50]. MR-proADM represents the amino acids 45-92 of the precursor protein preproadrenomedullin. It is secreted in equimolar amounts to ADM and it is known a major stability in plasma [51]. It has been reported that MR-proADM is superior to natriuretic peptides for short term prognosis in HF and a prognostic value in risk stratification of patients with HF was reported [50]. Moreover, MR-proADM plasma levels are elevated in some patients with acute pulmonary embolism and reflect the severity of pulmonary embolism [52]. This recent study concluded that although the levels of NT-proBNP and MR-proADM have similar predictive value in the assessment of outcomes, MR-proADM shows superiority in predicting all–cause mortality.

Copeptin

As we are discussing, it is recognized that an increased neurohormonal activation is a crucial hallmark of the HF pathophysiology. Indeed, the base of HF therapies relies on the restriction of the renin–angiotensin system and sympathetic nervous system [9, 53]. The vasopressin is a hormone especially involved in the osmoregulation, and seems to be related to HF pathophysiology. But more recently, copeptin was introduced as a representative biomarker for vasopressin and has been examined across diverse cardiovascular diseases including HF [53, 54]. Copeptin is the C-terminal part of pro-arginine vasopressin, a 39-amino acid glycolized glycopeptide with 5kDa of molecular mass. Copeptin and vasopressin are produced in equimolar quantity from the neurohypophysis. The inconvenient of vasopressin relies on an unstable behavior in plasma due to a short half-life and a pulsatile release pattern, that's why the research about copeptin have interested and demonstrated a better stability in plasma as a surrogate biomarker for vasopressin [54].

Different studies have been suggested that copeptin is a biomarker of physiological stress and has been correlated with disease severity in general. Copeptin has also been found to be an excellent biomarker in patients with community-acquired pneumonia, the same as in chronic obstructive pulmonary disease [55, 56]. Moreover, copeptin has been also reported as a strong biomarker of increased HF risk and death after MI [57]. Elevated copeptin levels in MI, reproduces an increased systemic stress and could represent a significant early biomarker to a MI diagnosis, in a early stage after ischemic onset in which cardiac troponins are often still negative [58]. Furthermore, copeptin is associated with post MI ventricular remodeling [59]. So, copeptin seems to carry on interesting prognostic information in most cardiovascular diseases.

Relating copeptin to HF, it is known how predicts the HF development in patients with previous MI and also has been observed how high copeptin levels associated with increased mortality in stable HF patients. This fact, promotes that copeptin could add prognostic information to other established HF predictors such as clinical variables or natriuretic peptides in some studies. Indeed, copeptin has even been found to be a better predictor biomarker compared to BNP and NT-proBNP in HF patients discharged after hospitalization [50, 53, 60, 61]. Nevertheless, copeptin has not demonstrated to be a good predictor of incident HF in patients without CV risk factors [62]. So, in the prediction of HF development, copeptin has been observed as a prognostic marker of HF in patients with previous MI as we commented, and hence there is evidence of the predictive potential of copeptin even in the early stages of HF [62]. Data of the studies appears contradictory, some studies presents how patients stratified according to elevated NT-proBNP and high copeptin levels were particularly predictors of poor outcome and define very high-risk group of patients [63]. In a sub-study from the OPTIMAAL trial, copeptin showed a predictive value stronger than BNP and NT-proBNP [60], so copeptin might also be used to predict HF in elderly patients. In the same way the combination of NT-proBNP elevated levels as well of high copeptin was associated with an increased risk of all-cause mortality after a long median follow-up of 13 years in other investigation [64]. However, not all studies have shown copeptin to be a predictive biomarker of HF development in general or in risk populations. Nevertheless, NT-proBNP was found to be superior to copeptin in predicting HF in a prospective study of elderly men in a follow-up of 11 years with no HF diagnosis at inclusion [65, 66].

Related to the prognostic role of copeptin in chronic HF patients, has repeatedly been observed that is an important prognostic biomarker of both mortality and morbidity [67]. Copeptin was first found to be a strong predictor of outcome in advanced HF [53]. Other data showed how copeptin was the single most potent predictor, and more accurate than BNP and NT-proBNP, in predicting mortality in patients in high NYHA class [61]. Moreover, copeptin could also be an indicator of need for transplantation in patients with chronic HF [68]. In other hand, copeptin was also been found as a prognostic biomarker for 90-day adverse events in patients with AHF, adding additional prognostic value to clinical predictors, plasma sodium levels and natriuretic peptides levels [49, 69].

In addition, it has been reported how copeptin could be useful as a tool in the biomarker-guided therapy. Analysis from the PEACE trial, who presented patients with stable coronary artery disease and preserved LVEF, found how high copeptin and ET-1 levels could predict a better effect of ACE-1 treatment [70]. Likewise, was found how low copeptin was a stronger biomarker of successful β-blocker up-titration than NT-proBNP in reduced LVEF patients [71]. These findings could support the use of copeptin in the monitoring of acute responses to therapy.

In conclusion, copeptin could be an accurate biomarker of HF development in patients with acute coronary syndrome, also in the HF prognosis. But, it is still unclear whether the prognostic information provided by copeptin levels in HF, can be applied into reduced morbidity or mortality of patients or into improve the utilization of healthcare resources. Before this, it seems unclear that this biomarker will replace the value of natriuretic peptides in HF. But, appears an interesting scenario in which raises the question of one single biomarker could fulfill all the pathophysiological roles of HF or if is possible that a combined multimarker strategy could enrich a clinical approach to all aspects of HF patient care [62].

Neprilysin

Natriuretic peptides are representative biomarkers of the neurohormonal activation pathway, but, soluble neprilysin form (sNEP) has recently arises as a potential alternative. Neprilysin (NEP) is a membrane-bound enzyme that cleaves several vasoactive peptides, including natriuretic peptides, adrenomedullin, angiotensin-I and II, bradykinin, and substance P [72]. NEP has been demonstrated as a important bio target in HF in PARADIGM-HF trial [73]. Moreover, the circulating sNEP was recently investigated in HF patients showing a median concentration of 0.642 ng/ml (0.385-1.219 ng/ml), in stable chronic HF patients; and sNEP was associated with short- and long-term outcomes, independently of NT-proBNP [74]. Also, sNEP seems to be an independent predictor of cardiovascular death and HF hospitalizations [74]. Another study from the same group showed median sNEP levels at admission were 0.67 ng/ml (0.37- 1.29 ng/ml) in a pilot multicenter study of AHF patients [75]. By virtue of its central role in neurohormonal regulation, sNEP provides prognostic value on the status of several pathophysiological pathways involved in HF. Circulating NEP is active and retains catalytic activity. Vodovar *et al.*, also showed that NEP activity varies in AHF and chronic HF as a function of circulating BNP levels. The higher the circulating BNP (>916 pg/mL), the lower the activity of circulating NEP and *vice versa*. This data suggests BNP as a substrate of NEP and could act as an endogenous inhibitor of circulating NEP [76].

Nevertheless, the results reported about this recent biomarker showed how sNEP provided independent information to other biomarkers that have been commonly used for HF risk stratification and could arise as an alternative to natriuretic peptides. Nevertheless, appears now a limitation in the assay used for sNEP determination, because is not yet approved for clinical use and requires ad hoc fine-tuning, although presented good intra and inter-assay variation coefficients [77].

In a recent study this group performed a head-to-head comparison of biomarkers within a multimarker strategy that also included natriuretic peptides, neprilysin, soluble suppression of tumorigenicity-2 (sST2) and high sensitivity troponin T (hs-TnT) for HF prognostication. They concluded that both sNEP and NT-proBNP behaved in a similar way in risk stratification in a large cohort with long-term follow-up of real-life HF patients. In other hand, sNEP was unaffected by renal dysfunction and BMI. Finally in a multimarker analysis context, particularly with the incorporation of sST2 and hs-TnT, both of which have been shown consistently to be strong prognosticators, only sNEP retained its prognostic value in ambulatory patients with HF [77].

So, in the way of managing HF patients, applying biomarker-guided therapy plasma levels of sNEP reflect heart condition and could be used to titrate HF medications during hospitalization and later during outpatient visits, but more studies have to confirm this recent data.

Biomarkers Related to Renal Dysfunction

Renal dysfunction is one of the best characterized biomarkers in HF but despite this, remains an underappreciated indicator of prognosis. Renal dysfunction consistently predicts risk in a wide of cardiovascular pathologies including HF [78]. The relationship between cardiac and renal function is a complex entity. Renal dysfunction usually appears related to HF and is associated with high ratios of hospitalization and mortality [79]. Into account, cardiorenal syndrome (CRS) is being used to describe the complex interaction whereby acute or chronic cardiac dysfunction can become to acute kidney injury or chronic kidney disease (CKD) [80]. After that a development of moderate or severe renal dysfunction marks an advanced HF stage. It is important to note that renal dysfunction and CKD could lead to promote cardiac remodeling, increasing the risk of adverse events [80]. Therefore, alterations in renal structure and function become relevant in HF pathophysiology. Commonly, renal function is generally defined by an increase in serum creatinine or reduction in glomerular filtration rate (eGFR), which reflects a late decline in renal function and precludes early identification. But, different biomarkers have been studied in this area, and are discussed in other chapter of this book, we will focus on C-type natriuretic peptide and its role in HF.

C-Type Natriuretic Peptide

Due to the consistent role of cardiac natriuretic peptides in HF diagnosis and the confirmed sensitivity of NT-proBNP for cardiac stress, injury and remodeling, there has been increasing interest in the investigation of renal-derived C-type natriuretic peptide (CNP) as a urinary biomarker of renal dysfunction and chronic renal remodeling in HF and CRS [81].

Highest levels of CNP expression and production occurs in the kidney [82], but CNP expression has also been found in cardiomyocytes, vascular endothelium, and bone, and it is known how the cardiac expression of CNP is typically low compared to ANP or BNP [83]. CNP lacks significant diuretic or natriuretic effects at physiological levels [84]. It is also known that CNP presents significant anti-proliferative and anti-fibrotic characteristics as suppression of the proliferation of fibroblasts and collagen production [85], inhibition of vascular smooth muscle cell proliferation, and could speed up the regeneration of endothelial cells [83, 86].

Talking about its relationship to HF, recently have been observed how high CNP levels were determined in urine from AHF patients *versus* healthy controls, and also showed poor correlation with plasma CNP levels [81, 86]. These founding expose an activation of renal natriuretic peptide system in HF, in which elevated urinary excretion of CNP is maintained to reflect increased renal interstitial pressure, renal tubular injury, hypoxia, and potentially renal fibrosis [81]. Moreover, CNP has been localized in renal tubules, representing another biomarker that reflects renal tubular integrity. CNP and its positive association with AHF prognosis, suggests a major sensitivity of another type of renal dysfunction in HF, thus better characterizing the evolution of CRS [81].

It is interesting the clear evidence for a reno-protective effect of CNP and deserves further investigation as a potential therapeutic mechanism. Additional investigation is also needed to clarify the relative biological significance and relationship between urinary CNP in HF with a reduced and preserved ejection fraction in CRS [83].

Biomarkers Related to Inflammation

The pathophysiology of HF is characterized by an immune activation with pro-inflammatory cytokines over-expressed in systemic circulation and locally myocardium muscle. Some efforts have expanded the application of inflammation markers to predict the HF new onset and the risk to stratify patients with established HF.

Cytokines are pleiotropic proteins that regulate leukocyte activity [87]. Several cytokines have been involved in the HF pathophysiology with major evidence in ***interleukin (IL-6)***, ***tumor necrosis factor (TNF)*** and its soluble receptor and the adhesion molecules and cell-surface ligands which mediate the inflammatory cascade. A promising role for the acute phase protein, ***C-reactive protein (CRP)***, showed up as a clinically relevant biomarker. It is caused by cytokines such as IL-1 and IL-6 drive production of reactant proteins, including CRP, in the acute-phase response [87, 88].

Taken from the Framingham study, participants with high inflammatory biomarkers, including TNF, IL-6 and CRP, presented a substantial future risk of HF [89]. Both elevated levels, TNF and IL-6, have been associated with worsening symptoms, and correlated with NYHA functional class. Moreover, both can be expressed in the myocardium under different classes of stress and could modulate cardiac function by a wide type of mechanisms including the induction of LV hypertrophy, cardiomyopathy, and apoptosis in cardiac myocytes [87].

CRP is an acute phase protein synthesized in the liver in response to stimulation by pro-inflammatory cytokines including IL-6 [90]. CRP stimulates the classical complement pathway, and is implicated as a mediator and a biomarker of the inflammatory process [90]. The role of CRP in HF is not well studied, the same as in atherosclerosis disease. In patients with acute MI, CRP behaves as a marker of the inflammatory intensity response in the damaged myocardium, and the future HF development [91]. One of the strongest evidence of the behavior of CRP levels in HF came from the Val-HeFT study. This investigation evidenced how plasma CRP levels determined at baseline, and showing CRP levels above the median, were associated with a quartile-dependent increase in adverse events, even after adjustment for other prognosis predictors, including BNP levels [92]. CRP has also been studied in pathologies derived from HF, as HCM. For example, Dimitrow *et al.*, showed that C-reactive protein was higher in HCM patients than in a control cohort [93] and Lamparter *et al.*, studied high sensitivity CRP (hs-CRP) values in cardiomyopathies, exactly in idiopathic dilated cardiomyopathy in a large cohort [94]. But, this study presented significant limitations because serum hs-CRP levels are non specific, reflecting a broad range of pathologic processes, including local or systemic inflammation, any kind of trauma, tissue injury or infection. Therefore it is demonstrated that hs-CRP levels cannot be interpreted without a complete medical history and physical examination [94]. It also known, that hs-CRP values could vary from day to day, so it would be convenient a kinetic study.

Growth Factors:

<u>Growth Differentiation Factor-15 (GDF-15)</u>

GDF-15 is a member of the transforming growth factor β (TGF-β) superfamily. This is a large family of related proteins that can be subdivided into two principal groups: the TGF-β/Activin and bone morphogenetic protein and the GDF branches, based on their sequence analogy [95]. The clinical literature have shown that cardiovascular diseases such as ischemia, inflammation, or injury [96] notoriously up-regulate GDF-15 production in the heart. These pathologies could express that this factor acts as a stress sign for the cardiomyocytes. In this way many clinical and experimental reports have evidenced a link between GDF-15 and vascular disorders [97]. Additionally, increased levels of GDF-15 were associated with poor prognosis in ACS [97, 98] and more recently in HF [99], remarking a vinculum with severity of the disease and even more, suggesting GDF-15 as a new prognostic biomarker in HF [100]. These observations also are suggested to identify high-risk patients across a broad spectrum of cardiovascular diseases and that it might display a regulatory role in the hypertrophy development [101].

GDF-15 levels were reported as lower in stable patients (NYHA class I) presenting a stronger power as a biomarker of changing status, since the cytokine was only highly increased in worse functional class [102]. By this way, GDF-15 could be useful to conclude in myocardium depth damage. Indeed, this study also observed how GDF-15 levels in NYHA class II were slightly raised, but flagrantly increased in class III compared to functional class I patients. In the same line, another interesting report found that GDF-15 levels presages non-cardiovascular death in older patients alone or combined with NT-proBNP levels [103].

Regarding to the prognostic role of GDF-15, the Val-HeFT study showed both basal levels and those of 1-year of follow-up, independently associated with risk of death, in a multivariate analysis with clinical variables and other biomarkers [99]. Other studies also confirmed the interesting role of GDF-15 in HF with preserved systolic function and data compared to NT-proBNP, ST2 and hs-TnT. GDF-15 was the biomarker with greater capacity for discrimination between control subjects and patients HF with preserved systolic function independently [104].

ST2

The biology of suppression of tumorigenicity 2 (ST2) suggests a pluripotent role, with important participation in both immunologic processes and fibrotic response to injury. ST2 is a member of the interleukin 1 receptor family, also known as interleukin 1 receptor-like 1 (IL1RL-1) and presents two main isoforms: transmembrane or cellular (ST2L) and soluble or circulating (sST2) forms [105]. ST2 is the receptor for interleukin-33 (IL-33), which is an IL-1-like cytokine secreted by living cells in response to cell damage. IL-33 exerts its effects by binding to the transmembrane receptor ST2L isoform. The interaction of IL-33 and ST2L has been proved to be cardioprotective, action that exclusively appears through the ST2L. The IL-33/sST2 system is upregulated in cardiomyocytes and fibroblasts in response to cardiac injury. sST2 avidly binds to IL-33 competing with ST2L. The interaction of this soluble receptor with IL-33 blocks the IL-33/ST2L system and, as a result, eliminates the cardioprotective effects, that's why sST2 is considered a decoy receptor in myocardium binding and blocking the antihypertrophic effects of IL-33 [106, 107]. ST2 is also associated with inflammatory and immune processes, especially those related to the regulation of mast cells and type 2 CD4 T-helper cells, and the production of Th2-associated cytokines [108].

It has been proved the prognostic sST2 role in decompensated AHF, being a useful in the risk stratification HF patients. sST2 was firstly studied in PRIDE cohort in emergency department [109] showing levels of sST2 significantly higher in patients with decompensated AHF than non-HF patients, being a powerful predictor of mortality. Patients who died at 1 year had higher values than survivors and there was a clear association between sST2 levels and mortality rates of risk. The multivariate analysis showed how sST2 remains being a strong predictor of 1-year mortality in patients with and without HF. Furthermore, the prognostic utility of sST2 added to NT-proBNP marked that patients with elevation of both markers had the highest 1-year mortality rate (almost 40%) [109]. Other studies corroborated similar results, relating how high sST2 levels at admission correlated with NYHA functional class [110, 111] and high sST2 levels reclassified risk of death in patients with low natriuretic peptides levels [112]. These studies used the more sensitive Presage ST2 assay (Critical Diagnostics, San Diego, CA, USA) who consider that a cutoff ≥ 35 ng/mL was associated with worse prognosis in HF patients [113]. The cutoff proposed in PRIDE study for AHF patients was 42.7 ng/mL [109]. Moreover, compared to other biomarkers, sST2 emerged as the strongest one with the ability to reclassify death risk in AHF, beyond a clinical mode being the best predictor of both 30-day and 1-year mortality [114]. Although baseline sST2 values at admission have been proved to predict outcomes, serial measurements levels, could be a greater prognostic biomarker. Boisot *et al.*, [115] measured sST2 on a daily basis in patients admitted with AHF and demonstrated that this biomarker quickly changes in response to treatment. Recently, similar results were obtained by Manzano-Fernandez *et al.*, also founding that patients with persistent elevation on day 4 of sST2 had a higher risk of death [116]. In AHF, was recently examined whether sST2 levels improves HF risk stratification relative to other representative biomarkers of multiple pathogenic pathways and they proved that sST2 provides valuable long-term risk stratification information in HF beyond that reported by other biomarkers of stretch, inflammation, necrosis, and remodeling [117]. So, they considered sST2 as a 3-in-1 prognosis biomarker in HF. Conceptually, therefore, ST2 sits at an intersection of strain, inflammation, cell death, and fibrosis with remodeling and reports the magnitude of the divergence of survival curves according to ST2 concentration above and below the median for every tertile of the other biomarkers is quite remarkable [117].

Furthermore, sST2 has been proven to be useful as a prognostic marker in chronic HF. Some initials contradictory studies did not all tested independent association of a single baseline measure of sST2 and adverse outcomes [118, 119]. But data from a recent study confirmed that sST2 and NT-proBNP significantly predicted death beyond conventional risk factors, and was also seen that sST2 was not influenced by renal function, as observed with NT-proBNP [120]. A posterior head-to-head comparison by the same group revealed that sST2 was superior to galectin-3 in risk stratification associated with cardiovascular mortality [121]. sST2 has also been shown to be a real sudden death predictor in patients with mild to moderate systolic HF [122], showing elevations of sST2 and NT-proBNP above the cutoff value associated with a high rate of sudden death (71%). All these data was confirmed in multicenter randomized studies confirming sST2 as the strongest predictor of cardiovascular death in HF [123, 124]. Indeed the CORONA study after initial adjustments for conventional variables, showed how baseline sST2 was a significant predictor of all endpoints, including the primary endpoint, death, worsening HF, and hospitalization for HF [123]. Moreover, the PROTECT study showed a multimarker approximation, concluding that serial measures of sST2, but neither hs-TnT nor GDF-15, changed significantly over a median of 10 months of follow-up as compared with baseline. So, baseline sST2 < 35 ng/mL was associated with longer time to first cardiovascular event [125]. They also studied the effects of medication on sST2 serial measurements proposing targeting with sST2. Those with elevated baseline sST2 levels who achieved higher beta blocker doses had significantly lower risk of events [126]. Finally sST2 baseline levels added significant information regarding first morbid event, death, but no HF hospitalization in the previously mentioned VAL-HeFT study [127].

Biomarkers Related to Myocardial Damage

Cardiac Troponins

It is know how serum levels of cardiac troponin T or I, became a specific and highly sensitive biomarker of myocardial injury and presented diagnostic and prognostic values, perfectly established in acute coronary syndromes [128]. On the other hand, cardiac troponins have been studied in many kind of patients, such as in idiopathic dilated cardiomyopathy with particularly poor prognosis, in which increased serum levels of troponin T (TnT) in absence of significant coronary stenosis, have been observed [129]. In outpatients with established chronic HF of a non-ischaemic feature, the revelation of TnT, even at low levels, could means an increased risk of adverse cardiac events, which seems to be independent of other clinical, analytical and ECG factors [130]. These higher levels of TnT could suggest ongoing myocardial damage or leaking of myofibrillar components, reflecting a loss of viable cardiac myocyte, that is a clear characteristic of progressive HF [131]. All this knowledge based on classical TnT determination procedures, reveal strong discrepancies on the positive TnT patients account. It is due to a probably heterogeneous detection limit of the analytical technique performed and the selected cutoff point. Moreover, the hemodynamic state (acute or stable patients) in the determination of TnT levels also adds difficulty to the evaluation of this biomarker. But nowadays, the new developed analytical techniques based on the relevance of the myocardial damage in cardiologic patients, such as high sensitive (hs-TnT) determination, mostly have shown an improvement in the diagnosis of ACS, due to its increased sensitivity for troponin T detection and quantification. This new scenario, also could help in the diagnosis of HF, but it has to be taken into account, to assess the renal function and the levels of hs-TnT in the same way with natriuretic peptides.

In those patients that present acute decompensated HF, a positive cardiac troponin test is related with mortality, independently of other predictive variables [132]. Omland *et al.*, presented a confirmation of hs-TnT elevations in stable coronary artery disease patients and its association with the incidence of cardiovascular death and HF [128]. Moreover, an increase in hs-TnT levels could be used as a biomarker of myocardial damage and remodeling, and this has been proposed prognostic marker in HCM [133, 134] closely related to advanced HF. This could reflect subclinical ongoing myocyte damage in HCM and, consequently, could take influence in functional status [133]. The prognostic value of hs-TnT is closely related to a worse NYHA class, fact this subclinical continuous damage [135]. In prognosis, addition of BNP/NT-proBNP and hs-TnT significantly improved prognostic information compared with clinical risk factors and when they are combined had the greatest discriminatory accuracy [136, 137]. This suggests that hs-TnT increment the prognostic information when is combined with BNP/NT-proBNP as part of a multimarker strategy, as was proved in the Val-HeFT analysis [135]. Finally, the addition of hs-TnT to others biomarkers may provide incremental and superior prognostic information than either marker alone, and would prove useful as part of a multi-marker approach for both acute and chronic HF, as we are being discussing related to others biomarkers cited previously.

Biomarkers Related to Cardiac Remodeling and Fibrosis Biomarkers

Collagen Peptides

Added to a myocyte injury, coronary microvascular dysfunction, the collagen turnover also plays a significant role in cardiac remodeling. A continuous extracellular matrix remodeling occurs in HF, especially in HCM, go through increased interstitial fibrosis, due to a deposition of raised amounts of collagen type I/III [138]. The collagen network is a metabolically vital structure in the fact of an important balance between the synthesis and degradation of collagen that determines its turnover, which is estimated to be from 80 to 120 days [139]. Primary, procollagen types I and III are synthesized and secreted by fibroblasts and myofibroblasts as a form of triple-helix and this procollagen precursor contained terminal propeptides. These propeptides are divided in block by specific procollagen proteinases, so, the new molecules get to the bloodstream and can be detected in blood by analytical techniques. That's why, if the propeptides are cleaved in every molecule of collagen, the amount of theirs quantified in serum could be proportional to the quantity of collagen formed, and could become as indexes of collagen synthesis. This propeptides are the carboxyterminal propeptide of procollagen type I (PICP) and likely for the amino-terminal propeptide of procollagen type I (PINP) [139].

The investigation for clear biomarkers of collagen metabolism has provided a large number of candidates molecules and can be categorized into 2 main groups: molecules related to the synthesis form new collagen fibers and molecules related to degradation of mature collagen fibers [139]. In this role, is important to note the importance of Collagen type III that is derived from a larger protein, procollagen type III, with two end extensions on both sides of the molecule. Many of the aminoterminal propeptide (PIIINP: N-terminal propeptide of procollagen type III) are produced during the collagen type III synthesis and deposition, another molecules are also detained on the forming part of the collagen fibers. Into these metabolism is very important the role of interstitial matrix metalloproteinases (MMP) who initiate the digestion of collagen by hydrolysis of the peptide bond after a glycine residue. So, the procollagen carboxyterminal telopeptide type I (CITP) is droped by the resultant action of matrix metalloproteinase-1 (MMP-1). Here, appears a relationship between the numbers of molecules of type I collagen degraded and the CITP molecules released. Thus, the quantity of CITP that could be determined in serum would be proportional to the quantity of degraded fibrillar collagen. So, this peptide could be associated as a biomarker of collagen lysis [139].

Lin *et al.* [140], explored the interstitial remodeling explained before, measuring the serum levels of PIIINP in HF patients. In this investigation, PIIINP was proposed as a biomarker of cardiac autonomic control and a risk factor of SCD. Interestingly, a recent report by Ho *et al.* [141], also examined different biomarkers of fibrosis and interstitial remodeling promoting that serum PICP levels could value increased myocardial collagen synthesis in sarcomere-mutation carriers without overt disease. They proposed that this profibrotic state precede the future development of left ventricular hypertrophy or fibrosis [141]. Fibrosis is a feature that provide electrical heterogeneity and is a substrate for arrhythmogenicity, which could cause SCD, that is one of the dangerous first symptom that could appear in the onset of pathologies as HF or HCM [140, 141].

But these findings appear contradictory, other studies could not corroborate only finding a higher ratio PICP/CITP in patients compared to controls [142].

Due to the conflicting data mentioned and the difficulty of matching cohorts where fibrosis could be similarly evaluated by collagen peptides, obtaining fully substantiated information, has made that these lines of investigation remain braked. This is mainly due to the appearance of other related cardiac fibrosis biomarkers as ST-2, previously discussed and especially the appearance of Galectin-3.

Galectin-3

Galectin-3 (GAL-3) is a member of the family of soluble β-galactoside-binding lectins; a protein secreted by activated macrophages who plays a crucial role in several diverse biological processes and diseases [143]. Recently, has been interestingly reported the role of GAL-3 in the pathophysiology of HF promoting cardiac fibroblast proliferation, collagen deposition, and ventricular dysfunction [143]. This leads to progressive inflammation and cardiac fibrosis [144]. GAL-3 is released in the myocardium, *via* a paracrine effect, stimulating proliferation of myofibroblasts and procollagen-1 deposition. The activation of fibroblasts and myofibroblasts and the deposition of procollagen into the extracellular matrix is what ultimately lead to cardiac fibrosis [144 - 146].

GAL-3 has been recently proposed as a new independent biomarker for diagnosis of AHF [145]. Peacock *et al.*, presented an algorithm in which GAL-3 may be used in this setting [147] (Fig. **2**). Moreover, several AHF studies have performed the value of GAL-3 levels in risk assessment. By the way, the PRIDE study demonstrated that GAL-3 was greater prognostic factor for 60-day mortality compared to NT-proBNP levels [148] and GAL-3 also correlated with parameters of diastolic function on echocardiogram. This finding was ratified in the COACH study that enrolled AHF patients at discharge [149]. GAL-3 rise as an independent predictor of unfavorable outcomes as mortality and HF hospitalizations and presented a clear promising value in HFpEF, because currently, the diagnosis and therapy of HFpEF are difficult and additional tools are demanded [150]. Furthermore, the predictive value of GAL-3 is clearly useful for short term outcomes and so, higher GAL-3 levels >17.8 ng/mL were associated with near-term re-hospitalization [151]. Other studies have confirmed the GAL-3 value of repeated measurements in acute setting (PROTECT) [152].

Regarding to chronic HF, Lok *et al.*, were the first to report the prognostic value of GAL-3 to this pathology and GAL-3 remained an independent prognostic biomarker for long-term all-cause mortality [153], and also evidenced an independent relationship between GAL-3 and left ventricular remodeling determined by echocardiography tools. Interestingly, in the CORONA trial after a median follow-up of 33 months, GAL-3 predicted the combined endpoint of cardiovascular death, nonfatal myocardial infarction and stroke, but after adjustment for NT-proBNP, the statistically significance was lost [154]. Contrarily, in the Val-HeFT trial, GAL-3 remained significantly associated with mortality and HF hospitalization, also after addition of eGFR and NT-proBNP to the regression model [155]. Interestingly, a recent study recommended by the American College of Cardiology/American Heart Association HF guidelines for their prognostic abilities in chronic HF assessed sST2 and GAL-3. This study confirmed both elevated biomarkers were associated with an increased risk for all-cause mortality but ST2 only for cardiovascular mortality. Meanwhile, GAL-3 might confer stronger prognostic information in early disease as compared to progressed disease [121]. Also, GAL-3 was recently recommended in the same guidelines for target for therapy in HF [9].

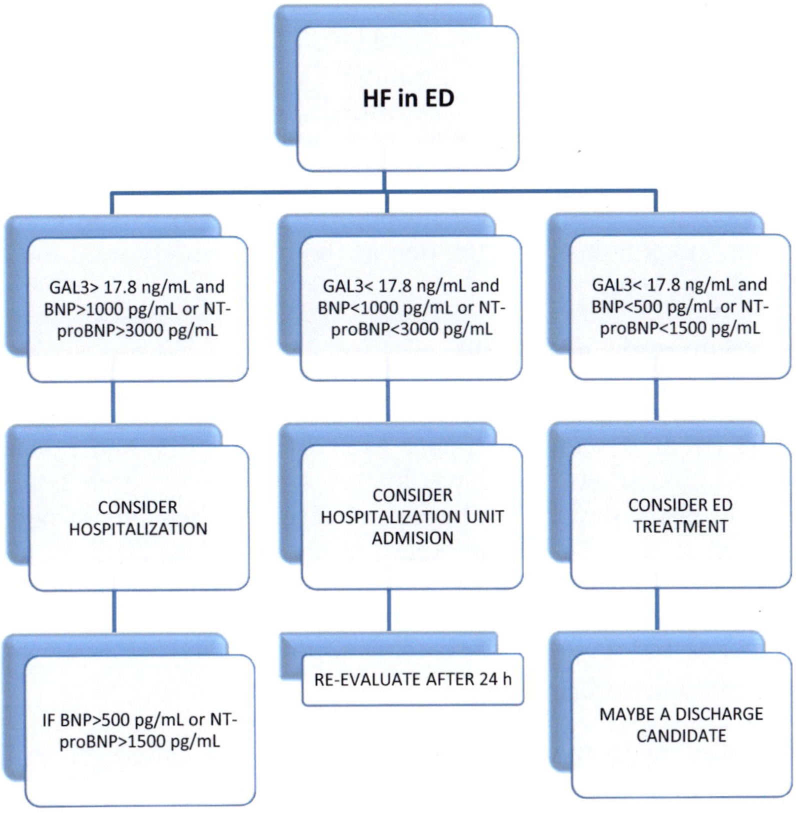

Fig. (2). Proposed algorithm for the emergency department, hypothetical flow-chart for clinical decision making, considering GAL-3 and B-type natriuretic peptides levels. Adapted from Peacok *et al.* (2014) [147].

Biomarkers Related to Oxidative Stress

HF is enrolled in the setting of a cascade of processes that conclude in an imbalance between bio-available nitric oxide and harmful reactive oxygen species (ROS) (so-called nitroso-redox imbalance). The pro-oxidant state mentioned contributes to myocyte apoptosis and necrosis, endothelial dysfunction and remodeling [88]. Some biomarkers of ROS burden are modified in HF, such as: urinary byopyrrins (derived from the oxidative bilirubin metabolism); urinary isoprostanes, 8-epi prostaglandin-F-a and plasma malondialdehyde (biomarkers of lipid peroxidation), *etc.* [156, 157]. But is important to note that urinary byopyrrins highlighted to be in higher proportion in the severity of LVSD and NYHA functional class (and correlated with B-type natriuretic peptides) [157].

Urate

Uric acid, as the final product of purine metabolism, has been reported as a potential biomarker of oxidative stress and HF. The high production appears as a consequence of increased activity of xanthine oxidase, a critical component of nitroso-redox balance. Urate was confirmed as an independent predictor of outcome moderate to severe HF in different studies [158]. To highlight one study, pre-discharge elevated urate was associated with a higher rate of death and combined endpoint of death or new hospitalization [159]. Notably, the urate is included in the index risk of Seattle, one of the most widely used to stratify the risk of HF patients [160].

Future Directions and Conclusions

Despite wide range of biomarkers revised in this review about HF, only natriuretic peptides have become standard-of-care biomarkers in HF and nowadays still being the only biomarkers that have intercrossed the research boundary and become used in every day medical practice. Recently, natriuretic peptides also were recognized as class IA biomarkers of prognosis in the 2013 ACC/AHA HF guidelines [9]. But these same guidelines are the one to have made, for the first time, a recommendation for fibrosis biomarkers, such as sST2 and GAL-3, in both acute and chronic HF. On top of that, the additive prognostic value to that of natriuretic peptides is emphasized [9].

Moreover, according to the previously discussed line called HF multimarker strategy. This is born from the complex pathophysiology of this syndrome and multiple pathophysiological pathways involved, and maybe can afford an accurate diagnostic and prognosis of HF patients. For that reason, a successful election of the biomarkers to be used and their own combined information is crucial. Highlighting, the way to understand and prove concise prognostic information of the combined biomarkers compared to a solo use.

Regarding to future directions in the setting of HF biomarkers studies, have to be noting the probably use of circulating microRNA (miRNA) who are small non-coding RNAs considered as regulatory molecules in many biological and pathological processes, including those related to cardiovascular disease and HF.

CONFLICT OF INTEREST

Dr. J. Ordoñez-Llanos has accepted honoraria for conferences/consultation from Abbott Laboratories, BioMerieux, BioRad, Roche Diagnostics, Stat Diagnostics and Thermo Fisher and support for research from Abbott Laboratories, Alere Diagnostics, Astute Diagnostics, Critical Diagnostics, Roche Diagnostics, Siemens Healthcare Diagnostics and Thermo Fisher. Dr Marín has received funding for research, consultancy and lecturing from Abbott, Boston Scientific, Bayer, Astra Zeneca, Daiichi-Sankyo, BMS/Pfizer and Boehringer Ingelheim. The rest of authors declare none conflict of interest.

ACKNOWLEDGEMENTS

Declared none.

REFERENCES

[1] Jessup M, Brozena S. Heart failure. N Engl J Med 2003; 348: 2007-18.

[2] McMurray JJ, Stewart S. Epidemiology, aetiology, and prognosis of heart failure. Heart 2000; 83: 596-602.

[3] Neumann T, Biermann J, Erbel R, *et al.* Heart failure: the commonest reason for hospital admission in Germany: medical and economic perspectives. Dtsch Arztebl Int 2009; 106: 269-75.

[4] McCullough PA. Practical experience using galectin-3 in heart failure. Clin Chem Lab Med 2014; 52: 1425-31.

[5] Vigen R, Maddox TM, Allen LA. Aging of the United States population: impact on heart failure. Curr Heart Fail Rep 2012; 9: 369-74.

[6] Sharma K, Kass DA. Heart failure with preserved ejection fraction: mechanisms, clinical features, and therapies. Circ Res 2014; 115: 79-96.

[7] McMurray JJ, Adamopoulos S, Anker SD, *et al.* ESC guidelines for the diagnosis and treatment of acute and chronic heart failure 2012: The Task Force for the Diagnosis and Treatment of Acute and Chronic Heart Failure 2012 of the European Society of Cardiology. Developed in collaboration with the Heart. Eur J Heart Fail 2012; 14: 803-69.

[8] Ho JE, Liu C, Lyass A, *et al.* Galectin-3, a marker of cardiac fibrosis, predicts incident heart failure in the community. J Am Coll Cardiol 2012; 60: 1249-56.

[9] Yancy CW, Jessup M, Bozkurt B, *et al.* 2013 ACCF/AHA guideline for the management of heart failure: a report of the American College of Cardiology Foundation/American Heart Association Task Force on Practice Guidelines. J Am Coll Cardiol 2013; 62: e147-239.

[10] Vasquez C, Benamer N, Morley GE. The cardiac fibroblast: functional and electrophysiological considerations in healthy and diseased hearts. J Cardiovasc Pharmacol 2011; 57: 380-8.

[11] Soman P, Lahiri A, Mieres JH, *et al.* Etiology and pathophysiology of new-onset heart failure: evaluation by myocardial perfusion imaging. J Nucl Cardiol Jan; 16: 82-91.

[12] Michels VV, Moll PP, Miller FA, *et al.* The frequency of familial dilated cardiomyopathy in a series of patients with idiopathic dilated cardiomyopathy. N Engl J Med 1992; 326: 77-82.

[13] Hutchinson KR, Stewart JA, Lucchesi PA. Extracellular matrix remodeling during the progression of volume overload-induced heart failure. J Mol Cell Cardiol 2010; 48: 564-9.

[14] Doust JA, Pietrzak E, Dobson A, Glasziou P. How well does B-type natriuretic peptide predict death and cardiac events in patients with heart failure: systematic review. BMJ 2005; 330: 625.

[15] Hughes MF, Saarela O, Blankenberg S, *et al.* A multiple biomarker risk score for guiding clinical decisions using a decision curve approach. Eur J Prev Cardiol 2012; 19: 874-84.

[16] Sturm HB, Haaijer-Ruskamp FM, Veeger NJ, Baljé-Volkers CP, Swedberg K, van Gilst WH. The relevance of comorbidities for heart failure treatment in primary care: A European survey. Eur J Heart Fail 2006; 8: 31-7.

[17] Arteaga E, Araujo AQ, Buck P, Ianni BM, Rabello R, Mady C. Plasma amino-terminal pro-B-type natriuretic peptide quantification in hypertrophic cardiomyopathy. Am Heart J 2005; 150: 1228-32.

[18] Køber L, Torp-Pedersen C, McMurray JJ, *et al.* Increased mortality after dronedarone therapy for severe heart failure. N Engl J Med 2008; 358: 2678-87.

[19] Abraham WT, Adamson PB, Bourge RC, *et al.* Wireless pulmonary artery haemodynamic monitoring in chronic heart failure: a randomised controlled trial. Lancet (London, England) 2011; 377: 658-66.

[20] Guazzi M, Samaja M, Arena R, Vicenzi M, Guazzi MD. Long-term use of sildenafil in the therapeutic management of heart failure. J Am Coll Cardiol 2007; 50: 2136-44.

[21] Abraham WT, Adams KF, Fonarow GC, *et al.* In-hospital mortality in patients with acute decompensated heart failure requiring intravenous vasoactive medications: an analysis from the Acute Decompensated Heart Failure National Registry (ADHERE). J Am Coll Cardiol 2005; 46: 57-64.

[22] Masip J, Roque M, Sánchez B, Fernández R, Subirana M, Expósito JA. Noninvasive ventilation in acute cardiogenic pulmonary edema: systematic review and meta-analysis. JAMA 2005; 294: 3124-30.

[23] Magga J, Sipola P, Vuolteenaho O, *et al.* Significance of plasma levels of N-terminal Pro-B-type natriuretic peptide on left ventricular remodeling in non-obstructive hypertrophic cardiomyopathy attributable to the Asp175Asn mutation in the alpha-tropomyosin gene. Am J Cardiol 2008; 101: 1185-90.

[24] Thaman R, Esteban MT, Barnes S, *et al.* Usefulness of N-terminal pro-B-type natriuretic peptide levels to predict exercise capacity in hypertrophic cardiomyopathy. Am J Cardiol 2006; 98: 515-9.

[25] Elliott P, McKenna WJ. Hypertrophic cardiomyopathy. Lancet (London, England) 2004; 363: 1881-91.

[26] Wu AH, Smith A, Wieczorek S, *et al.* Biological variation for N-terminal pro- and B-type natriuretic peptides and implications for therapeutic monitoring of patients with congestive heart failure. Am J Cardiol 2003; 92: 628-31.

[27] Coats CJ, Gallagher MJ, Foley M, *et al.* Relation between serum N-terminal pro-brain natriuretic peptide and prognosis in patients with hypertrophic cardiomyopathy. Eur Heart J 2013; 34: 2529-37.

[28] Januzzi JL, van Kimmenade R, Lainchbury J, *et al.* NT-proBNP testing for diagnosis and short-term prognosis in acute destabilized heart failure: an international pooled analysis of 1256 patients: the International Collaborative of NT-proBNP Study. Eur Heart J 2006; 27: 330-7.

[29] Maisel A, Xue Y, Greene SJ, *et al.* The potential role of natriuretic peptide-guided management for patients hospitalized for heart failure. J Card Fail 2015; 21: 233-9.

[30] Pascual-Figal DA, Manzano-Fernández S, Boronat M, *et al.* Soluble ST2, high-sensitivity troponin T-

and N-terminal pro-B-type natriuretic peptide: complementary role for risk stratification in acutely decompensated heart failure. Eur J Heart Fail 2011; 13: 718-25.

[31] Hildebrandt P, Collinson PO, Doughty RN, *et al.* Age-dependent values of N-terminal pro-B-type natriuretic peptide are superior to a single cut-point for ruling out suspected systolic dysfunction in primary care. Eur Heart J 2010; 31: 1881-9.

[32] Verdú JM, Comín-Colet J, Domingo M, *et al.* Rapid point-of-care NT-proBNP optimal cut-off point for heart failure diagnosis in primary care. Rev Esp Cardiol (Engl Ed) 2012; 65: 613-9.

[33] Masson S, Latini R, Anand IS, *et al.* Direct comparison of B-type natriuretic peptide (BNP) and amino-terminal proBNP in a large population of patients with chronic and symptomatic heart failure: the Valsartan Heart Failure (Val-HeFT) data. Clin Chem 2006; 52: 1528-38.

[34] Hartmann F, Packer M, Coats AJ, *et al.* Prognostic impact of plasma N-terminal pro-brain natriuretic peptide in severe chronic congestive heart failure: a substudy of the Carvedilol Prospective Randomized Cumulative Survival (COPERNICUS) trial. Circulation 2004; 110: 1780-6.

[35] Bettencourt P, Friões F, Azevedo A, *et al.* Prognostic information provided by serial measurements of brain natriuretic peptide in heart failure. Int J Cardiol 2004; 93: 45-8.

[36] Latini R, Masson S, Wong M, *et al.* Incremental prognostic value of changes in B-type natriuretic peptide in heart failure. Am J Med 2006; 119: 70.e23-30.

[37] Balion C, McKelvie R, Don-Wauchope AC, *et al.* B-type natriuretic peptide-guided therapy: a systematic review. Heart Fail Rev 2014; 19: 553-64.

[38] Januzzi JL. Use of biomarkers to "guide" care in chronic heart failure: what have we learned (so far)? J Card Fail 2011; 17: 622-5.

[39] Porapakkham P, Porapakkham P, Zimmet H, Billah B, Krum H. B-type natriuretic peptide-guided heart failure therapy: A meta-analysis. Arch Intern Med 2010; 170: 507-14.

[40] Krum H, Goldsmith R, Wilshire-Clement M, Miller M, Packer M. Role of endothelin in the exercise intolerance of chronic heart failure. Am J Cardiol 1995; 75: 1282-3.

[41] Pousset F, Isnard R, Lechat P, *et al.* Prognostic value of plasma endothelin-1 in patients with chronic heart failure. Eur Heart J 1997; 18: 254-8. [Internet].

[42] Selvais PL, Robert A, Ahn S, *et al.* Direct comparison between endothelin-1, N-terminal proatrial natriuretic factor, and brain natriuretic peptide as prognostic markers of survival in congestive heart failure. J Card Fail 2000; 6: 201-7.

[43] Gardner RS, Chong V, Morton I, McDonagh TA. N-terminal brain natriuretic peptide is a more powerful predictor of mortality than endothelin-1, adrenomedullin and tumour necrosis factor-alpha in patients referred for consideration of cardiac transplantation. Eur J Heart Fail 2005; 7: 253-60.

[44] Nishimura J, Seguchi H, Sakihara C, *et al.* The relaxant effect of adrenomedullin on particular smooth muscles despite a general expression of its mRNA in smooth muscle, endothelial and epithelial cells. Br J Pharmacol 1997; 120: 193-200.

[45] Tsuruda T, Kato J, Kitamura K, *et al.* Enhanced adrenomedullin production by mechanical stretching in cultured rat cardiomyocytes. Hypertension 2000; 35: 1210-4. [Internet].

[46] Bunton DC, Petrie MC, Hillier C, Johnston F, McMurray JJ. The clinical relevance of adrenomedullin: a promising profile? Pharmacol Ther 2004; 103: 179-201.

[47] Pousset F, Masson F, Chavirovskaia O, *et al.* Plasma adrenomedullin, a new independent predictor of prognosis in patients with chronic heart failure. Eur Heart J 2000; 21: 1009-14. [Internet].

[48] Khan SQ, O'Brien RJ, Struck J, *et al.* Prognostic value of midregional pro-adrenomedullin in patients with acute myocardial infarction: the LAMP (Leicester Acute Myocardial Infarction Peptide) study. J Am Coll Cardiol 2007; 49: 1525-32.

[49] Richards AM, Doughty R, Nicholls MG, *et al.* Plasma N-terminal pro-brain natriuretic peptide and

adrenomedullin: prognostic utility and prediction of benefit from carvedilol in chronic ischemic left ventricular dysfunction. Australia-New Zealand Heart Failure Group. J Am Coll Cardiol 2001; 37: 1781-7.

[50] Maisel A, Mueller C, Nowak RM, *et al.* Midregion prohormone adrenomedullin and prognosis in patients presenting with acute dyspnea: results from the BACH (Biomarkers in Acute Heart Failure) trial. J Am Coll Cardiol 2011; 58: 1057-67.

[51] Morgenthaler NG, Struck J, Alonso C, Bergmann A. Measurement of midregional proadrenomedullin in plasma with an immunoluminometric assay. Clin Chem 2005; 51: 1823-9.

[52] Pedowska-Włoszek J, Kostrubiec M, Kurnicka K, Ciurzynski M, Palczewski P, Pruszczyk P. Midregional proadrenomedullin (MR-proADM) in the risk stratification of patients with acute pulmonary embolism. Thromb Res 2013; 132: 506-10.

[53] Stoiser B, Mörtl D, Hülsmann M, *et al.* Copeptin, a fragment of the vasopressin precursor, as a novel predictor of outcome in heart failure. Eur J Clin Invest 2006; 36: 771-8.

[54] Morgenthaler NG, Struck J, Jochberger S, Dünser MW. Copeptin: clinical use of a new biomarker. Trends Endocrinol Metab 2008; 19: 43-9.

[55] Krüger S, Ewig S, Kunde J, *et al.* C-terminal provasopressin (copeptin) in patients with community-acquired pneumonia--influence of antibiotic pre-treatment: results from the German competence network CAPNETZ. J Antimicrob Chemother 2009; 64: 159-62.

[56] Stolz D, Christ-Crain M, Morgenthaler NG, *et al.* Copeptin, C-reactive protein, and procalcitonin as prognostic biomarkers in acute exacerbation of COPD. Chest 2007; 131: 1058-67.

[57] Mueller C. Biomarkers and acute coronary syndromes: an update. Eur Heart J 2014; 35: 552-6.

[58] Keller T, Tzikas S, Zeller T, *et al.* Copeptin improves early diagnosis of acute myocardial infarction. J Am Coll Cardiol 2010; 55: 2096-106.

[59] Reichlin T, Hochholzer W, Stelzig C, *et al.* Incremental value of copeptin for rapid rule out of acute myocardial infarction. J Am Coll Cardiol 2009; 54: 60-8.

[60] Voors AA, von Haehling S, Anker SD, *et al.* C-terminal provasopressin (copeptin) is a strong prognostic marker in patients with heart failure after an acute myocardial infarction: results from the OPTIMAAL study. Eur Heart J 2009; 30: 1187-94.

[61] Neuhold S, Huelsmann M, Strunk G, *et al.* Comparison of copeptin, B-type natriuretic peptide, and amino-terminal pro-B-type natriuretic peptide in patients with chronic heart failure: prediction of death at different stages of the disease. J Am Coll Cardiol 2008; 52: 266-72.

[62] Balling L, Gustafsson F. Copeptin as a biomarker in heart failure. Biomarkers Med 2014; 8: 841-54.

[63] Khan SQ, Dhillon OS, O'Brien RJ, *et al.* C-terminal provasopressin (copeptin) as a novel and prognostic marker in acute myocardial infarction: Leicester Acute Myocardial Infarction Peptide (LAMP) study. Circulation 2007; 115: 2103-10.

[64] Alehagen U, Dahlström U, Rehfeld JF, Goetze JP. Association of copeptin and N-terminal proBNP concentrations with risk of cardiovascular death in older patients with symptoms of heart failure. JAMA 2011; 305: 2088-95.

[65] Mason JM, Hancock HC, Close H, *et al.* Utility of biomarkers in the differential diagnosis of heart failure in older people: findings from the heart failure in care homes (HFinCH) diagnostic accuracy study. PLoS One 2013; 8: 535-60.

[66] Wannamethee SG, Welsh P, Whincup PH, Lennon L, Papacosta O, Sattar N. N-terminal pro brain natriuretic peptide but not copeptin improves prediction of heart failure over other routine clinical risk parameters in older men with and without cardiovascular disease: population-based study. Eur J Heart Fail 2014; 16: 25-32.

[67] Masson S, Latini R, Carbonieri E, *et al.* The predictive value of stable precursor fragments of

vasoactive peptides in patients with chronic heart failure: data from the GISSI-heart failure (GISSI-HF) trial. Eur J Heart Fail 2010; 12: 338-47.

[68] Miller WL, Grill DE, Struck J, Jaffe AS. Association of hyponatremia and elevated copeptin with death and need for transplantation in ambulatory patients with chronic heart failure. Am J Cardiol 2013; 111: 880-5.

[69] Maisel A, Xue Y, Shah K, *et al.* Increased 90-day mortality in patients with acute heart failure with elevated copeptin: secondary results from the Biomarkers in Acute Heart Failure (BACH) study. Circ Heart Fail 2011; 4: 613-20.

[70] Sabatine MS, Morrow DA, de Lemos JA, *et al.* Evaluation of multiple biomarkers of cardiovascular stress for risk prediction and guiding medical therapy in patients with stable coronary disease. Circulation 2012; 125: 233-40.

[71] Loncar G, von Haehling S, Tahirovic E, *et al.* Effect of beta blockade on natriuretic peptides and copeptin in elderly patients with heart failure and preserved or reduced ejection fraction: results from the CIBIS-ELD trial. Clin Biochem 2012; 45: 117-22.

[72] Braunwald E. The path to an angiotensin receptor antagonist-neprilysin inhibitor in the treatment of heart failure. J Am Coll Cardiol 2015; 65: 1029-41.

[73] McMurray JJ, Packer M, Desai AS, *et al.* Angiotensin–Neprilysin Inhibition *versus* Enalapril in Heart Failure. N Engl J Med 2014; 371: 993-1004.

[74] Bayés-Genís A, Barallat J, Galán A, *et al.* Soluble neprilysin is predictive of cardiovascular death and heart failure hospitalization in heart failure patients. J Am Coll Cardiol 2015; 65: 657-65.

[75] Bayés-Genís A, Barallat J, Pascual D, *et al.* Prognostic Value and Kinetics of Soluble Neprilysin in Acute Heart Failure: A Pilot Study. JACC Heart Fail 2015; 3: 641-4.

[76] Vodovar N, Séronde M-F, Laribi S, *et al.* Elevated plasma B-type natriuretic peptide concentrations directly inhibit circulating neprilysin activity in heart failure. JACC Heart Fail 2015; 3: 629-36.

[77] Bayes-Genis A, Barallat J, Galán A, *et al.* Multimarker Strategy for Heart Failure Prognostication. Value of Neurohormonal Biomarkers: Neprilysin *vs* NT-proBNP. Rev Esp Cardiol (Engl Ed) 2015; 68: 1075-84.

[78] Hillege HL, Nitsch D, Pfeffer MA, *et al.* Renal function as a predictor of outcome in a broad spectrum of patients with heart failure. Circulation 2006; 113: 671-8.

[79] McAlister FA, Ezekowitz J, Tarantini L, *et al.* Renal dysfunction in patients with heart failure with preserved *versus* reduced ejection fraction: impact of the new Chronic Kidney Disease-Epidemiology Collaboration Group formula. Circ Heart Fail 2012; 5: 309-14.

[80] Ronco C, Haapio M, House AA, Anavekar N, Bellomo R. Cardiorenal syndrome. J Am Coll Cardiol 2008; 52: 1527-39.

[81] Zakeri R, Sangaralingham SJ, Sandberg SM, Heublein DM, Scott CG, Burnett JC. Urinary C-type natriuretic peptide: a new heart failure biomarker. JACC Heart Fail 2013; 1: 170-7.

[82] Dean AD, Vehaskari VM, Greenwald JE. Synthesis and localization of C-type natriuretic peptide in mammalian kidney. Am J Physiol 1994; 266: 491-6.

[83] Zakeri R, Burnett JC, Sangaralingham SJ. Urinary C-type natriuretic peptide: an emerging biomarker for heart failure and renal remodeling. Clin Chim Acta 2015; 443: 108-13.

[84] Hunt PJ, Richards AM, Espiner EA, Nicholls MG, Yandle TG. Bioactivity and metabolism of C-type natriuretic peptide in normal man. J Clin Endocrinol Metab 1994; 78: 1428-35.

[85] Sangaralingham SJ, Huntley BK, Martin FL, *et al.* The aging heart, myocardial fibrosis, and its relationship to circulating C-type natriuretic Peptide. Hypertension 2011; 57: 201-7.

[86] Doi K, Ikeda T, Itoh H, *et al.* C-type natriuretic peptide induces redifferentiation of vascular smooth muscle cells with accelerated reendothelialization. Arterioscler Thromb Vasc Biol 2001; 21: 930-6.

[87] Buzás K, Megyeri K, Hõgye M, Csanády M, Bogáts G, Mándi Y. Comparative study of the roles of cytokines and apoptosis in dilated and hypertrophic cardiomyopathies. Eur Cytokine Netw 2004; 15: 53-9.

[88] Rocchiccioli JP, McMurray JJ, Dominiczak AF. Biomarkers in heart failure: a clinical review. Heart Fail 2010; 15: 251-73.

[89] Vasan RS, Sullivan LM, Roubenoff R, *et al.* Inflammatory markers and risk of heart failure in elderly subjects without prior myocardial infarction: the Framingham Heart Study. Circulation 2003; 107: 1486-91.

[90] Anker SD, von Haehling S. Inflammatory mediators in chronic heart failure: an overview. Heart 2004; 90: 464-70.

[91] Suleiman M, Khatib R, Agmon Y, *et al.* Early inflammation and risk of long-term development of heart failure and mortality in survivors of acute myocardial infarction predictive role of C-reactive protein. J Am Coll Cardiol 2006; 47: 962-8.

[92] Anand IS, Latini R, Florea VG, *et al.* C-reactive protein in heart failure: prognostic value and the effect of valsartan. Circulation 2005; 112: 1428-34.

[93] Dimitrow PP, Undas A, Bober M, Tracz W, Dubiel JS. Obstructive hypertrophic cardiomyopathy is associated with enhanced thrombin generation and platelet activation. Heart 2008; 94: e21.

[94] Lamparter S, Grimm W. Predictive value of high-sensitivity C-reactive protein in patients with idiopathic dilated cardiomyopathy. Pacing Clin Electrophysiol 2005; 28 (Suppl. 1): S233-6.

[95] de Caestecker M. The transforming growth factor-beta superfamily of receptors. Cytokine Growth Factor Rev 2004; 15: 1-11.

[96] Schlittenhardt D, Schober A, Strelau J, *et al.* Involvement of growth differentiation factor-15/macrophage inhibitory cytokine-1 (GDF-15/MIC-1) in oxLDL-induced apoptosis of human macrophages *in vitro* and in arteriosclerotic lesions. Cell Tissue Res 2004; 318: 325-33.

[97] Kempf T, Sinning J-M, Quint A, *et al.* Growth-differentiation factor-15 for risk stratification in patients with stable and unstable coronary heart disease: results from the AtheroGene study. Circ Cardiovasc Genet 2009; 2: 286-92.

[98] Eggers KM, Kempf T, Lagerqvist B, *et al.* Growth-differentiation factor-15 for long-term risk prediction in patients stabilized after an episode of non-ST-segment-elevation acute coronary syndrome. Circ Cardiovasc Genet 2010; 3: 88-96.

[99] Anand IS, Kempf T, Rector TS, *et al.* Serial measurement of growth-differentiation factor-15 in heart failure: relation to disease severity and prognosis in the Valsartan Heart Failure Trial. Circulation 2010; 122: 1387-95.

[100] Kempf T, von Haehling S, Peter T, *et al.* Prognostic utility of growth differentiation factor-15 in patients with chronic heart failure. J Am Coll Cardiol 2007; 50: 1054-60.

[101] Heger J, Schiegnitz E, von Waldthausen D, Anwar MM, Piper HM, Euler G. Growth differentiation factor 15 acts anti-apoptotic and pro-hypertrophic in adult cardiomyocytes. J Cell Physiol 2010; 224: 120-6.

[102] Montoro-García S, Hernández-Romero D, Jover E, *et al.* Growth differentiation factor-15, a novel biomarker related with disease severity in patients with hypertrophic cardiomyopathy. Eur J Intern Med 2012; 23: 169-74.

[103] Daniels LB, Clopton P, Laughlin GA, Maisel AS, Barrett-Connor E. Growth-differentiation factor-15 is a robust, independent predictor of 11-year mortality risk in community-dwelling older adults: the Rancho Bernardo Study. Circulation 2011; 123: 2101-10.

[104] Santhanakrishnan R, Chong JP, Ng TP, *et al.* Growth differentiation factor 15, ST2, high-sensitivity troponin T, and N-terminal pro brain natriuretic peptide in heart failure with preserved *vs.* reduced

ejection fraction. Eur J Heart Fail 2012; 14: 1338-47.

[105] Weinberg EO, Shimpo M, De Keulenaer GW, *et al.* Expression and regulation of ST2, an interleukin-1 receptor family member, in cardiomyocytes and myocardial infarction. Circulation 2002; 106: 2961-6.

[106] Sanada S, Hakuno D, Higgins LJ, Schreiter ER, McKenzie AN, Lee RT. IL-33 and ST2 comprise a critical biomechanically induced and cardioprotective signaling system. J Clin Invest 2007; 117: 1538-49.

[107] Villacorta H, Maisel AS. Soluble ST2 Testing: A Promising Biomarker in the Management of Heart Failure. Arq Bras Cardiol 2016; 106: 145-52.

[108] Pascual-Figal DA, Januzzi JL. The biology of ST2: the International ST2 Consensus Panel. Am J Cardiol 2015; 115: 3B-7B.

[109] Januzzi JL, Peacock WF, Maisel AS, *et al.* Measurement of the interleukin family member ST2 in patients with acute dyspnea: results from the PRIDE (Pro-Brain Natriuretic Peptide Investigation of Dyspnea in the Emergency Department) study. J Am Coll Cardiol 2007; 50: 607-13.

[110] Mueller T, Dieplinger B, Gegenhuber A, Poelz W, Pacher R, Haltmayer M. Increased plasma concentrations of soluble st2 are predictive for 1-year mortality in patients with acute destabilized heart failure. Clin Chem 2008; 54: 752-6.

[111] Rehman SU, Mueller T, Januzzi JL. Characteristics of the novel interleukin family biomarker ST2 in patients with acute heart failure. J Am Coll Cardiol 2008; 52: 1458-65.

[112] Manzano-Fernández S, Mueller T, Pascual-Figal D, Truong QA, Januzzi JL. Usefulness of soluble concentrations of interleukin family member ST2 as predictor of mortality in patients with acutely decompensated heart failure relative to left ventricular ejection fraction. Am J Cardiol 2011; 107: 259-67.

[113] Mueller T, Dieplinger B. The Presage(®) ST2 Assay: analytical considerations and clinical applications for a high-sensitivity assay for measurement of soluble ST2. Expert Rev Mol Diagn 2013; 13: 13-30.

[114] Lassus J, Gayat E, Mueller C, *et al.* Incremental value of biomarkers to clinical variables for mortality prediction in acutely decompensated heart failure: the Multinational Observational Cohort on Acute Heart Failure (MOCA) study. Int J Cardiol 2013; 168: 2186-94.

[115] Boisot S, Beede J, Isakson S, *et al.* Serial sampling of ST2 predicts 90-day mortality following destabilized heart failure. J Card Fail 2008; 14: 732-8.

[116] Manzano-Fernández S, Januzzi JL, Pastor-Pérez FJ, *et al.* Serial monitoring of soluble interleukin family member ST2 in patients with acutely decompensated heart failure. Cardiology 2012; 122: 158-66.

[117] Bayes-Genis A, Januzzi JL, Gaggin HK, *et al.* ST2 pathogenetic profile in ambulatory heart failure patients. J Card Fail 2015; 21: 355-61.

[118] Ky B, French B, McCloskey K, *et al.* High-sensitivity ST2 for prediction of adverse outcomes in chronic heart failure. Circ Heart Fail 2011; 4: 180-7.

[119] Weinberg EO, Shimpo M, Hurwitz S, Tominaga S, Rouleau J-L, Lee RT. Identification of serum soluble ST2 receptor as a novel heart failure biomarker. Circulation 2003; 107: 721-6. [Internet].

[120] Bayes-Genis A, de Antonio M, Galán A, *et al.* Combined use of high-sensitivity ST2 and NTproBNP to improve the prediction of death in heart failure. Eur J Heart Fail 2012; 14: 32-8.

[121] Bayes-Genis A, de Antonio M, Vila J, *et al.* Head-to-head comparison of 2 myocardial fibrosis biomarkers for long-term heart failure risk stratification: ST2 *versus* galectin-3. J Am Coll Cardiol 2014; 63: 158-66.

[122] Pascual-Figal DA, Ordoñez-Llanos J, Tornel PL, *et al.* Soluble ST2 for predicting sudden cardiac death in patients with chronic heart failure and left ventricular systolic dysfunction. J Am Coll Cardiol 2009; 54: 2174-9.

[123] Broch K, Ueland T, Nymo SH, *et al.* Soluble ST2 is associated with adverse outcome in patients with heart failure of ischaemic aetiology. Eur J Heart Fail 2012; 14: 268-77.

[124] Gruson D, Lepoutre T, Ahn SA, Rousseau MF. Increased soluble ST2 is a stronger predictor of long-term cardiovascular death than natriuretic peptides in heart failure patients with reduced ejection fraction. Int J Cardiol 2014; 172: e250-2.

[125] Gaggin HK, Szymonifka J, Bhardwaj A, *et al.* Head-to-head comparison of serial soluble ST2, growth differentiation factor-15, and highly-sensitive troponin T measurements in patients with chronic heart failure. JACC Heart Fail 2014; 2: 65-72.

[126] Gaggin HK, Motiwala S, Bhardwaj A, Parks KA, Januzzi JL. Soluble concentrations of the interleukin receptor family member ST2 and β-blocker therapy in chronic heart failure. Circ Heart Fail 2013; 6: 1206-13.

[127] Anand IS, Rector TS, Kuskowski M, Snider J, Cohn JN. Prognostic value of soluble ST2 in the Valsartan Heart Failure Trial. Circ Heart Fail 2014; 7: 418-26.

[128] Omland T, de Lemos JA, Sabatine MS, *et al.* A sensitive cardiac troponin T assay in stable coronary artery disease. N Engl J Med 2009; 361: 2538-47.

[129] Sato Y, Yamada T, Taniguchi R, *et al.* Persistently increased serum concentrations of cardiac troponin t in patients with idiopathic dilated cardiomyopathy are predictive of adverse outcomes. Circulation 2001; 103: 369-74.

[130] Pascual-Figal DA, Manzano-Fernández S, Pastor F, *et al.* Troponin-T monitoring in outpatients with nonischemic heart failure. Rev española Cardiol 2008; 61: 678-86.

[131] Sato Y, Taniguchi R, Yamada T, Matsumori A. Measurements of serum cardiac troponin T in patients with heart failure. Am Heart J 2003 Apr; 145: e18.

[132] Peacock WF, De Marco T, Fonarow GC, *et al.* Cardiac troponin and outcome in acute heart failure. N Engl J Med 2008; 358: 2117-26.

[133] Moreno V, Hernández-Romero D, Vilchez JA, *et al.* Serum levels of high-sensitivity troponin T: a novel marker for cardiac remodeling in hypertrophic cardiomyopathy. J Card Fail 2010; 16: 950-6.

[134] Sato Y, Taniguchi R, Nagai K, *et al.* Measurements of cardiac troponin T in patients with hypertrophic cardiomyopathy. Heart 2003; 89: 659-60.

[135] Latini R, Masson S, Anand IS, *et al.* Prognostic value of very low plasma concentrations of troponin T in patients with stable chronic heart failure. Circulation 2007; 116: 1242-9.

[136] Sakhuja R, Green S, Oestreicher EM, *et al.* Amino-terminal pro-brain natriuretic peptide, brain natriuretic peptide, and troponin T for prediction of mortality in acute heart failure. Clin Chem 2007; 53: 412-20.

[137] de Antonio M, Lupon J, Galan A, Vila J, Urrutia A, Bayes-Genis A. Combined use of high-sensitivity cardiac troponin T and N-terminal pro-B type natriuretic peptide improves measurements of performance over established mortality risk factors in chronic heart failure. Am Heart J 2012; 163: 821-8.

[138] Pujadas S, Carreras F, Arrastio X, *et al.* Detection and quantification of myocardial fibrosis in hypertrophic cardiomyopathy by contrast-enhanced cardiovascular magnetic resonance. Rev española Cardiol 2007 Jan; 60: 10-4.

[139] López B, González A, Díez J. Circulating biomarkers of collagen metabolism in cardiac diseases. Circulation 2010; 121: 1645-54.

[140] Lin Y-H, Lin C, Lo M-T, *et al.* The relationship between aminoterminal propeptide of type III procollagen and heart rate variability parameters in heart failure patients: a potential serum marker to evaluate cardiac autonomic control and sudden cardiac death. Clin Chem Lab Med 2010; 48: 1821-7.

[141] Ho CY, López B, Coelho-Filho OR, *et al.* Myocardial fibrosis as an early manifestation of

hypertrophic cardiomyopathy. N Engl J Med 2010; 363: 552-63.

[142] Vílchez JA, Hernández-Romero D, Ruiz-Espejo F, *et al.* Collagen peptides, interstitial remodelling and sudden cardiac death in hypertrophic cardiomyopathy. Clin Chem Lab Med 2011; 49: 1569-71.

[143] Liu F-T, Rabinovich GA. Galectins: regulators of acute and chronic inflammation. Ann N Y Acad Sci 2010; 1183: 158-82.

[144] Morrow DA, O'Donoghue ML. Galectin-3 in cardiovascular disease: a possible window into early myocardial fibrosis. J Am Coll Cardiol 2012; 60: 1257-8.

[145] Hrynchyshyn N, Jourdain P, Desnos M, Diebold B, Funck F. Galectin-3: a new biomarker for the diagnosis, analysis and prognosis of acute and chronic heart failure. Arch Cardiovasc Dis 2013; 106: 541-6.

[146] Lin Y-H, Lin L-Y, Wu Y-W, *et al.* The relationship between serum galectin-3 and serum markers of cardiac extracellular matrix turnover in heart failure patients. Clin Chim Acta 2009; 409: 96-9.

[147] Peacock WF, DiSomma S. Emergency department use of galectin-3. Crit Pathw Cardiol 2014; 13: 73-7.

[148] van Kimmenade RR, Januzzi JL, Ellinor PT, *et al.* Utility of amino-terminal pro-brain natriuretic peptide, galectin-3, and a pelin for the evaluation of patients with acute heart failure. J Am Coll Cardiol 2006; 48: 1217-24.

[149] de Boer RA, Lok DJ, Jaarsma T, *et al.* Predictive value of plasma galectin-3 levels in heart failure with reduced and preserved ejection fraction. Ann Med 2011; 43: 60-8.

[150] Filipe MD, Meijers WC, Rogier van der Velde A, de Boer RA. Galectin-3 and heart failure: prognosis, prediction & clinical utility. Clin Chim Acta 2015; 443: 48-56.

[151] Meijers WC, Januzzi JL, DeFilippi C, *et al.* Elevated plasma galectin-3 is associated with near-term rehospitalization in heart failure: a pooled analysis of 3 clinical trials. Am Heart J 2014; 167: 853-60.

[152] Motiwala SR, Szymonifka J, Belcher A, *et al.* Serial measurement of galectin-3 in patients with chronic heart failure: results from the ProBNP Outpatient Tailored Chronic Heart Failure Therapy (PROTECT) study. Eur J Heart Fail 2013; 15: 1157-63.

[153] Lok DJ, Lok SI, Bruggink-André de la Porte PW, *et al.* Galectin-3 is an independent marker for ventricular remodeling and mortality in patients with chronic heart failure. Clin Res Cardiol 2013; 102: 103-10.

[154] Gullestad L, Ueland T, Kjekshus J, *et al.* The predictive value of galectin-3 for mortality and cardiovascular events in the Controlled Rosuvastatin Multinational Trial in Heart Failure (CORONA). Am Heart J 2012; 164: 878-83.

[155] Anand IS, Rector TS, Kuskowski M, Adourian A, Muntendam P, Cohn JN. Baseline and serial measurements of galectin-3 in patients with heart failure: relationship to prognosis and effect of treatment with valsartan in the Val-HeFT. Eur J Hear Fail 15: 511-8.

[156] McMurray J, Chopra M, Abdullah I, Smith WE, Dargie HJ. Evidence of oxidative stress in chronic heart failure in humans. Eur Heart J 1993; 14: 1493-8.

[157] Hokamaki J, Kawano H, Yoshimura M, *et al.* Urinary biopyrrins levels are elevated in relation to severity of heart failure. J Am Coll Cardiol 2004; 43: 1880-5.

[158] Cicoira M, Zanolla L, Rossi A, *et al.* Elevated serum uric acid levels are associated with diastolic dysfunction in patients with dilated cardiomyopathy. Am Heart J 2002; 143: 1107-11.

[159] Pascual-Figal DA, Hurtado-Martínez JA, Redondo B, Antolinos MJ, Ruiperez JA, Valdes M. Hyperuricaemia and long-term outcome after hospital discharge in acute heart failure patients. Eur J Heart Fail 2007; 9: 518-24.

[160] Levy WC, Mozaffarian D, Linker DT, *et al.* The Seattle Heart Failure Model: prediction of survival in heart failure. Circulation 2006; 113: 1424-33.

CHAPTER 12

Emerging Biomarkers in Peripheral Artery Disease

Isabel Fort-Gallifa, Anna Hernández-Aguilera, Vicente Martín-Paredero, Jorge Joven and **Jordi Camps**[*]

Unitat de Recerca Biomèdica, and Laboratori de Referència SUD, Hospital Universitari Sant Joan, Institut d'Investigació Sanitària Pere Virgili, Universitat Rovira i Virgili, Reus, Spain Carrer Sant Llorenç 21, 43201-Reus, Spain

Abstract: Peripheral artery disease is probably the most common cardiovascular disease. It affects extensive areas of arteries in the lower extremities. Almost invariably, it is the result of a sustained and asymptomatic progression of the disease, and effective preventive measures are applied too late. In peripheral artery disease, the degree of atherosclerosis is greater than in patients with adverse events in other arterial territories. Different aspects of atherosclerosis may also be represented, and designing strategies for prompt identification could facilitate therapeutic goals. The main objective of this chapter is to describe biomarkers that can be measured in serum or plasma, which help diagnose, monitor and/or predict the course of the disease. This chapter defines and describes established biomarkers, emerging biomarkers and candidate biomarkers. This chapter includes short descriptions of their main characteristics and their usefulness in the clinical evaluation of peripheral artery disease. Our aim is to give reader a perspective on the advances made in the research laboratory over the recent years in the analysis of these patients.

Keywords: Arginine, Atherosclerosis, β-hydroxybutyrate, β-2-melatonin, Biomarkers, C-reactive protein, CCL2, Cysteine, Galectin-3, Homocysteine, Hydroxycholesterol, Inflammation, Isoprostanes, Oxidative stress, Peripheral artery disease, Microglobulin, PON1, Protein carbonyls, Tricarboxylic acid cycle products, 8-hydroxy-2-deoxy guanosine.

INTRODUCTION

Peripheral Artery Disease: Clinical and Epidemiological Characteristics

Lower extremity peripheral artery disease (PAD) is a major health concern since it could result in catastrophic events in different arterial territories such as the extracranial carotid, upper extremity arteries, mesenteric and renal circulation, and

[*] **Corresponding author Jordi Camps:** Unitat de Recerca Biomèdica, and Laboratori de Referència SUD, Hospital Universitari Sant Joan, Institut d'Investigació Sanitària Pere Virgili, Universitat Rovira i Virgili, Reus, Spain Carrer Sant Llorenç 21, 43201-Reus, Spain; Tel/Fax: +34-977.310.300; E-mail: jcamps@grupsagessa.com

arteries of the lower extremities [1].

The incidence of this disease increases with age and, in patients >55 years old, the prevalence of PAD in the legs is approximately 20% [2, 3]. The burden of atherosclerosis is much higher than in patients with adverse events in other arterial territories, and preventive measures are only effective in the initial stages of the disease.

Unfortunately, the initial stages of PAD are often asymptomatic, and appropriate preventive measures are applied too late. In 2010, this disorder affected between 8 and 12 million people in the USA; an increasing prevalence in Europe and Asia is being expected in the near future [4]. A recent meta-analysis estimated that 202 million people worldwide are affected with the disease; the analysis indicating an increase in PAD prevalence of 23.5% in the first decade of the 21st century [5].

Total or partial obstruction of lower extremity arteries is caused by poor circulation resulting in intermittent claudication, characterised by pain in leg muscles during walking; the pain is relieved within a few minutes after coming to rest. It is the initial symptom of the disease, and should not be confused with pain caused by pinched or damaged spinal nerves *i.e.* a pain which does not disappear when walking is suspended [1]. Progression of the illness causes pain even at rest, and it is accentuated when legs are raised. The pain can appear in calf muscles, or feet. The most severe stages of PAD are associated with tissue hypo-perfusion resulting in difficulty in wound healing that can progress to ischaemia, ulceration and gangrene; in one third of the patients, this situation can lead to amputation of the limb [1]. Presence of pain at rest or tissue loss, known as critical limb ischaemia, is associated with about 20% mortality/year.

Clinical evaluation of PAD needs to consider different aspects of the disease: history, epidemiology, and risk factors [6] which are factors predictive of illness progression if not addressed. Risk factors of atherosclerosis in PAD are similar to those in coronary artery disease (CAD). The most important risk factors are: race (being Black is associated with twice the risk for developing PAD compared to White race); age (older members of the population have a higher risk of suffering this disease than do younger members); sex (being male carries a higher risk than being female) [7]. Other risk factors of premature PAD include smoking, diabetes, hypertension, higher circulating concentrations of cholesterol, triglycerides, and homocysteine. High blood fibrinogen concentration is a thrombosis risk factor, and an elevated haematocrit is identified as a factor promoting hyper-coagulation [8]. High-density lipoprotein cholesterol to total cholesterol ratio is, currently, the best biochemical predictor of PAD, according to the consensus document (TASMCII) for the management of PAD [6].

Epidemiological studies indicate that the prevalence of PAD is between 3 and 10% in the general population, but this range increases up to 20% if 70 years of age is used as the cut-off point. These results from the presence of different comorbidities associated with age; essentially increased oxidative stress [1, 6]. The ankle-brachial index (ABI) has been, and still is, used as a screening method for assessing severity of PAD. ABI compares systolic blood pressure in the upper limbs *versus* lower limbs. Using Doppler ultrasound and a sphygmomanometer, the blood pressure is measured in the arm and the corresponding leg. The index so obtained reflects the extent of PAD. ABI is an effective and objective classification; reduced values of ABI confirm the diagnosis and detect the presence of PAD in asymptomatic patients. It also identifies patients with impaired limb function (difficulty to walk a defined distance, or at a standard walking speed). Finally, it provides important information on prognosis; an ABI < 0.9 is associated with a 3-6 fold increased risk of death from cardiovascular disease [9, 10] (Table **1**).

Table 1. ABI classification.

ABI value	Interpretation
>1.3	Abnormal, non-compressible vessel (calcified)
1.0-1.29	Normal
0.91-0.99	Indeterminate
0.89-0.91	Mild peripheral artery disease
0.41-0.89	Mild-moderate peripheral artery disease
<0.4	Severe peripheral artery disease

However, ABI has a drawback in diabetic patients, or patients with renal failure, which can have non-compressible and calcified arteries in the lower limbs and, therefore, the ABI is falsely elevated. In these cases the arterial pressure needs to be measured in the toes and related to the measurement in the arm. There are several classifications for PAD, the most widely employed being the Fontaine classification and the Rutherford classification (Table **2**).

Table 2. Fontaine and Rutherford classification.

Fontaine classification		Rutherford classification		
Stage	**Clinical**	**Grade**	**Stage**	**Clinical**
I	Asymptomatic	0	0	Asymptomatic
II a	Mild claudication (at distances > 200 meters)	I	1	Mild claudication
II b	Moderate to severe claudication (at distances < 200 meters)	I	2	Moderate claudication

(Table 2) contd.....

Fontaine classification		Rutherford classification		
Stage	**Clinical**	**Grade**	**Stage**	**Clinical**
III	Ischaemic pain at rest	I	3	Severe claudication
IV	Ulceration or gangrene	II	4	Ischaemic pain at rest
		III	5	Minor tissue loss
		III	6	Major tissue loss

Arteriography is still the "gold standard" technique, and all classifications need to be compared with it. However, arteriography is a semi-invasive technique and would only be indicated for patients who are candidates for surgery or percutaneous interventions. In addition, the technique cannot be applied to patients with kidney disease because the contrast medium used in the arteriography is contra-indicated in patients with impaired renal function, especially those who have an advanced stage renal disease [11, 12].

Biochemistry and Cell Biology of Atherosclerosis Onset and Development

Atherosclerosis is initiated by an increase in the inflammation of endothelial cells of the vessel wall associated with low-density lipoprotein (LDL) particles retained within the cells. This phenomenon may be a cause or an effect of the underlying inflammatory process. Atherosclerosis is considered to commence with damage to vascular endothelium caused by a variety of insults such as high blood pressure, tobacco smoking, high cholesterol concentration, diabetes and hyper-homocysteinaemia [13 - 15].

Plaque formation is initiated by endothelial damage (or dysfunction) in the lumen of the arterial wall. In a second stage, LDL passes through the damaged endothelium and deposits within the artery wall. A hallmark in the development of atherosclerosis is LDL oxidation generating reactive oxygen species (ROS) (Fig. **1**). ROS are chemical species containing one unpaired electron in the outer orbital of the oxygen molecule. ROS have high biological reactivity and are produced in normal aerobic metabolism in processes such as food metabolism, breathing or physical exercise. ROS production is strictly controlled by the organism with the help of endogenous and exogenous antioxidants. ROS, if uncontrolled, can oxidise and damage essential biomolecules such as polyunsaturated fatty acids (PUFA), DNA, proteins and carbohydrates leading to cellular structure and function damage and, ultimately, cell death [16]. Disequilibrium between ROS production and antioxidant control mechanisms is termed oxidative stress. Direct measurement of ROS is difficult due, in general, to the short half-life of the molecule. Usually, oxidative stress is assessed indirectly as a function of measurable markers.

PUFA present in LDL molecules are particularly sensitive to oxidation. Lipid peroxidation of PUFA has identical general characteristics as that described for oxidative processes in lipids in cell membranes. Fatty acid peroxidation is a chain reaction consisting of three stages. In the 1st, stage a fatty acid radical is produced when fatty acids react with ROS. The 2nd stage is known as propagation and is due to fatty acid radicals being unstable molecules that react with molecular oxygen leading to a peroxyl radical fatty acid. This is also a highly reactive species that can react with another fatty acid to initiate a chain reaction. The 3rd stage, which only occurs when ROS concentration is high enough, consists of a reaction between two radicals to produce a stable species [17]. This process plays a fundamental role in the onset and development of atherosclerosis [18].

When LDL crosses the damaged endothelium, it is partially oxidised by ROS, and is termed minimally oxidised LDL (MO-LDL). MO-LDL stimulates endothelial cells and smooth muscle cells, which generate the chemokine (C-C motif) ligand 2 (CCL2), formerly termed monocyte chemo attractant protein-1 (MCP1) [19]. ROS also stimulate the expression of adhesins such as intercellular adhesion molecule-1 (ICAM-1 or CD54) and vascular cell adhesion molecule-1 (VCAM-1) and leukocyte adhesion molecules which enhance adhesion of leukocytes to endothelial cells. When monocytes adhere to the endothelial cell surface, MO-LDL is oxidised to OX-LDL. Complete LDL oxidation from MO-LDL is catalysed mainly by myeloperoxidase (an enzyme which is expressed in macrophages found in atherosclerotic lesions) and glycosylases. Some studies have suggested that enzymatic LDL oxidation is possible *in vivo* by interaction with the artery wall, blood cells, plasma constituents and components from the artery wall matrix which are able to hydrolyse cholesterol esters, phosphoglycerides and triglycerides.

PUFA are oxidised by lipo-oxygenases from macrophages, leading to formation of hydro-peroxides from LDL fatty acids. Similarly, plasma cholesterol oxidases are able to generate LDL that is enriched with products from oxidised cholesterol [9, 18]. MO-LDL induces an inflammatory response leading to apoptosis *via* activation of nuclear factors, coagulation stimulation, peroxidation induction in lesions, and inhibition of nitric oxide (NO) production. Similarly, OX-LDL produces an inflammatory reaction resulting in lymphocyte T infiltration. Both oxidised molecules, MO-LDL and OX-LDL, are immunogenic and are able to stimulate the release of auto-antibodies while, in addition, platelet aggregation can be altered. OX-LDL and CCL2 promote monocyte migration into the sub-endothelium. Further, OX-LDL is an inducer of pro-inflammatory molecules such as VCAM-1, ICAM-1, selectin P, and CCL2. The inflammatory process depends, mainly, on two adhesion molecules: selectins (implicated in the deposition of monocytes and lymphocytes in the endothelium) and immunoglobulins (responsible for leukocyte adhesion). Currently, however, there is no clear evidence regarding the factors responsible for increasing local expression of adhesion molecules and cytokines, but the more LDL molecules are oxidised the more pro-inflammatory molecules appear to be generated. Differentiation from monocytes to macrophages is due to monocyte colony stimulating factor release from endothelial cells by MO-LDL. Differentiated macrophages develop a receptor to OX-LDL and these molecules are captured by macrophages *via* phagocytosis to form foam cells. Eventually, tissue necrosis results [18]. OX-LDL exacerbates intra-cellular endothelial damage while, at the time, the lack of NO generates vasoconstriction which regulates vascular tone and inhibits platelet aggregation [20]. NO derives from L-arginine and is released by endothelial cells. NO functions at the transcriptional level by modulating the signalling pathway of nuclear factor kappa B (NF-κB) and by inhibiting the expression of the *VCAM-1* gene in endothelial cells [1]. Macrophages generate many growth factors involved in the formation of collagen, elastic fibres and proteins. Smooth muscle cell proliferation, monocyte migration together with the synthesis of connective tissue and extracellular matrix are major factors in atherosclerosis progression [9].

Despite these disruptive mechanisms, the organism has several systems to counteract the oxidative stress-mediated alterations. Under laminar blood flow conditions, NO synthase is increased, together with NO levels; its anti-inflammatory and vasodilator actions are enhanced. Increasing NO levels is the first defence mechanism against plaque formation; the two essential functions of NO being to relax the vessels and to inhibit platelet aggregation [9, 21]. Under turbulent flow conditions, such as in arterial branches or when vascular endothelium is dysfunctional, there is an inhibition of NO synthesis, increased VCAM-1 expression, and transformation of monocytes into foam cells [1, 22]. It is not only turbulent flow that induces vascular contraction; ROS have a similar effect. Oxygen (O_2) and superoxide anion (H_2O_2) produce contractions in vessels *via* two possible mechanisms; directly [23], or indirectly mediated by endothelin-1 or by NO breakdown [24-26]. ROS can induce turbulent flow directly by damaging endothelial cells and reducing NO production. O_2 can react with NO to produce more active ROS, such as peroxynitrite (ONO_2^-). ROS can reduce prostacyclin production and decrease the production of vasodilators in the endothelium. The result is an increase in circulating catecholamines causing vasoconstriction of blood vessels and an increase in blood pressure. Another mechanism is mediated by glucose, which increases sympathetic activity and the increase in blood pressure [9, 18]. All these data imply that ROS, *via* several different mechanisms, can induce hypertension and atherosclerosis.

The second defence mechanism is provided by HDL which penetrates into the sub-endothelium and is instrumental in binding and removal of intra-cellular cholesterol molecules. In atherogenesis, this mechanism fails because the HDL is insufficient to remove the excess LDL entering the cell. Several studies [19, 27] have demonstrated that HDL particles protect LDL from oxidation by means of, at least, two apolipoproteins with enzyme activity: paraoxonase 1 (PON1) and acetylhydrolase platelet activating factor [19, 27, 28]. PON1 is a hydrolase capable of degrading a wide range of different substrates. It circulates in plasma bound to HDL. The enzyme is able to hydrolyse phospholipids present in oxidised LDL and, as has been suggested, is responsible for many of the observed beneficial effects (*e.g.* athero-protective) of HDL [18].

The third defence mechanism is provided by platelets, which adhere to damaged endothelium and release growth factors leading to connective tissue formation.

Cholesterol debris and foam cell accumulation in the sub-endothelium is what leads to the core of atherosclerotic plaque. In a more advanced atherosclerotic plaque, accumulation of foam cells in the intimal endothelium layer occurs, and plaque gradually becomes fibrous as smooth muscle cells accumulate within the lesion and produce extra-cellular proteins that form a fibrous matrix. Calcium also accumulates in atherosclerotic plaque. This plaque progresses with protein production by vascular smooth muscle cells.

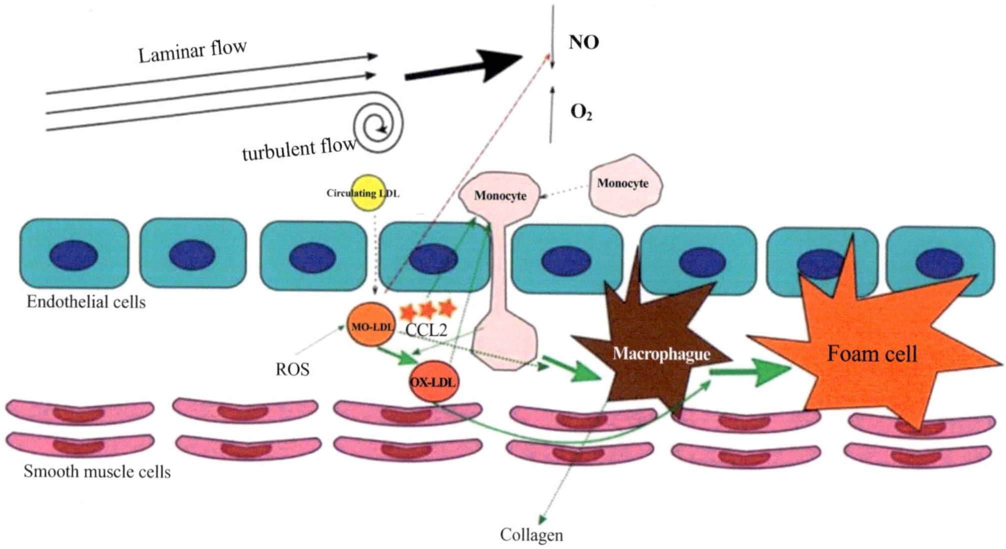

Fig. (1). Entry of circulating LDL, MO-LDL transformation, OX-LDL transformation and atherosclerotic plaque formation.

BIOMARKERS

Biomarkers of Peripheral Artery Disease

Definition of "Biomarker"

A biomarker is a molecule in the circulation with biological characteristics that can be measured and evaluated; the objective being to improve the physician's ability to assess different aspects of a disease beyond that indicated by the established clinical signs and symptoms. When a biomarker indicates susceptibility to the disease, it is termed a risk factor; when it indicates a capacity to discriminate the affected from the non-affected population, it is a screening biomarker; when it estimates the progression of the disease, it is a staging biomarker; when it estimates the prognosis of the disease, it is a prognosis biomarker; and finally, when it evaluates the efficacy of a treatment, it is a monitoring biomarker.

The accuracy of a biomarker is usually analysed by Receiver Operating Characteristics (ROC) plots. The ROC plot represents the values of 1-specificity for each data point on the x-axis, and the sensitivity on the y-axis. The area-under-the-curve (AUROC) may be between 0.5 and 1.0, and the diagnostic accuracy is higher when this parameter is closer to 1.0 [29, 30].

The search for specific biomarkers for PAD is difficult because this disease and heart disease share risk-factors in common, and have common mechanisms of illness initiation and progression. Comparisons of these diseases indicate similar alterations in: chemokines and inflammatory cytokines; markers of endothelial dysfunction; vascular regeneration mediators; lipoproteins; oxidative stress biomarkers; metabolic modulators and clotting factors. There is no specific biomarker for PAD since all the above are elevated in CAD and other cardiovascular diseases, as well [4]. This is not entirely unexpected but, since the burden of atherosclerosis is much higher in PAD than in CAD *i.e.* the extent of the affected arteries is much higher, greater alterations would be expected in patients with PAD, or even specific biomarkers of this condition would be identified.

Established Biomarkers

C-Reactive Protein

C-reactive protein (CRP) is an acute phase reactant synthesized by the liver. The serum concentrations of this protein increase with inflammation or infection, and also following myocardial infarction, surgical intervention, or trauma [4, 31].

The high-sensitivity method enables the measurement of very low levels of this protein (hs-CRP), and is very useful in establishing presence of inflammation and/or cardiovascular disease (CVD) risk. This analyte is used in combination with other markers such as cholesterol, triglycerides or lipoprotein-a (Lp_a) in determining CVD or PAD risk status. The American Heart Association (AHA) and the Center for Disease Control and Prevention (CDC), defined the following risk groups with respect to CRP [32]:

Low risk: Values <1.0 mg/L

Medium risk: Values between 1.0 and 3.0 mg/L

High risk: Values >3.0 mg/L

β-2-Microglobulin

B-2-microglobulin (B2M) is a low-molecular-weight protein found on the plasma membrane of all nucleated cells. It belongs to the major histocompatibility complex class I. B2M is released into the circulation by nucleated cells, particularly by B lymphocytes and tumour cells. Under normal conditions it is released in very small quantities into the urine; 99% of the filtered B2M being reabsorbed by the renal tubules. High urinary concentrations of B2M are indicative of renal tubule disease, while increased serum B2M concentrations are found in systemic infection and/or in autoimmune diseases due, in these cases, to increased production rather than decreased clearance. Patients with PAD have elevated serum B2M concentrations due to arterial stiffness, vessel inflammation, decreased renal clearance [11], repeated bouts of ischaemia and reperfusion in the legs [33]. Arterial stiffening reduces the buffering capacity of the elastic arteries which increases systolic and pulse pressure, promotes left ventricular hypertrophy and dysfunction, and impairs the capacity for myocardial perfusion. As such, it plays an important role in atherogenesis and PAD [34].

Emerging Biomarkers

This section focuses on describing the results of research into new biomarkers related to oxidative stress and inflammation. Specifically, these molecules such as isoprostanes and carbonylated proteins involve the coordinated roles of PON1 and CCL2 and indices of oxidative stress processes.

Isoprostanes

Isoprostanes are a family of eicosanoids derived from arachidonic acid, and produced by random oxidation. These molecules are generated by oxidation by ROS of polyunsaturated acyl groups *in situ* in membrane phospholipids.

Isoprostanes have powerful effects on the vascular system [35]. At least one of the isoprostanes, 8-isoprostane/15-F2t-isoprostane/8-iso-15(S)-Prostaglandin $F_{2\alpha}$/8-iso-15(S)-Prostaglandin $F_{2\alpha}$/8-*iso*$PGF_{2\alpha}$/$iPF_{2\alpha}$-III/8-*epi* $PGF_{2\alpha}$ (Fig. **2**) is a powerful pulmonary and renal vasoconstrictor, and has been implicated in the patho-physiology of hepato-renal syndrome and pulmonary oxygen toxicity. This molecule has been proposed as a marker of antioxidant deficiency, and oxidative stress [36, 37].

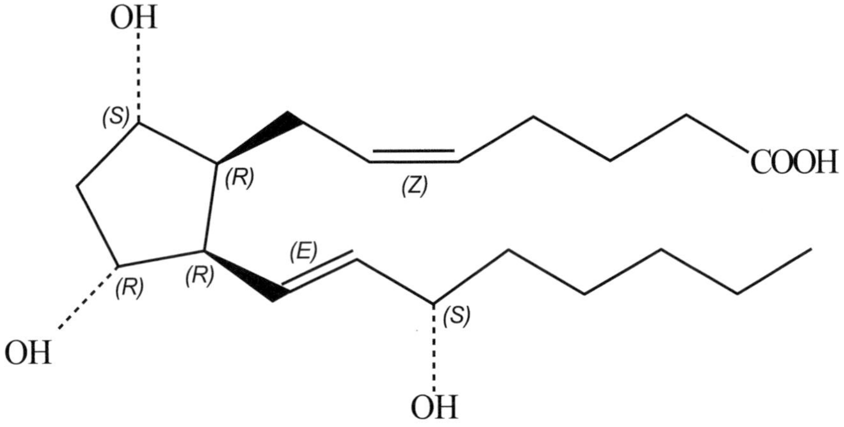

Fig. (2). 8-isoprostane (chemdraw.cdx Perkin Elmer software).

Isoprostanes have several modes-of-action: 8-*iso*$PGF_{2\alpha}$ inhibits angiogenesis by blocking vascular endothelial growth factor-induced endothelial cell migration, tube formation and cardiac vessel sprouting. It stimulates tumour growth factor-β that can promote nephropathy associated with type I diabetes. The activity is blocked by antagonists of thromboxane A_2 (TXA_2) receptor, indicating that, in some cases, isoprostanes stimulate TXA_2 biosynthesis.

Plasma from healthy individuals contains small amounts of 8-isoprostane, about 4 to 10 pg/mL. Isoprostanes are considered the most sensitive and reproducible analytes in quantifying lipid peroxidation *in vivo* [38].

These molecules are almost ubiquitously distributed in plasma, cerebrospinal fluid, urine, bile, broncho-alveolar lavage fluid, and in several tissues. The most convenient matrixes for quantification are plasma and urine because they can be obtained without invasive procedures. The values in these fluids are a good reflex of their endogenous production and, thus, provide an accurate index of *in vivo* oxidative stress [39]. Elevation of isoprostanes has been found in several human diseases including atherosclerosis, diabetes, obesity, and pulmonary disease.

There are several methods to quantify isoprostanes in body fluids. These include gas chromatographic/negative-ion chemical ionisation mass spectrometric (GC/MS); liquid chromatographic/mass spectrometric (LC/MS); and immunoassays [40]. One important advantage of LC/MS methods is that the sample preparation is simpler than for GC/MS. Immunoassays are even simpler but have a potential drawback in that there is limited information available regarding precision and accuracy. In addition, there is a paucity of data comparing immunoassays to MS. Despite the limitations of immunoassays, they are successful because of their low cost and ease of methodology. Results obtained by our group have shown that patients with PAD have significantly increased levels of isoprostanes relative to control individuals; AUROC curve analysis of 0.999 demonstrating the high level of accuracy of this measurement in the diagnosis of PAD.

PON1

PON1 belongs to an enzyme family composed of PON1, PON2 and PON3. These three enzymes derive from a common precursor, and their genes are located adjacent to each other on chromosome 7 arm q (21.3-22.1) [41 - 43].

In humans, PON1 and PON3 expression are observed in most tissues, especially liver, kidney and epithelium. They are found in plasma linked to HDL. PON2 expression is practically ubiquitous in tissues, but it is an intracellular enzyme and cannot be measured in the circulation [13, 44]. PON1 has lactonase, paraoxonase, and arylesterase activities and has been implicated in defence against xenobiotics. PON2 and PON3 have lactonase activity, but not esterase activity. PON1, PON2 and PON3 are able of degrading lipid peroxides in LDL and HDL. PON2 counteracts cellular oxidative stress and apoptosis in endothelial vascular cells.

The enzymes degrade oxidised phospholipids derived from LDL and HDL, as well as from cellular membranes. These features highlight their anti-inflammatory and athero-protective function [19, 45].

In spite the importance of PON1 in the pathophysiology of many diseases, the measurement of this enzyme is difficult to apply in routine clinical chemistry laboratories, due to methodological difficulties. The substrates used for PON1 measurement are not physiological, and some of them are toxic or unstable. These drawbacks preclude automation of PON1 measurement and, as such, it is not included yet in panels of standard biochemical tests in the clinical chemistry laboratory. The most widely used method for the measurement of PON1 activity is the based on the hydrolysis of paraoxon, measured by the increase in absorbance at 412 nm. However, paraoxon has two major drawbacks: it is extremely toxic and very unstable. The solution to the latter problem is to prepare the reagent immediately before use. The solution to the former problem requires that the substrate must be handled in an air-extraction cupboard and the operator has to take safety precautions such as wearing masks and gloves to protect against accidental contact or inhalation of the toxic fumes. The recommended reaction medium is a glycine buffer (0.05 M, pH = 10.5) with 1 mM $CaCl_2$, or a Tris-HCl buffer (0.1 M, pH = 8.5) with 2 mM $CaCl_2$. Some investigators add NaCl (1 or 2 M) to increase PON1 activity. The temperature of reaction can be 37ºC or 25ºC [46 - 48]. These differences imply that values obtained from different centres vary considerably, and this precludes inter-laboratory comparisons.

Another widely employed substrate is phenylacetate. The reaction buffer is Tris-HCl (9 mM, pH = 8.5) with 0.9 mM $CaCl_2$. The reaction is conducted at 25ºC and monitored at 270 nm [46, 47, 49]. The toxicity or phenylacetate is lower to that of paraoxon, but it also needs to be handled under safety conditions. As we have already described, there is evidence showing that native activity of PON1 is as a lactonase, and that lipophylic lactones constitute the native substrates. Two tests based on this capacity of PON1 were proposed some years ago [50, 51]. One of the methods uses 5-thiobutyl butyrolactone (TBBL) as a substrate. TBBL is a synthetic lactone that resembles the natural substrate of PON1. The method allows PON1 activity to be analysed using a more 'physiologically-akin' substrate. The lactonase measurement correlates well with the levels of PON1-HDL complex and provides an estimate of the quality of the HDL particles to which the enzyme is attached [51, 52]. The incubation reagent contains 1 mM $CaCl_2$, 0.25 mM TBBL, and 0.5 mM 5,5'-dithio-bis-nitrobenzoic acid in 0.05 mM Tris-HCl buffer (pH = 8.0). The temperature of reaction is 25ºC and the change in absorbance is monitored at 412 nm. The other method employs 7-*O*-diethyl phosphoryl 3-cyano 4-methyl 7-hydroxycoumarin (DEPCyMC) as substrate, and this method is a good estimate of the total PON1 protein mass, whether bound or not to HDL particles. The incubation reagent contains 1 mM $CaCl_2$ and 1 mM DEPCyMC, in 50 mM bis-trispropane (pH = 9.0). The temperature of reaction is 25ºC and the change in absorbance is monitored at 400 nm. These methods have two great advantages over the esterase assays. First, they measure a type of PON1 activity that is closer to the physiology. Second, the TBBL and DEPCyMC substrates are not toxic. During the 1990s, some groups were able to generate monoclonal or polyclonal antibodies against purified PON1 [53, 54]. These antibodies were employed in developing ELISA methods for the measurement of serum PON1 concentration and, as well, in immunohistochemical studies [55 - 57].

The problem associated with these methods was that the quality of the antibody was very variable, and depended on the quality of the initial purified PON1 preparation. More recently, we developed antibodies against synthetic peptides which are specific of PON1, PON2 and PON3 mature sequences. These peptides are CRNHQSSYQTRLNALREVQ (PON1 specific), CKEEKPRARELRISR GFDL (PON2 specific) and CRVNASQEVEPVEPEN (PON3 specific). These antibodies enabled us to develop immunohistochemical methods to study the distribution of the three PON in tissues [58], including human arteries [59], and ELISA methods to measure the serum concentrations of PON1 and PON3 [58 - 61].

Our studies showed that PAD patients have decreased PON1 lactonase and paraoxonase activities, and decreased PON1 concentrations. The best diagnostic accuracy was observed for PON1 concentration, with an AUROC value of >0.95.

CCL2

Chemokines, or chemo attractant cytokines, are a family of small secreted proteins that are globular in structure and which function as messengers to activate and to direct migration of specific subtypes of leukocytes from the bloodstream into injured tissues. Chemokines, in particular CCL2 (chemokine (C-C motif) ligand 2) also termed MCP-1 (monocyte chemo attractant protein 1), play a significant role in the regulation of the inflammatory response [19, 62]. Chemokines are associated with the migration of immune cells and, as well, are of considerable importance in the correct functioning of metabolic pathways [62]. The most-known function of CCL2 is to stimulate monocyte migration to the inflammation sites and, subsequently, to induce their differentiation into macrophages [63]. CCL2 binds to CC chemokine receptor 2 (CCR2), which is expressed on the cell surface, and promotes accumulation of monocytes on the vascular wall [64, 65].

Circulating chemokines cause metabolic perturbations which can be reversed by anti-inflammatory drugs [66]. Both, CCL2 protein and mRNA, are expressed in the majority of tissues, suggesting that there is a systemic production and response to inflammatory stimuli [67]. Thus, CCL2 synthesis is induced in several diseases that feature monocyte-rich cellular infiltrates. These diseases include atherosclerosis, congestive heart failure and rheumatoid arthritis [68]. Many stimuli, particularly oxidative stress, cause the production of CCL2 in vascular cells and stimulation of foam cell formation, inflammation, and progression of the atherosclerosis process [65, 69]. Increased levels of CCL2 have been observed in PAD, and it has been suggested that CCL2 concentration in plasma can be an index of the severity of the disease. However, this hypothesis needs confirmation [63]. The study of the interrelations between CCL2 and PON1 may provide clues to the understanding of PAD. For example, the increase in the production of CCL2 by endothelial cells may be significantly inhibited by the action of PON1.

These results suggest inter-related roles for CCL2 and PON1 molecules [44].

Early studies from our group showed that serum PON1 activity and concentration were significantly lower, and CCL2 concentration higher, in PAD patients compared to control individuals. Our results also showed that the combination of plasma CCL2 and PON1-related variables was capable of almost complete discrimination of control individuals from patients [67, 70, 71].

Protein Carbonyls

The most widely used indicator (or marker) of protein oxidation is carbonylated protein content [16]. Carbonylation represents the most frequent oxidative modification of proteins. This modification is chemically stable and is a feature of particular importance in protein storage and in the methodology for the detection of protein carbonyls [16, 17].

Cations generated by the redox cycle, as Fe^{2+} or Cu^{2+}, can bind to cation binding sites of proteins which, when binding to H_2O_2 or O_2, can transform to carbonyls, the amine groups from the protein lateral chain of several amino acids. Protein oxidation catalysed by metals is not the only mechanism by which carbonyls are introduced inside the protein structure. For example, cigarette smoke and aldehydes are also implicated in protein oxidation. Many analytical methods have been developed to detect and quantify the extent of protein carbonylation in protein preparations. The most used method consists of protein carbonyl derivatisation with 2,4-dinitrophenylhydrazine (DNPH) and subsequent spectrophotometric assay. This method enables an overall quantification of protein carbonyl content due to the ability of DNPH to react with carbonyls to produce an adduct that absorbs light at 366nm. These methods has been sophisticated including a high-performance liquid chromatography to separate molecules and parallel measurement of absorbing adducts. Subsequent, immunological techniques, such as Western blot or ELISA, have been developed leading to an increase in sensitivity of protein carbonyl detection [72].

Currently, these methods are extensively employed in evaluating changes in total protein carbonyls and, eventually, to identify the specific proteins undergoing oxidation. More recently, many mass-spectrometry methods have been developed for the identification of protein carbonyls, and the associated amino acid residues that have been modified to carbonyl derivatives. Although these mass-spectrometry methods are detailed and accurate, due to their ability to identify the amino acid residues undergoing carbonylation, they require expensive equipment, and which limits their general availability [73]. Despite the potential usefulness of this marker in processes in which oxidative stress is implicated, results from our laboratory show that the measurement of protein carbonyls has no greater accuracy than do the measurements of hs-CRP and β2-microglobulin in the diagnosis of PAD.

Homocysteine

Homocysteine is an intermediate metabolite of methionine resulting from the action of the enzyme methionine adenosyl transferase (MAT). It is a sulphur amino acid, the importance of which lies in the transference of methyl groups in cellular metabolism [74]. Methionine is generated from homocysteine re-methylation and enzyme catalysis by homocysteinemethyltransferase (HMT); this enzyme needing vitamin B12 and 5,10-methylenetetrahydrofolate for its function. The 5,10-methylenetetrahydrofolate acts as co-substrate in the course of its conversion to 5-methylenetetrahydrofolate by the enzyme methylene tetra-hydrofolate reductase (MTHFR). Vitamins B6 and B12 act as co-enzymes for methionine and homocysteine metabolism. Hence, there is a direct relationship between these two vitamins and homocysteine levels. Homocysteine contributes to cardiovascular disease through mechanisms that include endothelial dysfu-nction, increased lipid permeability, and vascular inflammation. The association between hyper-homocysteinaemia and increased oxidative stress has been demonstrated [75]; clinical studies have shown an incidence of hyper-homocysteinaemia of between 28 and 30% in patients with premature PAD. Thus, hyper-homocysteinaemia appears to be an independent risk factor for the development of this disease [76]. Mutations in the genes for MTHFR and cystathionine beta synthase (CBS) enzymes may alter the metabolism of homocysteine, promoting its increased concentration [77]. The *MTHFR* gene has received attention due to the relationship between folic acid intake and cardiovascular diseases. This gene has several polymorphisms; the C677T mutation being one of them. Patients bearing the 677T allele have lower enzyme activity than those patients with the 677C allele. The 677T allele is associated with an *in vitro* thermo-lability and a reduction of 50% in the enzyme's activity [78, 79].

There are three genotypes: 677CC (homozygote wild type), 677CT (heterozygote mutant) and 677TT (homozygote mutant). Subjects with 677TT have higher circulating homocysteine concentrations. Those with the 677TT allele represent about 10% of the population [80]. A second mutation in *MTHFR* in exon 7 is A1298C which results in a mutation that reduces MTHFR enzyme activity to a lesser extent than does C677T. However, compound heterozygous MTHFR A1298C/C677T carriers may develop hyper-homocysteinaemia.

CBS enzymes can alter the metabolism of homocysteine in favour of its increase [78]. Several mutations have been identified in the *CBS* gene, and the presence of these mutations in the homozygous form is associated with hyper-homocysteinaemia. The prevalence of these mutations varies among populations and ethnic groups; 844ins68 and T833C mutations being the most commonly reported to occur in the cis position in carriers of the 844ins68 mutation. The 844ins68 mutation in exon 8 causes the insertion of 68 base pairs in position 844, while the T833C mutation occurs as a replacement of a thymine by a cytosine nucleotide in position 833 [77, 78].

Homocysteine was first measured using amino acid analysers, or by radio-immunoassay. However, these techniques have been replaced with more convenient ones that are more rapid, and do not require the radioactivity of radio-immunoassay. More recently, high resolution liquid chromatography, with electrochemical or fluorescent detectors, as well as enzyme-immunoassays, have been developed. Immunoassay has the advantage of easy automation, enabling the measurement of homocysteine in routine clinical chemistry laboratories. Liquid chromatography linked to a mass-mass detector is the reference method, but it is time-consuming and the instrumentation is not readily available.

Collection and conservation is one of the most important features in homocysteine measurement. To pre-empt transfer of homocysteine from red blood cells into plasma, one of the most important features is to centrifuge the blood sample quickly at 4ºC, and separate the plasma from the rest of the constituents. There are no significant differences between serum and plasma samples with respect to homocysteine measurements. Overall, these data suggest that hyper-homocysteinaemia is an independent risk factor for PAD and CAD [80].

Metabolomics and the Search for New Biomarker Candidates

Metabolomics techniques are revolutionising clinical chemistry analyses, particularly in the search for circulating metabolites that are significantly altered in various diseases; their altered concentrations can be used as biomarkers of disease presence and/or disease status. In this section we propose some metabolites which, although not yet established unequivocally in the evaluation of peripheral arterial disease, there is sufficient evidence to warrant further research.

β-hydroxybutyrate

Ketones such as beta-hydroxybutyrate and acetoacetate are produced in states of negative energy catabolism, or in decreased glucose utilisation such as low carbohydrate ketogenic diets, starvation, or high-intensity exercise. Caloric restriction or fasting reduces inflammation as the organism adapts to low glucose intake. Energy metabolism then switches to mitochondrial fatty acid oxidation, ketogenesis and ketolysis [81]. In the liver, β-hydroxybutyrate levels are increased when there is an excess production of acetyl coenzyme A (acetyl CoA). Acetyl-CoA is enzymatically transformed to acetoacetate which, subsequently, is reduced to beta-hydroxybutyrate. Production of acetoacetate increases during carbohydrate deprivation (fasting, digestion perturbations, vomiting), decreased carbohydrate utilisation (glycogen storage diseases, diabetes mellitus, and alkalosis). The increase in acetoacetate may exceed the capacity of tissues and, as acetoacetate accumulates in the blood, a small amount is converted to acetone by decarboxylation. The remaining, and greater, portion of acetoacetate is converted to β-hydroxybutyrate. This compound is indirectly measured in human serum or urine (ketone bodies analysis) as a specific indicator of diabetic ketoacidosis in diabetes mellitus [82]. Ketone body metabolism comprises three pathways which are non-oxidative metabolic fates of ketone bodies: ketogenesis within hepatic mitochondria (the primary source of circulating ketone bodies); terminal oxidation within mitochondria of extrahepatic tissues *via* CoA transferase (primary metabolic fate of ketone bodies); and cytoplasmic *de novo* lipogenesis and cholesterol synthesis [81].

β-hydroxybutyrate levels are likely to be altered in PAD, since artery diseases modify fatty acid metabolism causing an accelerated rate of lactate formation from glucose. Moderately ischaemic tissue from oxidation causes a disruption in cell homeostasis resulting in lactate accumulation and decrease in pH and ATP. This situation can be inhibited by metabolites that reduce fatty acid oxidation and increase the combustion of glucose and lactate [83]. In addition, β-hydroxybutyrate inhibits the NLRP3 inflammasome in response to several NLRP3 activators. The NLRP3 inflammasome is an innate immune sensor that becomes activated in response to damage-associated molecular patterns (DAMPs) such as toxins, excess glucose, ATP, amyloids, ceramides, urate and cholesterol crystals. NLRP3 controls the activation of caspase-1 and the release of the pro-inflammatory cytokines IL-18 and IL-1β in macrophages [84, 85]. From a mechanistic point of view, β-hydroxybutyrate inhibits NLRP3 inflammasome without undergoing oxidation in the TCA cycle. It has other roles which implicate Sirt2 and the receptor Gpr109a. Inhibition of NLRP3 did not correlate with the magnitude of histone acetylation in macrophages; the inflammasome-mediated IL-1β and IL-18 production in human monocytes being inhibited. *In vivo*, this molecule attenuates caspase-1 activation and IL-1β secretion in mice. These data suggest that the anti-inflammatory effects of caloric restriction (or ketogenic diets) are mechanistically linked to inhibition of NLRP3 inflammasome *via* β-hydroxybutyrate. Hence, there is a potential therapeutic use of β-hydroxybutyrate against pro-inflammatory diseases, by acting against NLRP3 inflammasome [86].

Tricarboxylic Acid Cycle Products

LDL uptake by endothelial cells results in the generation of acetyl-CoA [37], which drives flux through the tricarboxylic acid cycle (TCA cycle). Increased acetyl-CoA alleviates the need for carbohydrate-derived precursors, thereby inhibiting glycolytic flux in the TCA cycle, and elevating the 3-carbon intermediates (Fig. **3**).

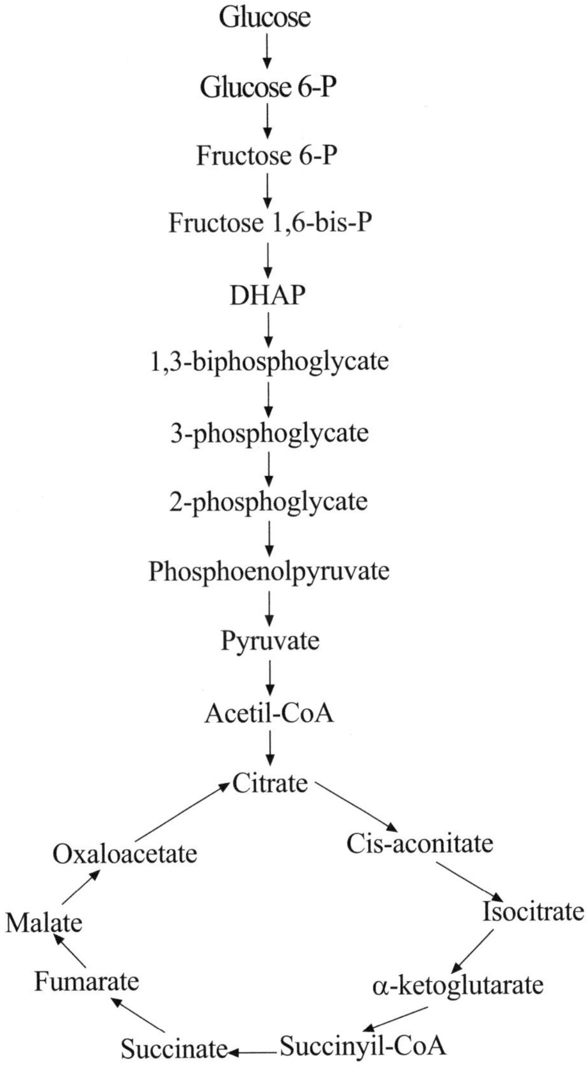

Fig. (3). Krebs' cycle (or "citric acid cycle"). Fig. adapted from Lehninger. Principios de Bioquímica.

Studies conducted by our group showed that treatment of endothelial cells with oxidised LDL could induce levels of oxidative stress sufficient to impair normal energy pathways. In cells treated with oxidised LDL there was an accumulation of 6-carbon intermediates and a decrease of 3-carbon intermediates. The hypothesis is that this situation may be due to changes in glyceraldehyde-3-phosphate dehydrogenase (GADPH; whether in concentration or activity) in response to oxidised LDL, since superoxide overproduction inhibits GADPH *via* a mechanism that involves poly-ADP-ribose polymerase (PARP) activation [60]. Following on from this possible explanation, it is necessary to highlight that the TCA cycle intermediates are lowered, probably due to the attenuated conversion of 6-carbon glycolytic intermediates to 3-carbon compounds. This situation feeds into the TCA cycle *via* two compounds, pyruvate and acetyl-CoA. When normal HDL is added to oxidised LDL-treated cells, there is a partial reversion of energy metabolism pathways.

One could conclude that oxidised LDL promotes the accumulation of 6-carbon glycolytic intermediates, and decreases 3-carbon TCA intermediates due to changes in GADPH *via* PARP activation. HDL affects this *via* PON-1. Our group has reported that the addition of normal HDL to oxidised LDL in cells, partially reverses the impact on energy metabolism pathways and, therefore, the concentrations of 3-carbon intermediates, and the TCA cycle, are normalised. However, the addition of HDL from PON1-deficient mice was shown not to normalise these alterations. These results suggest that PON1 has a beneficial role by curtailing the changes in energy metabolism often leading to endothelial dysfunction [27].

Amino Acids Arginine and Cysteine Related to TCA Cycle

These are glycogenic amino acids which are converted into TCA cycle intermediates, and their concentrations in the circulation have been observed to be increased in PAD. Arginine and its metabolite, homoarginine, increases endothelial NO production; the lack of NO increases vasoconstriction and inhibits platelet aggregation [9]. Several studies have investigated the role of arginine in PAD and its therapeutic supplementation on vascular and metabolic functions. However, results have been inconsistent. Some studies showed that short term complementation of arginine in the diet has positive effects in patients with PAD, but others showed that long-term supplementation could even be harmful. Symmetric dimethyl-arginine (SDMA), a structural analogue of asymmetric dimethyl-arginine (ADMA), inhibits NO formation by inhibiting cellular uptake of L-arginine [9, 87 - 89]. More studies are needed to define the status and function of arginine and its derivatives in PAD.

Cysteine is a non-essential amino acid, and is usually synthesized from serine and methionine. Cysteine plays a role, as a component of glutathione, in an endogenous defence mechanism to decrease oxidative stress. But cysteine is also a component of CCL2, a cysteine-cysteine protein, and its plasma concentration is decreased in PAD, probably due to increased CCL2 synthesis [90].

Other Biochemical Candidates

Hydroxycholesterol

The 'Oxysterol Hypothesis' was formulated by Kandutsch and his team in 1978. The assertion was that the negative feed-back of cholesterol on its own synthesis is mediated not by cholesterol itself, but by oxygenated forms of cholesterol *i.e.* the so called oxysterols. Currently, oxysterols are considered end-products of cholesterol metabolism; they are oxygenated sterol derivatives and can be biologically potent. However, the half-life is short [91, 92].

The field of oxysterol research has been revitalised recently since these molecules have been implicated in the development and progression of atherosclerosis. However, there is no consensus regarding their involvement in atherosclerotic lesions. Currently, it is known that 27-hydroxy cholesterol (27HC) is increased in atherosclerotic lesions, but little is known of its role in atherogenesis. Other potentially interesting hydroxycholesterols are 24-hydroxy cholesterol (24HC) which eliminates excess cholesterol from brain. 25HC in humans is exclusively produced in the brain and is involved in immune function together with 27HC which is a selective modulator; both are precursors of bile acid synthesis [92].

Galectin-3

Galectins were identified after the work by Ashwell and Morrell who characterised, in the 1960s, the asialoglycoprotein receptor. Galectins are a family of lectins that bind β-galactosides. They are highly conserved mammalian lectins that regulate inflammation. Galectins are classified into three distinct groups based on their quaternary structure: prototypical, chimeric, and tandem repeat [93].

Galectin-3 is a carbohydrate-binding protein which participates in a variety of biological process including proliferation, macrophage chemotaxis, phagocytosis, neutrophil extravasation, oxidative stress, apoptosis and angiogenesis. All of these mechanisms are involved in cardiovascular diseases [94]. The role of galectin-3 as a biomarker of heart failure has been demonstrated; it seems that galectin-3 has value in predicting heart failure [95, 96], but there a dearth of physiological evidence of its involvement in atherosclerosis. Galectin-3 has a key role in vascular fibrosis, and is known to be involved in monocyte to macrophage conversion.

The study by Madrigal-Matute *et al.* [94] showed that galectin-3 was implicated in PAD and had a relevant contribution to patient mortality. The authors observed that galectin-3 levels are a significant, and independent, risk factor associated with overall cardiovascular mortality in patients with PAD. Other studies showed that galectin-3 levels were positively associated with nephropathy, heart failure, and PAD, and that galectin-3 could be used as a marker of systemic inflammation *via* macrophage activation promotion and monocyte attraction. The action is, predominantly, pro-inflammatory. Neutrophil-endothelial interactions are promoted and numerous cell types involved in inflammatory and immune responses are activated, leading to fibrosis, but essential for tissue repair. Galectin-3 could be acting as an inflammatory factor in vascular disease, thereby accelerating adverse ventricular remodelling [97].

Melatonin

Melatonin is an endogenous hormone synthesized and secreted mainly by the pineal gland. It was isolated and chemically identified as *N*-acetyl-5-metox-ytryptamine in the 1960s. Melatonin derives from tryptophan which, in a first step is transformed into serotonin, and in a second step, in melatonin by two enzymatic reactions. Melatonin is a potent free radical scavenger, with antioxidant properties [98].

Experimental studies show that melatonin acts at multiple levels: direct scavenging of free radicals and singlet oxygen; stimulation of antioxidant enzymes activities; inhibition of pro-oxidant enzymes; reduction of radical formation in neurones by anti-excitatory actions; anti-inflammatory properties; reduction of electron leakage by support of mitochondrial electron flux; prevention of oxidative stress by optimising phasing and amplitudes of metabolic circadian rhythms; formation of redox reactive melatonin metabolites with radical-scavenging properties; and, finally, scavenging of low reactivity free radicals by melatonyl radicals, or mediated by catalysts.

Melatonin stimulates gene expression and activity of several antioxidant enzymes such as glutathione peroxidase (GSH-Px), catalase (CAT), and superoxide dismutase (SOD) [98 - 101]. These data highlight melatonin as a potentially useful biomarker of atherosclerosis, and a possible therapeutic tool for this disease [102].

8-Hydroxy-2-Deoxy Guanosine

8-hydroxy-2-deoxy guanosine (8-OhdG) is an oxidised derivative of deoxy-guanosine. It is one of the major products of DNA oxidation. The measurement of 8-OhdG concentration in serum or urine is a good index of oxidative stress. Endogenous oxidative damage to DNA can be measured as an indicator of aging, or as an indicator of oxidative stress; both being related. The extent of damage can be assessed by measuring the steady-state level of 8-OhdG [103]. 8-OHdG is used in measuring endothelial dysfunction in CAD [104]. However, there are few studies that have investigated the relationship of 8-OhdG with PAD [105].

CONCLUSION

The lack of specificity of currently-used established biomarkers such as CRP and B2M limit their usefulness in the study of patients with PAD. In particular, these molecules are often increased in serum when the disease is already clinically symptomatic, and they are not useful in discriminating between different stages of the disease, or of severity. This problem has encouraged investigators to focus on evaluating new markers that have sufficient sensitivity and specificity to identify asymptomatic PAD patients, alternatively, the patients may be diagnosed already and have had clinical intervention. In such cases, these markers may be used to monitor the course of the disease. Results from our group have shown that emerging biomarkers related to oxidative stress and inflammation may be useful in the early diagnosis of PAD. These biomarkers include 8-isoprostanes, CCL2 and PON1. Investigation of these biomarkers and alterations in energy dynamics and metabolism ("metabolomics") would provide new analytical tools with potential clinical and therapeutic importance.

LIST OF ABBREVIATIONS

ABI	ankle-brachial index
CAD	coronary artery disease
CCL2	chemokine (C-C motif) ligand 2
CVD	cardiovascular disease
HDL	high density lipoproteins
ICAM-1	intercellular adhesion molecule-1
LDL	low density lipoproteins
MCP1	monocyte chemo attractant protein-1
NO	nitric oxide
PAD	peripheral artery disease
PON	paraoxonase

ROS	reactive oxygen species
VCAM-1	vascular cell adhesion molecule-1
8-OHdG	8-hydroxy-deoxyguanosine
24-HC	24- hydroxy cholesterol
25-HC	25- hydroxy cholesterol
27-HC	27- hydroxy cholesterol

CONFLICT OF INTEREST

The author (editor) declares no conflict of interest, financial or otherwise.

ACKNOWLEDGEMENTS

Studies from the *Unitat de Recerca Biomèdica* reported in this manuscript have been supported by grants from the *Instituto de Salud Carlos III* (PI1102817, PI1100130, and PI15/00285) and the *Fondo Europeo de Desarrollo Regional* (FEDER), Madrid, Spain, and also from *Generalitat de Catalunya* (14SGR1227), Barcelona, Spain. Editorial assistance was provided by Dr. Peter R. Turner of *t-SciMed.*

REFERENCES

[1] Ouriel K. Peripheral arterial disease. Lancet 2001; 358(9289): 1257-64. [http://dx.doi.org/10.1016/S0140-6736(01)06351-6]

[2] Brevetti G, Giugliano G, Brevetti L, Hiatt WR. Inflammation in peripheral artery disease. Circulation 2010; 122(18): 1862-75. [http://dx.doi.org/10.1161/CIRCULATIONAHA.109.918417]

[3] Tzoulaki I, Murray GD, Lee AJ, Rumley A, Lowe GD, Fowkes FG. Inflammatory, haemostatic, and rheological markers for incident peripheral arterial disease: Edinburgh Artery Study. Eur Heart J 2007; 28(3): 354-62. [http://dx.doi.org/10.1093/eurheartj/ehl441]

[4] Cooke JP, Wilson AM. Biomarkers of peripheral arterial disease. J Am Coll Cardiol 2010; 55(19): 2017-23. [http://dx.doi.org/10.1016/j.jacc.2009.08.090]

[5] Conte MS, Pomposelli FB. Society for Vascular Surgery Practice guidelines for atherosclerotic occlusive disease of the lower extremities management of asymptomatic disease and claudication. Introduction. J Vasc Surg 2015; 61(3) (Suppl.): 1S. [http://dx.doi.org/10.1016/j.jvs.2014.12.006]

[6] Norgren L, Hiatt WR, Dormandy JA, *et al.* Inter-society consensus for the management of peripheral arterial disease (TASC II). J Vasc Surg 2007; (Suppl S)S5-67.

[7] Fowkes FG, Rudan D, Rudan I, *et al.* Comparison of global estimates of prevalence and risk factors for peripheral artery disease in 2000 and 2010: a systematic review and analysis. Lancet 2013; 382(9901): 1329-40. [http://dx.doi.org/10.1016/S0140-6736(13)61249-0]

[8] Kumakura H, Fujita K, Kanai H, *et al.* High-sensitivity C-reactive protein, lipoprotein(a) and

homocysteine are risk factors for coronary artery disease in Japanese patients with peripheral arterial disease. J Atheroscler Thromb 2015; 22(4): 344-54. [http://dx.doi.org/10.5551/jat.25478]

[9] Prasad K, Dhar I. Oxidative stress as a mechanism of added sugar-induced cardiovascular disease. Int J Angiol.: official publication of the International College of Angiology. Inc 2014; 23(4): 217-26.

[10] Guindo J, Gusi G, Punti J, Bermúdez P, Martínez-Rubio A. Métodos diagnósticos de la enfermedad arterial periférica. Rev Esp Cardiol 2009; 09(D): 11-7.

[11] Missouris CG, Buckenham T, Cappuccio FP, MacGregor GA. Renal artery stenosis: a common and important problem in patients with peripheral vascular disease. Am J Med 1994; 96(1): 10-4. [http://dx.doi.org/10.1016/0002-9343(94)90109-0]

[12] Sadat U, Walsh SR, Norden AG, Gillard JH, Boyle JR. Does oral N-acetylcysteine reduce contrast-induced renal injury in patients with peripheral arterial disease undergoing peripheral angiography? A randomized-controlled study. Angiology 2011; 62(3): 225-30. [http://dx.doi.org/10.1177/0003319710377078]

[13] Camps J, Marsillach J, Joven J. The paraoxonases: role in human diseases and methodological difficulties in measurement. Crit Rev Clin Lab Sci 2009; 46(2): 83-106. [http://dx.doi.org/10.1080/10408360802610878]

[14] Kalampogias A, Siasos G, Oikonomou E, Tsalamandris S, Mourouzis K, Tsigkou V, *et al.* Basic Mechanisms in Atherosclerosis: The Role of Calcium. Med Chem 2016; 12(2): 103-13. [http://dx.doi.org/10.2174/1573406411666150928111446]

[15] Williams KJ, Tabas I. Atherosclerosis--an inflammatory disease. N Engl J Med 1928; 340(24)

[16] Rani V, Deep G, Singh RK, Palle K, Yadav UC. Oxidative stress and metabolic disorders: Pathogenesis and therapeutic strategies. Life Sci 2016; 148: 183-93. [http://dx.doi.org/10.1016/j.lfs.2016.02.002]

[17] Chisolm GM, Steinberg D. The oxidative modification hypothesis of atherogenesis: an overview. Free Radic Biol Med 2000; 28(12): 1815-26. [http://dx.doi.org/10.1016/S0891-5849(00)00344-0]

[18] Stocker R, Keaney JF Jr. Role of oxidative modifications in atherosclerosis. Physiol Rev 2004; 84(4): 1381-478. [http://dx.doi.org/10.1152/physrev.00047.2003]

[19] Mackness B, Hine D, Liu Y, Mastorikou M, Mackness M. Paraoxonase-1 inhibits oxidised LDL-induced MCP-1 production by endothelial cells. Biochem Biophys Res Commun 2004; 318(3): 680-3. [http://dx.doi.org/10.1016/j.bbrc.2004.04.056]

[20] Offredo L, Pignatelli P, Cangemi R, *et al.* Imbalance between nitric oxide generation and oxidative stress in patients with peripheral arterial disease: effect of an antioxidant treatment. J Vasc Surg 2006; 44(3): 525-30. [http://dx.doi.org/10.1016/j.jvs.2006.05.023]

[21] Fry BC, Edwards A, Layton AT. Impact of nitric-oxide-mediated vasodilation and oxidative stress on renal medullary oxygenation: a modeling study. Am J Physiol Renal Physiol 2016; 310(3): F237-47. [http://dx.doi.org/10.1152/ajprenal.00334.2015]

[22] van Hinsbergh VW. Endothelium--role in regulation of coagulation and inflammation. Semin Immunopathol 2012; 34(1): 93-106. [http://dx.doi.org/10.1007/s00281-011-0285-5]

[23] Bharadwaj L, Prasad K. Mediation of H_2O_2-induced vascular relaxation by endothelium-derived relaxing factor. Mol Cell Biochem 1995; 149-150: 267-70. [http://dx.doi.org/10.1007/BF01076587]

[24] Hughes AK, Stricklett PK, Padilla E, Kohan DE. Effect of reactive oxygen species on endothelin-1

production by human mesangial cells. Kidney Int 1996; 49(1): 181-9.
[http://dx.doi.org/10.1038/ki.1996.25]

[25] Prasad K, Kalra J, Chan WP, Chaudhary AK. Effect of oxygen free radicals on cardiovascular function at organ and cellular levels. Am Heart J 1989; 117(6): 1196-202.
[http://dx.doi.org/10.1016/0002-8703(89)90396-7]

[26] Prasad K, Kalra J, Chaudhary AK, Debnath D. Effect of polymorphonuclear leukocyte-derived oxygen free radicals and hypochlorous acid on cardiac function and some biochemical parameters. Am Heart J 1990; 119(3 Pt 1): 538-50.
[http://dx.doi.org/10.1016/S0002-8703(05)80276-5]

[27] Garcia-Heredia A, Marsillach J, Rull A, *et al.* Paraoxonase-1 inhibits oxidized low-density lipoprotein-induced metabolic alterations and apoptosis in endothelial cells: a nondirected metabolomic study. Mediators of inflamm 2013; 2013: 156053.
[http://dx.doi.org/10.1155/2013/156053]

[28] Mackness MI, Arrol S, Durrington PN. Paraoxonase prevents accumulation of lipoperoxides in low-density lipoprotein. FEBS Lett 1991; 286(1-2): 152-4.
[http://dx.doi.org/10.1016/0014-5793(91)80962-3]

[29] Zweig MH, Campbell G. Receiver-operating characteristic (ROC) plots: a fundamental evaluation tool in clinical medicine. Clin Chem 1993; 39(4): 561-77.

[30] Campbell G. Advances in statistical methodology for the evaluation of diagnostic and laboratory tests. Stat Med 1994; 13(5-7): 499-508.
[http://dx.doi.org/10.1002/sim.4780130513]

[31] Murabito JM, Keyes MJ, Guo CY, *et al.* Cross- sectional relations of multiple inflammatory biomarkers to peripheral arterial disease: The Framingham Offspring Study. Atherosclerosis 2009; 203(2): 509-14.
[http://dx.doi.org/10.1016/j.atherosclerosis.2008.06.031]

[32] Buckley DI, Fu R, Freeman M, Rogers K, Helfand M. C-reactive protein as a risk factor for coronary heart disease: a systematic review and meta-analyses for the U.S. Preventive Services Task Force. Ann Intern Med 2009; 151(7): 483-95.
[http://dx.doi.org/10.7326/0003-4819-151-7-200910060-00009]

[33] Real de Asua D, Puchades R, Garcia-Polo I, Suarez C. A study on the relationship between serum beta 2-microglobulin levels, underlying chronic kidney disease, and peripheral arterial disease in high-vascular-risk patients. Int Cardiovasc Res J 2012; 6(4): 107-12.

[34] Kals J, Zagura M, Serg M, *et al.* beta2-microglobulin, a novel biomarker of peripheral arterial disease, independently predicts aortic stiffness in these patients. Scand J Clin Lab Invest 2011; 71(4): 257-63.
[http://dx.doi.org/10.3109/00365513.2011.558108]

[35] Morrow JD, Roberts LJ II. The isoprostanes. Current knowledge and directions for future research. Biochem Pharmacol 1996; 51(1): 1-9.
[http://dx.doi.org/10.1016/0006-2952(95)02072-1]

[36] Evans AR, Junger H, Southall MD, *et al.* Isoprostanes, novel eicosanoids that produce nociception and sensitize rat sensory neurons. J Pharmacol Exp Ther 2000; 293(3): 912-20.

[37] Czerska M, Zielinski M, Gromadzinska J. Isoprostanes - A novel major group of oxidative stress markers. Int J Occup Med Environ Health 2016; 29(2): 179-90.
[http://dx.doi.org/10.13075/ijomeh.1896.00596]

[38] Liu W, Morrow JD, Yin H. Quantification of F2-isoprostanes as a reliable index of oxidative stress *in vivo* using gas chromatography-mass spectrometry (GC-MS) method. Free Radic Biol Med 2009; 47(8): 1101-7.
[http://dx.doi.org/10.1016/j.freeradbiomed.2009.07.028]

[39] Milne GL, Yin H, Morrow JD. Human biochemistry of the isoprostane pathway. J Biol Chem 2008;

283(23): 15533-7.
[http://dx.doi.org/10.1074/jbc.R700047200]

[40] Milne GL, Yin H, Brooks JD, Sanchez S, Jackson Roberts L II, Morrow JD. Quantification of F2-isoprostanes in biological fluids and tissues as a measure of oxidant stress. Methods Enzymol 2007; 433: 113-26.
[http://dx.doi.org/10.1016/S0076-6879(07)33006-1]

[41] Primo-Parmo SL, Sorenson RC, Teiber J, La Du BN. The human serum paraoxonase/arylesterase gene (PON1) is one member of a multigene family. Genomics 1996; 33(3): 498-507.
[http://dx.doi.org/10.1006/geno.1996.0225]

[42] Sorenson RC, Primo-Parmo SL, Camper SA, La Du BN. The genetic mapping and gene structure of mouse paraoxonase/arylesterase. Genomics 1995; 30(3): 431-8.
[http://dx.doi.org/10.1006/geno.1995.1261]

[43] Mackness MI, Mackness B, Durrington PN. Paraoxonase and coronary heart disease. Atheroscler Suppl 2002; 3(4): 49-55.
[http://dx.doi.org/10.1016/S1567-5688(02)00046-6]

[44] Hine D, Mackness B, Mackness M. Coincubation of PON1, APO A1, and LCAT increases the time HDL is able to prevent LDL oxidation. IUBMB Life 2012; 64(2): 157-61.
[http://dx.doi.org/10.1002/iub.588]

[45] Hernandez-Aguilera A, Sepulveda J, Rodriguez-Gallego E, *et al.* Immunohistochemical analysis of paraoxonases and chemokines in arteries of patients with peripheral artery disease. Int J Mol Sci 2015; 16(5): 11323-38.
[http://dx.doi.org/10.3390/ijms160511323]

[46] Eckerson HW, Romson J, Wyte C, La Du BN. The human serum paraoxonase polymorphism: identification of phenotypes by their response to salts. Am J Hum Genet 1983; 35(2): 214-27.

[47] Eckerson HW, Wyte CM, La Du BN. The human serum paraoxonase/arylesterase polymorphism. Am J Hum Genet 1983; 35(6): 1126-38.

[48] Richter RJ, Jampsa RL, Jarvik GP, Costa LG, Furlong CE. Determination of paraoxonase 1 status and genotypes at specific polymorphic sites 2004. Chapter 4:Unit4.12
[http://dx.doi.org/10.1002/0471140856.tx0412s19]

[49] Haagen L, Brock A. A new automated method for phenotyping arylesterase (EC 3.1.1.2) based upon inhibition of enzymatic hydrolysis of 4-nitrophenyl acetate by phenyl acetate. European journal of clinical chemistry and clinical biochemistry. Eur J Clin Chem Clin Biochem 1992; 30(7): 391-5.

[50] Khersonsky O, Tawfik DS. Chromogenic and fluorogenic assays for the lactonase activity of serum paraoxonases. ChemBioChem 2006; 7(1): 49-53.
[http://dx.doi.org/10.1002/cbic.200500334]

[51] Gaidukov L, Tawfik DS. The development of human sera tests for HDL-bound serum PON1 and its lipolactonase activity. J Lipid Res 2007; 48(7): 1637-46.
[http://dx.doi.org/10.1194/jlr.D600045-JLR200]

[52] Gaidukov L, Rosenblat M, Aviram M, Tawfik DS. The 192R/Q polymorphs of serum paraoxonase PON1 differ in HDL binding, lipolactonase stimulation, and cholesterol efflux. J Lipid Res 2006; 47(11): 2492-502.
[http://dx.doi.org/10.1194/jlr.M600297-JLR200]

[53] Kawai H, Yomoda S, Inoue Y. ELISA using monoclonal antibody to human serum arylesterase. Clin Chim Acta 1991; 202(3): 219-25.
[http://dx.doi.org/10.1016/0009-8981(91)90052-E]

[54] Mackness B, Hunt R, Durrington PN, Mackness MI. Increased immunolocalization of paraoxonase, clusterin, and apolipoprotein A-I in the human artery wall with the progression of atherosclerosis. Arterioscler Thromb Vasc Biol 1997; 17(7): 1233-8.

[http://dx.doi.org/10.1161/01.ATV.17.7.1233]

[55] Ferre N, Marsillach J, Camps J, *et al.* Paraoxonase-1 is associated with oxidative stress, fibrosis and FAS expression in chronic liver diseases. J Hepatol 2006; 45(1): 51-9. [http://dx.doi.org/10.1016/j.jhep.2005.12.018]

[56] James RW, Blatter Garin MC, Calabresi L, *et al.* Modulated serum activities and concentrations of paraoxonase in high density lipoprotein deficiency states. Atherosclerosis 1998; 139(1): 77-82. [http://dx.doi.org/10.1016/S0021-9150(98)00058-6]

[57] Deakin SP, James RW. Genetic and environmental factors modulating serum concentrations and activities of the antioxidant enzyme paraoxonase-1. Clin Sci (Lond) 2004; 107(5): 435-47. [http://dx.doi.org/10.1042/CS20040187]

[58] Marsillach J, Mackness B, Mackness M, *et al.* Immunohistochemical analysis of paraoxonases-1, 2, and 3 expression in normal mouse tissues. Free Radic Biol Med 2008; 45(2): 146-57. [http://dx.doi.org/10.1016/j.freeradbiomed.2008.03.023]

[59] Marsillach J, Camps J, Beltran-Debon R, *et al.* Immunohistochemical analysis of paraoxonases-1 and 3 in human atheromatous plaques. Eur J Clin Invest 2011; 41(3): 308-14. [http://dx.doi.org/10.1111/j.1365-2362.2010.02411.x]

[60] Marsillach J, Aragones G, Mackness B, *et al.* Decreased paraoxonase-1 activity is associated with alterations of high-density lipoprotein particles in chronic liver impairment. Lipids Health Dis 2010; 9: 46. [http://dx.doi.org/10.1186/1476-511X-9-46]

[61] Aragones G, Guardiola M, Barreda M, *et al.* Measurement of serum PON-3 concentration: method evaluation, reference values, and influence of genotypes in a population-based study. J Lipid Res 2011; 52(5): 1055-61. [http://dx.doi.org/10.1194/jlr.D014134]

[62] Rodriguez-Gallego E, Riera-Borrull M, Hernandez-Aguilera A, Marine-Casado R, Rull A. Ubiquitous transgenic overexpression of C-C chemokine ligand 2: a model to assess the combined effect of high energy intake and continuous low-grade inflammation. Mediators Inflamm 2013; 2013: 953841.

[63] Satiroglu O, Uydu HA, Demir A, Bostan M, Atak M, Bozkurt E. Association between plasma monocyte chemoattractant protein-1 levels and the extent of atherosclerotic peripheral artery disease. Tohoku J Exp Med 2011; 224(4): 301-6. [http://dx.doi.org/10.1620/tjem.224.301]

[64] Rull A, Hernandez-Aguilera A, Fibla M, *et al.* Understanding the role of circulating chemokine (C-C motif) ligand 2 in patients with chronic ischemia threatening the lower extremities. Vasc Med 2014; 19(6): 442-51. [http://dx.doi.org/10.1177/1358863X14554034]

[65] Charo IF, Taubman MB. Chemokines in the pathogenesis of vascular disease. Circ Res 2004; 95(9): 858-66. [http://dx.doi.org/10.1161/01.RES.0000146672.10582.17]

[66] Rius B, Lopez-Vicario C, Gonzalez-Periz A, *et al.* Resolution of inflammation in obesity-induced liver disease. Front Immunol 2012; 3: 257. [http://dx.doi.org/10.3389/fimmu.2012.00257]

[67] Rodriguez-Sanabria F, Rull A, Beltran-Debon R, *et al.* Tissue distribution and expression of paraoxonases and chemokines in mouse: the ubiquitous and joint localisation suggest a systemic and coordinated role. J Mol Histol 2010; 41(6): 379-86. [http://dx.doi.org/10.1007/s10735-010-9299-x]

[68] Rull A, Camps J, Alonso-Villaverde C, Joven J. Insulin resistance, inflammation, and obesity: role of monocyte chemoattractant protein-1 (or CCL2) in the regulation of metabolism. Mediators Inflamm 2010; 2010: 326580.

[http://dx.doi.org/10.1155/2010/326580]

[69] Camps J, Rodriguez-Gallego E, Garcia-Heredia A, Triguero I, Riera-Borrull M. Paraoxonases and chemokine (C-C motif) ligand-2 in noncommunicable diseases. Adv Clin Chem 2014; 63: 247-308. [http://dx.doi.org/10.1016/B978-0-12-800094-6.00007-8]

[70] Camps J, Joven J. Chemokine ligand 2 and paraoxonase-1 in non-alcoholic fatty liver disease: The search for alternative causative factors. World J Gastroenterol 2015; 21(10): 2875-82. [http://dx.doi.org/10.3748/wjg.v21.i10.2875]

[71] Rull A, Garcia R, Fernandez-Sender L, *et al.* The role of combined assessment of defense against oxidative stress and inflammation in the evaluation of peripheral arterial disease. Curr Mol Med 2011; 11(6): 453-64. [http://dx.doi.org/10.2174/156652411796268713]

[72] Colombo G, Clerici M, Garavaglia ME, *et al.* A step-by-step protocol for assaying protein carbonylation in biological samples. J Chromatogr B Analyt Technol Biomed Life Sci 2016; 1019: 178-90. [http://dx.doi.org/10.1016/j.jchromb.2015.11.052]

[73] Wong CM, Cheema AK, Zhang L, Suzuki YJ. Protein carbonylation as a novel mechanism in redox signaling. Circ Res 2008; 102(3): 310-8. [http://dx.doi.org/10.1161/CIRCRESAHA.107.159814]

[74] Finkelstein JD, Martin JJ. Homocysteine. Int J Biochem Cell Biol 2000; 32(4): 385-9. [http://dx.doi.org/10.1016/S1357-2725(99)00138-7]

[75] Welch GN, Upchurch GR Jr, Loscalzo J. Homocysteine, oxidative stress, and vascular disease. Hosp Pract (1995) 1997 Jun 15; 32(6): 81-2. 85, 88-92

[76] Santos ME. Mutations in methylenetetrahydrofolate reductase and in cysthationine beta synthase: is there a link to homocysteine levels in peripheral arterial disease? Mol Biol Rep 2011; 38(5): 3361-6. [http://dx.doi.org/10.1007/s11033-010-0443-1]

[77] Alessio AC, Siqueira LH, Bydlowski SP, Hoehr NF, Annichino-Bizzacchi JM. Polymorphisms in the CBS gene and homocysteine, folate and vitamin B12 levels: association with polymorphisms in the MTHFR and MTRR genes in Brazilian children. Am J Med Genet A 2008; 146A(20): 2598-602. [http://dx.doi.org/10.1002/ajmg.a.32496]

[78] Khandanpour N, Willis G, Meyer FJ, *et al.* Peripheral arterial disease and methylenetetrahydrofolate reductase (MTHFR) C677T mutations: A case-control study and meta-analysis. J Vasc Surg 2009; 49(3): 711-8. [http://dx.doi.org/10.1016/j.jvs.2008.10.004]

[79] Anderson CA, Beresford SA, McLerran D, *et al.* Response of serum and red blood cell folate concentrations to folic acid supplementation depends on methylenetetrahydrofolate reductase C677T genotype: results from a crossover trial. Mol Nutr Food Res 2013; 57(4): 637-44. [http://dx.doi.org/10.1002/mnfr.201200108]

[80] Sheweita SA, Baghdadi H, Allam AR. Role of genetic changes in the progression of cardiovascular diseases. Int J Biomed Sci 2011; 7(4): 238-48.

[81] Cotter DG, Schugar RC, Crawford PA. Ketone body metabolism and cardiovascular disease. Am J Physiol Heart Circ Physiol 2013; 304(8): H1060-76. [http://dx.doi.org/10.1152/ajpheart.00646.2012]

[82] Nakamura A, Osonoi T, Terauchi Y. Relationship between urinary sodium excretion and pioglitazone-induced edema. J Diabetes Investig 2010; 1(5): 208-11. [http://dx.doi.org/10.1111/j.2040-1124.2010.00046.x]

[83] Wolff AA, Rotmensch HH, Stanley WC, Ferrari R. Metabolic approaches to the treatment of ischemic heart disease: the clinicians' perspective. Heart Fail Rev 2002; 7(2): 187-203. [http://dx.doi.org/10.1023/A:1015384710373]

[84] Lamkanfi M, Dixit VM. Mechanisms and functions of inflammasomes. Cell 2014; 157(5): 1013-22. [http://dx.doi.org/10.1016/j.cell.2014.04.007]

[85] Franchi L, Eigenbrod T, Munoz-Planillo R, Nunez G. The inflammasome: a caspase-1-activation platform that regulates immune responses and disease pathogenesis. Nat Immunol 2009; 10(3): 241-7. [http://dx.doi.org/10.1038/ni.1703]

[86] Youm YH, Nguyen KY, Grant RW, *et al.* The ketone metabolite beta-hydroxybutyrate blocks NLRP3 inflammasome-mediated inflammatory disease. Nat Med 2015; 21(3): 263-9.

[87] Schneider JY, Rothmann S, Schroder F, *et al.* Effects of chronic oral L-arginine administration on the L-arginine/NO pathway in patients with peripheral arterial occlusive disease or coronary artery disease: L-Arginine prevents renal loss of nitrite, the major NO reservoir. Amino Acids 2015; 47(9): 1961-74. [http://dx.doi.org/10.1007/s00726-015-2031-0]

[88] Staniszewska A, Rajagopalan S, Al-Shaheen A, Thies F, Brittenden J. Increased levels of symmetric dimethyl-arginine are associated with all-cause mortality in patients with symptomatic peripheral arterial disease. J Vasc Surg 2015; 61(5): 1292-8. [http://dx.doi.org/10.1016/j.jvs.2015.01.002]

[89] Vogl L, Pohlhammer J, Meinitzer A, *et al.* Serum concentrations of L-arginine and L-homoarginine in male patients with intermittent claudication: a cross-sectional and prospective investigation in the CAVASIC Study. Atherosclerosis 2015; 239(2): 607-14. [http://dx.doi.org/10.1016/j.atherosclerosis.2015.02.019]

[90] Zagura M, Kals J, Kilk K, *et al.* Metabolomic signature of arterial stiffness in male patients with peripheral arterial disease. Hypertens Res 2015; 38(12): 840-6. [http://dx.doi.org/10.1038/hr.2015.71]

[91] Luu W, Sharpe LJ, Capell-Hattam I, Gelissen IC, Brown AJ. Oxysterols: old tale, new twists. Annu Rev Pharmacol Toxicol 2016; 56: 447-67. [http://dx.doi.org/10.1146/annurev-pharmtox-010715-103233]

[92] Gill S, Chow R, Brown AJ. Sterol regulators of cholesterol homeostasis and beyond: the oxysterol hypothesis revisited and revised. Prog Lipid Res 2008; 47(6): 391-404. [http://dx.doi.org/10.1016/j.plipres.2008.04.002]

[93] Arnoys EJ, Ackerman CM, Wang JL. Nucleocytoplasmic shuttling of galectin-3. Methods Mol Biol 2015; 1207: 465-83. [http://dx.doi.org/10.1007/978-1-4939-1396-1_30]

[94] Madrigal-Matute J, Lindholt JS, Fernandez-Garcia CE, *et al.* Galectin-3, a biomarker linking oxidative stress and inflammation with the clinical outcomes of patients with atherothrombosis. J Am Heart Assoc 2014; 3(4): e000785. [http://dx.doi.org/10.1161/JAHA.114.000785]

[95] de Boer RA, Yu L, van Veldhuisen DJ. Galectin-3 in cardiac remodeling and heart failure. Curr Heart Fail Rep 2010; 7(1): 1-8. [http://dx.doi.org/10.1007/s11897-010-0004-x]

[96] Peacock WF. How galectin-3 changes acute heart failure decision making in the emergency department. Clin Chem Lab Med 2014; 52(10): 1409-12. [http://dx.doi.org/10.1515/cclm-2014-0221]

[97] Jin QH, Lou YF, Li TL, Chen HH, Liu Q, He XJ. Serum galectin-3: a risk factor for vascular complications in type 2 diabetes mellitus. Chin Med J (Engl) 2013; 126(11): 2109-15.

[98] Sanchez A, Calpena AC, Clares B. Evaluating the oxidative stress in inflammation: role of melatonin. Int J Mol Sci 2015; 16(8): 16981-7004. [http://dx.doi.org/10.3390/ijms160816981]

[99] Emamgholipour S, Hossein-Nezhad A, Ansari M. Can melatonin act as an antioxidant in hydrogen peroxide-induced oxidative stress model in human peripheral blood mononuclear cells? Biochem Res Int 2016; 2016: 5857940.
[http://dx.doi.org/10.1155/2016/5857940]

[100] Rodriguez C, Mayo JC, Sainz RM, *et al.* Regulation of antioxidant enzymes: a significant role for melatonin. J Pineal Res 2004; 36(1): 1-9.
[http://dx.doi.org/10.1046/j.1600-079X.2003.00092.x]

[101] Menendez-Pelaez A, Rodriguez C, Dominguez P. 5-aminolevulinate synthase mRNA levels in the Harderian gland of Syrian hamsters: correlation with porphyrin concentrations and regulation by androgens and melatonin. Mol Cell Endocrinol 1991; 80(1-3): 177-82.
[http://dx.doi.org/10.1016/0303-7207(91)90154-K]

[102] Stasiak M, Zasada K, Lewinski A, Karbownik-Lewinska M. Melatonin restores the basal level of lipid peroxidation in rat tissues exposed to potassium bromate *in vitro.* Neuroendocrinol Lett 2010; 31(3): 363-9.

[103] Fraga CG, Shigenaga MK, Park JW, Degan P, Ames BN. Oxidative damage to DNA during aging: 8-hydroxy-2'-deoxyguanosine in rat organ DNA and urine. Proc Natl Acad Sci USA 1990; 87(12): 4533-7.
[http://dx.doi.org/10.1073/pnas.87.12.4533]

[104] Maschirow L, Khalaf K, Al-Aubaidy HA, Jelinek HF. Inflammation, coagulation, endothelial dysfunction and oxidative stress in prediabetes--Biomarkers as a possible tool for early disease detection for rural screening. Clin Biochem 2015; 48(9): 581-5.
[http://dx.doi.org/10.1016/j.clinbiochem.2015.02.015]

[105] Loffredo L, Marcoccia A, Pignatelli P, *et al.* Oxidative-stress- mediated arterial dysfunction in patients with peripheral arterial disease. Eur Heart J 2007; 28(5): 608-12.
[http://dx.doi.org/10.1093/eurheartj/ehl533]

CHAPTER 13

Early Neonatal Sepsis Biomarkers

Natalia Sancho-Rodríguez[1,*], Marta Sancho-Rodríguez[2], Irene De-Migue--Elízaga[3] and Ana Martínez-Ruiz[4]

[1] *Santa Lucía General University Hospital, Cartagena, Murcia, Spain*

[2] *Pharmaceutical Production Team in MSD, Oss, The Netherlands*

[3] *Unilabs Laboratory. Torrevieja-Salud Hospital, Alicante, Spain*

[4] *Reina Sofía General University Hospital, Murcia, Spain*

Abstract: Early-onset neonatal sepsis is currently a major cause of morbidity and mortality in neonatal period and its rapid diagnosis can help to establish an effective antibiotic treatment. The suspicion diagnosis of neonatal sepsis is based on a number of risk factors and non-specific clinical and laboratory parameters, therefore in many cases it is difficult to assess when it is the proper period to initiate antibiotic treatment. Confirmatory diagnosis depends on the results of blood cultures in the neonatal period, hence the importance of a biochemical marker that allows predicting the likelihood infection, as well as supporting the diagnosis of sepsis. Therefore identifying tools for rapid detection of neonatal sepsis is an objective of great importance in perinatal medicine, as an early and accurate diagnosis leads to an appropriate treatment thus potentially improving the final prognosis of these patients. The objective of this work is to study different markers of early neonatal sepsis, biochemical and haematological, particularly in cord blood; and establish its potential clinical usefulness. New techniques of molecular biology in cord blood are being studied in different types of samples, both blood and neonatal cord blood.

Keywords: C-Protein Reactive (CRP), Cord blood, Early-onset neonatal sepsis (EONS), Interleukin-6 (IL-6), Procalcitonin (PCT), Septifast.

INTRODUCTION

Infectious diseases are considered a major cause of morbidity and mortality in neonatal period. Among others, early sepsis is considered one of the factors which most affect neonatal period.

The suspicion diagnosis of neonatal sepsis is based on a number of risk factors and non-specific clinical and laboratory parameters, therefore in many cases it is

* **Corresponding author Natalia Sancho-Rodríguez:** Santa Lucía General University Hospital, Cartagena, Murcia, Spain; Tel/Fax: 968128600; E-mail: natsancho11@gmail.com

Juan Antonio Vílchez (Ed.)

difficult to assess when it is the proper period to initiate antibiotic treatment [1].

Furthermore, confirmatory diagnosis depends on the results of blood cultures in the neonatal period. These are less profitable because the extraction volume is often insufficient to detect bacteraemia, because in some cases they are performed in an intermittent way. Thus, importance of a biochemical marker that allows us to predict the probability of infection and support sepsis diagnosis [2].

Therefore, identification of tools for rapid detection of early sepsis is a highly relevant objective within perinatal medicine, since an early and accurate diagnosis leads to appropriate treatment and potentially improves the final outcome of these patients.

Over the last decades many studies of different markers in cord blood have been performed in order to anticipate sepsis diagnosis in cases of suspected neonatal sepsis. This chapter will be focused on the study of these cord blood biomarkers for the early diagnosis of neonatal sepsis.

Human body has many ways of protecting. Some are simply physical barriers such as the external hard layer of skin keratin, which protects the cells placed below from a hostile environment [3]. Human body also produces a great number of a different group of serum proteins during intense infections. They are known as "acute phase proteins" which have antimicrobial effects; for example, C reactive protein (CRP) is fixed to protein C on the surface of *Pneumococci* bacteria and consequently their destruction is activated by means of the complement cascade [4].

But more complex, dynamic and effective defence strategies are carried out by specialized cells that travel through the body to destroy microorganisms and other foreign substances.

In humans there are three main groups of cells that provide this type of defence. The neutrophils and a number of monocytes-macrophages are phagocytic cells which act mainly by digesting bacteria, cell remains and other particles. The third group, lymphocytes and related elements have low phagocytic capacity, but instead they are involved in a large number of other protective reactions, which are collectively known as immune responses. Both phagocytes and lymphocytes are essential for human health, often acting together and depending on each other in order to reach maximum efficiency [5].

Inflammation process can be defined as the initial non-specific response to tissue injury caused by a mechanical, chemical or microbial stimulation [6]. Inflammation is a quick, humoral and cellular response, which is highly amplified but controlled process, wherein the cytokine cascade, complement, coagulation and fibrinolytic cascade are triggered together by bacterial elements by means of activation of macrophages and endothelial cells.

This local response is considered a benign process as long as the inflammatory process is properly regulated. The reaction has pro-inflammatory and anti-inflammatory components which sometimes are equivalent or higher than pro-inflammatory ones.

Cytokines are small molecules which are produced as endogenous mediators in immune response against bacterial infection, *i.e.* are physiological messengers of the inflammatory response. They are glycoprotein molecules whose fundamental roles are to intervene in the transmission of information or signals from one cell to another. They bind to specific receptors on their target cells causing changes in these cells and consequently leading to the synthesis and release of secondary mediators [7].

Although many cytokines play a possible role in the pathogenesis process and all of these have been isolated and characterized, only five of them have a clinically important role as pro-inflammatory cytokines: tumour necrosis factor alpha (TNF-alpha), interleukin-1 beta, interleukin-6 (IL-6) and interleukin-8 (IL-8) and interferon.

Infection process is the main stimulus for releasing cytokines by means of the action of bacterial molecules such as endotoxin (LPS), which are recognized by cells of the innate immune system. Polymorphic-nuclear (PMN), monocyte-macrophages and endothelial cells are the cellular effectors of inflammatory response [8].

In most cases, beneficial effect of pro-inflammatory mediators overcomes negative effects. They remove damaged tissue, promote tissue growth and fight pathogens, neoplastic cells and foreign antigens. In order to avoid that these harmful effects mediators develop over-stimulation, human body quickly develops an anti-inflammatory response [9].

This reaction involves compensatory anti-inflammatory cytokines such as interleukin 4 (IL-4), 10 (IL-10) and 11 (IL-11), soluble receptors and receptor antagonists. Their effect is less well known than one of pro-inflammatory mediators, but apparently they alter the function of monocytes and reduce the ability of cells to produce pro-inflammatory cytokines.

Endothelium is another key component in the system. Normally endothelial cells express an anticoagulant, anti-cell adhesion and vasodilator phenotype. When activated as in inflammation process, these cells express pro-coagulant properties and pro-cell adhesion [10].

The opposing forces between inflammation and anti-inflammation processes may tend to a balance or imbalance states. The imbalance state with increased inflammatory response produces a deep shock state, with high mortality. The process results fulminant as an example in the sepsis caused by meningococcus bacteria.

In the other hand, the prevalence of anti-inflammation leads the patient after the first days of sepsis to an energy state or "immune paralysis". This situation defines a term called Compensatory Anti-inflammatory Response Syndrome (CARS) which explains the increase of secondary infections such as those caused by catheters or injury [11].

The definition of sepsis was established in 1991 when the American College of Critical Care Medicine and Society of Critical Care Medicine (ACCM-SCCM) consensus took place to unify criteria regarding the sepsis definitions and new sepsis definitions and related processes were proposed.

A new conference of AMCM-SCCM took place in 1992 [12]. Systemic inflammatory response syndrome (SIRS) term was reintroduced into the common language, defined as some clinical manifestations of the inflammatory response caused by infectious and non-infectious causes. Subsequently, Bone *et al.* defined sepsis as systemic inflammatory response to infection [13].

An important fact in this new terminology is that it recognizes the fundamental role that systemic inflammation plays in sepsis, accepting that clinical manifestations are not caused only by factors related to microbial pathogenicity, but involves a conceptual change in the assessment of critical infected patients; that is a change in perspective and not a new clinical entity.

Therefore, neonatal sepsis is defined as an acute systemic infection with toxic manifestations caused by the invasion and proliferation of bacteria into the bloodstream and various organs, which occurs within the first four weeks of life and is shown by blood culture positive. There are two differences considered in neonatal sepsis period regarding the way of contamination [14].

Early-Onset Neonatal Sepsis (EONS)

Early neonatal sepsis or "vertical transmission", on which this paper is focused; it is usually presented as a fulminate and multisystem disease during the first three days of life through vertical transmission [15], either by germs located in the mother's birth canal that contaminate the fetus *via* upward progressing through the birth canal until it reaches the amniotic fluid; or by direct contact with contaminated fetus to pass through birth canal secretions [16].

For this reason possibility of sepsis should be considered when a positive culture by pathogenic bacteria in vaginal channel exudates during the two weeks before delivery is obtained, since there is growing evidence that the infection begins in the uterus in a significant number of newborns with early sepsis.

Sepsis of Late Submission

Late submission of sepsis can be classified according to the acquisition place of the etiologic agent; therefore it is named nosocomial sepsis and sepsis of Community origin.

The "nosocomial sepsis" is produced by microorganisms from hospital environment especially in neonatal intensive care units, which colonize the infant by contacting the health workers or from contaminated material [17].

However "Sepsis of Community origin" is caused due to microorganisms that contaminate the newborn at home and these are very unusual. The clinic process starts after 72 hours of life although it can begin earlier. The pathogens responsible for infection are different from the vertical sepsis; they are predominantly entero-bacteria and other gram-negative organisms [17].

EPIDEMIOLOGY

Sepsis remains as an important cause of unacceptably high mortality and morbidity in neonatal units despite advances in antibiotic therapy, support measures and knowledge of risk factors for infection.

Regarding the epidemiology of these types of neonatal sepsis, with estimation by the World Health Organization (WHO), approximately 20% of all live births in developing countries develop an infection and 1% dies due to neonatal sepsis [18]. The incidence in developed countries ranges from 1/500 to 1/1600 live births.

In specialized hospitals these rates of early onset sepsis or perinatal sepsis are close to 1/1000 term newborn, 1/230 in low birth weight and 164/1000 in premature births from 1000 to 1500 grams [19]. Late or nosocomial sepsis affects 2-5% of all newborns hospitalized and up to 15% of newborns admitted more than 48 hours in the neonatal ICU (Intensive Care Unit).

ETIOLOGY

Pathogens can contaminate the newborn at skin and/or respiratory level or digestive mucosa level and according to their characteristics pathogens are divided and able to cross the cutaneous and mucosal barrier and reach the bloodstream [17]. Once in the blood, if there is a competent immune defence, germs can start dividing logarithmically leading to a neonatal sepsis in the newborn.

Until the 70s the main etiological agents were gram-negative bacilli and it was later replaced by *Streptococcus agalactiae group B*, which explained the approximately 50% of early sepsis [20].

Main pathogens and risk factors associated to neonatal sepsis are shown in Table **1**.

Table 1. Main pathogens and risk factors associated to neonatal sepsis [21].

Neonatal sepsis	Microbial pathogens	Risk factors
Early-onset	Group B stretococci *Escherichia coli* *Streptococcus viridans* *Enterococci* *Staphylococcus aureus* *Pseudomonas aeruginosa* Other gram-negative bacilli	Maternal Group B streptococcal colonization Chorioamnionitis Premature rupture of membranes Prolonged rupture of membranes (>18h) Preterm birth (<37 weeks) Multiple gestation
Late-onset	Coagulase-negative Staph. *Staphylococcus aureus* *Candida albicans* *Enterococci* *Escherichia coli* *Klebsiella pneumoniae* *Pseudomonas aeruginosa* Group B streptococci	Prematurity Low birth weight Prolonged indwelling catheter use Invasive procedures Ventilator associated pneumonia Prolonged antibiotics

B-hemolytic *Streptococcus* group B (GBS) or *Streptococcus agalactiae*

It is most frequent bacterium associated (50-60%) to early onset neonatal sepsis [21]. GBS sepsis is associated to a mortality rate between 5 and 20% in developed countries and a significant percentage of sequelae (30%) among survivors [22].

It is a gram-positive *coccus* facultative anaerobic microorganism which is surrounded by a capsule composed mainly of polysaccharides with anti-phagocytic properties. According to its composition can be classified in 7 serotypes: Ia, Ib, II, III, IV, V, VI, and other temporary additional serotypes.

One of the major virulence factors of these bacteria is the presence of sialic acid component of the capsule. The sialic acid decreases activation of the alternative complement pathway, as well as presence of neuraminidase, whose pathogenic role is not well known [23] and further it is characterized by a high minimum inhibitory concentration (MIC) to penicillin. This gram-positive diplococcal bacterium in contact with the blood agar medium (type β haemolysis) turns to complete haemolysis.

Among the different serotypes, type III is the most frequently associated to positive cultures in the mother (26%) and illness in the newborn (64%) [20]. Type III is usually acquired during labour and is rarely observed in infants born *via* caesarean section with intact membranes [24].

The SGB is nowadays the most common cause of perinatal bacterial infection of vertical transmission in the Western world in absence of prevention. The exposure to GBS during birth is common resulting in colonization of 1 out of 10 infants. The process incidence (without prevention) reaches 3% of live births, with mortality in 1970 of 50%, which is currently reduced to 4-5% as a result of advances in neonatology [25].

Measures to prevent vertical transmission have been implemented due to high morbidity and mortality associated to sepsis caused by this bacterium. In 1992 American Academy of Paediatrics (AAP) [26] recommended a screening in pregnant women between 26 and 28 weeks gestation in order to find those women colonized by GBS.

Later, American College of Obstetricians and Gynaecologists (ACOG) recommended based on risk factors screening to include other factors as preterm birth (gestational age less than 37 weeks), history of previous child with GBS sepsis, GBS bacteraemia in current pregnancy, premature rupture of membranes more than 18 hours or intrapartum fever (greater than or equal to 38°C), as it was associated to a mortality rate between 5 and 20% [27].

Escherichia coli

Through the implemented protocols based on prevention antepartum GBS antibiotic prophylaxis, the importance of GBS as etiologic agent of early sepsis has been reduced since 2001. It has been replaced by other bacterium Escherichia coli, which is the causal agent who heads the new caseloads, returning gram-negative bacilli to occupy the first places in the ranking as in previous decades [22].

This organism is a Gram-negative bacillus facultative anaerobic microorganism present in most of the intestines of healthy mammals; whose function is necessary for the proper functioning of the digestive process, acting as commensally gut flora, helping the absorption of nutrients and B and K vitamins production.

In humans *Escherichia coli* bacterium colonizes the gastrointestinal tract of the newborn, by means of adhesion to the mucus of the large intestine within 48 hours after the first meal. It is associated with neonatal meningitis, acquired in the birth canal and to a lesser extent for nosocomial infection. In the WHO study Young Infant Study where 360 infants were included 26% of them had positive cultures where the largest growth was due to *Escherichia coli* [28].

Staphylococcus Epidermidis

Coagulase-negative staphylococci bacteria (CNS) are an essential part of the commensal flora skin and mucosa of human beings. They are considered classic opportunistic microorganisms which only have a limited potential pathogen in immune-competent people.

Staphylococcus epidermidis bacterium is consistent positive cocci bacterial specie aggregated in groups, responsible for 70-80% of infections caused by CNS. Furthermore, bacteria of these species are probably the most common ones found in laboratory analyses due to contamination.

The infections increase by this bacterium is clearly associated to increased survival of preterm infants with very low weight, longer stay in the Neonatal Intensive Care Unit (NICU) and exposure to intravascular devices as well as support nutrition. These are factors leading these patients to acquired sepsis by this organism [29].

DIAGNOSIS

Diagnosis in a newborn must consider clinical supports, being helpful the assessment of the presence of maternal risk factors in the vertical sepsis and one or more identifiable risk factors present usually on these infants.

A previous infectious check before starting empirical antibiotic therapy is necessary in order to clarify the diagnosis aetiology and address to sepsis.

Difficulty in Diagnosing

When the cause of etiologic agent sepsis invades the newborn body, immune system is activated to fight back. However, sometimes system can be physiologically immature in the premature newborn; the most important defence mechanism of the phagocytic deficit in new-born is a failure in the production and neutrophil chemotaxis to the infection site. Newborns have a very restricted capacity to accelerate neutrophils production in response to an infectious process and the deposit in the bone marrow is very limited so it is quickly over [29].

In addition there is no evidence either in clinic or laboratory with sufficient predictive value to confirm or rule out sepsis; thus different laboratory tests have been combined in order to enhance the diagnostic capacity. Neonatal sepsis is a disease of high severity; therefore examination which does not fail in any case is needed. That means high sensitivity tests are required and these should detect sepsis when the disease is present, that is, with high positive predictive value. The objective of a lab test is to help medical personnel to determine the start of antibiotic treatment or not. Therefore the ideal test would be one that has a high sensitivity and a high positive predictive value.

Considering that bacteriological results take several days to provide reliable results, a rapid screening procedure is needed in order to guide the establishment of newborn antibiotic treatment. To discard sepsis with a positive predictive value of 100%, infants with risk of infection are evaluated. This value of 100% is obtained at least 12-24 hours after birth, using the combination of WBC and CRP determination.

Etiologic Diagnosis

Etiologic diagnosis is based on the isolation of a pathogen on a normally sterile body fluid.

In case of suspected neonatal sepsis [30], studies to be performed are:

- Blood cultures considered as "gold standard" for sepsis diagnosis.
- Cerebrospinal fluid (CSF): whose importance consists on the fact that 20-25% of neonatal sepsis is associated to meningitis.
- Urine culture: the sample must be taken by supra-pubic catheterization or bladder puncture in the urethra. Urine culture in neonates less than 72 hours old has low sensitivity (less than 0.5%) and it is not recommended in case of absence of known anatomic abnormalities.
- Cultivation of peripheral exudates: they have indicative value of the aetiology in vertical sepsis. Positive result indicates contamination but the positivity of 3 or more of the same seed exudates with presence of clinical sepsis and negative blood cultures is a great value to address to the responsible bacterium.

Blood Microbiologic Culture

The cause of infection in bacterial neonatal sepsis is identified by blood culture, considered as gold standard test which allows determining also the bacterial susceptibility to antibiotics.

Bacterial culture is the most used and valued method in the clinic; however, it has low sensitivity and specificity for detection of the causal agents of infection as a diagnostic test in neonatal blood samples. This is related to the low volume (less than 1 mL) of blood to be drawn from the newborn. An additional limitation of blood culture is that it takes up to four days to obtain a laboratory result and cannot analyze the total content of bacteria that may be of interest to identify the disease aetiology.

Cultivation techniques automated or semi-automated based on the detection of CO_2 produced by bacterial metabolism enable to report positive blood culture within 24 hours.

Despite this, positivity of blood cultures in neonatal sepsis does not exceed 80-85%; thus a negative result in presence of risk factors and compatible symptoms does not discard infection [31].

For these reasons, methodologies based on a quick, accurate and highly sensitive analysis are required to help neonatologists and the responsible medical personnel to achieve a timely evaluation and better treatment of clinical and subclinical cases of neonatal sepsis.

Sepsis Biomarkers

Silbergeld and Davis [32], consider biomarkers to refer to physiological signals induced by a reflecting xenobiotic exposure, early cellular response or inherent or acquired susceptibility, providing a strategy for solving these problems. Thus, the proposal of a biomarker must be based on response mechanism knowledge.

Biomarkers should present also proper reproducibility, sensitivity, specificity, reversibility, applicability, usability and cost/effectiveness capacities. It is essential that response time of the biomarker is short, thus it can be used as "early warning system"; in case it shows diagnostic value, it can then be used predictively.

Therefore a marker of neonatal sepsis should allow early sensitive diagnosis (differentiate between infectious cause or not to an inflammation) and if possible report the prognosis [33].

Some parameters have been evaluated in recent decades to early diagnosis of neonatal sepsis. The most studied parameters so far are the white blood cell count, total number of neutrophils, immature neutrophils ratio, C-reactive protein, procalcitonin and cytokines.

Therefore, principal markers widely used in the evaluation of suspected neonatal sepsis and on which this study is focused are:

Haematological Parameters

Haematological parameters have a high specificity and measurement of peripheral blood leukocytes and differential study probably are the most rapid and useful nonspecific tests to obtain.

Several decades ago it was demonstrated that leukocyte count could be normal in a third part of infected newborns [34].

Numerous studies show that one isolated factor for sepsis diagnosis cannot be considered. The White Blood Cells (WBC) count and absolute neutrophil immature/mature ratio, changes in morphology or degeneration such as vacuolization, intracellular bacteria and toxic granulation neutrophils, should be studied and analyzed individually and altogether. In these analyzes the sensitivity is variable, reporting data from 40-90% without being directly related to the type of seed, which among other factors could affect the results [35].

Because of this variability in sensitivity of hematological parameters considered individually, several authors have proposed strategies in obtaining scores ratios or combining some of these parameters. The performance of these scoring systems is shown in Table **2**, depending on the hours of neonatal life (1-7 hours and 12-24 hours after birth) [36].

Table 2. Performance hematologic scoring systems according to various authors for the diagnosis of early neonatal GBS sepsis in the first 24 hours of life of newborn (Values of Sensitivity (S), Specificity (SP), Positive Predictive Value (PPV) and Negative Predictive Value (NPV) [36].

	S (%)	SP (%)	PPV (%)	NPV (%)
Initial blood count (1-7 hours after birth)				
Manroe *et al.*	68	45	43	70
Rodwell *et al.*	63	55	46	71
Spector *et al.*	31	90	67	68
Blood count 12-24 hours after birth				
Manroe *et al.*	100	50	56	100
Rodwell *et al.*	100	73	73	100
Spector *et al.*	53	83	67	73

These data suggest that early blood counts provide limited diagnostic information in evaluating a newborn with or without symptoms with a possible bacterial infection.

Furthermore total absolute neutropenia and a high ratio between immature and total neutrophilic forms strongly suggest a bacterial infection in absence of maternal hypertension, severe asphyxia, periventricular hemorrhage, maternal fever or haemolytic disease [22].

Presence of neutropenia is associated to a poor prognosis because it usually indicates a reserve decrease of neutrophils in bone marrow, disturbances in releasing them to the peripheral system and inefficiency of stem cells to respond to demands [37].

Thrombocytopenia (platelet count less than 150,000/mL) is associated to severe infections, especially caused by fungi and in less proportion by gram-negative bacteria and viruses and rarely with Gram-positive. It is related to a combination of diffuse injury of the endothelial cell, bacterial and fungal toxins, platelet activation increase and intravascular coagulation dissemination, which implicates an increase in consumption of platelets [38], and a limited response of production and thrombopoietin, being predictors of poor prognosis [39, 40].

In addition, other parameters can be used to assess the bacterial infection type.

Immature granulocytes ratio indicates the number of immature cells in peripheral blood which is usually low and unnoticed in the leukocyte count (observed from 100 to 200 cells). Some of these cells (promyelocytes, myelocytes or meta-myelocytes) can be observed under physiological conditions such as pregnancy and in neonatal period, pathological as bacterial infections, solid tumours or myeloproliferative disorders, or conditions associated with surgery.

The immature neutrophils ratio and total number of neutrophils is also a parameter commonly used to diagnose sepsis, considered a normal value more than 0.2 (or more than 0.3 in the first 24 hours of life).

C-Reactive Protein

C-reactive protein (CRP) is a protein found in the blood as response to inflammation; it is an acute phase protein. It is produced in the intestinal cells and also by adipocytes in response to an IL-6 increase (produced predominantly by macrophages and adipocytes) in plasma concentration after the inflammatory stimulus.

The exact function is not known but the activate complement function is known:

interaction with phagocytic cells. Apparently CRP acts on the first line of immune defence, serves on lipid metabolism and contributes to pathogenesis of diseases such as arteriosclerosis and myocardial infarction [41].

Normal levels of this protein are increased in 6 hours after the inflammatory stimulus, presenting a peak within 48 hours [42] with normal values of about 1.5 mg/mL in these first 48 hours. Its half-life is constant and therefore its level is mainly determined by the production rate and the severity of the case.

Nowadays it is one of the most widely used tests for diagnosis and monitoring response to treatment in the early diagnosis of neonatal sepsis. It is a laboratory test that has high sensitivity but low specificity [43].

It should be considered that the protein is not only altered by infection but also by neonatal asphyxia, meconium aspiration, prolonged labour, hypertensive syndrome Severe Pregnancy Induced or cases of intracranial haemorrhage [44].

Regarding the relationship between CRP levels and intra-partum risk factors mentioned above, studies show that some of the major risk factors can cause a protein increase in absence of infection [45].

This may be caused because until the PCR does not cross the placenta, elevated levels are raised due to the production of CRP in the neonate; also there are few longitudinal studies to assess changes in CRP levels in healthy infants with intra-partum infection risk factors. In this same study it was observed that the cut-off of CRP levels in cord blood sample may differ to cord and serum samples for 24 hours.

Procalcitonin

Procalcitonin (PCT) is a pro-hormone which is produced in C-cells of the thyroid gland. It is the Calcitonin precursor and systemic levels are undetectable or less than 0.1 ng/mL under normal conditions in humans.

PCT has a serum half-life of 25-30 hours and it is important to take into account the physiological elevation of this marker in the first 48 hours of life, which can complicate the interpretation of results during this period [46].

During inflammation caused by infection, PCT rises quicker than the PCR. This indicates an earlier prediction factor of severity and mortality; however this is not applied to infants because they have a physiological peak within 48 hours, reason by which the specificity does not exceed 65% and consequently having a poor positive predictive value [33].

In late or nosocomial neonatal sepsis PCT measurement is important because several studies have found a high sensitivity and specificity. This should be requested for the biochemical profile of these patients [47]. The PCT increase is shown in infections of bacterial, fungal or parasitic origin; while processes caused by viruses do not exhibit elevated PCT [48].

PCT is a marker of severe infection, often generalized infection, and it is in these cases when levels of this marker can be increased above 100 ng/mL, occurring also in other tissues of the thyroid in different cases of sepsis.

The PCT effects are not clearly defined but are known as part of systemic inflammatory response and its production or relation to cytokines are increased in response to endotoxins [45].

PCT increases usually occur in bacterial, parasitic and fungal infections with systemic manifestations. Unlike other biochemical markers, in severe viral infections or severe systemic inflammation without infection, there is usually PCT increase. In case it increases, it does it minimally and without relation to the magnitude of the overall response of the body [49]. As a conclusion, PCT is considered an early marker for severe sepsis.

In several studies it has been found that serum levels of PCT were correlated to severity of microbial invasion and quickly decrease after appropriate antibiotic therapy [50]. The authors conclude that the PCT and CRP concentrations in infants with neonatal sepsis are increased: increase in PCT levels is higher.

Interleukins

In the past decade several studies have been published about various cytokines [51]. Most studied ones are interleukins 1-β, 6, 8 and tumour necrosis factor alpha (TNF-α). Cytokines receptors have also been used as soluble markers of infection which are raised in infectious blood. The soluble receptor IL-2 (IL-2Rs) is the most studied interleukin [52].

IL-6 is a pleiotropic cytokine or glycoprotein secreted by macrophages, T-cells, endothelial cells and fibroblasts, presenting a wide range of functions. It is located on chromosome 7 and its release is induced by IL-1 and its increase is activated in response to TNF-α. Therefore, it is a pro-inflammatory and anti-inflammatory cytokine.

It is an endogenous pyrogen that stimulates ACTH production in the pituitary and whose functions are: immunoglobulin production, differentiation of B lymphocytes, activation of cytotoxic T lymphocytes and plasma cells, haematopoiesis modulation and synthesis of acute phase proteins such as PCR in the liver together with IL-1 [43].

It reaches its maximum peak at 2 hours of the onset of bacterial stimulus, thus its levels may be elevated before the onset of symptoms and before the markers used routinely as PCR are increased [53]. IL-6 production is induced quickly during acute inflammatory reactions associated to injuries, trauma, stress, infection, neoplasms and other causes.

As previous studies have already shown [54], use of sepsis biomarkers as PCT, CRP and IL-6 serum facilitates the sepsis diagnosis, even before onset of clinical manifestations. However, none of these markers is absolutely reliable and accurate marker for early diagnosis.

Tumour Necrosis Factor (TNF-α) presents a very similar kinetic as IL-6 and it has been observed in multiple studies where this cytokine concentrations are significantly increased in infected neonates *versus* non-infected neonates [55]. Furthermore, authors Silveira *et al.*, observed a diagnostic accuracy equivalent to PCT [56]. Sensitivity and specificity capacities increase from 60 to 100% respectively when TNF-α and IL-6 are combined for the diagnosis.

IL-8 is a pro-inflammatory cytokine that helps in neutrophil activation and chemotaxis. Not only works as a sepsis marker but also is associated to the infection severity. It is produced as a result of infection by monocytes, macrophages and endothelial cells with a similar kinetics to IL-6. The IL-8 sensitivity and specificity are defined from 80 to 91% and from 76 to 100% respectively [57].

In a study performed by Boskabadi *et al.*, 93 infants aged 72 hours, IL-8 serum concentrations in infants who died were 3.3 times higher than in surviving infants [58]. The combination of IL-8 and CRP is more reliable for early diagnosis of neonatal sepsis, with a sensitivity and specificity of 91 and 73% respectively [59]. Therefore the combination of CRP and IL-8 can reduce the overuse of antibiotics.

Regarding to cytokines, it has to be considered that although IL-6 and IL-8 levels increase very quickly with bacterial invasion, the levels are quickly normalized in serum (within 24 hours) which limits their ability to be used as an ideal marker. Therefore this is a use limitation in daily clinical practice when cases of operational difficulties of cytokines detection, cytokine elevation in non-specific circumstances and unavailability in many centres. Thus additional studies and methods are needed.

Inter-inhibitory protein alpha family (IAIP) is a group of proteins that inhibit serine protease and provide increased protease activity, associated to systemic activation of the immune system that accompanies inflammation and sepsis. They are involved in processes such as extracellular matrix stabilization, inflammation, wound healing and play an important role in anti-inflammation and infection regulations [60].

The hetero IAIP is a trimer composed by two heavy chains and one light which are held together by glycosaminoglycans links. The light chain has a molecular weight of 30 kDa and is the active component of anti-protease molecule. In the presence of serine proteases this light chain is released providing a protective effect. Its half-life is very short and is rapidly excreted by the kidneys.

IAIP concentrations are independent to gestational age and similar to concentrations in adults. However IAIP levels are significantly lower in septic neonates than in non-septic neonates. COR-curve analysis has shown promising clinical characteristics of sensitivity and specificity obtained from a pilot study on 573 infants. IAIP levels were decreasing not only in neonatal sepsis, but also increased with antibiotic treatment and apparently they are mortality predictive [61].

Some authors describe the kinetics of several inflammatory markers in response to a pathogenic microorganism, being the first marker to appear TNF-α (at 2-3 hours), followed by IL6 (at 3-6 hours), PCT (at 6-12 hours) and finally, CRP (at 48-72 hours) [62].

CORD BLOOD SEPSIS BIOMARKERS

Determination of biochemical markers in cord blood may be useful in the diagnosis of early neonatal sepsis, as demonstrated in several studies in the last decade. The clinical relevance of these consists on their measurements on the birth day avoiding the new-born venipuncture. Therefore the congenital sepsis risk can be assessed even if the mother had no amniocentesis and the results could be available before the placental pathology report is generated.

Regarding to CRP, cord plasma concentrations have resulted higher in patients with microbial invasion from the amniotic cavity and funisitis. According to several studies, CRP levels in cord blood from 0.02 mg/dL could be used to modify the index of suspicion that the infant has been exposed to a microorganism before birth or had an inflammatory response [63].

Furthermore a high level of CRP, results of the total WBC count, differential and total count of platelets in blood from umbilical cord, could be informative in assessing the risk of neonatal sepsis by finding a significant relation between concentrations of CRP and IL-6 [64].

Studies on cord blood where PCT is measured show that this is a sensitive and specific marker of infection antenatal with high positive and negative predictive value. It presents a good positive and negative probability ratio and appears more reliable than CRP [65].

Some studies regarding to IL-6 show that cord blood levels reached a sensitivity of 87-100% and a negative predictive value between 93 and 100%. At the beginning of the infection it had a high sensitivity (89%) and negative predictive value of 91%; however, levels in blood decrease quickly after initiating treatment because IL-6 has a very short half-life. Conversely the blood levels in many patients are undetectable after 24 hours of newborn life. It can be observed that the sensitivity decreases after 24 and 48 hours (67% and 58% respectively) of the beginning of sepsis [66, 67].

Elevated levels of IL-6 in maternal and umbilical cord blood in premature births have been also associated to increased morbidity and mortality in neonatal period, both early neonatal sepsis as in other non-infectious diseases. Furthermore, presence of chorioamnionitis and elevated levels of IL-6 in cord blood may occur in non-infected neonates [45, 68].

Moreover, IL-6 in cord blood appears to be a predictive marker of early sepsis in premature rupture of membranes neonates (PROM); therefore IL-6 should be considered a highly sensitive marker of early infection when it is interpreted together with the PCR in the first 48 hours of infection [69].

Currently there are different investigations focused on the study of these biochemical markers in cord blood for diagnosis of neonatal sepsis [70, 71]. Table **3** summarizes the results of the evaluation of IL-6 and CRP in cord blood by Cernada *et al.* [72] observed in 128 infants diagnosed with early neonatal sepsis with positive blood culture confirmed in 5 cases of sepsis. IL-6 was considered the best predictor of sepsis with higher positive and negative probability ratios than PCR.

And a second study performed by Joram *et al.* [65], concluded a diagnosed early neonatal sepsis in 16 neonates out of 197 neonates with suspected infection. Nowadays, many studies are being conducted predominantly in cord blood, whose objective is the observation of these biochemical markers behaviours in the diagnosis of early neonatal sepsis.

Table 3. Values of Sensitivity (S), Specificity (SP), Positive Predictive Value (PPV) and Negative Predictive Value (NPV) of IL-6 and CRP in cord blood of newborns with prenatal risk factors [72] and the PCT [65].

Sepsis biomarkers in cord blood	S (%)	SP (%)	PPV (%)	NPV (%)
IL-6 (Cut-off: 255 pg/mL)	90	87,4	37,5	99
CRP (Cut-off: 0,1 mg/dL)	60	79	19,4	95,9
PCT (Cut-off: 0,5 ng/mL)	87,5	98,7	87,5	98,7

Cell Surface Markers

With advances in technology like flow cytometry, cell surface antigens are detectable in blood cells. These tests are easy to perform and they require very low volume (0.05 mL) of blood. Clusters of neutrophil differentiation (CD) CD11-β and CD64 have been postulated as good markers for the detection of early and late neonatal sepsis respectively, with high sensitivity and specificity [73].

Its expression increases within minutes after exposure to bacterial products. Moreover due to biological activities of cytokines cannot always be detected by their circulating levels; measurement of cellular response to cytokines may be a better way to recognize an early immune response to infection. However cost, need of sophisticated equipment and processing time are barriers to use these markers in clinical practice.

Molecular Biology

The amplified nucleic acid segments defined by polymerase chain reaction (PCR) started in basic research but its application has been incorporated into the clinical microbiology laboratories and health, where a significant progress has been observed when identifying the etiology of infectious processes.

Unlike bacterial culture, PCR requires minimal concentrations of DNA (pg/mL). Recently bacterial initiators have been synthesized for the amplification of the small region of 16S ribosomal (16S rDNA). 16S rDNA is present in all bacteria, highly conserved and with infinite sequences, allowing distinguish between species of bacteria [74].

Use of 16S rDNA can explore at once all bacteria contained in various biological samples, becoming a better marker compared to other bacterial genes [75].

Although PCR has advantages in bacterial culture, it has several limitations.

Firstly it requires a good quality DNA target without fragmentations and biological samples containing enzymes such as DNases and endonucleases, which degrade the DNA. Secondly, it requires blood samples containing different structural components such as rings porphyrins which interfere with the activity of the TaqDNA polimerasa. And thirdly when obtaining the sample and extracting the DNA, there may be contamination with other bacteria that do not necessarily correspond to the sample source.

In recent years, studies have been conducted evaluating a new technique for detecting microorganisms by real-time PCR as a tool in the microbiological diagnosis of sepsis.

SeptiFast Test (Roche Diagnostics®) is a technique "multiplex" by real-time PCR able to detect 25 microorganisms which are the most frequently found in blood infections. Whilst microbiological methods detect only viable microorganisms, this test detects viable, non-viable and non-microbial DNA [76].

When to Perform the Diagnosis

Infectious diseases are a major cause of morbidity and mortality in neonatal period. A proper early diagnosis leading to an appropriate treatment could prevent the final outcome of these patients. Therefore, any tool that quickly identifies an early neonatal sepsis is highly important objective in perinatal medicine.

There is evidence that infection begins in the womb in a great number of infants with early neonatal sepsis. Many of them are asymptomatic in the early hours of life. It has been observed that mothers with bacterial colonization and/or other infectious risk factors are more likely to give birth to children with bacteraemia already present at birth [77].

As mentioned previously, many biological parameters or sepsis markers have been evaluated in the past decades, including haematological parameters or proteins such as CRP or PCT and some cytokines. However, despite the great number of tests currently available, none of them is completely reliable. Thus many neonatologists often start antibiotic treatment in neonates with risk factors infectious, exposing many of these patients to unnecessary treatment.

For all this reasons it becomes important the diagnosis time because an early and accurate diagnosis will lead to appropriate treatment and avoid unnecessary treatments to newborns.

FUTURE AREAS OF RESEARCH

There is no doubt that early diagnosis of infection is crucial for a proper therapeutic approach and a favourable prognosis in the infected patient. However in clinical practice, conventional methods of diagnosing infectious diseases confront difficulties related to lack of precision, slowness and technical problems. These limitations have not only led to development of more effective techniques and automated diagnosis, as mentioned above, but also a stimulating challenge in finding new biomarkers of infection in the scientific community.

The progress made in recent years in different biomics disciplines (genomics, proteomics and metabolomics, *etc.*) enhanced by bioinformatics as well, offer currently an opportunity for the discovery of molecular profiles and useful markers in diagnosing early neonatal sepsis.

Notably proteomics is perhaps the methodology which has raised greater expectations in the field of infectious diseases. The reason is because its object of study is the systematic and large-scale analysis of the proteome (set of all proteins expressed by the genome of a cell, tissue and organism) in a certain time and under determined conditions of time and environment.

In order to consolidate this relatively recent science, it has been necessary the development of mass spectrometry as a technique applied to the analysis of biological molecules, combined with the use of methods of fractionation and separation of peptides and proteins such as polyacrylamide two-dimensional electrophoresis and high performance liquid chromatography (HPLC).

It is noteworthy that protein microarrays (microarrays or biochips) allow the analysis and simultaneous identification of large panels of probes for specific genomic regions of different biomarkers. Due to versatility and diversity of microarray systems, antibiotic resistance genes and virulence markers can also be evaluated.

Some studies [78, 79], assessed the bacterial DNA detection (16S rRNA region) by PCR amplification followed by DNA microarray hybridization in blood samples of 172 infants with suspected sepsis. In both, microarray was very specific in some of the samples with positive detection of certain microorganisms and in the evaluation of antibiotic resistance genes related to blood culture.

The low sensitivity of the microarray obtained can be due to the low amount of pathogen DNA in patient samples, which is enhanced by PCR amplification prior to microarray hybridization.

In a more recent study Tissari *et al.*, evaluated a microarray platform for the identification of positive blood cultures, where gyr B, par E and mec A genes from 50 bacterial species were amplified [80]. This test was quicker after 18 hours than conventional methods with a sensitivity and specificity of 94.7 and 98.8% respectively.

Therefore, it should be considered to pay special attention to these new techniques advance and the validation of this new generation of proteomic markers as future areas of research in the diagnosis of neonatal sepsis.

CONFLICT OF INTEREST

The author (editor) declares no conflict of interest, financial or otherwise.

ACKNOWLEDGEMENTS

Declared none.

REFERENCES

[1] Lam HS, Ng PC. Biomarkers in neonatal infection. Biomarkers Med 2007; 1(1): 133-43. [http://dx.doi.org/10.2217/17520363.1.1.133]

[2] Ng PC, Lam HS. Diagnostic markers for neonatal sepsis. Curr Opin Pediatr 2006; 18(2): 125-31. [http://dx.doi.org/10.1097/01.mop.0000193293.87022.4c]

[3] Haake A, Holbrook K. The structure and development of skin. Dermatology in General Medicine. 5th ed. New York: McGraw-Hill 1999; pp. 70-107.

[4] Nielsen MB, Odum N, Gerwien J, *et al. Staphylococcal enterotoxin-A* directly stimulates signal transduction and interferon-gamma production in psoriatic T-cell lines. Tissue Antigens 1998; 52(6): 530-8.
[http://dx.doi.org/10.1111/j.1399-0039.1998.tb03083.x]

[5] Hack CE, Zeerleder S. The endothelium in sepsis: source of and a target for inflammation. Crit Care Med 2001; 29(7) (Suppl.): S21-7.
[http://dx.doi.org/10.1097/00003246-200107001-00011]

[6] Opal SM. The uncertain value of the definition for SIRS. Systemic inflammatory response syndrome. Chest 1998; 113(6): 1442-3.
[http://dx.doi.org/10.1378/chest.113.6.1442]

[7] Alper CA. The human immune response to hepatitis B surface antigen. Exp Clin Immunogenet 1995; 12(3): 171-81.

[8] Salinas J, Fica A. Immunoglobulins in sepsis and septic shock. Rev Chilena Infectol 2005; 22(1): 21-31.
[http://dx.doi.org/10.4067/S0716-10182005000100003]

[9] Johnston JB, Rahman MM, McFadden G. Strategies that modulate inflammasomes: insights from host-pathogen interactions. Semin Immunopathol 2007; 29(3): 261-74.
[http://dx.doi.org/10.1007/s00281-007-0080-5]

[10] Affara NI, Robertson FM. Vascular endothelial growth factor as a survival factor in tumor-associated angiogenesis. In Vivo 2004; 18(5): 525-42.

[11] Carrillo-Esper R, Nunez-Monroy FN. Systemic inflammatory response syndrome: new concepts. Gac Med Mex 2001; 137(2): 127-34.

[12] Bone RC, Balk RA, Cerra FB, *et al.* Definitions for sepsis and organ failure and guidelines for the use of innovative therapies in sepsis. The ACCP/SCCM Consensus Conference Committee. American College of Chest Physicians/Society of Critical Care Medicine. Chest 1992; 101(6): 1644-55. [http://dx.doi.org/10.1378/chest.101.6.1644]

[13] Bone R, Grodzin C, Balk R. Sepsis: a new hypothesis for pathogenesis of the disease process. Chest 1998; 112: 235-43. [http://dx.doi.org/10.1378/chest.112.1.235]

[14] Lopez Sastre JB, Cotallo GD, Ramos Aparicio M. Sepsis Neonatal. Protocolos diagnósticos y terapéuticos en pediatría 2007; 36: 307-14.

[15] Afroza S. Neonatal sepsis-- a global problem: an overview. Mymensingh Med J 2006; 15(1): 108-14.

[16] Adair CE, Kowalsky L, Quon H, *et al.* Risk factors for early-onset group B streptococcal disease in neonates: a population-based case-control study. CMAJ 2003; 169(3): 198-203.

[17] Lopez Sastre JB, Perez SD. Definitions of neonatal sepsis: a long way to go. An Pediatr (Barc) 2006; 65(6): 525-8. [http://dx.doi.org/10.1157/13095843]

[18] Vergnano S, Sharland M, Kazembe P, Mwansambo C, Heath PT. Neonatal sepsis: an international perspective. Arch Dis Child Fetal Neonatal Ed 2005; 90(3): F220-4. [http://dx.doi.org/10.1136/adc.2002.022863]

[19] Escobar GJ, Li DK, Armstrong MA, *et al.* Neonatal sepsis workups in infants >/=2000 grams at birth: A population-based study. Pediatrics 2000; 106(2 Pt 1): 256-63. [http://dx.doi.org/10.1542/peds.106.2.256]

[20] Gibbs RS, Schrag S, Schuchat A. Perinatal infections due to group B streptococci. Obstet Gynecol 2004; 104(5 Pt 1): 1062-76. [http://dx.doi.org/10.1097/01.AOG.0000144128.03913.c2]

[21] Edwards MS, Baker CJ. Group B streptococcal infections in elderly adults. Clin Infect Dis 2005; 41(6): 839-47. [http://dx.doi.org/10.1086/432804]

[22] Tapia IJ, Reichhard TC, Saldias RM, *et al.* Neonatal sepsis in the era of antenatal antibiotic prophylaxis. Rev Chilena Infectol 2007; 24(2): 111-6. [http://dx.doi.org/10.4067/S0716-10182007000200004]

[23] Juncosa T, Bosch J, Dopico E, *et al.* Neonatal infection by *Streptococcus agalactiae*. Multicenter study in the area of Barcelona, Spain. Enferm Infecc Microbiol Clin 1998; 16(7): 312-5.

[24] Coto Cotallo GD, Ibáñez Fernández A. Protocolo diagnóstico-terapéutico de la sepsis neonatal. Bol Pediatr 2006;46 (Suplem 1):125-34.26. Commitee on Infectious Diseases and Commitee on Fetus and Newborn. Guidelines for prevention of group B streptococcal (GBS) infection by chemoprophylaxis. Pediatrics 1992; 90: 775-8.

[25] Schrag SJ, Zywicki S, Farley MM, *et al.* Group B streptococcal disease in the era of intrapartum antibiotic prophylaxis. N Engl J Med 2000; 342(1): 15-20. [http://dx.doi.org/10.1056/NEJM200001063420103]

[26] Commitee on Infectious Diseases and Commitee on Fetus and Newborn. Guidelines for prevention of group B streptococcal (GBS) infection by chemoprophylaxis. Pediatrics 1992; 90: 775-8.

[27] Schrag S, Gorwitz R, Fultz-Butts K, Schuchat A. Prevention of perinatal group B streptococcal disease. Revised guidelines from CDC. MMWR Recomm Rep 2002; 51(RR-11): 1-22.

[28] Duke T. Neonatal pneumonia in developing countries. Arch Dis Child Fetal Neonatal Ed 2005; 90(3):

F211-9.
[http://dx.doi.org/10.1136/adc.2003.048108]

[29] Stoll BJ, Hansen N, Fanaroff AA, *et al.* Late-onset sepsis in very low birth weight neonates: the experience of the NICHD Neonatal Research Network. Pediatrics 2002; 110(2 Pt 1): 285-91.
[http://dx.doi.org/10.1542/peds.110.2.285]

[30] Anwer SK, Mustafa S, Pariyani S, Ashraf S, Taufiq KM. Neonatal sepsis: an etiological study. J Pak Med Assoc 2000; 50(3): 91-4.

[31] Sundaram V, Kumar P, Dutta S, *et al.* Blood culture confirmed bacterial sepsis in neonates in a North Indian tertiary care center: changes over the last decade. Jpn J Infect Dis 2009; 62(1): 46-50.

[32] Silbergeld EK, Davis DL. Role of biomarkers in identifying and understanding environmentally induced disease. Clin Chem 1994; 40(7 Pt 2): 1363-7.

[33] Sheu JR, Hung WC, Wu CH, *et al.* Reduction in lipopolysaccharide-induced thrombocytopenia by triflavin in a rat model of septicemia. Circulation 1999; 99(23): 3056-62.
[http://dx.doi.org/10.1161/01.CIR.99.23.3056]

[34] Manroe BL, Weinberg AG, Rosenfeld CR, Browne R. The neonatal blood count in health and disease. I. Reference values for neutrophilic cells. J Pediatr 1979; 95(1): 89-98.
[http://dx.doi.org/10.1016/S0022-3476(79)80096-7]

[35] Weber TR, West KW, Grosfeld JL. Broviac central venous catheterization in infants and children. Am J Surg 1983; 145(2): 202-4.
[http://dx.doi.org/10.1016/0002-9610(83)90062-4]

[36] Greenberg DN, Yoder BA. Changes in the differential white blood cell count in screening for group B streptococcal sepsis. Pediatr Infect Dis J 1990; 9(12): 886-9.
[http://dx.doi.org/10.1097/00006454-199012000-00006]

[37] Brady MT. Health care-associated infections in the neonatal intensive care unit. Am J Infect Control 2005; 33(5): 268-75.
[http://dx.doi.org/10.1016/j.ajic.2004.11.006]

[38] Cornette L. Fetal and neonatal inflammatory response and adverse outcome. Semin Fetal Neonatal Med 2006; 11: 459-70.

[39] Chiesa C, Panero A, Rossi N, *et al.* Reliability of procalcitonin concentrations for the diagnosis of sepsis in critically ill neonates. Clin Infect Dis 1998; 26(3): 664-72.
[http://dx.doi.org/10.1086/514576]

[40] Griffin MP, Lake DE, O'Shea TM, Moorman JR. Heart rate characteristics and clinical signs in neonatal sepsis. Pediatr Res 2007; 61(2): 222-7.
[http://dx.doi.org/10.1203/01.pdr.0000252438.65759.af]

[41] Bamonde L, Caamaño B, Alonso M. La Procalcitonina como marcador de infección. Una revisión desde atención primaria. Rev ped de atención primaria 2002; 4(16): 617-30.

[42] Hawk M. C-reactive protein in neonatal sepsis. Neonatal Netw 2008; 27(2): 117-20.
[http://dx.doi.org/10.1891/0730-0832.27.2.117]

[43] Santana RC, Garcia-Munoz F, Reyes D, Gonzalez G, Dominguez C, Domenech E. Role of cytokines (interleukin-1beta, 6, 8, tumour necrosis factor-alpha, and soluble receptor of interleukin-2) and C-reactive protein in the diagnosis of neonatal sepsis. Acta Paediatr 2003; 92(2): 221-7.

[44] Kawczynski P, Piotrowski A. Procalcitonin and C-reactive protein as a markers of neonatal sepsis. Ginekol Pol 2004; 75(6): 439-44.

[45] Chiesa C, Pellegrini G, Panero A, *et al.* C-reactive protein, interleukin-6, and procalcitonin in the immediate postnatal period: influence of illness severity, risk status, antenatal and perinatal complications, and infection. Clin Chem 2003; 49(1): 60-8.
[http://dx.doi.org/10.1373/49.1.60]

[46] Santuz P, Soffiati M, Dorizzi RM, Benedetti M, Zaglia F, Biban P. Procalcitonin for the diagnosis of early-onset neonatal sepsis: a multilevel probabilistic approach. Clin Biochem 2008; 41(14-15): 1150-5.
[http://dx.doi.org/10.1016/j.clinbiochem.2008.05.015]

[47] Farmaki E, Evdoridou J, Pouliou T, *et al.* Fungal colonization in the neonatal intensive care unit: risk factors, drug susceptibility, and association with invasive fungal infections. Am J Perinatol 2007; 24(2): 127-35.
[http://dx.doi.org/10.1055/s-2007-970078]

[48] Peidró P, González de Dios J, Urán M, García B, de la Morena Campillo A, Mora M. Utilidad de la Procalcitonina como prueba diagnóstica precoz de sepsis neonatal en recién nacidos con factores de riesgo de infección. An Pediatr (Barc) 2007; 67(6): 530-5.
[http://dx.doi.org/10.1016/S1695-4033(07)70799-3]

[49] Guibourdenche J, Bedu A, Petzold L, *et al.* Biochemical markers of neonatal sepsis: value of procalcitonin in the emergency setting. Ann Clin Biochem 2002; 39(Pt 2): 130-5.
[http://dx.doi.org/10.1258/0004563021901874]

[50] Monneret G, Labaune JM, Isaac C, Bienvenu F, Putet G, Bienvenu J. Procalcitonin and C-reactive protein levels in neonatal infections. Acta Paediatr 1997; 86(2): 209-12.
[http://dx.doi.org/10.1111/j.1651-2227.1997.tb08870.x]

[51] Harris MC, D'Angio CT, Gallagher PR, Kaufman D, Evans J, Kilpatrick L. Cytokine elaboration in critically ill infants with bacterial sepsis, necrotizing entercolitis, or sepsis syndrome: correlation with clinical parameters of inflammation and mortality. J Pediatr 2005; 147(4): 462-8.
[http://dx.doi.org/10.1016/j.jpeds.2005.04.037]

[52] Satar M, Turhan E, Yapicioglu H, Narli N, Ozgunen FT, Cetiner S. Cord blood cytokine levels in neonates born to mothers with prolonged premature rupture of membranes and its relationship with morbidity and mortality. Eur Cytokine Netw 2008; 19(1): 37-41.

[53] Volante E, Moretti S, Pisani F, Bevilacqua G. Early diagnosis of bacterial infection in the neonate. J Matern Fetal Neonatal Med 2004; 16 (Suppl. 2): 13-6.
[http://dx.doi.org/10.1080/jmf.16.2.13.16]

[54] Chiesa C, Pellegrini G, Panero A, *et al.* Umbilical cord interleukin-6 levels are elevated in term neonates with perinatal asphyxia. Eur J Clin Invest 2003; 33(4): 352-8.
[http://dx.doi.org/10.1046/j.1365-2362.2003.01136.x]

[55] Shah BA, Padbury JF. Neonatal sepsis: an old problem with new insights. Virulence 2014; 5(1): 170-8.
[http://dx.doi.org/10.4161/viru.26906]

[56] Silveira RC, Procianoy RS. Evaluation of interleukin-6, tumour necrosis factor-alpha and interleukin-1beta for early diagnosis of neonatal sepsis. Acta Paediatr 1999; 88(6): 647-50.
[http://dx.doi.org/10.1080/08035259950169314]

[57] Wang Z, Yu J. Recent progress in the diagnosis of neonatal septicemia. Zhongguo Dang 2013; 15: 236-41.

[58] Boskabadi H, Maamouri G, Afshari JT, Ghayour-Mobarhan M, Shakeri MT. Serum interleukin 8 level as a diagnostic marker in late neonatal sepsis. Iran J Pediatr 2010; 20(1): 41-7.

[59] Mehr S, Doyle LW. Cytokines as markers of bacterial sepsis in newborn infants: a review. Pediatr Infect Dis J 2000; 19(9): 879-87.
[http://dx.doi.org/10.1097/00006454-200009000-00014]

[60] Kasper DC, Altiok I, Mechtler TP, *et al.* Molecular detection of late-onset neonatal sepsis in premature infants using small blood volumes: proof-of-concept. Neonatology 2013; 103(4): 268-73.
[http://dx.doi.org/10.1159/000346365]

[61] Chaaban H, Singh K, Huang J, Siryaporn E, Lim YP, Padbury JF. The role of inter-alpha inhibitor proteins in the diagnosis of neonatal sepsis. J Pediatr 2009; 154(4): 620-2. [http://dx.doi.org/10.1016/j.jpeds.2008.10.008]

[62] Courdec R, Mary R, Veinberg F. Marcadores de inflamación en pediatría. Acta Bioquim Clin Latinoam 2004; 38: 513-7.

[63] Yoon BH, Romero R, Shim JY, Shim SS, Kim CJ, Jun JK. C-reactive protein in umbilical cord blood: a simple and widely available clinical method to assess the risk of amniotic fluid infection and funisitis. J Matern Fetal Neonatal Med 2003; 14(2): 85-90. [http://dx.doi.org/10.1080/jmf.14.2.85.90]

[64] Dollner H, Vatten L, Austgulen R. Early diagnostic markers for neonatal sepsis: comparing C-reactive protein, interleukin-6, soluble tumour necrosis factor receptors and soluble adhesion molecules. J Clin Epidemiol 2001; 54(12): 1251-7. [http://dx.doi.org/10.1016/S0895-4356(01)00400-0]

[65] Joram N, Boscher C, Denizot S, *et al.* Umbilical cord blood procalcitonin and C reactive protein concentrations as markers for early diagnosis of very early onset neonatal infection. Arch Dis Child Fetal Neonatal Ed 2006; 91(1): F65-6. [http://dx.doi.org/10.1136/adc.2005.074245]

[66] Khassawneh M, Hayajneh WA, Kofahi H, Khader Y, Amarin Z, Daoud A. Diagnostic markers for neonatal sepsis: comparing C-reactive protein, interleukin-6 and immunoglobulin M. Scand J Immunol 2007; 65(2): 171-5. [http://dx.doi.org/10.1111/j.1365-3083.2006.01878.x]

[67] Perafán M. Fisiopatología de la sepsis. Tópicos en Medicina Intensiva 2003; 3: 71-9.

[68] Hatzidaki E, Gourgiotis D, Manoura A, *et al.* Interleukin-6 in preterm premature rupture of membranes as an indicator of neonatal outcome. Acta Obstet Gynecol Scand 2005; 84(7): 632-8. [http://dx.doi.org/10.1111/j.0001-6349.2005.00747.x]

[69] Weitkamp J, Ascher JL. Diagnostic use of C-reactive protein (CRP) in assessment of neonatal sepsis. NeoReviews 2005; 6(11): 508-15.

[70] Campos DP, Silva MV, Machado JR, Castellano LR, Rodrigues V, Barata CH. Early-onset neonatal sepsis: cord blood cytokine levels at diagnosis and during treatment. J Pediatr (Rio J) 2010; 86(6): 509-14.

[71] Santana C, Guindeo MC, Gonzalez G, Garcia-Munoz F, Saavedra P, Domenech E. Cord blood levels of cytokines as predictors of early neonatal sepsis. Acta Paediatr 2001; 90(10): 1176-81. [http://dx.doi.org/10.1111/j.1651-2227.2001.tb03250.x]

[72] Cernada M, Badia N, Modesto V, *et al.* Cord blood interleukin-6 as a predictor of early-onset neonatal sepsis. Acta Paediatr 2012; 101: 203-07.

[73] Ng PC, Lam HS. Biomarkers for late-onset neonatal sepsis: cytokines and beyond. Clin Perinatol 2010; 37(3): 599-610. [http://dx.doi.org/10.1016/j.clp.2010.05.005]

[74] Jordan JA, Durso MB. Real-time polymerase chain reaction for detecting bacterial DNA directly from blood of neonates being evaluated for sepsis. J Mol Diagn 2005; 7(5): 575-81.

[75] Dark PM, Dean P, Warhurst G. Bench-to-bedside review: the promise of rapid infection diagnosis during sepsis using polymerase chain reaction-based pathogen detection. Crit Care 2009; 13(4): 217. [http://dx.doi.org/10.1186/cc7886]

[76] Westh H, Lisby G, Breysse F, *et al.* Multiplex real-time PCR and blood culture for identification of bloodstream pathogens in patients with suspected sepsis. Clin Microbiol Infect 2009; 15(6): 544-51. [http://dx.doi.org/10.1111/j.1469-0691.2009.02736.x]

[77] Philip AG, Hewitt JR. Early diagnosis of neonatal sepsis. Pediatrics 1980; 65(5): 1036-41.

[78] Cleven BE, Palka-Santini M, Gielen J, Meembor S, Kronke M, Krut O. Identification and characterization of bacterial pathogens causing bloodstream infections by DNA microarray. J Clin Microbiol 2006; 44(7): 2389-97.
[http://dx.doi.org/10.1128/JCM.02291-05]

[79] Shang S, Chen G, Wu Y. DU L, Zhao Z. Rapid diagnosis of bacterial sepsis with PCR amplification and microarray hybridization in 16S rRNA gene. Pediatr Res 2005; 58(1): 143-8.
[http://dx.doi.org/10.1203/01.PDR.0000169580.64191.8B]

[80] Tissari P, Zumla A, Tarkka E, *et al.* Accurate and rapid identification of bacterial species from positive blood cultures with a DNA-based microarray platform: an observational study. Lancet 2010; 375(9710): 224-30.
[http://dx.doi.org/10.1016/S0140-6736(09)61569-5]

CHAPTER 14

Sepsis: Traditional and Emergent Biomarkers for Diagnosis and Prognosis

Luis García de Guadiana-Romualdo, Patricia Esteban-Torrella and **María Dolores Albaladejo-Otón***

Clinical Analysis Department, Santa Lucia General University Hospital, Cartagena, Spain

Abstract: Sepsis and its related complications are associated with significant morbidity and mortality among populations worldwide. Early diagnosis and prompt initial management are keys to improve sepsis outcome. Therefore, biological markers (biomarkers) can be useful for identifying or ruling out sepsis, identifying patients who may benefit from specific therapies or assessing the response to therapy. Although numerous biomarkers have been investigated, only two biomarkers are currently used in the clinical practice, C-reactive protein (CRP) and procalcitonin (PCT). Both were included in 2001 in the revised definition of sepsisas variables to diagnose sepsis, but even these have limited ability to distinguish sepsis from other inflammatory conditions or to predict outcome, and there is a continuous search for better biomarkers of sepsis. Moreover, the recent Third Consensus Definitions for Sepsis and Septic Shock does not include the use of biomarkers as tools for management of sepsis and, probably, to redefine the role of these biomarkers is necessary. The purpose of this review is to describe the most relevant sepsis biomarkers used currently in the clinical practice and discuss the future role of some emergent biomarkers for the management of sepsis.

Keywords: C-reactive protein (CRP), Cytokines, Definitions, Diagnosis, Endocan, Guide antibiotic decisions, Lipopolysaccharide-binding protein (LBP), Mid-regional pro-Adrenomedullin (pro-AMP), Pancreatic Stone Protein (PSP), Pentraxin 3, Presepsin, Procalcitonin (PCT), Prognosis, Sepsis, Septic shock, Soluble E-selectin (sE-selectin), Soluble Intercellular Adhesion Molecule (sICAM-1), Soluble Triggering Receptor Expressed on Myeloid Cells-1 (sTREM-1), Soluble Vascular Cell Adhesion Molecule (sVCAM-1), Soluble urokinase-type Plasminogen Activator Receptor (suPAR).

INTRODUCTION

Sepsis is a major challenge to clinicians and a global burden to healthcare systems because the resources needed for sepsis care are high [1]. Its reported incidence

* **Corresponding author María Dolores Albaladejo-Otón:** Clinical Analysis Department, Santa Lucia General University Hospital, Cartagena, Murcia, Spain; Tel/Fax: +34 96812 86 00; E-mail: mariad.albaladejo@carm.es

has been increasing [2, 3], and due to the increase in age of patient, with more comorbidities, greater recognition [4] and, in some cases, reimbursement-favorable coding [5]. The actual incidence of sepsis is unknown, but a conservative estimation indicate that sepsis is a one the most important cause of morbidity and mortality around the world [6, 7], despite the use of modern antibiotic and resuscitation therapies [8]. In Spain, mortality reached 28 and 46% in patients with severe sepsis and septic shock, respectively [9]. In fact, the number of people dying of sepsis each year is similar to the number of deaths caused by acute myocardial infarction, and far exceeds those by HIV, breast cancer or stroke [10]. Moreover, the survivors of sepsis suffer from physical and cognitive disabilities leading to reduced long-term quality of life [11].

Sepsis is a syndrome of physiologic, pathologic, and biochemical abnormalities due to infection and is a result of complex chain of events that implies inflammatory and anti-inflammatory processes, humoral and cellular reactions and circulatory abnormalities, including activation of complement system, coagulation cascades and the vascular endothelial system [12, 13]. In addition, the immune response of host depends on several individual factors, such as age, underlying diseases, nutritional condition and genetic variability and indeed pathogens factors vary by patient, with Gram-positive bacterial pathogens being the most common cause of sepsis [10].

The diagnosis of sepsis and evaluation of severity are initially based on clinical findings but are complicated by the highly variable and non-specific nature of the sign of symptoms, particularly in populations such as elderly and children [14]. An early diagnosis of sepsis, prompt therapy and identification of high risk patients contribute to reduce the mortality associated with sepsis [4]; therefore, the availability of tools such as the biological markers (biomarkers) can help in the diagnostic process indicating the presence or absence of severity of sepsis [15, 16], and can differentiate bacterial of fungal or viral infection, and sepsis from local infection. Besides, biomarkers could be useful for prognosis, evaluating the response to therapy and predicting the progression to Multiple Organ Dysfunction Syndrome [17]. In this chapter, the traditional and emergent biomarkers used for diagnosis and prognosis of sepsis and guide antibiotic therapy are described.

Pathophysiology of Sepsis

Sepsis is the result of host response to infection due to microbial pathogens. Traditionally, sepsis has been considered as result of uncontrolled inflammatory response resulting in shock or organ dysfunction [12], divided into two phases: following to infection an early hyper-inflammatory stage mediated by cytokines as tumor necrosis factor-alpha (TNF-α), interleukine (IL)-1β and IL-6 are responsible for the features of systemic inflammatory response syndrome (SIRS). This first phase often progress as to a second stage called by Bone *et al.* "compensatory anti-inflammatory response syndrome" (CARS) [18], mainly in those patients with organ dysfunction (Fig. **1** [19]). In 1999, van der Poll *et al.* [20] proposed a model alternative to the "sequential model" originally proposed by Bone for the progression of sepsis to organ dysfunction in which the CARS begins while the pro-inflammatory SIRS is still present. This discrepancy needs to be resolved [19].

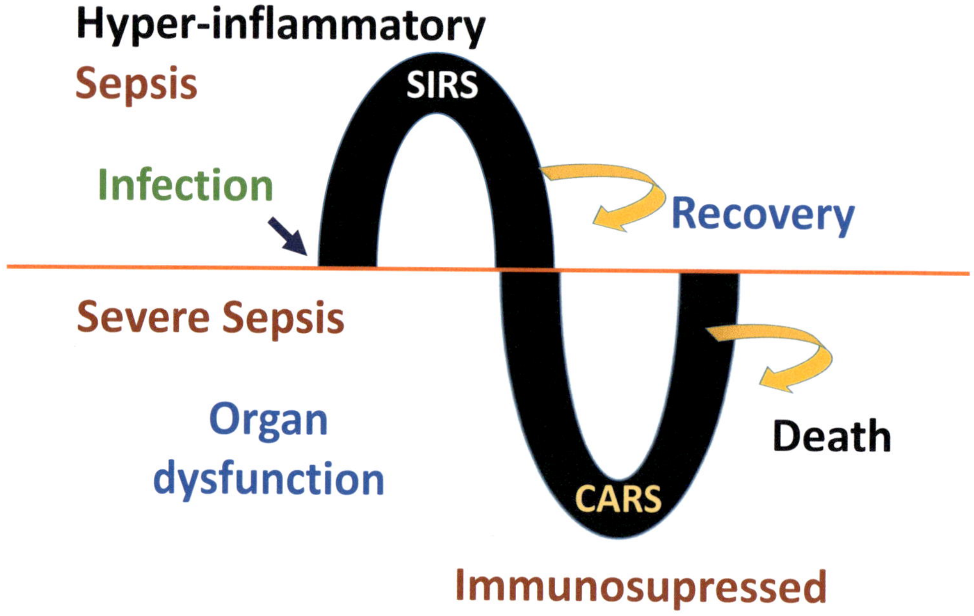

Fig. (1). Adapted of reference [19]. Sepsis includes two phases. Following infection, a hyper-inflammatory phase is characterized by SIRS and this can evolved to sepsis with organ dysfunction. During this phase, there is evidence of CARS with immunosuppression and multiple organ dysfunction.

However, the paradigm of understanding sepsis has shifted, the focus towards immunosuppressive effects [21]. For example, elderly patients with sepsis show lack of typical signs or symptoms as fever, associated with poor outcomes. The immunosuppression is considered a key pathogenesis associated with sepsis mortality which leads to nosocomial infections, with a worse prognosis. Therefore, early hyper-inflammatory and late compensatory anti-inflammatory responses are included in current immunological models of sepsis, but are not a simple biphasic model in most cases [22]. The degree and duration of immune response differ among patients depending on different conditions: age, comorbidities, pathogen virulence, pathogen burden and genetic factors [23].

During the course of sepsis, the degree and duration of immunosuppression could affect the outcome, which leads us to consider tailored immunomodulatory therapy.

Sepsis Definitions

The word "sepsis" derives from the Greek "σηψιζ", referred to the "decomposition of animal, or vegetable or organic matter in the presence of bacteria" [24]. The first use of "sepsis" term in the medical setting occurred over 2700 years ago in the Homer´s poems. In this use, the term "sepsis"' derives directly from the word "σηπω", which means "I rot". This term is also found in the writings of the physician and philosopher Hippocrates; without using the same term, he was probably the first to describe the clinical course of septic shock ("when continuing fever is present, it is dangerous if the outer parts are cold, but the inner parts are burning hot").

Moreover, Hippocrates was one of the first to examine antisepsis properties of potential medicinalcompounds including alcohol in wine and vinegar. Later, Galen and Celsus described the sign of inflammation as peripheral vasodilatation (*rubor*), fever (*calor*), increased capillary permeability (*tumor*) and organ dysfunction (*function laesa*). In 1914, Hugo Schottmüller provided the first scientific definition of sepsis: "*sepsis is a state caused by microbial invasion from a local infectious source into the bloodstream which leads to signs of systemic illness in remote organs*" [25].

According to this definition, bacteremia was a needed condition for the identification of sepsis. This notion did not change significantly over the years and both terms were often used interchangeably.

The modern concept of sepsis focuses on the human response to infection. William Osler (1849 to 1919) was the first to recognize the important role of the host response in sepsis and in 1904 he cited: "*It appears that patients are dying not from their infection but rather their reaction to them*". A 1991 consensus conference developed the first definitions that focused on the view that sepsis was the result from a host's systemic inflammatory response syndrome (SIRS) to infection (Table **1**) [26].

Sepsis with organ dysfunction was termed as severe sepsis, which could progress to septic shock, defined as "*sepsis-induced hypotension persisting despite adequate fluid resuscitation*". These simple clinical criteria allowed researchers to identify patients to include in sepsis trials and were rapidly adopted.

Table 1. SIRS (Systemic Inflammatory Response Syndrome). From Bone *et al.* [26].

Two or more of:
Temperature > 38 ºC or < 36 ºC
Heart rate > 90/min
Respiratory rate > 20/min and/or PaCO2 < 32 mm Hg (4.3 kPa)
White blood cell count > 12000/mm3 or < 4000/mm3 or > 10% immature bands

Table 2. Diagnostic criteria for sepsis.

Sepsis: Infection, documented or suspected, and some of the following:
General variables
Fever (> 38.3 ºC)
Hypothermia (core temperature <36 ºC)
Heart rate > 90 min-1 or more than two SD above the normal value for age
Tachypnea
Altered mental status
Significant edema or positive fluid balance (> 20 mL/kg over 24 h)
Hyperglycemia (plasma glucose > 140 mg/dL or 7.7 mmol/L) in the absence of diabetes
Inflammatory variables
Leukocytosis (WBC count > 12000 μL-1)
Leukopenia (WBC count < 4,000 μL-1)
Normal WBC count with greater than 10% immature forms
Plasma C-reactive protein more than two SD above the normal value
Plasma procalcitonin more than two SD above the normal value
Hemodynamic variables
Arterial hypotension (SBP > 90 mmHg, MAP <70 mmHg, or an SBP decrease > 40 mmHg in adults or less than two SD below normal for age)
Organ dysfunction variables
Arterial hypoxemia (PaO2/FiO2 < 300)
Acute oliguria (urine output < 0.5 mL/kg/h for at least 2 h despite adequate fluid resuscitation)
Creatinine increase > 0.5 mg/dL or 44.2 lmol/L
Coagulation abnormalities (INR > 1.5 or aPTT> 60 s)
Ileus (absent bowel sounds)
Thrombocytopenia (platelet count < 100000 μL-1)
Hyperbilirubinemia (plasma total bilirubin> 4 mg/dLor 70 mmol/L)
Tissue perfusion variables
Hyperlactatemia (> 1 mmol/L)
Decreased capillary refill or mottling
WBC = white blood cell; SBP = systolic blood pressure; MAP = mean arterial pressure; INR = international normalized ratio; aPTT = activated partial thromboplastin time Adapted from [4]

However, the sepsis approach has some important limitations. The most important is that the SIRS criteria are so sensitive that a high percentage of patients admitted in an Intensive Care Unit (ICU) and indeed hospitalized in ward meet the criteria

[27]. Other non-infectious clinical processes such as severe trauma, burns, pancreatitis can cause SIRS.

Table 3. Definitions of severe sepsis and septic shock.

Severe sepsis definition: sepsis-induced tissue hypoperfusion or organ dysfunction (any of the following thought to be due to the infection)
Sepsis-induced hypotension Lactate above upper limits laboratory normal Urine output< 0.5 mL/kg/h for more than 2 hours despite adequate fluid resuscitation Acute lung injury with $PaO_2/FiO_2 < 250$ in the absence of pneumonia as infection source Acute lung injury with $PaO_2/FiO_2 < 200$ in the presence of pneumonia as infection source Creatinine > 2.0 mg/dL (176.8 μmol/L) Bilirubin > 2 mg/dL (34.2 μmol/L) Platelet count < 100000 μL-1 Coagulopathy (international normalized ratio > 1.5
Septic shock: sepsis-induced hypotension persisting despite adequate fluid resuscitation
Adapted from [4]

A second consensus conference in 2001 [28], recognizing limitations with these definitions, expanded the list of diagnostic criteriabut did not offer alternatives because of the lack of supporting evidence. In 2012 a new consensus conference updated these criteria (Tables **2** and **3**) [4]. The definitions of sepsis, septic shock, and organ dysfunction contained in these documents have remained largely unchanged for more than one decade.

In recent years some authors have discussed about the need to change the sepsis definition. So, for Vincent *et al.* [29] and Lin HY [30] the evidence of organ dysfunction should be included in the criteria for sepsis. Hence, sepsis should be defined as a systemic response to infection with the presence of some degree of organ dysfunction. Recognizing this need to reexamine the current definitions theEuropean Society of Intensive Care Medicine and the Society of Critical Care Medicine convened a task force to update sepsis definition.

Recently, the Third International Consensus Definitions for Sepsis and Septic Shock (Sepsis-3) have been published [31]. In this document excessive focus on inflammation, the misleading model that sepsis follows a continuum through severe sepsis to shock, and inadequate specificity and sensitivity of the SIRS criteria are cited as limitations of previous definitions. The main differences regarding previous definitions include the elimination of SIRS criteria to identify sepsis because these criteria do not necessarily indicate a dysregulated, life-threatening response, and the update of sepsis definition including organ dysfunction, defined by Sequential Organ Failure Assessment (SOFA) ≥ 2

(Table **4**). Therefore, sepsis involves organ dysfunction, indicating a pathobiology more complex than infection plus an accompanying inflammatory response alone. This consensus recommends a new score called quick SOFA (*qSOFA*), including altered mentation, systolic blood pressure of 100 mm Hg or less, and respiratory rate ≥ of 22/min, as a simple bedside tool in the out-of-hospital, emergency department and ward settings to identify adult patients with suspected infection who are likely to have poor outcomes and suggests that *qSOFA* be used to prompt clinicians to further investigate for organ dysfunction, to initiate or escalate therapy as appropriate, and to consider referral to critical care or increase the frequency of monitoring. Finally, the biomarkers included in former definitions, such as C-reactive protein (CRP) and procalcitonin (PCT) have not been in the updated definitions.

Table 4. New terms and definitions (Sepsis-3).

• Sepsis is defined as life-threatening organ dysfunction caused by a dysregulated host response to infection. • Organ dysfunction can be identified as an acute change in total SOFA score ≥ 2 points consequent to the infection. • The baseline SOFA score can be assumed to be zero in patients not known to have preexisting organ dysfunction. • In lay terms, sepsis is a life-threatening condition that arises when the body's response to an infection injures its own tissues and organs. •Septic shock is a subset of sepsis in which underlying circulatory and cellular/metabolic abnormalities are profound enough to substantially increase mortality. • Patients with septic shock can be identified with a clinical construct of sepsis with persisting hypotension requiring vasopressors to maintain MAP ≥ 65mmHg and having a serum lactate level >2 mmol/L (18mg/dL) despite adequate volume resuscitation.
Adapted from [30]

SEPSIS BIOMARKERS

Definition

Overall, a biomarker has been defined as "*an indicator of normal biological processes, pathogenic processes, or pharmacologic responses to a therapeutic intervention*" [15]. Julián-Jiménez *et al.* define a biomarker of inflammation and/or infection as "*that molecule objectively measurable and with reproducibility, which levels are an indicator of a normal or pathogenic process and are useful to guide antibiotic therapy*". Moreover, a biomarker must be able to provide additional information to that obtained with clinical findings [32].

The usefulness of biomarkers for management of septic patients was first addressed systematically in a colloquium organized by the International Sepsis Forum in 2005 [33]. In this report, it was concluded that: "*First, sepsis is a concept-that of disease arising from the host response to infection-rather than a measurable pathological process. Second, that concept is a complex one that hinges on documentation of both infection and a response. Third, that response is nonspecific, defined by consensus criteria that emphasize physiologic changes in vital parameters that are common to a number of disparate processes*" and that "*biomarkers promise to transform sepsis from a physiologic syndrome to a group of distinct biochemical disorders*".

Sepsis biomarkers could be for three objectives: (1) to rule out infection and establish an early diagnosis, contributing to prompt therapy (2) to assess the severity of sepsis, which is information useful in patient triage and mainly to make a decision about Intensive Care Unit (ICU) or ward admission, and (3) to evaluate the clinical course and suggest the need to change or withdraw the treatment on the basis of the biomarker levels [32, 34].

Classification of Sepsis Biomarkers

By definition, sepsis biomarkers should reflect the biology of sepsis and evidence the characteristic biochemical changes of the host response to infection. Searching for sepsis biomarkers is focused on the biochemical changes at the plasma level (complement system, coagulation system, and kallikrein-kinin system) and the indicators of the activation or downregulation of cellular elements implied in host response, which may lead to the release of a high number of mediators and molecules, including cytokines, chemokines, and acute-phase proteins [35]. Some of these have been proposed to be sepsis biomarkers [17, 36]. However, it is unlikely that one single biomarker will be able to satisfy all the characteristics for an "ideal" sepsis biomarker.

In a recent review, Pierrakos *et al.* identified 178 different biomarkers for potential use in sepsis. They and other authors have categorized the sepsis biomarkers according to their pathophysiological role in sepsis [17, 32, 37]. Coagulation, complement, contact system activation, inflammation and apoptosis are all involved in sepsis process and separate markers for each system have been proposed. Additionally, the systemic nature of sepsis and the large numbers of cells type, tissues and organs involved increase the number of potential biomarkers candidates. In this chapter we discuss those biomarkers that are measurable by assays which have been approved for clinical use.

Acute-Phase Protein Biomarkers

PCT and CRP are both proteins in response to infection and/or inflammation and they are the most widely used biomarkers for management of sepsis.

C-Reactive Protein

CRP was discovered by Tillet *et al.*, in a patient with lobar pneumonia in 1930 as a protein responsible for precipitating C polysaccharide during the acute phase of *Streptococcus pneumoniae* infection [38]. CRP is an acute phase protein released by the liver after the onset of inflammation or tissue damage. Although IL-6 is the prototypical stimulus for the induction of CRP, other cytokines, such as IL-8, also play a role in its production [39]. CRP has been traditionally the reference biomarker for inflammatory systemic response, but its specificity has been challenged, because CRP levels also increase in non-infectious conditions [40]. Its low specificity and inability to differentiate bacterial infections from non-infectious causes of inflammation makes CRP of limited diagnostic value. Other limitation for an early diagnosis in acute setting is its kinetic because CRP levels increase with a delay of up to 24 h compared to cytokines or PCT [37] and remain elevated for up to several days, even when infection has been eliminated [41] (Fig. (**2**) [42]). Perhaps, in infection setting, its measurement only should be recommended in patients with suspected infection when PCT measurement is not available [32]. Finally, its measurement is recommended for assessment of severity of acute pancreatitis [43].

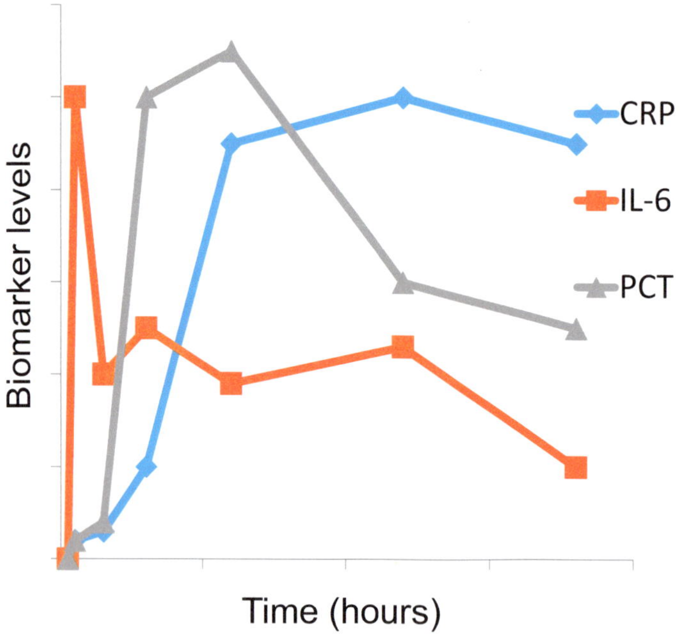

Fig. (2). Adapted from [42]. Kinetic profile of infection and/or inflammation markers.

Procalcitonin

PCT, the 116-amino acid protein precursor of the hormone calcitonin, is normally synthesized by neuro-endocrine C-cells of the thyroid gland and is cleaved into mature calcitonin, which plays a role in calcium homeostasis [44]. Its complete sequence has been known since 1984 [45, 46]. Because PCT cleavage into calcitonin occurs before secretion, circulating PCT level is very low in normal subjects; so, Becker *et al.*, using a highly sensitivity assay, described normal PCT in non-infected people as 0.033 ± 0.003 ng/mL [47], and levels < 0,05 ng/mL are considered as normal [48]. However, systemic inflammation can trigger extra thyroidal non-neuroendocrine calcitonin gene expression. In this context, different tissues can secrete immature PCT into the bloodstream, whilst calcitonin remains undetectable, in response to several pro-inflammatory signals such as endotoxin, IL-1β, IL -6 or TNF-α [44, 49 - 51], and its level is associated to bacterial load [52]. Conversion from PCT to calcitonin is inhibited by the cytokines and endotoxins secreted during bacterial infections; hence, in these conditions, PCT levels increases specifically. The induction of PCT can be attenuated by interpheron-gamma (IFN-γ), a cytokine released during viral infection [51]; therefore, PCT levels can be useful to differentiate bacterial infections from viral illness [53].

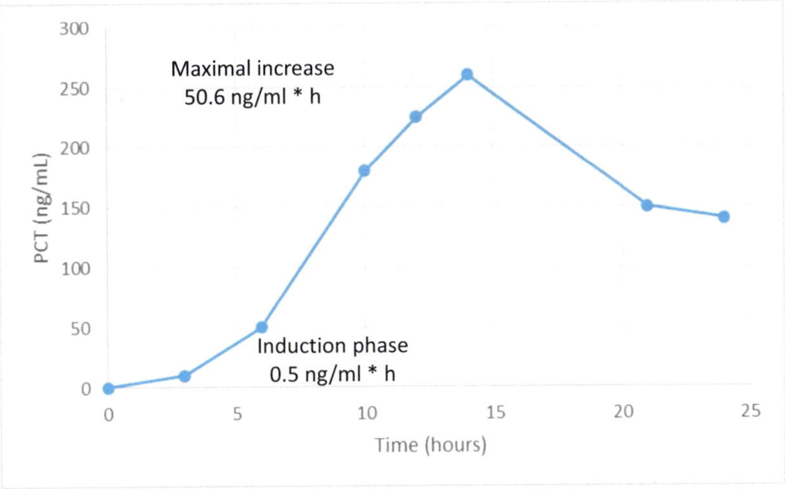

Fig. (3). Adapted from [55]. PCT plasma concentrations (ng/ml) following infusion of an accidentally bacterially (*Acinetobacter baumanii*) contaminated infusion solution to a 76 year-old female patient. The induction period can be described according to 2 types of kinetics: during the first phase (<6h), PCT increased by approximately 0.5 ng/ml per hour after a latency phase of about 2-3 hours (first measurable value recorded at time = 3h) followed by massive PCT production at the rate of approximately 50 ng/ml per hour over subsequent hours.

The use of PCT as a potential biomarker for sepsis and infection was first described in 1993 [54]. PCT shows a favorable kinetic profile for use as a clinical marker (Figs. (**2**) [42] and (**3**) [55]): after induction, *e.g.* by endotoxin, PCT increase is observed within 2-3 hours. Levels then rise rapidly, reaching a plateau after 6-12 hours. PCT levels remain high for up to 48 hours, falling to their baseline, in absence of other stimulation [56].

PCT as a Diagnostic Marker

Numerous studies have investigated the diagnostic usefulness of PCT to differentiate sepsis from others forms of SIRS and in different clinical situations and different types and sites of infection, usually comparing with CRP. It is of interest to note that, as with any diagnostic tool, PCT should be embedded in clinical algorithms adapted to the type of infection and the clinical context and setting.

In the review and meta-analysis by Simon *et al.* [57], the diagnostic accuracy of PCT was higher than that of CRP among patients hospitalized for suspected bacterial infections. PCT level was more sensitive (88% *vs.* 75%) and specific (81% *vs.* 67%) than CRP level to differentiate bacterial from non-infective origins of inflammation. Besides, the sensitivity for differentiating bacterial from viral infections was also higher for PCT (92% *vs.* 86%); the specificities were comparable (73% *vs.* 70%). In a Spanish study including patients admitted in an Emergency Department, Julián-Jiménez *et al.* reported that PCT was more sensitive and specific than PCR to distinguish sepsis from non-infectious SIRS [58]. In critically ill patients, a recent systematic review and meta-analysis showed that PCT yielded a pooled sensitivity and specificity, for a cutoff of 1.1 ng/mL, of 0.77 and 0.79, respectively, and an area under the receiver characteristic operating curve (AUC ROC) of 0.85 [59], concluding that the results of PCT must be interpreted in the context of medical history, physical examination and microbiological assessment.

Diagnostic accuracy of PCT has also been investigated in other infectious diseases:

- In acute meningitis, several studies have shown that serum PCT allows to distinguish between bacterial and viral etiologies in children and adultswith suspected meningitis, with a higher performance than CRP and other markers such as white blood count (WBC) [60, 61], although the optimal cutoffs vary depending on the authors [32]. It is also a good indicator to evaluate the efficacy of therapy in bacterial meningitis [60]. In children, a recent meta-analysis demonstrates that serum PCT assay is a highly accurate and powerful test for rapidly differentiating the etiology of pediatric meningitis with pooled sensitivity, specificity and AUC ROC of 0.96, 0.89 and 0.97, respectively [62]. In adults, other recently published systematic review and meta-analysis, Vikse *et al.* conclude that serum PCT is a highly accurate test for diagnosing meningitis, with a pooled sensitivity and specificity of 0.90 and 0.98, respectively [63].

- As predictor of bacteremia, PCT has demonstrated to be a useful tool to predict bloodstream infection in patients with SIRS admitted in an Emergency Department [64, 65] and in patients with community-acquired pneumonia (CAP) [66] and urinary tract infection (UTI) [52] and to distinguish blood contamination from true bloodstream infection in patients with growth of coagulase-negative *staphylococci* in blood cultures [67].
- In UTI, the most of evidence has been obtained in pediatric populations. Some observational studies suggest than PCT is an accurate marker for early diagnosis of acute pyelonephritis, with a similar sensitivity but higher specificity than CRP, and determination of the severity of renal parenchymal involvement, allowing prediction of high risk patients for permanent parenchymal renal lesions [68, 69]; these findings have also been confirmed in a meta-analysis by Leroy *et al.* [70]. However, these conclusions have been recently questioned by Shaikh *et al.* They concluded that the limited number of studies and the marked heterogeneity between studies prevents us from reaching definitive conclusions [71].
- In patients with infective endocarditis PCT levels were increased compared to non-infected patients in two studies [72, 73], but Yu *et al.* meta-analysis conclude that the current evidence does not support the routine use of serum PCT or CRP to rule in or rule out infective endocarditis in patients suspected to have infective endocarditis [74].
- In patients with febrile neutropenia, the value of PCT for diagnosis of infection is controversial. García de Guadiana *et al.* describe PCT levels higher in infected patients than in those with fever of unknown origin, with an AUC ROC for diagnosing infection of 0.88 [75]. Results are controversial in different reviews and meta-analysis. Sak *et al.* concluded that PCT is useful as a diagnostic and prognostic tool in patients with febrile neutropenia. However, due to differences in the studies design and the definitions for classifications of patients, further research is needed [76]. Most recently, Wu *et al.* concluded that PCT can be used to confirm the diagnosis of bacterial infections in patients with febrile neutropenia but due to high negative likelihood ratio (0.40) the decision for stopping antibiotics based on PCT levels alone in this type of patients is not be possible.
- In acute pancreatitis, PCT is a helpful tool to predict the severity of acute pancreatitis and for predicting and early diagnosis of infected pancreatic necrosis [77, 78]. Its measurement is recommended in patients with acute

pancreatitis, being a level ≥ 0.5 ng/mL during the first 24 hours an pancreatic alarm sign [43].

- In community-acquired pneumonia (CAP), PCT is useful to the making decision about the need and optimal duration of antibiotic therapy (Fig. **4**) [32, 79] and to obtain prognostic information, evaluated together clinical rules such as FINE, PSI or CURB-65 [80, 81]. Therefore, some guidelines recommends PCT measurement in patients with CAP because it may guide shorter treatment duration [82]. Moreover, PCT contributes to identify a bacterial etiology in patients with lower respiratory tract infection symptoms, avoiding antibiotic treatment [83].

PCT (ng/mL)	Bacterial infection	Recommendations to start antibiotics	Considerations
< 0,1	Very unlikely	NO	• If antibiotic therapy is withheld, control PCT after 6-24 h • Antibiotic teraphy can be considered in case of: ✓ Respiratory or hemodinamic instability, severe comorbidities or requirement of ICU management ✓ Sepsis with organ dysfunction or septic shock ✓ Local infection (empyema) ✓ Neutropenic or inmunossuppresed patients ✓ PCT < 0,1: CAP with PSI V or CURB > 3, COPD with GOLD IV ✓ PCT < 0,25: CAP with PSI IV/V or CURB >2, COPD with GOLD III/IV
0,1-0,24	Unlikely	No	
0,25-0,5	Likely	Yes	• Consider the course of PCT • If antibiotic teraphy is initiated: ✓Measure PCT on days 3, 5 and 7; stop antibiotic teraphy using the same thresholds. ✓If peak PCT levels are very high, then stop when 80-90% decrease of peak ✓If PCT remains high, value treatment failure
> 0,5	Very likely	YES	

Fig. (4). Adapted from [32] and [79]. PCT algorithm for patients with respiratory tract infections in the Emergency Department.

PCT as a Guide for Antibiotic Decisions

Several randomized-controlled studies have evaluated the use of PCT to assist in decisions about initiation and/or duration of antibiotic therapy (antibiotic stewardship). Because PCT is associated with bacterial infection, it has been proposedthat this biomarker might be helpful to decide whether or not treat patients with suspected infection. Thereby the benefit of PCT was measured by clinical outcomes, assuming that if the patient recovers without antibiotic treatment, there was not relevant bacterial illness in need of antibiotics. The published studies on antibiotic stewardship used algorithms with recommendations for or against antibiotic treatment based on PCT thresholds. An example of these algorithms is shown in Fig. (**6**) [32, 79] and it is based on the results of Christ-Crain M *et al.* study, including patients with suspected lower respiratory tract infections [84]. Clinical outcomes were similar in both groups, but PCT guidance substantially reduced antibiotic use in lower respiratory tract infections in all diagnostic subgroups, particularly in patients with acute bronchitis and acute exacerbation of chronic obstructive pulmonary disease. Similar results were reported in a subsequent multicenter study [85], showing besides that in patients with CAP the decrease to exposure to antibiotics was associated to a reduction in duration of therapy. In a systematic review Schuetz el al. assessed the safety and efficacy of PCT for starting or stopping antibiotics over a large range of patients with varying severity of acute respiratory infections and from different clinical settings. They concluded that its use to guide initiation and duration of antibiotic treatment in patients with acute respiratory infection was not associated with higher mortality rates or treatment failure. Besides, antibiotic consumption was reduced across different clinical settings and acute respiratory infection diagnoses [86].

The role of PCT for guidance of antibiotic stewardship has been widely investigated in more high risk patients in ICU setting, mainly for discontinuation of antibiotic therapy. Hence, the decrease in PCT levels has been proposed to guide stopping antibiotics, allowing their administration duration to become "customized" for each patient. PRORATA, a large multicenter randomized trial in France including 621 unselected patients ICU with suspected bacterial infection, tested the usefulness of a PCT-based algorithm (Fig. **5**) to guide initiation, continuation and stopping of antibiotics [87]. Notably, the initial prescription rate did not differ between PCT-guided and control groups and in PCT-guided patientshad similar rates of relapses and 30-day and 60-day mortality but significantly more days without antibiotics than did those in the control group (14.3 *vs* 11.6 days). Similar findings were reported in postsurgical ICU patients with severe sepsis [88], in patients with ventilator-associated pneumonia [89] and in newborns with suspected neonatal early-onset sepsis [90]. A recent systematic review concluded that PCT levels may be effective and cost-effective when they are used to guide discontinuation of antibiotics in adults being treated for suspected or confirmed sepsis in ICU settings, and initiation of antibiotics in adults presenting to the ED with respiratory symptoms and suspected bacterial infection. In children, further research is needed to assess the effectiveness of adding PCT algorithms to the information used to guide antibiotic treatment [91].

PCT as a Prognostic Marker in ICU Patients with Severe Sepsis or Septic Shock

Numerous studies have evaluated the value of PCT and its kinetics, expressed as clearance of PCT in the first 42, 48 or 72 hours, in ICU patients with severe sepsis or septic shock as a marker for prognosis of mortality [92 - 100]. In most studies, a single value of PCT measured on admission did not predict the prognosis of the critically ill septic patient. Although changes in PCT levels in the first 24 [95, 100], 48 [94, 97] and 72 [96, 98, 99] hours were predictors of mortality, the added value to clinical scores such as is controversial.

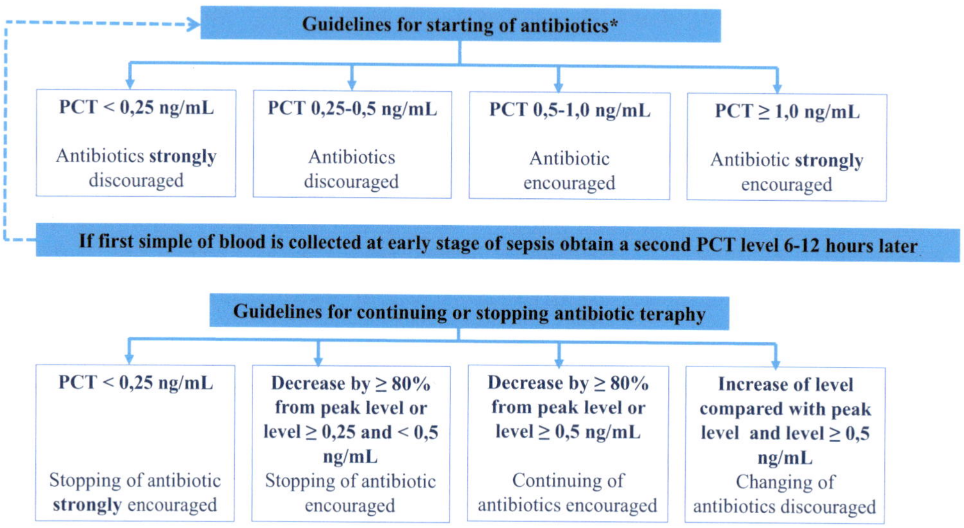

Fig. (5). PCT-based algorithm used in the PRORATA trial for starting, continuing, or stopping of antibiotics [87]. *Excludes situations requiring immediate antibiotic treatment (*e.g.*, septic shock, purulent meningitis).

Limitations of PCT

Although PCT is the biomarker most widely used for diagnosing sepsis, interpretation of its levels requires the knowledge of some limitations:

- In healthy term neonates circulating concentrations of PCT physiologically increase gradually from birth to reach peak values at about 24 h of age and then decrease gradually by 48 h of life [101]. Moreover, during the early neonatal period, the healthy preterm baby has an earlier, higher, and longer PCT response compared with that found in the healthy term baby, demonstrating an inverse relationship between stage of development and magnitude of neonatal PCT response [102].
- Different pathogens might induce different responses, causing a variable up-regulation of PCT levels. In Krüger *et al.* study, in patients with typical bacterial CAP, levels of PCT were significantly higher compared to CAP of atypical or viral etiology [103].

- Previous antimicrobial treatment may influence the PCT level resulting in lower levels [104].
- Moreover, PCT can be elevated also in absence of bacterial infections in conditions such as severe trauma and surgery [78] or after cardiac shock [105]. Elevated PCT levels have also been reported in patients with medullary thyroid carcinoma [106] and lung cancer [107]. Other conditions with increased serum concentrations of PCT include different types of immunotherapies [108, 109], and some autoimmune diseases [110].

Lipopolysaccharide-Binding Protein

Lipopolysaccharide (LPS)-binding protein (LBP) is an acute phase reactant of hepatic origin that plays an important role in innate immunity mechanisms. It forms a complex with LPS and this complex binds to CD14 and to the Toll-like receptor 4 (TLR-4)/MD2-complex resulting in transcription of cytokines and other pro-inflammatory mediators, such as IL-6 and IL-1 and TNFα [111].

Because LBP levels increase 3 to 4-fold in sepsis, it makes LBP promising forthe diagnosis of sepsis and indeed, a good discrimination between SIRS and sepsis was reported [112]. However, further studies did not confirm these findings showing that LBP is a rather non-specific marker of the inflammatory response [113] and is not predictive of outcome [97, 114]. Currently, LBP does not appear to have a role in the diagnosis of sepsis [22, 110].

Pentraxin 3

Pentraxin 3 (PTX-3) is a secretory protein classified as a long pentraxin subfamily member of the pentraxin family, recognized as key components of humoral innate immunity [115]. It can be rapidly produced and released by mononuclear phagocytes, polymorphonuclear neutrophils (PMNs), epithelial and endothelial cells in response to primary inflammatory signals such as IL-1 and TNF-α [116]. PTX-3 plays a central role in the early phases of inflammation: it recognizes microbial components, activates the classical pathway of complement and facilitates recognition by macrophages and dendritic cell [116]. PTX-3 levels increase rapidly during multiple inflammatory conditions similarly to other acute-phase reactants; therefore, PTX3 is not a specific marker for bacterial infection and it should be considered a biomarker of disease severity. An increase in plasma PTX-3 levels have been described in several inflammatory conditions. High PTX-3 levels have been correlated with unfavorable outcomes in several conditions such as cardiovascular diseases [117], lung cancer [118] and polymyalgia rheumatic [119] and were predictors of severity of different diseases: dengue virus infection [120], leptospirosis [121] and epidemic nephropathy [122].

In the context of infection and sepsis, in emergency department high PTX-3 levels has been shown to predict the severity of disease in febrile patients [123] and severe sepsis, with a similar AUC ROC (0.73) than PCT (0.77) and higher than that of CRP (0.60), and 28 and 90-days mortality and other adverse outcomes (admission in ICU, need for mechanical ventilation and vasopressors, acute renal failure) in patients with suspected infection [124]. In critically ill ICU-patients, PTX-3 levels have correlated with the severity of disease and infection [125]. In ICU patients receiving mechanical ventilation, PTX-3 levels shown a moderate diagnostic accuracy for ventilator-associated pneumonia (VAP) (AUC ROC: 0,790) and correlated with the severity of sepsis and mortality [126].

In a recent review Liu *et al.* concluded that the diagnostic value of PTX3 is low in patients with sepsis and does not improve to other biomarkers as PCT and that circulating levels of PTX3 have prognosticvalue and may add to prognostication of patients with SIRS or sepsis, complementing severity of disease classification scores and other biological markers [127].

Pancreatic Stone Protein

Pancreatic stone protein (PSP), also called PSP/reg or lithostathine, is a polypeptide belonging to the family of lectin-binding proteins [128]. PSP historically has been reported mainly regarding to the pancreas, and, indeed, pancreaticacinar cells are considered its main source although PSP was reported to be expressed in various other tissues, including stomach and kidney [128]. Its serum elevation seemed to be correlated strongly with severe acute pancreatitis and less tightly withchronic pancreatitis and pancreatic cancer [129]. In recent years, PSP/reg investigations focused on infection and inflammation [130], from the result of Keel *et al.*, who concluded that that PSP/reg is regulated in blood after trauma and related to the severity of inflammation. Furthermore, PSP/reg binds to and activates neutrophils. Therefore, PSP might be an acute-phase reactant that responds to injury during the early phase of infection [131].

PSP is useful for differentiating SIRS from sepsis (AUC ROC: 0.93), with a similar performance than that of PCT (AUC ROC: 0.84) and to assess the severity of sepsis in ICU patients [132]. PSP levels, measured on admission or within 24 hours of ICU admission, are a prognostic marker in patients with severe sepsis or septic shock (ACU ROC for in-hospital mortality: 0.65) [133], improving the diagnostic accuracy when is combined with PCT and APACHE II score (AUC ROC: 0.710) [134]. Moreover, PSP is related to development of organ failure in patients with VAP [135].

Cytokine And Chemokines Biomarkers

Cytokines

IL-1β, IL-6 or TNF-α are cytokines responsible for mediation of the initial immune system response to injury or infection, contributing to fever, activate endothelial cells and attract PMNs. Therefore, they mediators responsible in the pathophysiology of sepsis, but their levels also increase in other non-infectious inflammatory conditions such as trauma, surgery, stroke or autoimmune diseases. Other limitations are the lack of specificity, the kinetic profile of these mediators and other cytokines, such as IL-10 or IL-8 (Fig. **3**), peaking in 2-3 hours, the short plasma half-life and the low biological stability [32].

IL-6 and IL-8 shown a moderate performance to differentiate sepsis from SIRS, lower than PCT [136, 137]. This finding was confirmed by Tsalik *et al.* [138]. In patients with febrile neutropenia IL-6 shown a similar performance than PCT for diagnosis of infection [97]. Moreover, serum levels of IL-6 are closely related to the severity andoutcome of sepsis in patients IL-6 is a predictor for mortality [139, 140]. Recent studies have proposed that the measurement of multiples cytokines correlates with disease severity and prognosis [141, 142].

While IL-1β, IL-6 and TNF-α are proinflammatory mediators released within minutes from exposure to bacterial lipopolysaccharides and are classified as markers for early response to sepsis, there are two well-known inflammatory mediators, *high-mobility-group box 1 (HMGB1) and macrophage migration inhibitory factor (MIF)*, which are important in late phase of severe infection [143]; hence, these biomarkers are indicators of severity and poor outcome, and both could predict sepsis prognosis [22, 144].

Cell Surface Markers and Soluble Receptors Biomarkers

Soluble Cluster of Differentiation 14 Subtype (Presepsin)

Cluster of differentiation (CD) 14 is a glycoprotein expressed on the membrane surface of various cells, including monocytes, macrophages and PMNs (mCD14) and serves as a specific high-affinity receptor for complexes of LPS, a compound from the outer cell wall of Gram-negative bacteria, and LBP. Upon binding of the LBP complex, CD14 activates the TLR4-specific pro-inflammatory signaling cascade, thereby starting the inflammatory reaction of the host against infectious agents (Fig. **6**) [145]. Besides mCD14 may function as a receptor for peptidoglycan, the major cell wall component of Gram-positive bacteria, and other microbial products with similar structural features. CD14 is also found in a circulating soluble (sCD14) state and circulating plasma proteases activate a cleavage of sCD14, generating a truncated form of named sCD14 subtype (sCD14-ST), re-named as presepsin. sCD14-ST reduces the mortality rate caused by endotoxin shock and the severity of Gram-negative bacterial infections [146].

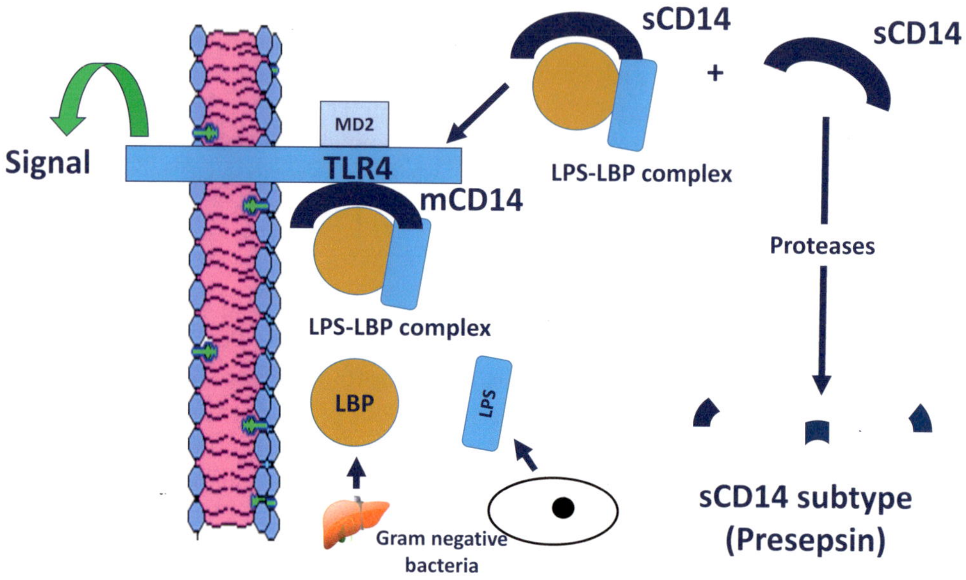

Fig. (6). Mechanism of secretion of presepsin. Abreviatures: mCD14: membrane CD14; sCD14: soluble CD14; sCD14-ST: soluble CD14 subtype (Presepsin); LPS: lipopolysaccharide; LBP: lipopolysaccharide binding protein; TLR4: toll-like-receptor 4; MD2: Co-protein of TLR4.

Several recent studies have pointed out the value of presepsin for diagnosing sepsis, assessing the severity of sepsis and providing a prognostic evaluation of patient outcomes, including those with severe sepsis and septic shock requiring management in an ICU.

Regarding to diagnostic value, several reviews and meta-analyses have recently evaluated the diagnostic accuracy of presepsin for sepsis with pooled ROC AUCs ranging from 0.86 to 0.89 and pooled sensitivity from 77% to 86%, specificity from 73% to 81%, positive like hood ratio from 3.8 to 4.8 and negative like hood ratio from 0,18 to 0,21. However, the comparison of results among the published studies has several limitations due to the great heterogeneity found in the included studies, with confounding potential sources such as the study design, clinical setting (emergency departments, ICU), type of patients (medical, surgical, burn patients), criteria for sepsis definition and even the type of sample (plasma, serum or whole blood) for measurement of presepsin [147 - 151].

As prognostic marker, Ulla *et al.* reported, in septic patients admitted in an emergency department, that presepsin levels on admission correlated with 60-day mortality [152], similarly to results reported by Liu *et al.* [153] and Carpio *et al.* [154] for 28-day mortality and 30-day mortality, respectively, indicating the possible prognostic role of presepsin for predicting mortality and identifying high-risk patients. In ICU patients with severe sepsis and septic shock presepsin reveals valuable diagnostic capacity to differentiate sepsis severity compared to PCT, IL-6 and CRP and prognostic value for 30- and 6 months all-cause mortality throughout the first week of ICU treatment [155]. In Masson *et al.* the prognostic accuracy of presepsin yielded AUCs for ICU survival of 0.69, 0.70 and 0.74 on days 1, 2 and 7, respectively, higher than PCT and similar than SOFA score [156].

Soluble Triggering Receptor Expressed on Myeloid Cells-1 (sTREM-1)

The triggering receptor expressed on myeloid cells-1 (TREM-1) is a glycopeptide receptor expressed on the surface of myeloid cells such as PMNs, mature monocytes and macrophages and its expression is upregulated on phagocytes after exposure to bacteria and fungi [157]. During sepsis, activated phagocytes release a soluble form of TREM-1 (sTREM-1) which among other body fluids can be found in blood. Gibot. *et al* reported a higher performance than that of PCT and CRP to differentiate patients with sepsis from those with systemic inflammation of non-infectious cause in ICU patients, with an AUC ROC of 0.97 [158], finding confirmed in Jiyong *et al.* meta-analysis, reporting a pooled AUC ROC, sensitivity and specificity of 0.86, 0.82 and 0.86, respectively, for diagnosis of bacterial infection [159]. However, a more recent meta-analysis including 1795 patients shown than sTREM-1 had a sensitivity of 0.79 and specificity of 0.8 for differentiating sepsis from non-infectious SIRS; the authors concluded that sTREM-1 has only a moderate diagnostic accuracy for differentiating sepsis from SIRS although sensitivityand specificity were comparable to the values found for PCT [160]. Regarding to its prognostic value, a prospective study in Korea in severe sepsis patients reported that sTREM-1 levels, measured on admission, correlated with worst outcome and could be used as a marker to identify patients with a poor prognosis [161]. The role of sTREM-1 as a biomarker in management of sepsis remains undefined and larger studies are necessary to clarify this issue [22, 110].

Neutrophil CD64

The CD64 antigen is the high affinity receptor for the Fcγ part of the IgG heavy chain and can bind monomeric IgG1 and IgG3 as well as aggregated IgG. Phagocytosis of bacteria and other microorganisms is mediated by this receptor [162]. In contrast to monocytes where CD64 antigen is constitutively expressed, resting neutrophils have very low levels of CD64 antigen on their membrane. CD64 expression is increased upon activation of PMNs by pro-inflammatory cytokines within 4-6 hours and can reach more than 10-fold higher levels than in resting conditions, allowing good discrimination between resting and activated neutrophils [163, 164].

Neutrophil CD64 has several characteristics that make is well suited for clinical application: on resting PMNs CD64 expression is low and after activation it is significantly upregulated within few hours. Once the activation stimulus disappears, CD64 expression returns to its basal level in few days. Moreover, CD64 is relatively stable after blood collection and the assay is straightforward and requires only small sample volume. Moreover, neutrophil CD64 expression represents a physiological process which plays a key role in the innate immune response: PMNs acting as phagocytes [165].

A recent meta-analysis conclude that neutrophil CD64 expression could be a promising biomarker for diagnosing bacterial infection, with a summary ROC of 0,92 and summary estimate of 0,76 for sensitivity and 0,85 for specificity [166]. However, the calculations in this and other meta-analysis [167, 168] were performed using studies with different designs and populations: both adults and neonates were included, patients with sepsis and local infections were mixed, different criteria inclusion were used, generally used small numbers of patients and there was substantial variation in analytical methodology. Taken together, there is evidently a need for much larger, multicenter studies with uniform inclusion criteria and standardized analytical methods in order to appreciate the clinical utility of neutrophil CD64 as a sepsis biomarker [165, 166].

Neutrophil CD64 expression is also useful as indicator of sepsis severity and to predict poor outcome [169], although the available data about its prognostic value are highly controversial, because other studies indicated that higher expression of CD64 was related to survival of the patients [170]. Most likely the small patient groups and the design of the studies explain these conflicting findings [165].

Soluble Urokinase Type Plasminogen Activator Receptor (suPAR)

The urokinase-type plasminogen activator (uPA) system consists of a protease, inhibitors and a receptor (uPAR), expressed on various cell types, including PMNs, lymphocites, monocytes/macrophages, endothelial and tumor cellsand is involved in themigration of inflammatory cells from the bloodstream into tissues. Its soluble form (suPAR) was identified in 1991 [171]. After cleavage from the cell surface, suPARcan be found in blood and other organic fluids, including urine, cerebrospinal fluid, saliva and bronchial washing fluid, in three forms with different properties due to structural differences. Increased inflammation secondary to activation of theimmune system thereby produces increased concentrations of suPAR in body fluids.

Levels of suPARare increased in acutely ill patients, but this increase is not specific for sepsis. Therefore, suPAR is of limited value as a diagnostic biomarker of sepsis [22, 110, 172]. Alternatively, however, several studies have indicated that suPAR levels may reflect the severity of infection and they are related with a poor outcome in non-infectious and infectious diseases. Several studies have confirmed the prognostic value of suPAR. In Giamarellos-Bourboulis *et al.* study, in ICU patients, suPARlevels >12 ng/ml were associated with an unfavorable outcome, mainly in patients with an APACHE II score >17 [173]. Recently, Donadello *et al.* reported a diagnostic accuracy for sepsis weaker than that of CRP in critically ill patients, but high suPAR levels on admission were a strong independent predictor for ICU and 28-day mortality in global population, although the addition of suPAR to the APACHE II and SOFA scores did not significantly increase the predictive power [174], similarly to results obtained by Suverbiola *et al.* in patients with severe sepsis [175].

In summary, suPAR might have better prognostic value to predict mortality instead of diagnosing sepsis [22].

Biomarkers of Endothelial Activation

A hallmark of sepsis is a change in microvascular function. Widespread endothelial damage and apoptosisappears to be directly involved, withnumerous associations observed between sepsis and endothelial cell (EC) activation [176, 177]. Therefore, there is a strong biologic rationale for targeting markers of endothelial activation as biomarkers of sepsis. A large number of EC-active molecules have been investigatedas potential biomarkers for the early diagnosis and prognosis of sepsis.

Recently Xing *et al.* reviewed the utility of biomarkers of endothelial cell activation in sepsis and they classified them in the following groups [178]:

The Angiopoietin System

<u>The Leukocyte Adhesion Pathway</u>, including *soluble intercellular adhesion molecule* (*sICAM-1*), *soluble vascular cell adhesion molecule* (*sVCAM-1*), *soluble E-selectin*(*sE-selectin*) and *Endocan*

<u>Mediators of Permeability and Vasomotor Tone</u>, including *soluble vascular endothelial growth factor* (*VEGF*), *soluble fms-like tyrosine kinase*(*s-1FlT-1*) and *endothelin-1*

Mediators of Coagulation, including *Von Willebrand factor* and *ADAMTS13*

Although recent studies have confirmed the association among the biomarkers of endothelial activation and severity of sepsis, organ dysfunction and mortality [179, 180] the clinical utility or application of these molecules as biomarkers in sepsis is limited due to a lack of standardization in analytical assays, a lack of data regarding ROC curves, in the few cases where thresholds have been reported, and a lack of validation in representative patient populations [178].

Biomarkers Related to Vasodilatation

Pro-Adrenomedullin (proADM)

Adrenomedullin (ADM) is a 52-amino acid peptide with a strong vasodilatory activity [181, 182]. Its circulating levelshave been described for several disease states, including dysfunction of the cardiovascular system [183] and sepsis [184]. Therefore, quantification of ADM would be helpful in the diagnosis, monitoring, and prognosis of various cardiovascular diseases and sepsis. However, the reliable measurement of ADM release in the circulation is difficult. In addition to immediate binding of ADM to receptors in the vicinity of its production, peripheral measurement is also hampered by the existence of a binding protein, the short half-life of ADM (22 min), and technical difficulties [185 - 187]. ADM is derived from a larger precursor peptide (preproADM; 185 amino acids) by posttranslational processing. During the processing of preproADM, other peptides are generated: another biologically active peptide termed proadrenomedullin N-terminal 20 peptide (PAMP) with a suggested hypotensive effect, and 2 peptides flanking ADM: one midregional part of proADM (proADM 45-92) and the COOH terminus of the molecule (proADM 153-185) (Fig. **7**). Because of its probable stoichiometric generation, the released amounts of MR-proADM may directly reflect those of ADM and PAMP [184].

In the context of sepsis and infection, the main usefulness of MR-proAMP is due to its prognostic value. Most studies published about MR-proADM evaluated it predictive value for mortality in patients with CAP, independently or combined with prognostic scores such as PSI or CURB-65 [188 - 191] and concluded that MR-proADM is significantly associated with short and long-term mortality and complications in patients with CAP. Moreover, the addition of MR-proADM improved the discriminant ability of prognostic scores. In conclusion, MR-proADM represent a promising tool for risk stratification [192].

In sepsis, there is limited evidence on its potentialclinical role. In studies evaluating the prognostic value of MR-proAMP in ICU patients with sepsis the results are less conclusive. MR-proAMP on admission shown a moderate performance for mortality and the combination of severity scores and MR-proAMP did not provide superior AUCs [175, 193]. However, more recently, Ojeda *et al*. concluded that MR-proADM is a consistent marker of mortality risk along time in sepsis,which provides a selective advantage of this moleculeover other biomarkers as prognostic tool, although this study didn´t compare the value of MR-proAMP with scoring systems [194].

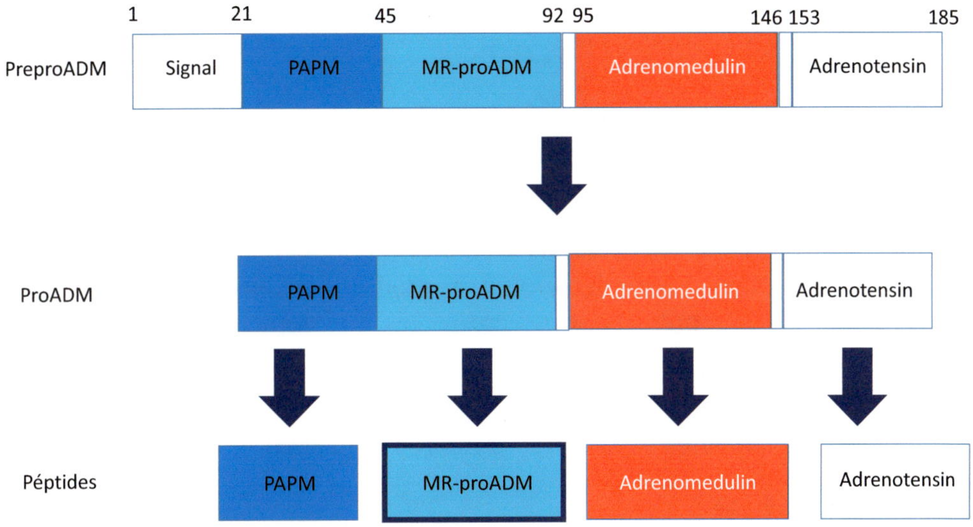

Fig. (7). Sequence of prepro-ADM.

CONCLUSIONS AND FUTURE DIRECTIONS

This chapter reviews the traditional and some emergent biomarkers available biomarkers for the diagnosis and prognosis of sepsis, most of them under investigation.

Despite the wide number of biomarkers investigated for use in different settings related with sepsis (diagnosis, prognosis and biomarker-guided antibiotic therapy), none has sufficient specificity and sensitivity to be routinely employed in clinical practice. PCT and CRP have been most widely used and indeed they were included as criteria for diagnosis of sepsis in 2001 sepsis definition [28].

Moreover, nowadays the role of biomarkers for diagnosis of sepsis is not clear. Recently, an update definition for sepsis has been published and it has not included the biomarkers to identify patients with sepsis [31]. The role of biomarkers in the initial management of patients with sepsis should probably be reviewed and large and multicentric studies, using designs similar to that used in Seymour *et al.* study [195]. Could biomarkers improve or increase the performance of *qSOFA* score to identify adult patients with suspected infection who are likely to have poor outcomes? Currently, this question remains unanswered.

In any case, in view of the complexity of the sepsis response, it is unlikely that a single ideal biomarker will be found. Probably, a combination of several biomarkers or of biomarkers with score systems may be more effective [196 - 198], but this approach requires further evaluation, because these result have not been confirmed in other studies [199, 200].

Regarding to future directions, there is a growing interest in the potential utility of molecular diagnostics, such as real-time polymerase chain reaction (PCR),to improve the detection of life-threatening infection by detecting pathogen deoxyribonucleic acid (DNA) in blood samples within a few hours [201, 202] or in the use of circulating cell-free DNA as a potential biomarker in critically ill patients [203]. Further research is also needed to evaluate the influence of polymorphisms in susceptibility to severe sepsis and septic shock [204].

CONFLICT OF INTEREST

The author (editor) declares no conflict of interest, financial or otherwise.

ACKNOWLEDGEMENTS

Declared none.

REFERENCES

[1] Torio CM, Andrews RM. National inpatient hospital costs: the most expensive conditions by payer Statistical Brief #160 Healthcare Cost and Utilization Project (HCUP) Statistical Briefs August 2011. Available at: http://www.ncbi.nlm.nih.gov/books/NBK169005/. Accessed October 31

[2] Iwashyna TJ, Cooke CR, Wunsch H, Kahn JM. Population burden of long-term survivorship after severe sepsis in older Americans. J Am Geriatr Soc 2012; 60: 1070-7. [http://dx.doi.org/10.1111/j.1532-5415.2012.03989.x]

[3] Gaieski DF, Edwards JM, Kallan MJ, Carr BG. Benchmarking the incidence and mortality of severe sepsis in the United States. Crit Care Med 2013; 41: 1167-74. [http://dx.doi.org/10.1097/CCM.0b013e31827c09f8]

[4] Dellinger RP, Levy MM, Rhodes A, *et al.* Surviving Sepsis Campaign: international guidelines for management of severe sepsis and septic shock, 2012. Crit Care Med 2013; 41: 580-637. [http://dx.doi.org/10.1097/CCM.0b013e31827e83af]

[5] Rhee C, Gohil S, Klompas M. Regulatory mandates for sepsis care-reasons for caution. N Engl J Med 2014; 370: 1673-6. [http://dx.doi.org/10.1056/NEJMp1400276]

[6] Vincent JL, Marshall JC, Namendys-Silva SA, *et al.* ICON Investigators. Assessment of the worldwide burden of critical illness: the Intensive Care Over Nations (ICON) audit. Lancet Respir Med 2014; 2: 380-6. [http://dx.doi.org/10.1016/S2213-2600(14)70061-X]

[7] Fleischmann C, Scherag A, Adhikari NK, *et al.* Assessment of global incidence and mortality of hospital-treated sepsis: current estimates and limitations. Am J Respir Crit Care Med 2016; 193: 259-72.
[http://dx.doi.org/10.1164/rccm.201504-0781OC]

[8] Angus DC, Linde-Zwirble WT, Lidicker J, Clemont G, Carcillo J, Pinsky MR. Epidemioology of severe sepsis in the United States: analysis of incidence, outcome and associated cost of care. Crit Care Med 2001; 29: 1303-10.
[http://dx.doi.org/10.1097/00003246-200107000-00002]

[9] Esteban A, Frutos-Vivar F, Ferguson ND, *et al.* Sepsis incidence and outcome: Contrasting the intensive care unit with the hospital ward. Crit Care Med 2007; 35: 1284-9.
[http://dx.doi.org/10.1097/01.CCM.0000260960.94300.DE]

[10] Martin GS. Sepsis, severe sepsis and septic shock: changes in incidence, pathogens and outcomes. Expert Rev Anti Infect Ther 2012; 10: 701-6.
[http://dx.doi.org/10.1586/eri.12.50]

[11] Iwashyna TJ, Ely EW, Smith DM, Langa KM. Long-term cognitive impairment and functional disability among survivors of severe sepsis. JAMA 2010; 304: 1787-94.
[http://dx.doi.org/10.1001/jama.2010.1553]

[12] Hotchkiss RS, Karl IE. The pathophysiology and treatment of sepsis. N Engl J Med 2003; 348: 138-50.
[http://dx.doi.org/10.1056/NEJMra021333]

[13] Gullo A, Bianco N, Berlot G. Management of severe sepsis and septic shock: challenges and recommendations. Crit Care Clin 2006; 22: 489-501.
[http://dx.doi.org/10.1016/j.ccc.2006.03.006]

[14] Lever A, Mackenzie I. Sepsis definition, epidemiology, and diagnosis. BMJ 2007; 335: 879-83.
[http://dx.doi.org/10.1136/bmj.39346.495880.AE]

[15] Biomarkers and surrogate endpoints: prefered definitions and conceptual framework. Clin Pharmacol Ther 2001; 69: 89-95.
[http://dx.doi.org/10.1067/mcp.2001.113989]

[16] Marshall JC, Reinhart K. Biomarkers of sepsis. Crit Care Med 2009; 37: 2290-8.
[http://dx.doi.org/10.1097/CCM.0b013e3181a02afc]

[17] Pierrakos C, Vincente JL. Sepsis biomarkers: a review. Crit Care 2010; 14: R15.
[http://dx.doi.org/10.1186/cc8872]

[18] Bone RC, Grodzin CJ, Balk RA. Sepsis: a new hypothesis for pathogenesis of the disease process. Chest 1997; 112: 235-43.
[http://dx.doi.org/10.1378/chest.112.1.235]

[19] Faix JD. Biomarkers of sepsis. Crit Rev Lab Clin 2013; 50: 23-36.
[http://dx.doi.org/10.3109/10408363.2013.764490]

[20] van der Poll T, van Deventer SJ. Cytokines and anticytokines in the pathogenesis of sepsis. Infect Dis Clin North Am 1999; 13: 413-26.
[http://dx.doi.org/10.1016/S0891-5520(05)70083-0]

[21] Hotchkiss RS, Monneret G, Payen D. Immunosupression in sepsis: a novel understanding of the disorder and a new therapeutic approach. Lancet Infect Dis 2013; 13: 260-8.
[http://dx.doi.org/10.1016/S1473-3099(13)70001-X]

[22] Cho SY, Coi JH. Biomarkers of sepsis. Infect Chemother 2014; 46: 1-12.
[http://dx.doi.org/10.3947/ic.2014.46.1.1]

[23] Garnacho J, Garnacho MC, Ortiz C, Aldabó T. Polimorfismos genéticos en la sepsis. Med Intensiva 2005; 29: 185-9.

[http://dx.doi.org/10.1016/S0210-5691(05)74226-1]

[24] Geroulanos S, Douka ET. Historical perspective of the word "sepsis.". Intensive Care Med 2006; 32: 2077.
[http://dx.doi.org/10.1007/s00134-006-0392-2]

[25] Wesen SH, der Sepsis B. Wesen und Behandlung der Sepsis. Inn Med 1914; 31: 257-80.

[26] Bone RC, Balk RA, Cerra FB, *et al.* American College of Chest Physicians/Society of Critical Care Medicine Consensus Conference Definitions for sepsis and organ failure for the use of innovative therapies in sepsis. Crit Care Med 1992; 20: 864-74.
[http://dx.doi.org/10.1097/00003246-199206000-00025]

[27] Sprung Cl, Sakr Y, Vincent JL, Le Gall JR, Reinhart K, Ranieri VM. An evaluation of systemic inflammatory response syndrome signs in the Sepsis Occurrence In Acutely Ill Patients (SOAP) study. Intensive Care Med 2006; 32: 421-7.
[http://dx.doi.org/10.1007/s00134-005-0039-8]

[28] Levy MM, Fink MP, Marshall JC, *et al.* International Sepsis Definitions Conference. 2001 SCCM/ESICM/ACCP/ATS/SIS International Sepsis Definitions Conference. Intensive Care Med 2003; 29: 530-8.
[http://dx.doi.org/10.1007/s00134-003-1662-x]

[29] Vincent JL, Opal SM, Marshall JC, Tracey KJ. Sepsis definitions: time for a change. Lancet 2013; 381: 774-5.
[http://dx.doi.org/10.1016/S0140-6736(12)61815-7]

[30] Lin HY. Why we need to revise the definition and diagnostic criteria for sepsis. Chin J Traumatol 2015; 18: 249-50.
[http://dx.doi.org/10.1016/j.cjtee.2015.11.009]

[31] Singer M, Deutschman CS, Seymour CW, *et al.* The Third International Consensus Definitions for Sepsis and Septic Shock (Sepsis-3). JAMA 2016; 315: 801-10.
[http://dx.doi.org/10.1001/jama.2016.0287]

[32] Julián-Jiménez A, Candel-González FJ, González del Castillo J. Utilidad de los biomarcadores de inflamación e infección en los servicios de urgencias. Enferm Infecc Microbiol Clin 2014; 3: 117-90.

[33] Marshall JC, Reinhart K. Biomarkers of sepsis. Crit Care Med 2009; 37: 2290-8.
[http://dx.doi.org/10.1097/CCM.0b013e3181a02afc]

[34] Vincent JL, Teixeira L. Sepsis biomarkers. Value and limitations. Am J Respir Crit Care Med 2014; 190: 1081-2.
[http://dx.doi.org/10.1164/rccm.201410-1895ED]

[35] Karzai W, Meisner M, Reinhart K. New approaches to the diagnosis of sepsis. Curr Opin Crit Care 1999; 5: 357-62.
[http://dx.doi.org/10.1097/00075198-199910000-00004]

[36] Marshall JC, *et al.* Measures, markers, and mediators: toward a staging system for clinical sepsis. A report of the Fifth Toronto Sepsis Roundtable, Toronto, Ontario, Canada, October 25-26, 2000. Crit Care Med 2003; 31: 1560-7.
[http://dx.doi.org/10.1097/01.CCM.0000065186.67848.3A]

[37] Reinhart K, Bauer M, Riedemann NC, Hartoga CS. New approaches to sepsis: molecular diagnostics and biomarkers. Clin Microbiol Rev 2012; 25: 609-34.
[http://dx.doi.org/10.1128/CMR.00016-12]

[38] Tillet WS, Francis T. Serological reactions in pneumonia with a non-protein somatic fraction of pneumococcus. J Exp Med 1930; 22: 309-13.

[39] Gabay C, Kushner I. Acute-phase proteins and other systemic responses to inflammation. N Engl J Med 1999; 340: 448-54.

[http://dx.doi.org/10.1056/NEJM199902113400607]

[40] Povoa P, Coelho L, Almeida E, *et al.* C-reactive protein as a marker of infection in critically ill patients. Clin Microbiol Infect 2005; 11: 101-8. [http://dx.doi.org/10.1111/j.1469-0691.2004.01044.x]

[41] Meisner M, Tschaikowsky K, Palmaers T, Schmidt J. Comparison of procalcitonin (PCT) and C-reactive protein (CRP) plasma concentrations at different SOFA scores during the course of sepsis and MODS. Crit Care 1999; 3: 45-50. [http://dx.doi.org/10.1186/cc306]

[42] Meissner M. Procalcitonin: experience with a new diagnostic tool for bacterial infection and systemic inflammation. J Lab Med 1999; 23: 263-72.

[43] Maraví Poma E, Zubia Olascoaga F, Petrov MS, *et al.* SEMICYUC 2012. Recommendations for intensive care management of acute pancreatitis. Med Intensiva 2013; 37: 163-79. [http://dx.doi.org/10.1016/j.medin.2013.01.007]

[44] Becker KL, Nylen ES, White JC, Muller B, Snider RH. Clicical review 167: Procalcitonin and the calcitonin gene family of peptides inflammation, infection, and sepsis.: a journey from calcitonin back to its precursors. J Clin Endocrinol Metab 2004; 89: 1512-25. [http://dx.doi.org/10.1210/jc.2002-021444]

[45] Jacobs JW, Lund PK, Potts JT, Bell NH, Habener JF. Procalcitonin is a glycoprotein. J Biol Chem 1981; 256: 2803-7.

[46] Le Moullec JM, Jullienne A, Chenais J, *et al.* The complete sequence of human preprocalcitonin. FEBS Lett 1984; 167: 93-7. [http://dx.doi.org/10.1016/0014-5793(84)80839-X]

[47] Becker KL, Snider R, Nylen ES. Procalcitonin assay in systemic inflammation, infection, and sepsis: clinical utility and limitations. Crit Care Med 2008; 36: 941-52. [http://dx.doi.org/10.1097/CCM.0B013E318165BABB]

[48] Barassi A, Pallotti F, Melzi GV. Biological variation of procalcitonin in healthy individuals. Clin Chem 2004; 50: 1878. [http://dx.doi.org/10.1373/clinchem.2004.037275]

[49] Müller B, White JC, Nylén ES, *et al.* Ubiquitous expression of the calcitonin-i gene in multiple tissues in response to sepsis. J Clin Endocrinol Metab 2001; 86: 396-404.

[50] Linscheid P, Seboek D, Nylen ES, *et al. In vitro* and *in vivo* calcitonin I gene expression in parenchymal cells: a novel product of human adipose tissue. Endocrinology 2003; 144: 5578-84. [http://dx.doi.org/10.1210/en.2003-0854]

[51] Linscheid P, Seboek D, Schaer DJ, Zulewski H, Keller U, Muller B. Expression and secretion of procalcitonin and calcitonin gene-related peptide by adherent monocytes and by macrophage-activated adipocytes. Crit Care Med 2004; 32: 1715-21. [http://dx.doi.org/10.1097/01.CCM.0000134404.63292.71]

[52] Van Nieeuwkoop C, Bonten TN, van't Wout JW, Kuijper EJ, Groenevald GH, Becker MJ. Procalcitonin reflects bacteremia and bacterial load in urosepsis syndrome: a prospective observational study. Crit Care 2010; 14: R206. [http://dx.doi.org/10.1186/cc9328]

[53] Moulin F, Raymond J, Lorrot M, *et al.* Procalcitonin in children admitted to hospital with community acquired pneumonia. Arch Dis Child 2001; 84: 332-6. [http://dx.doi.org/10.1136/adc.84.4.332]

[54] Assicot M, Gendrel D, Carsin H, Raymond J, Guilbaud J, Bohuon C. High serum procalcitonin concentrations in patients with sepsis and infection. Lancet 1993; 341: 515-8. [http://dx.doi.org/10.1016/0140-6736(93)90277-N]

[55] Brunkhorst FM, Heinz U, Forycki ZF. Kinetics of procalcitonin in iatrogenic sepsis. Intensive Care Med 1998; 24: 888-92. [http://dx.doi.org/10.1007/s001340050683]

[56] Meisner M. Procalcitonin - Biochemistry and Clinical Diagnosis. 1st ed., UNI-MED Science 2010.

[57] Simon L, Gauvin F, Amre DK, Saint-Louis P, Lacroix J. Serum procalcitonin and C-reactive protein levels as markers of bacterial infection: a systematic review and meta-analysis. Clin Infect Dis 2004; 39: 206-17. [http://dx.doi.org/10.1086/421997]

[58] Julián A, Palomo MJ, Ortiz R, Pedrosa A, Parejo R, Salcedo R. Utilidad de la procalcitonina y de la proteína C reactiva en el paciente con sepsis en urgencias. Emergencias 2009; 21: 23-7.

[59] Wacker C, Prkno A, Brunkhorst FM, Schlattmann P. Procalcitonin as a diagnostic marker for sepsis: a systematic review and meta-analysis. Lancet Infect Dis 2013; 13: 426-35. [http://dx.doi.org/10.1016/S1473-3099(12)70323-7]

[60] Alkholi UM, Al-monem NA, Abd El-Azim AA, Sultan MH. Serum procalcitonin in viral and bacterial meningitis. J Glob Infect Dis 2011; 3: 14-8. [http://dx.doi.org/10.4103/0974-777X.77290]

[61] Shen HY, Gao W, Cheng JJ, *et al.* Direct comparison of the diagnostic accuracy between blood and cerebrospinal fluid procalcitonin levels in patients with meningitis. Clin Biochem 2015; 48: 1079-82. [http://dx.doi.org/10.1016/j.clinbiochem.2015.06.017]

[62] Henry BM, Roy J, Ramakrishnan PK, Vikse J, Tomaszewski KA, Walocha JA. Procalcitonin as a serum biomarker for differentiation of bacterial meningitis from viral meningitis in children: evidence from a meta-analysis. Clin Pediatr (Phila) 2016; 55: 749-64.

[63] Vikse J, Henry BM, Roy J, Ramakrishnan PK, Tomaszewski KA, Walocha JA. The role of serum procalcitonin in the diagnosis of bacterial meningitis in adults: a systematic review and meta-analysis. Int J Infect Dis 2015; 38: 68-76. [http://dx.doi.org/10.1016/j.ijid.2015.07.011]

[64] Romualdo LG, Torrella PE, González MV, *et al.* Diagnostic accuracy of presepsin (soluble CD14 subtype) for prediction of bacteremia in patients with systemic inflammatory response syndrome in the Emergency Department. Clin Biochem 2014; 47: 505-8. [http://dx.doi.org/10.1016/j.clinbiochem.2014.02.011]

[65] Tromp M, Lansdorp B, Bleeker-Rovers CP, Klein JM, Kullberg BJ, Pickkers P. Serial and panel analyses of biomarkers do not improve the prediction of bacteremia compared to one procalcitonin measurement. J Infect 2012; 65: 292-301. [http://dx.doi.org/10.1016/j.jinf.2012.06.004]

[66] Julián-Jiménez A, Timón Zapata J, Laserna Mendieta EJ, Parejo Miguez R, Flores Chacartegui M, Gallardo Schall P. Ability of procalcitonin to predict bacteremia in patients with community acquired pneumonia. Med Clin (Barc) 2014; 142: 285-92.

[67] Schuetz P, Mueller B, Trampuz A. Serum procalcitonin for discrimination of blood contamination from bloodstream infection due to coagulase-negative staphylococci. Infection 2007; 35: 352-5. [http://dx.doi.org/10.1007/s15010-007-7065-0]

[68] Pecile P, Miorin E, Romanello C, *et al.* Procalcitonin: a marker of severity of acute pyelonephritis among children. Pediatrics 2004; 114: e249-54. [http://dx.doi.org/10.1542/peds.114.2.e249]

[69] Smolkin V, Koren A, Raz R, Colodner R, Sakran W, Halevy R. Procalcitonin as a marker of acute pyelonephritis in infants and children. Pediatr Nephrol 2002; 17: 409-12. [http://dx.doi.org/10.1007/s00467-001-0790-1]

[70] Leroy S, Fernandez-Lopez A, Nikfar R, *et al.* Association of procalcitonin with acute pyelonephritis

and renal scars in pediatric UTI. Pediatrics 2013; 131: 870-9. [http://dx.doi.org/10.1542/peds.2012-2408]

[71] Shaikh N, Borrell JL, Evron J, Leeflang MM. Procalcitonin, C-reactive protein, and erythrocyte sedimentation rate for the diagnosis of acute pyelonephritis in children. Cochrane Database Syst Rev 2015; 1: CD009185.

[72] Mueller C, Huber P, Laifer G, Mueller B, Perruchoud AP. Procalcitonin and the early diagnosis of infective endocarditis. Circulation 2004; 109: 1707-10. [http://dx.doi.org/10.1161/01.CIR.0000126281.52345.52]

[73] Knudsen JB, Fuursted K, Petersen E, *et al.* Procalcitonin in 759 patients clinically suspected of infective endocarditis. Am J Med 2010; 123: 1121-7. [http://dx.doi.org/10.1016/j.amjmed.2010.07.018]

[74] Yu CW, Juan LI, Hsu SC, *et al.* Role of procalcitonin in the diagnosis of infective endocarditis: a meta-analysis. Am J Emerg Med 2013; 31: 935-41. [http://dx.doi.org/10.1016/j.ajem.2013.03.008]

[75] García de Guadiana-Romualdo L, Español-Morales I, Cerezuela-Fuentes P, *et al.* Value of lipopolysaccharide binding protein as diagnostic marker of infection in adult cancer patients with febrile neutropenia: comparison with C-reactive protein, procalcitonin, and interleukin 6. Support Care Cancer 2015; 23: 2175-82. [http://dx.doi.org/10.1007/s00520-014-2589-1]

[76] Sakr Y, Sponholz C, Tuche F, Brunkhorst F, Reinhart K. The role of procalcitonin in febrile neutropenic patients: review of the literature. Infection 2008; 36: 396-407. [http://dx.doi.org/10.1007/s15010-008-7374-y]

[77] Mofidi R, Suttie SA, Patil PV, Ogston S, Parks RW. The value of procalcitonin at predicting the severity of acute pancreatitis and development of infected pancreatic necrosis: systematic review. Surgery 2009; 146: 72-81. [http://dx.doi.org/10.1016/j.surg.2009.02.013]

[78] Domínguez-Comesaña E, Ballinas-Miranda JR. La procalcitonina como marcador de infección intraabdominal. Cir Cir 2014; 82: 231-9.

[79] Schuetz P, Albrich W, Mueller B. Procalciton for diagnosis of infection and guide to antibiotic decisions: past, present and future. BMC Med 2011; 9: 107. [http://dx.doi.org/10.1186/1741-7015-9-107]

[80] Julián-Jiménez A, Palomo De Los Reyes MJ, Laín-Terés N, Estebaran-Martín J. ¿Es posible mejorar la decisión de ingreso en el hospital en la neumonía adquirida en la comunidad? Rev Clin Esp 2012; 212: 52-3. [http://dx.doi.org/10.1016/j.rce.2011.05.001]

[81] Huang DT, Weissfeld LA, Kellum JA, *et al.* Risk prediction with procalcitonin and clinical rules in community-acquired pneumonia. Ann Emerg Med 2008; 52: 48-58. [http://dx.doi.org/10.1016/j.annemergmed.2008.01.003]

[82] Woodhead M, Blasi F, Ewig S, *et al.* Guidelines for the management of adult lower respiratory tract infections-summary. Clin Microbiol Infect 2011; 17 (Suppl. 6): 1-24. [http://dx.doi.org/10.1111/j.1469-0691.2011.03602.x]

[83] Fowler CL. Procalcitonin for triage of patients with respiratory tract symptoms: a case study in the trial design process for approval of a new diagnostic test for lower respiratory tract infections. Clin Infect Dis 2011; 52: S351-6. [http://dx.doi.org/10.1093/cid/cir058]

[84] Christ-Crain M, Jaccard-Stolz D, Bingisser R, *et al.* Effect of procalcitonin-guided treatment on antibiotic use and outcome in lower respiratory tract infections: cluster-randomised, single-blinded intervention trial. Lancet 2004; 21(363): 600-7.

[http://dx.doi.org/10.1016/S0140-6736(04)15591-8]

[85] Schuetz P, Christ-Crain M, Thomann R, *et al.* Effect of procalcitonin-based guidelines *vs* standard guidelines on antibiotic use in lower respiratory tract infections: the ProHOSP randomized controlled trial. JAMA 2009; 302: 1059-66.
[http://dx.doi.org/10.1001/jama.2009.1297]

[86] Schuetz P, Müller B, Christ-Crain M, *et al.* Procalcitonin to initiate or discontinue antibiotics in acute respiratory tract infections. Cochrane Database Syst Rev 2012; 9: CD007498.

[87] Bouadma L, Luyt CE, Tubach F, Cracco C, Alvarez A, Schwebel C. Use of procalcitonin to reduce patients' exposure to antibiotics in intensive care units (PRORATA trial): a multicentre randomised controlled trial. Lancet 2010; 375: 463-74.
[http://dx.doi.org/10.1016/S0140-6736(09)61879-1]

[88] Schroeder S, Hochreiter M, Koehler T, *et al.* Procalcitonin (PCT)-guided algorithm reduces length of antibiotic treatment in surgical intensive care patients with severe sepsis: results of a prospective randomized study. Langenbecks Arch Surg 2009; 394: 221-6.
[http://dx.doi.org/10.1007/s00423-008-0432-1]

[89] Stolz D, Smyrnios N, Eggimann P, *et al.* Procalcitonin for reduced antibiotic exposure in ventilator-associated pneumonia: a randomised study. Eur Respir J 2009; 34: 1364-75.
[http://dx.doi.org/10.1183/09031936.00053209]

[90] Stocker M, Fontana M, El Helou S, Wegscheider K, Berger TM. Use of procalcitonin-guided decision-making to shorten antibiotic therapy in suspected neonatal early-onset sepsis: prospective randomized intervention trial. Neonatology 2010; 97: 165-74.
[http://dx.doi.org/10.1159/000241296]

[91] Westwood M, Ramaekers B, Whiting P, *et al.* Procalcitonin testing to guide antibiotic therapy for the treatment of sepsis in intensive care settings and for suspected bacterial infection in emergency department settings: a systematic review and cost-effectiveness analysis. Health Technol Assess 2015; 19: 1-236.
[http://dx.doi.org/10.3310/hta19960]

[92] Georgeopoulou A, Savva A, Giamarello-Bourboulis EJ, *et al.* Early changes of procalcitonin may advise about prognosis and appropriateness of antimicrobial therapy in sepsis. J Crit Care 2011; 26: 331.e1-7.
[http://dx.doi.org/10.1016/j.jcrc.2010.07.012]

[93] Karlsson S, Heikinnen M, Pettila V, *et al.* Predictive value of procalcitonin decrease in patients with severe sepsis: A prospective observational study. Crit Care Med 2010; 14: R205.

[94] Suberviola B, Castellanos-Ortega A, González-Castro A, García-Astudillo LA, Fernández Miret B. Valor pronóstico del aclaramiento de procalcitonina, PCR y leucocitos en el shock séptico. Med Intensiva 2012; 36: 177-84.
[http://dx.doi.org/10.1016/j.medin.2011.09.008]

[95] de Azevedo JR, Torres OJ, Czeczko NG, Tuon FF, Nassif PA, Souza GD. Procalcitonin as a prognostic biomarker of severe sepsis and septic shock. Rev Col Bras Cir 2012; 39: 456-61.

[96] Schuetz P, Maurer P, Punjabi V, Desai A, Amin DN, Gluck E. Procalcitonin decrease over 72 hours in US critical care units predicts fatal outcome in sepsis patients. Crit Care 2013; 17: R115.
[http://dx.doi.org/10.1186/cc12787]

[97] García de Guadiana-Romualdo LM, Rebollo-Acebes S, Esteban-Torrella P, *et al.* Valor pronóstico de la proteína fijadora de lipopolisácaridos y la procalcitonina en pacientes con sepsis grave y shock séptico ingresados en una unidad de cuidados intensivos. Med Intensiva 2015; 39: 207-12.
[http://dx.doi.org/10.1016/j.medin.2014.04.005]

[98] Poddar B, Gurjar M, Singh S, *et al.* Procalcitonin kinetics as a prognostic marker in severe sepsis/septic shock. Indian J Crit Care Med 2015; 19: 140-6.

[http://dx.doi.org/10.4103/0972-5229.152755]

[99] Pierallia F, Vannucchia V, Mancinia A, *et al.* Procalcitonin kinetics in the first 72 hours predicts 30-Day mortality in severely ill septic patients admitted to an Iintermediate Care Unit. J Clin Med Res 2015; 7: 706-13.
[http://dx.doi.org/10.14740/jocmr2251w]

[100] de Azevedo JR, Torres OJ, Beraldi RA, Ribas CA, Malafaia O. Prognostic evaluation of severe sepsis and septic shock: procalcitonin clearance *vs* Δ Sequential Organ Failure Assessment. J Crit Care 2015; 30: 219.e9-219.e12.
[http://dx.doi.org/10.1016/j.jcrc.2014.08.018]

[101] Chiesa C, Panero A, Rossi N, *et al.* Reliability of procalcitonin concentrations for the diagnosis of sepsis in critically ill neonates. Clin Infect Dis 1998; 26: 664-72.
[http://dx.doi.org/10.1086/514576]

[102] Chiesa C, Natale F, Pascone R, *et al.* C reactive protein and procalcitonin: reference intervals for preterm and term newborns during the early neonatal period. Clin Chim Acta 2011; 412: 1053-9.
[http://dx.doi.org/10.1016/j.cca.2011.02.020]

[103] Krüger S, Ewig S, Papassotiriou J, *et al.* Inflammatory parameters predict etiologic patterns but do not allow for individual prediction of etiology in patients with CAP: results from the German competence network CAPNETZ. Respir Res 2009; 10: 65.
[http://dx.doi.org/10.1186/1465-9921-10-65]

[104] Krüger S, Ewig S, Kunde J, *et al.* Assessment of inflammatory markers in patients with community-acquired pneumonia--influence of antimicrobial pre-treatment: results from the German competence network CAPNETZ. Clin Chim Acta 2010; 411: 1929-34.
[http://dx.doi.org/10.1016/j.cca.2010.08.004]

[105] Schuetz P, Affolter B, Hunziker S, *et al.* Serum procalcitonin, C-reactive protein and white blood cell levels following hypothermia after cardiac arrest: a retrospective cohort study. Eur J Clin Invest 2010; 40: 376-81.
[http://dx.doi.org/10.1111/j.1365-2362.2010.02259.x]

[106] Machens A, Lorenz K, Dralle H. Utility of serum procalcitonin for screening and risk stratification of medullary thyroid cancer. J Clin Endocrinol Metab 2014; 99: 2986-94.
[http://dx.doi.org/10.1210/jc.2014-1278]

[107] Avrillon V, Locatelli-Sanchez M, Folliet L, *et al.* Lung cancer may increase serum procalcitonin level. Infect Disord Drug Targets 2015; 15: 57-63.
[http://dx.doi.org/10.2174/1871526515666150320162950]

[108] Brodska H, Drabek T, Malickova K, *et al.* Marked increase of procalcitonin after the administration of anti-thymocyte globulin in patients before hematopoietic stem cell transplantation does not indicate sepsis: a prospective study. Crit Care 2009; 13: R37.
[http://dx.doi.org/10.1186/cc7749]

[109] Robier C, Neubauer M, Reicht G. Marked elevation of procalcitonin in a patient with a drug related infusion reaction to rituximab. Clin Chem Lab Med 2016; 54: e101-3.
[http://dx.doi.org/10.1515/cclm-2015-0772]

[110] Bloos F, Reinhart K. Rapid diagnosis of sepsis. Virulence 2014; 5: 154-60.
[http://dx.doi.org/10.4161/viru.27393]

[111] Schumann R, Zweigner J. A novel acute-phase marker: lipopolysaccharide binding protein (LBP). Clin Chem Lab Med 1999; 37: 271-4.
[http://dx.doi.org/10.1515/CCLM.1999.047]

[112] Gaïni S, Koldkjaer OG, Pedersen C, Pedersen SS. Procalcitonin, lipopolysaccharide-binding protein, interleukin-6 and C-reactive protein in community acquired infections and sepsis: a prospective study. Crit Care 2006; 10: R53.

[http://dx.doi.org/10.1186/cc4866]

[113] Sakr Y, Burgett U, Nacul FE, Reinhart K, Brunkhorst F. Lipopolysaccharide binding protein in a surgical intensive care unit: a marker of sepsis? Crit Care Med 2008; 36: 2014-22. [http://dx.doi.org/10.1097/CCM.0b013e31817b86e3]

[114] Prucha M, Herold I, Zazula R, *et al.* Significance of lipopolysaccharide-binding protein (an acute phase protein) in monitoring critically ill patients. Crit Care 2003; 7: R154-9. [http://dx.doi.org/10.1186/cc2386]

[115] Bottazzi B, Doni A, Garlanda C, Mantovani A. An integrated view of humoral innate immunity: pentraxins as a paradigm. Annu Rev Immunol 2010; 28: 157-83. [http://dx.doi.org/10.1146/annurev-immunol-030409-101305]

[116] Bottazzi B, Vouret-Craviari V, Bastone A, *et al.* Multimer formation and ligand recognition by the long pentraxin PTX3. Similarities and differences with the shorter pentraxins C-reactive protein and serum amyloid P component. J Biol Chem 1997; 272: 32817-23. [http://dx.doi.org/10.1074/jbc.272.52.32817]

[117] Garlanda C, Bottazzi B, Moalli F, Deban L, Molla F, *et al.* Pentraxins and atherosclerosis: the role of PTX3. Curr Pharm Des 2011; 17: 38-46. [http://dx.doi.org/10.2174/138161211795049750]

[118] Diamandis EP, Goodglick L, Planque C, Thornquist MD. Pentraxin-3 is a novel biomarker of lung carcinoma. Clin Cancer Res 2011; 17: 2395-9. [http://dx.doi.org/10.1158/1078-0432.CCR-10-3024]

[119] Pulsatelli L, Peri G, Macchioni P, *et al.* Serum levels of long pentraxin PTX3 in patients with polymyalgia rheumatica. Clin Exp Rheumatol 2010; 28: 756-8.

[120] Mairuhu AT, Peri G, Setiati TE, *et al.* Elevated plasma levels of the long pentraxin, pentraxin 3, in severe dengue virus infections. J Med Virol 2005; 76: 547-52. [http://dx.doi.org/10.1002/jmv.20397]

[121] Wagenaar JF, Goris MG, Gasem MH, *et al.* Long pentraxin PTX3 is associated with mortality and disease severity in severe leptospirosis. J Infect 2009; 58: 425-32. [http://dx.doi.org/10.1016/j.jinf.2009.04.004]

[122] Outinen TK, Makela S, Huhtala H, *et al.* High pentraxin-3 plasma levels associate with thrombocytopenia in acute Puumala hantavirus-induced nephropathia epidemica. Eur J Clin Microbiol Infect Dis 2011; 31: 957-3. [http://dx.doi.org/10.1007/s10096-011-1392-x]

[123] de Kruif MD, Limper M, Sierhuis K, *et al.* PTX3 predicts severe disease in febrile patients at the emergency department. J Infect 2010; 60: 122-7. [http://dx.doi.org/10.1016/j.jinf.2009.11.010]

[124] Uusitalo-Seppala R, Huttunen R, Aittoniemi J, *et al.* Pentraxin 3 (PTX3) is associated with severe sepsis and fatal disease in emergency room patients with suspected infection: a prospective cohort study. PLoS One 2013; 8: e53661. [http://dx.doi.org/10.1371/journal.pone.0053661]

[125] Mauri T, Bellani G, Patroniti N, *et al.* Persisting high levels of plasma pentraxin 3 over the first days after severe sepsis and septic shock onset are associated with mortality. Intensive Care Med 2010; 36: 621-9. [http://dx.doi.org/10.1007/s00134-010-1752-5]

[126] Lin Q, Fu F, Shen L, Zhu B. Pentraxin 3 in the assessment of ventilator-associated pneumonia: An early marker of severity. Heart Lung 2013; 42: 139-45. [http://dx.doi.org/10.1016/j.hrtlng.2012.11.005]

[127] Liu S, Qu X, Liu F, Wang C. Pentraxin 3 as a prognostic biomarker in patients with systemic inflammation or infection. Mediators Inflamm 2014; 2014: 421-29.

[http://dx.doi.org/10.1155/2014/421429]

[128] Watanabe T, Yonekura H, Terazono K, Yamamoto H, Okamoto H. Complete nucleotide sequence of human reg gene and its expression in normal and tumoral tissues. The reg protein, pancreatic stone protein, and pancreatic thread protein are one and the same product of the gene. J Biol Chem 1990; 265: 7432-9.

[129] Hayakawa T, Kondo T, Shibata T, *et al.* Serum pancreatic stone protein in pancreatic diseases. Int J Pancreatol 1993; 13: 97-103.

[130] Cash HL, Whitham CV, Behrendt CL, Hooper LV. Symbiotic bacteria direct expression of an intestinal bactericidal lectin. Science 2006; 313: 1126-30. [http://dx.doi.org/10.1126/science.1127119]

[131] Keel M, Härter L, Reding T, *et al.* Pancreatic stone protein is highly increased during posttraumatic sepsis and activates neutrophil granulocytes. Crit Care Med 2009; 37: 1642-8. [http://dx.doi.org/10.1097/CCM.0b013e31819da7d6]

[132] Llewelyn MJ, Berger M, Gregory M, *et al.* Sepsis biomarkers in unselected patients on admission to intensive or high-dependency care. Crit Care 2013; 17: R60. [http://dx.doi.org/10.1186/cc12588]

[133] Que YA, Delodder F, Guessous I, *et al.* Pancreatic stone protein as an early biomarker predicting mortality in a prospective cohort of patients with sepsis requiring ICU management. Crit Care 2012; 16: R114. [http://dx.doi.org/10.1186/cc11406]

[134] Que YA, Guessous I, Dupuis-Lozeron E, *et al.* Prognostication of Mortality in Critically Ill Patients With Severe Infections. Chest 2015; 148: 674-82. [http://dx.doi.org/10.1378/chest.15-0123]

[135] Boeck L, Graf R, Eggimann P, *et al.* Pancreatic stone protein: a marker of organ failure and outcome in ventilator-associated pneumonia. Chest 2011; 40: 925-32.

[136] Harbarth S, Holeckova K, Froidevaux C, *et al.* Diagnostic value of procalcitonin, interleukin-6, and interleukin-8 in critically ill patients admitted with suspected sepsis. Am J Respir Crit Care Med 2001; 164: 396-402. [http://dx.doi.org/10.1164/ajrccm.164.3.2009052]

[137] Meynaar IA, Droog W, Batstra M, Vreede R, Herbrink P. In Critically Ill Patients, Serum procalcitonin is more useful in differentiating between sepsis and SIRS than CRP, IL-6, or LBP. Crit Care Res Pract 2011; 2011: 594645.

[138] Tsalik EL, Jaggers LB, Glickman SW, *et al.* Discriminative value of inflammatory biomarkers for suspected sepsis. J Emerg Med 2012; 43: 97-106. [http://dx.doi.org/10.1016/j.jemermed.2011.05.072]

[139] Pettilä V, Hynninen M, Takkunen O, Kuusela P, Valtonen M. Predictive value of procalcitonin and interleukin 6 in critically ill patients with suspected sepsis. Intensive Care Med 2002; 28: 1220-5. [http://dx.doi.org/10.1007/s00134-002-1416-1]

[140] Miguel-Bayarri V, Casanoves-Laparra EB, Pallás-Beneyto L, *et al.* Prognostic value of the biomarkers procalcitonin, interleukin-6 and C-reactive protein in severe sepsis. Med Intensiva 2012; 36: 556-62. [http://dx.doi.org/10.1016/j.medin.2012.01.014]

[141] Bozza FA, Salluh JI, Japiassu AM, *et al.* Cytokine profiles as markers of disease severity in sepsis: a multiplex analysis. Crit Care 2007; 11: R49. [http://dx.doi.org/10.1186/cc5783]

[142] Andaluz-Ojeda D, Bobillo F, Iglesias V, *et al.* A combined score of pro- and anti-inflammatory interleukins improves mortality prediction in severe sepsis. Cytokine 2012; 57: 332-6. [http://dx.doi.org/10.1016/j.cyto.2011.12.002]

[143] Sundén-Cullberg J, Norrby-Teglund A, Rouhiainen A, *et al.* Persistent elevation of high mobility group box-1 protein (HMGB1) in patients with severe sepsis and septic shock. Crit Care Med 2005; 33: 564-73.
[http://dx.doi.org/10.1097/01.CCM.0000155991.88802.4D]

[144] Gibot S, Massin F, Cravoisy A, *et al.* High-mobility group box 1 protein plasma concentrations during septic shock. Intensive Care Med 2007; 33: 1347-53.
[http://dx.doi.org/10.1007/s00134-007-0691-2]

[145] Bas S, Gauthier B, Spenato U, Stingelin S, Gabay C. CD14 Is an Acute-Phase Protein. J Inmunol 2004; 172: 4470-9.

[146] Mussap M, Noto A, Fravega M, Fanos V. Soluble CD14 subtype presepsin (sCD14-ST) and lipopolysaccharide binding protein (LBP) in neonatal sepsis: new clinical and analytical perspectives for two old biomarkers. J Matern Fetal Neonatal Med 2011; 24 (Suppl. 2): 12-4.
[http://dx.doi.org/10.3109/14767058.2011.601923]

[147] Zheng Z, Jiang L, Ye L, Gao Y, Tang L, Zhang M. The accuracy of presepsin for the diagnosis of sepsis from SIRS: a systematic review and meta-analysis. Ann Intensive Care 2015; 5: 48.
[http://dx.doi.org/10.1186/s13613-015-0089-1]

[148] Zhang X, Liu D, Liu YN, Wang R, Xie LX. The accuracy of presepsin (sCD14-ST) for the diagnosis of sepsis in adults: a meta-analysis. Crit Care 2015; 19: 323.
[http://dx.doi.org/10.1186/s13054-015-1032-4]

[149] Wu J, Hu L, Zhang G, Wu F, He T. Accuracy of presepsin in sepsis diagnosis: a systematic review and meta-analysis. PLoS One 2015; 10: e0133057.
[http://dx.doi.org/10.1371/journal.pone.0133057]

[150] Zhang J, Hu ZD, Song J, Shao J. Diagnostic value of presepsin for sepsis: a systematic review and meta-analysis. Medicine (Baltimore) 2015; 94: e2158.
[http://dx.doi.org/10.1097/MD.0000000000002158]

[151] Tong X, Cao Y, Yu M, Han C. Presepsin as a diagnostic marker for sepsis: evidence from a bivariate meta-analysis. Ther Clin Risk Manag 2015; 11: 1027-33.

[152] Ulla M, Pizzolato E, Lucchiari M, *et al.* Diagnostic and prognostic value of Presepsin in the management of sepsis in the emergency department: a multicentre prospective study. Crit Care 2013; 17: R168.
[http://dx.doi.org/10.1186/cc12847]

[153] Liu B, Chen YX, Yin Q, Zhao YZ, Li CS. Diagnostic value and prognostic evaluation of Presepsin for sepsis in an Emergency Department. Crit Care 2013; 17: R24.
[http://dx.doi.org/10.1186/cc13070]

[154] Carpio R, Zapata J, Spanuth E, Hess G. Utility of presepsin (sCD14-ST) as a diagnostic and prognostic marker of sepsis in the Emergency Department. Clin Chim Acta 2015; 450: 169-75.
[http://dx.doi.org/10.1016/j.cca.2015.08.013]

[155] Behnes M, Bertsch T, Lepiorz D, *et al.* Diagnostic and prognostic utility of soluble CD 14 subtype (presepsin) for severe sepsis and septic shock during the first week of intensive care treatment. Crit Care 2014; 18: 507.
[http://dx.doi.org/10.1186/s13054-014-0507-z]

[156] Masson S, Caironi P, Spanuth E, *et al.* Presepsin (soluble CD14 subtype) and procalcitonin levels for mortality prediction in sepsis: data from the Albumin Italian Outcome Sepsis trial. Crit Care 2014; 18(1): R6.
[http://dx.doi.org/10.1186/cc13183]

[157] Bouchon A, Facchetti F, Weigand MA, Colonna M. TREM-1 amplifies inflammation and is a crucial mediator of septic shock. Nature 2001; 410: 1103-7.
[http://dx.doi.org/10.1038/35074114]

[158] Gibot S, Kolopp-Sarda MN, Béné MC, *et al.* Plasma level of a triggering receptor expressed on myeloid cells-1: its diagnostic accuracy in patients with suspected sepsis. Ann Intern Med 2004; 141: 9-15.
[http://dx.doi.org/10.7326/0003-4819-141-1-200407060-00009]

[159] Tiancha H, Wei C, Huahao S. Diagnostic value of the soluble triggering receptor expressed on myeloid cells-1 in bacterial infection: a meta-analysis. Intensive Care Med 2009; 35: 587-95.
[http://dx.doi.org/10.1007/s00134-008-1333-z]

[160] Wu Y, Wang F, Fan X, *et al.* Accuracy of plasma sTREM-1 for sepsis diagnosis in systemic inflammatory patients: a systematic review and meta-analysis. Crit Care 2012; 16: R229.
[http://dx.doi.org/10.1186/cc11884]

[161] Jeong SJ, Song YG, Kim CO, *et al.* Measurement of plasma sTREM-1 in patients with severe sepsis receiving early goal-directed therapy and evaluation of its usefulness. Shock 2012; 37: 574-8.
[http://dx.doi.org/10.1097/SHK.0b013e318250da40]

[162] Schiff DE, Rae J, Martin TR, Davis BH, Curnutte JT. Increased phagocyte Fc gammaRI expression and improved Fc gamma- receptor-mediated phagocytosis after *in vivo* recombinant human interferon-gamma treatment of normal human subjects. Blood 1997; 90: 3187-94.

[163] Davis BH. Improved diagnostic approaches to infection/sepsis detection. Expert Rev Mol Diagn 2005; 5: 193-207.
[http://dx.doi.org/10.1586/14737159.5.2.193]

[164] Song SH, Kim HK, Park MH, Cho HI. Neutrophil CD64 expression is associated with severity and prognosis of disseminated intravascular coagulation. Thromb Res 2008; 121: 499-507.
[http://dx.doi.org/10.1016/j.thromres.2007.05.013]

[165] Johannes JM. Hoffmann. Neutrophil CD64 as a sepsis biomarker. Biochem Med 2011; 21: 282-90.

[166] Li S, Huang X, Chen Z, *et al.* Neutrophil CD64 expression as a biomarker in the early diagnosis of bacterial infection: a meta-analysis. Int J Infect Dis 2013; 17: e12-23.
[http://dx.doi.org/10.1016/j.ijid.2012.07.017]

[167] Hoffmann JJ. Neutrophil CD64: a diagnostic marker for infection and sepsis. Clin Chem Lab Med 2009; 47: 903-16.
[http://dx.doi.org/10.1515/CCLM.2009.224]

[168] Cid J, Aguinaco R, Sánchez R, García-Pardo G, Llorente A. Neutrophil CD64 expression as marker of bacterial infection: a systematic review and meta-analysis. J Inf 2010; 60: 313-9.
[http://dx.doi.org/10.1016/j.jinf.2010.02.013]

[169] Livaditi O, Kotanidou A, Psarra A, *et al.* Neutrophil CD64 expression and serum IL-8: sensitive early markers of severity and outcome in sepsis. Cytokine 2006; 36: 283-90.
[http://dx.doi.org/10.1016/j.cyto.2007.02.007]

[170] Cid J, García-Pardo G, Aguinaco R, Sánchez R, Llorente A. Neutrophil CD64: diagnostic accuracy and prognostic value in patients presenting to the emergency department. Eur J Clin Microbiol Infect Dis 2011; 30: 845-52.
[http://dx.doi.org/10.1007/s10096-011-1164-7]

[171] Rønne E, Behrendt N, Jensen AL, Blasi F, Danø K. Cellular receptor for urokinase plasminogen activator. Carboxyl-terminal processing and membrane anchoring by glycosyl-phosphatidylinositol. J Biol Chem 1991; 266: 1926-33.

[172] Donadello K, Scolletta S, Covajes C, Vincent JL. suPAR as a prognostic biomarker in sepsis. BMC Med 2012; 10: 2.
[http://dx.doi.org/10.1186/1741-7015-10-2]

[173] Giamarellos-Bourboulis EJ, Norrby-Teglund A, Mylona V, *et al.* Risk assessment in sepsis: a new prognostication rule by APACHE II score and serum soluble urokinase plasminogen activator

receptor. Crit Care 2012; R16: 149.
[http://dx.doi.org/10.1186/cc11463]

[174] Donadello K, Scolletta S, Taccone FS, *et al.* Soluble urokinase-type plasminogen activator receptor as a prognostic biomarker in critically ill patients. J Crit Care 2014; 29: 144-9.
[http://dx.doi.org/10.1016/j.jcrc.2013.08.005]

[175] Suberviola B, Castellanos-Ortega A, Ruiz Ruiz A, Lopez-Hoyos M, Santibañez M. Hospital mortality prognostication in sepsis using the new biomarkers suPAR and proADM in a single determination on ICU admission. Intensive Care Med 2013; 39: 1945-52.
[http://dx.doi.org/10.1007/s00134-013-3056-z]

[176] Hotchkiss RS, Tinsley KW, Swanson PE, Karl IE. Endothelial cell apoptosis in sepsis. Crit Care Med 2002; 30: S225-8.
[http://dx.doi.org/10.1097/00003246-200205001-00009]

[177] Hack CE, Zeerleder S. The endothelium in sepsis: source of and a target for inflammation. Crit Care Med 2001; 29: S21-7.
[http://dx.doi.org/10.1097/00003246-200107001-00011]

[178] Xing X, Murthy S, Conrad Liles W, Singh JM. Clinical utility of biomarkers of endothelial activation in sepsis-a systematic review. Crit Care 2012; 16: R7.
[http://dx.doi.org/10.1186/cc11145]

[179] Skibsted S, Jones AE, Puskarich MA, *et al.* Biomarkers of endothelial cell activation in early sepsis. Shock 2013; 39: 427-32.
[http://dx.doi.org/10.1097/SHK.0b013e3182903f0d]

[180] Mihajlovic DM, Lendakb DF, Brkic SV, *et al.* Endocan is useful biomarker of survival and severity in sepsis. Microvasc Res 2014; 93: 92-7.
[http://dx.doi.org/10.1016/j.mvr.2014.04.004]

[181] Kitamura K, Kangawa K, Kawamoto M, *et al.* Adrenomedullin: a novel hypotensive peptide isolated from human pheochromocytoma. Biochem Biophys Res Commun 1993; 192: 553-60.
[http://dx.doi.org/10.1006/bbrc.1993.1451]

[182] Hinson JP, Kapas S, Smith DM. Adrenomedullin, a multifunctional regulatory peptide. Endocr Rev 2000; 21: 138-67.

[183] Nicholls MG, Charles CJ, Lainchbury JG, *et al.* Adrenomedullin in heart failure. Hypertens Res 2003; 200(26) (Suppl.): S135-40.

[184] Morgenthaler NG, Struck J, Alonso C, Bergmann A. Measurement of midregional proadrenomedullin in plasma with an immunoluminometric assay. Clin Chem 2005; 51: 1823-9.
[http://dx.doi.org/10.1373/clinchem.2005.051110]

[185] Pio R, Martinez A, Unsworth EJ, *et al.* Complement factor H is a serum-binding protein for adrenomedullin, and the resulting complex modulates the bioactivities of both partners. J Biol Chem 2001; 276: 12292-300.
[http://dx.doi.org/10.1074/jbc.M007822200]

[186] Meeran K, O'Shea D, Upton PD, *et al.* Circulating adrenomedullin does not regulate systemic blood pressure but increases plasma prolactin after intravenous infusion in humans: a pharmacokinetic study. J Clin Endocrinol Metab 1997; 82: 95-100.

[187] Lewis LK, Smith MW, Yandle TG, Richards AM, Nicholls MG. Adrenomedullin(1-52) measured in human plasma by radioimmunoassay: plasma concentration, adsorption, and storage. Clin Chem 1998; 44: 571-7.

[188] Julián-Jiménez A, Timón Zapata J, Laserna Mendieta EJ, *et al.* Poder diagnóstico y pronóstico de los biomarcadores para mejorar el manejo de la neumonía adquirida en la comunidad en los servicios de urgencias. Enferm Infecc Microbiol Clin 2014; 32: 225-35.
[http://dx.doi.org/10.1016/j.eimc.2013.04.015]

[189] Gordo-Remartínez S, Calderón-Moreno M, Fernández-Herranz J, *et al.* Usefulness of midregional proadrenomedullin to predict poor outcome in patients with community acquired pneumonia. PLoS One 2015; 10: e0125212. [http://dx.doi.org/10.1371/journal.pone.0125212]

[190] Suberviola B, Castellanos-Ortega A, Llorca J, Ortiz F, Iglesias D, Prieto B. Prognostic value of proadrenomedullin in severe sepsis and septic shock patients with community-acquired pneumonia. Swiss Med Wkly 2012; 142: w13542.

[191] Bello S, Lasierra AB, Mincholé E, *et al.* Prognostic power of proadrenomedullin in community-acquired pneumonia is independent of aetiology. Eur Respir J 2012; 39: 1144-55.

[192] Cavallazzi R, El-Kersh K, Abu-Atherah E, *et al.* Midregional proadrenomedullin for prognosis in community-acquired pneumonia: a systematic review. Respir Med 2014; 108: 1569-80. [http://dx.doi.org/10.1016/j.rmed.2014.09.018]

[193] Christ-Crain M, Morgenthaler NG, Struck J, Harbarth S, Bergmann A, Müller B. Mid-regional pro-adrenomedullin as a prognostic marker in sepsis: an observational study. Crit Care 2005; 9: R816-24. [http://dx.doi.org/10.1186/cc3885]

[194] Andaluz-Ojeda D, Cicuéndez R, Calvo D, *et al.* Sustained value of proadrenomedullin as mortality predictor in severe sepsis. J Infect 2015; 71: 136-9. [http://dx.doi.org/10.1016/j.jinf.2015.02.002]

[195] Seymour CW, Liu VX, Iwashyna TJ, *et al.* Assessment of Clinical Criteria for Sepsis: For the Third International Consensus Definitions for Sepsis and Septic Shock (Sepsis-3). JAMA 2016; 315: 762-74. [http://dx.doi.org/10.1001/jama.2016.0288]

[196] Naderi HR, Sheybani F, Sarvghad M, Nooghabi MJ. Can Procalcitonin Add to the Prognostic Power of the Severity Scoring System in Adults with Pneumonia? Tanaffos 2015; 14: 95-106.

[197] España PP, Capelastegui A, Mar C, *et al.* Performance of pro-adrenomedullin for identifying adverse outcomes in community-acquired pneumonia. J Infect 2015; 70: 457-66. [http://dx.doi.org/10.1016/j.jinf.2014.12.003]

[198] Alan M, Grolimund E, Kutz A, *et al.* Clinical risk scores and blood biomarkers as predictors of long-term outcome in patients with community-acquired pneumonia: a 6-year prospective follow-up study. J Intern Med 2015; 278: 174-84. [http://dx.doi.org/10.1111/joim.12341]

[199] Ratzinger F, Schuardt M, Eichbichler K, *et al.* Utility of sepsis biomarkers and the infection probability score to discriminate sepsis and systemic inflammatory response syndrome in standard care patients. PLoS One 2013; 8: e82946. [http://dx.doi.org/10.1371/journal.pone.0082946]

[200] Tromp M, Lansdorp B, Bleeker-Rovers CP, Gunnewiek JM, Kullberg BJ, Pickkers P. Serial and panel analyses of biomarkers do not improve the prediction of bacteremia compared to one procalcitonin measurement. J Infect 2012; 65: 292-301. [http://dx.doi.org/10.1016/j.jinf.2012.06.004]

[201] Dark P, Wilson C, Blackwood B, *et al.* Accuracy of LightCycler(R) SeptiFast for the detection and identification of pathogens in the blood of patients with suspected sepsis: a systematic review protocol. BMJ Open 2012; 2: e000392. [http://dx.doi.org/10.1136/bmjopen-2011-000392]

[202] Warhurst G, Dunn G, Chadwick P, *et al.* Rapid detection of health-care-associated bloodstream infection in critical care using multipathogen real-time polymerase chain reaction technology: a diagnostic accuracy study and systematic review. Health Technol Assess 2015; 19: 1-142. [http://dx.doi.org/10.3310/hta19350]

[203] Dwivedi DJ, Toltl LJ, Swystun LL, *et al.* Prognostic utility and characterization of cell-free DNA in patients with severe sepsis. Crit Care 2012; 16: R151.

[http://dx.doi.org/10.1186/cc11466]

[204] Gutiérrez Junco SM. Polimorfismos en genes de la inmunidad innata determinan la susceptibilidad y la mortalidad de la sepsis grave y el shock séptico. Tesis Doctoral Universidad de Valladolid 2014.

CHAPTER 15

Biomarkers in Saliva as a Tool for Health Diagnosis

Ana M. Moreno-Fuentes and **Carmen Nieto-Sánchez***

Clinical Analysis Department, Santa Lucía General University Hospital, Cartagena, Murcia, Spain

Abstract: Over the last years, there has been an increasing attention over the use of saliva as a proper diagnostic fluid. One of the most interesting aspects in saliva based diagnosis is the non-invasive method of obtaining oral samples. This is a safe method for both, the health worker and the patient, related to the management of saliva samples.

The utility of the whole saliva sample as an instrument of salivary diagnosis has been proven in almost all the research fields. The relation between saliva based components and blood based components has been proven and hence, salivary diagnosis could be a proper and accurate tool for disease diagnosis. The research of salivary biomarkers in viral and bacterial infections, autoimmune diseases, endocrine diseases, oncology, systemic diseases, stress assessment, medication detection and forensic science among others, is becoming a reality.

It is expected that, with the help of current technological advances, salivary analysis which owns the analytical sensitivity and specificity required for this technology, could be used as a valuable tool and a diagnostic medium to achieve successful findings. However, there is a lot of work to do due to the fact that, some results from certain investigations still remain contradictory.

Definitions of Words and Terms

According to the National Institutes of Health (NIH), a *Biomarker* is an objectively measured and evaluated indicator of normal biological processes, pathogenic processes or pharmacological responses to a therapeutic intervention [1].

Saliva is a complex fluid composed of different secretions from the major and minor salivary glands.

* **Corresponding author Carmen Nieto-Sanchez:** Clinical Analysis Department, Santa Lucía General University Hospital, Cartagena, Murcia; Tel/Fax: 0034968128600; E-mail: carmennietosanchez@yahoo.es

Keywords: Acini salivary, Acinus cell, Analyte, Biofluid, Biomarkers, Crevicular fluid, Early diagnosis, miRNA, mRNA, ncRNA, Oral cancer, Point of care, Salivary diagnostic, Salivary gland, Salivary metabolome, Salivary microbiome, Salivary proteome, Salivary transcriptome, Systemic diseases, Whole saliva.

INTRODUCTION

Traditionally, blood-based components analysis was the cornerstone of diagnosis procedures in the clinic laboratory. Nowadays, there is a great interest in diagnosis based on salivary analysis because the collection of salivary samples is a simple, fast, non-invasive method and could be easily collected by the patients themselves. Collecting oral samples is safe for the health worker as well as the patient, not to mention the simple and cost-efficient storage. This is one of the most important characteristics: the possibility of obtaining samples in patients who are not prone to collaborate, as children and elderly people. Saliva is important as a diagnostic tool because there is a direct relation between the basic biochemical parameters in both saliva and blood. This close connection between these two fluids would allow the screening, progression and even early detection of a wide range of diseases [2].

What was once considered simply a digestive juice is now being counted as a biological fluid with the capacity to express an individual's up-to-date health status. Great attempts in this domain must be made to the setting up of clinically acceptable tests and to integrate these methodologies into current clinical practice [3].

Nature of Human Saliva

Saliva is a clear, slightly acid (pH 6.0 to 7.0) and heterogeneous viscous biofluid, composed of water, ions, mucin, plasmatic proteins, leucocytes and cells debris [4]. The main function is the lubrication of the food to help the mastication and formation of the cud and hence, the beginning of the carbohydrates degradation due to the alpha-amylase enzyme found in this fluid. Besides, human saliva helps to maintain the health of teeth and the oral cavity and it is a proper defensive system, facing general pathogens.

Saliva and the Salivary Glands

Saliva is generated within the salivary glands (a kind of exocrine glands) by acinus cells, collected in small ducts, and subsequently released into the oral cavity [5]. The different salivary glands are: major glands, responsible for the main saliva production (92-95%) and minor salivary glands (labial, buccal, lingual and palatal glands) that supply the remainder [3]. The terms minor and major refer to the anatomic size of the glands. As said before, there is three pair of major glands:

- Parotid glands, are placed in front of each ear, being the largest of the three major salivary glands. They are responsible for the 30% of the amount produced.
- Submandibular glands, reside beneath the lower jaw just posterior and below the sublingual glands. They are responsible for the 55-65% of the amount produced.
- Sublingual glands are located under the tongue. They are responsible for the 5% of the amount produced.

The minor salivary glands can be found in mouth, pharynx, nostrils, larynx, paranasal sinuses and tracheal mucosa. The most important are labial, buccal, lingual and palatal glands. Those can be considered as mucous glands (Fig. **1**).

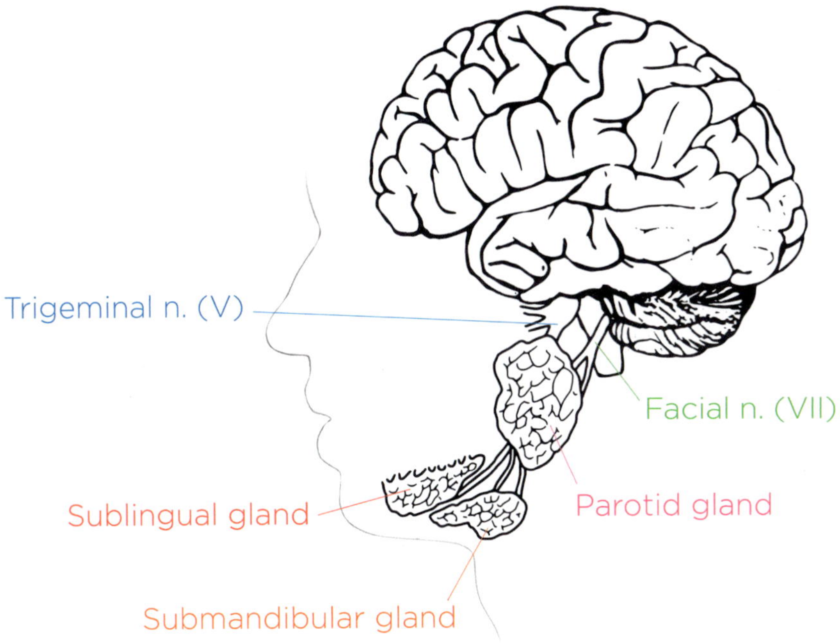

Fig. (1). Locations of salivary glands (parotid, submandibular, and sublingual) and the nerves charged with their innervation (adapted from Yoshizawa *et al*).

According to the nature of the salivary secretion, the acini can be characterized as serous, seromucous or mucous. A salivary gland can be composed by different kind of acini (mix glands) or an specific group or cells which produce only one type of secretion (pure glands). Serous glands are composed of serous cells that produce clear secretion rich in albumin and alpha-amylase, this kind of secretion is known as dilution saliva. Mucous glands produce a viscous, sticky and rich in mucin, salivary secretion. Salivary glands are constituted by acini cells, duct cells and myoepithelial cells surrounding the acini to contract and discharge the secretion into the duct system that drains into the oral cavity (Fig. **2**) [6].

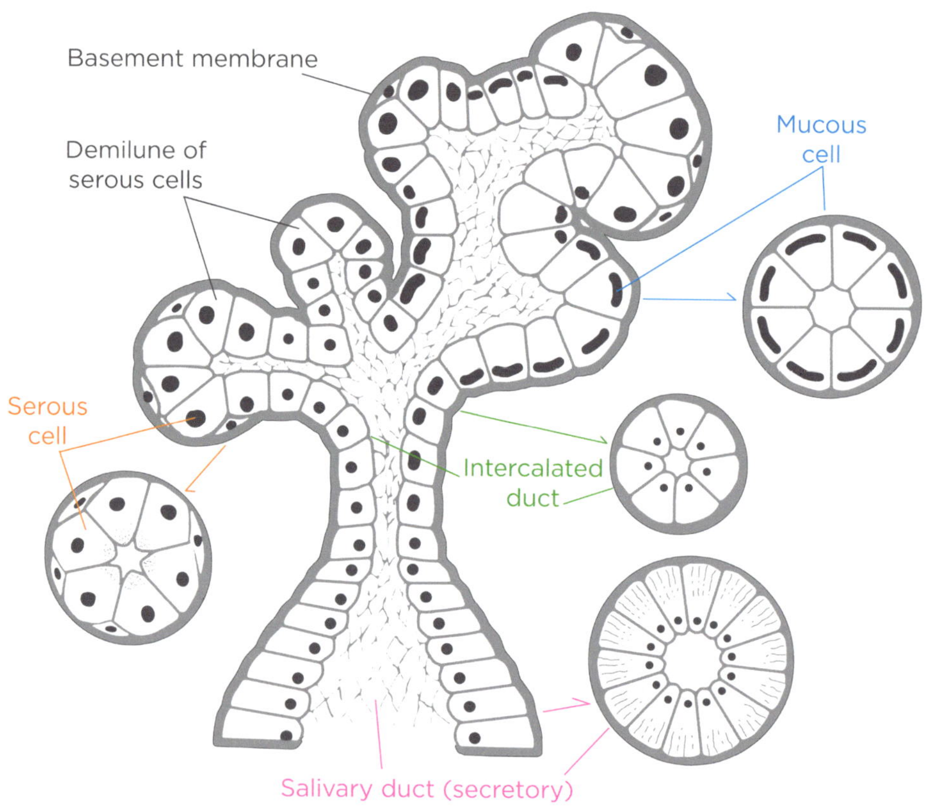

Fig. (2). Structure of salivary acini. Different kind of cells is shown. The secretion produced by the cells drained into intercalated ducts to salivary ducts (adapted from Keppel and Stanton).

Salivary glands are very permeable and highly vascularized. This characteristic permits the free exchange and replacement of blood-based molecules by ultrafiltration towards the contiguous acinus cells which are the ones that produce salivary secretions. Transport mechanisms to produce saliva from blood components are quite diverse [7]:

1. Blood molecules entering salivary tissues *via* transcellular passive diffusion, is the most common route for substances to migrate from blood to saliva, as lipophilic molecules, or with the use of protein channels, as peptides (IgA).
2. Extracellular ultrafiltration is provided by pores in the cellular membrane like water and electrolytes. The use of paths through the spaces between acinus and ductal cells is an alternative transportation for molecules as sulphated steroids. In addition, ultrafiltration through gap junctions can be observed in secretory units of acinus cells. Only molecules as water, ions, catecholamine's and steroids use this way of transport.
3. Active secretions of vesicles from cellular acini cytoplasm.

Any change involving blood biomarkers levels or molecules could be absorbed by salivary glands and therefore they could modify the complex biochemical composition of the saliva. As a result, salivary secretions may carry valuable molecular information able to communicate a patient´s current condition of health as an option to blood and tissue based diagnostic [7].

Saliva Composition

Saliva is composed of water nearly about 95-98% and other components as minerals, electrolytes, hormones, immunoglobulins, cytokines and proteins depending on the origin of the gland secretion. Whole saliva sample is a complex mixture of oral fluids coming from salivary glands and other secretions as gingival crevicular fluid, oral mucosa transudate, secretions from nasal and pharyngeal mucosa, microorganism, desquamated epithelial cells, keratin debris, bloods cell and even food or chemical substances.

The quantity and salivary composition depends on salivary flow, circadian rhythm, origin of gland secretion, secretion stimulation, diet, drugs or medication, sex, age and physiological state. The majority of molecules found in the saliva are coming from blood-based components and the rest are originated from the own salivary gland secretion. The composition of proteins found in salivary secretion is quite different among individuals and for hence, trying to explore the relations between saliva host proteins and systemic diseases, could be and opening vast field of knowledge. The main proteins found in saliva samples are listed in Table **1** [8].

The mucins are a group of proteins which can be secreted from all mucous and seromucous salivary glands and are involved in almost all functions of saliva. As a result of their specific nature, they are responsible for the rheological properties of saliva, as viscosity, elasticity and stickiness [9]. Mucins MUC5B and MUC7 can be found mainly in saliva. Viscosity and elasticity are characteristics provided by MUC5B, which has a high molecular weight and it is concerned in the protection of the tooth enamel surface against the proteolytic attack of pathogens. Stickiness is the capacity that saliva owns, to adhere to all kind of surfaces, including oral cavity, tissues and microorganism, and this effect is provided by MUC7.

Table 1. List of principal proteins components found in human saliva (adapted from Pfeiffer *et al*).

Protein percentage	Gland origin	Protein function
Proline Rich Peptides (PRPs), 37%	Parotid and Submandibular glands	Mineral homeostasis, neutralization of toxic substances, protection of the underlying tissue against proteolytic attack by microorganisms [10]
α-Amylase, 20%	Parotid and Submandibular glands	The first step in carbohydrate digestion. It breaks down starch into sugars, antibacterial function, involved in mouth lubrication [11]
Mucins, 20%	Parotid glands	Protection of the tissue beneath against enzymatic attack of microorganism, cellular protection, mouth lubrication, protection against dehydration, allowance of viscosity [12]
Cystatins, 8%	Parotid, Submandibular and Sublingual glands	Antibacterial and antiviral function, involved in protein metabolism regulation, protection from proteolytic attack by microorganism, aids teeth mineralization [12]
Human serum albumin (6%)	Parotid, Submandibular and Sublingual glands	Protein's transporter, negative acute-phase protein, pH buffer maintenance [13]
Soluble IgA (3%)	Parotid and Submandibular glands	Immunity [14]
IgG (2%)	Parotid and Submandibular glands	Secondary immune response, its major function is binding pathogens [14]
Stateliness, 1%	Present in Parotid glands	Inhibits hydroxyapatite crystal growth; protection tissue against enzymatic attack by microorganisms; cellular protection, lubrication; responsible for saliva viscoelasticity [15]
Histatins, NA[a]	Present in all types of glands	Antifungal and antimicrobial capacity; participation in the mineralization dynamics of oral fluids, its presence inhibits the release of histamine from mast cells, suggesting a role in regulation of oral inflammation [15, 16]

NA[a]: percentage not available.

Salivary Secretion Control

Salivary flow is under control of the autonomous nervous system, mainly controlled by the parasympathetic nervous system. The parotid gland is served by the parasympathetic side of glossopharyngeal nerve (cranial nerve IX). The parasympathetic side of the facial nerve (cranial nerve VII) provides the innervations of submandibular and sublingual glands, conducted by submandibular ganglion.

Each gland is autonomous. They are innervated and depend on parasympathetic and sympathetic stimulation, and they are deemed to be exocrine in function.

Spontaneous secretion discharge of minor salivary glands is produced by the absence of nervous stimulation; consequently, the oral mucous is protected during all day. A watery, rich in bicarbonate ions salivary secretion, is produced by parasympathetic activity, that persists whereas the glands are continually stimulated. On the other hand, sympathetic activity on the salivary glands results on salivary secretion increments but the effect is transitory and the saliva which is produced, has higher viscosity [7].

Currently, the oral cavity has an average volume of 1.1ml of saliva but daily secretion production ranges between 500 and 700ml. At resting periods, the saliva secretion ranges from 0.25 to 0.35ml/min and is mostly produced by the submandibular and sublingual glands. Sensory, electrical, or mechanical stimuli can raise the production rate to 1.5ml/min. The greatest volume of saliva is produced before, during and after meals, reaching its maximum peak at around 12 a.m., and falls considerably at night, when individuals are sleeping [17].

The Main Functions of the Saliva:

1. Protection of oral cavity tissues and upper respiratory tract against the microorganism attack and inflammations due to a great number of proteins presents in saliva, by mechanical cleaning and immunological defensive action. Furthermore, teeth protection is essential to maintenance the balance of organic-inorganic compounds in order to create and to preserve cell cement, as well as inhibit bacterial growth [9].
2. Digestive function of saliva is in charge of creating and maintains the bolus. At this point food begins digestion, salivary proteins (α-amylase) break food starch into oligosaccharides and glucose. Besides, saliva is responsible for gustatory sensation thanks to the humidification and lubrication of taste buds. (See Table **2**).

According to the multitude of molecules found in saliva, the reason why they appear in the oral cavity still remains unclear. The role that saliva might play in the maintenance of body homeostasis is not entirely clear [18].

Table 2. Components of saliva: molecules functions (Adapted from Amidogen *et al*).

Teeth	Remineralisation Inhibition of mineralization Lubrication and viscoelasticity Buffering effect	Proline Rich Peptides (PRPs), Slathering, Calcium Phosphate Mucins Mucins and PRPs Bicarbonate, Phosphate and proteins
Food	Digestion Taste Bolus	Lipasa, alpha-amylase, Proteasas, Donase, Rase Zinc Mucinas
Microorganism	Antiviral Antifungals Antibacterial	Mucins, Cystatins, Immunoglobulins Immunoglobulins, Mucins, Histatins Mucins, Histatins, Cystatins, Lactoferrin, Lacto peroxidase, Immunoglobulin, Lysozyme.

Collecting Saliva Samples

Saliva as a whole sample, presents many advantages and some limitations regarding any other biological samples. As some advantages:

1. Obtaining whole saliva samples is easy and safe. It is only needed basic tools and materials, so it is more economical [19].
2. Saliva collection is simple – it doesn´t need a specialized worker. The sample can be obtained by anybody, including self-collection.
3. The procedure is undemanding and non-invasive. Getting saliva sample is neither harmful nor painful. Most people do not feel discomfort unlike from biopsies samples and repeated blood collection, which are more invasive procedures. Besides, saliva samples can be collected under no stress situations and multiple samples can be obtained in different times if repeated measurements must be required. Furthermore, saliva samples do not clot so we can minimize testing errors.
4. Samples are safer to manage. The whole saliva sample contains biochemical factors that inhibit the infectivity of HIV. For hence, salivary secretions do not increase the rates of oral transmission [20].
5. The majority of analytes are exceptionally stable in this kind of samples.
6. Samples are easier to handle and to transport.

Despite this promising future, saliva has not yet become a mainstream diagnostic fluid due to some relevant limitations. As some disadvantages:

1. Most analytes which are detected in blood serum are also found in saliva secretions too, but their levels are essentially different [2]. Analytes concentrations depend on salivary flow and the type of molecular mechanism

of transport from serum into salivary gland ducts.

2. Sometimes the lack of correlation between salivary compounds and blood-based levels of constituents complicates the accuracy of the results.
3. Despite these two biofluids are individually separated, they are linked in a molecular level because salivary glands are very permeable and they are surrounded by multiple blood vessels, allowing the exchange of blood-based molecules.
4. Some technological limitations. The lack of sensibility and specificity of diagnosis tests used to measure the concentration of most analytes found in saliva.

Factors Affecting Salivary Flow Rate

Circadian rhythm, the kind of salivary gland, the type of salivary stimulus, diet, age, physiological status and method of collection are various factors that may affect daily salivary flow [21]. Excessive salivation may occur in certain neurological disorders, gastroesophageal reflux, side-effects of drugs that trigger the parasympathetic nervous system functions as pilocarpine, muscle relaxants, antiepileptic drugs, antipsychotics, over hydration and heavy metals poisoning, among others [22].

Hypo salivation or asialorrhea may be caused by Sjogren's syndrome, drugs that block the action of the chemical messenger acetylcholine (anticholinergic drugs), antihistamine medication, chemotherapy and head and neck radiation therapy between others, can caused changes in salivary flow too. Those situations that could affect salivary flow must be taken into account when there is a needed to collect saliva samples by gland stimulus. Chewing parafilm material or a drop of citric acid on the tip of the tongue, lead the most successful results [21].

The most desirable saliva collection method is the obtaining of whole saliva sample without any gland stimulus [22]. Stimulated whole saliva sample is less suitable for diagnostic applications because of the external substances used to stimulate saliva. These kind of agents tend to modulate the pH fluid and generally stimulate the water phase of saliva secretion, resulting in an over-dilution of targeted proteins concentrations [23]. A proper collection of saliva samples is the first step for an accuracy saliva-based diagnostic procedure. Most of the times they can be obtained by draining, spitting and suctioning oral fluids from the glands and the oral cavity [21].

A large variety of devices are currently commercialized to collect saliva samples. The most common of them are reviewed in Table **3** [3]. Not to interfere in the detection of the analyte concentrations found in saliva samples and the use of a validated material and method for the measurement of a concrete analyte are the most desirable characteristics of the collecting devices.

Occasionally, gathering oral fluids from specific glands must be required. Suctioning and cannulation of the glands are complex and invasive procedures and qualified trained personal must be needed. Glass is the material of reference due to its natural specific composition, free of saliva compounds absorption so it remains the election choice, but because of its fragility, should be rejected. Cotton swabs have also been used, but latest publications have shown that they may introduce undesirable errors because saliva analytes could be trapped into their material composition [24].

Each major salivary gland contributes with a different protein proportion and fluid in the formation of the whole salivary secretion. Parotid glands are mainly implicated in alpha-amylase secretion and mucins are essentially secreted by submandibular and sublingual glands. There is also an unequal contribution of each salivary gland depending on the state of stimulation. In an unstimulated situation, about 65% of the secretion is originated in the submandibular gland, about 25% from the parotid gland, 4% from the sublingual gland and the remaining percentage, form other general salivary glands [25].

It is highly recommended to use a validated device for every specific analyte detection. Although there is no a widespread technique or a validated used of protocols for sample collection, it is extremely important not to omit the instructions of use in the commercial collection device protocols [3]. Besides, it is mandatory a previous cleaning of the oral cavity in order to avoid the presence of diverse contaminants.

Table 3. Most Common saliva collection devices (adapted from Yoshizawa *et al*).

Brand company	Collection device
Salimetrics	Salimetrics Saliva collection aid
	Salimetrics oral swab
	Salimetrics children's swab
	Salimetrics infant's swab
Oasis Diagnostics	DNA·SAL
	UltraSal-2
	Super·SAL
Malvern Medical Developments	Oracol
DNA Genotek	ORA collect · DNA
	Oragene · DNA and RNA

(Table 3) contd.....

Brand company	Collection device
Sarstedt	Salivette
Immunalysis	Quantisal
Norgen	Saliva DNA collection and preservation device
Biomatrica	DNAgard

APPLICATIONS OF SALIVARY BIOMARKERS

According to the National Institutes of Health (NIH), "a biomarker is an objectively measured and evaluated indicator of normal biologic processes, pathogenic processes, or pharmacologic responses to therapeutic intervention" [1].

Taking into account that biomarkers are capable of providing decisive information regarding the current physiologic state of a living organism and the main fact that biomarkers are constituent elements of the body, it is assumed that biomarkers are efficient tools as evaluators of health [26]. There is a wide range of different biomarkers, including proteins, hormones, metabolites, antibodies, lipids, DNA, RNA and microbes. Changes in their intrinsic concentration values, structure, function, even in the manner they conform their actions, may be associated with the onset, development of a specific illness and how the organism may deal with it [27].

Furthermore, to assess an accurate diagnosis method for selecting biomarkers, researchers should be encouraged to follow Prospective specimen collection and Retrospective Blinded Evaluation (PRoBE), a study protocol where biological specimens are collected from a cohort population well characterized in which the biomarker will be measured. Besides, the second part of this methodology protocol required a blinded design to obtain, as much as possible, unbiased data regarding specificity and sensitivity of the biomarkers analysed [28].

Salivary Biomarkers in Cardiovascular Diseases

Currently, the diagnosis of AMI (Acute Myocardial Infarction) is based on clinical findings, electrocardiogram findings and concentrations of serum biomarkers. Cardiovascular disease is a leading cause of death worldwide.

The success in early diagnosis and effective intervention of an AMI depends on the proper sample management and analysis, therefore, new platforms and methodologies are sought for faster screening diagnosis. New biological specimens such as whole saliva secretions could potentially show faster results and for hence, faster successful interventions [29]. There are many analytes related to the ischemic cascade, which result from acute coronary syndromes that can be detected in saliva. For further information, see Table **4**.

The gold standard in AMI biochemical diagnosis is the Troponin. It has to be measured in blood analysis but it means that the blood sample must be obtained by invasive methods. In this case, salivary samples could be an option to avoid more invasive methods of sample collection, but despite successful advancements in AMI early diagnosis, there is not much available information about the diagnostic capability of troponin in saliva obtained from AMI patients [30]. Some researchers have developed salivary multiplexed tests that combined with electrocardiology findings, have demonstrated sensitivity values in the range of 90–100% for the detection of an AMI [30].

Those biomarkers are completely promising but in fact, salivary diagnostics related to cardiovascular diseases must be taken cautiously into account because saliva components like proteins, inhibitors and enzymes, could alter or hide antigenic determinants needed for the immunoassays tests.

Patients suffering from gingivitis and periodontitis show high concentrations of inflammatory biomarkers and proteases. These high levels of inflammatory biomarkers could act as potentially confounders in the screening of cardiovascular diseases because we could not discriminate them from non-specific inflammatory diseases. For example, elevated levels of salivary CRP, MMP-8, MMP-9 and IL-1β during periodontitis could confound the utility of these markers for AMI diagnoses and studies are needed to identify their discriminatory capacity [31, 32].

Table 4. Useful salivary biomarkers for the diagnosis of cardiovascular diseases (adapted from Miller *et al*).

Biomarker	Diagnosing AMI	Compound levels in disease
Myoglobin	Within 48h of chest-pain onset	Elevated [29]
CRP	Marker of systemic inflammation, and plays a central role in atherosclerosis	Elevated [30]
TNF-α	Marker of systemic inflammation.	Elevated [29]
MMP-9	Macrophages that release lytic enzymes like metalloproteinases present in acute coronary syndromes [33]. Within 24h after the onset of chest pain	Elevated [29]
Myeloperoxidase	Neutrophils that is secreted during cell activation contribute to tissue injury [34].	Elevated [29]
Soluble CD40 ligand	Plaque destabilization/rupture is associated with release of soluble CD40 ligand and specific adhesion molecules [35]	Decreased [30]

(Table 4) contd.....

Biomarker	Diagnosing AMI	Compound levels in disease
ICAM-1	Specific adhesion molecules [35]	Elevated [30]

Salivary Biomarkers for Renal Disease

Unfortunately, there is a lack of reports attending the potential capacity of saliva regarding renal disorders. It has been reported in patients suffering of renal disease, changes in salivary creatinine levels according to the renal disease phase with a high sensitivity and specificity [36].

Some researchers pointed out that saliva could be an useful sample in the detection of heavy metals poisoning, such as lead and cadmium because they can be detected in the saliva samples although there are originated from the molecules found in the soluble fraction of blood components [2]. Besides, advanced research is needed for the basement of any saliva diagnosis methodology related to renal diseases.

Salivary Biomarkers for Infectious Diseases

Saliva as a diagnostic fluid is also used for the detection of a variety of antibodies secreted against virus, fungus, bacteria, parasitic agent and allergy reactions. The *Helicobacter pylori* infection is related to a multitude of pathologies such as atrophic gastritis, gastric and duodenal ulcers and in some cases Mucosa Associated Lymphoid Tissue (MALT) lymphoma and gastric carcinomas, being the mucous of oral cavity a reservoir for this pathogen [37]. Some authors have proved the presence of *H. pylori* in oral samples from patients suffering from digestive pathologies with a *H. pylori* biopsy positive confirmatory [37]. Currently the gold standard to prove the presence of *H. pylori* in gastric ulcers is the performance of a biopsy. These methods are quite invasive and with no lack of inconveniences, that could be the reason why the PCR technology could be an excellent alternative. Detecting *H. pylori* DNA has been shown to be highly sensitive and specific in the case of saliva samples [38]. In addition, these bacteria specifically bind to salivary mucins, MUC-5B and MUC 7 secreted by salivary glands, so this mucins could be used as an indicator of the presence of this microorganism [39].

Regarding the presence of the Cytomegalovirus (CMV) in saliva samples, it has been proved the presence of the virus in the oral mucous, so salivary samples could be a good method for the early diagnosis of CMV infection [40].

Recent studies have proved the use of saliva for the diagnosis of Human

Immunodeficiency Virus (HIV), because the specific antibodies against the virus can be detected in this kind of samples [41]. Suspected patients can be screened for the 1-2 HIV virus using ELISA technology, this method has shown a high sensitivity and specificity in the detection of the virus in saliva samples. It is the reason why salivary diagnosis has demonstrated its usefulness as an alternative tool for the detection of confirmatory antibodies against HIV [42]. Recently, the US Centers for Disease Control and Prevention (CDC) have been using a saliva-based test for the detection of HIV that provides results in only 20 minutes. This is a quite interesting characteristic because almost 25% of the people recently infected by HIV virus are not aware of their condition, so these rapid and easy to use devices, could contribute to decrease the probability of infection. It is important to consider that many of these patients have their veins damaged by continuous inoculation of drugs in their blood and others patients are afraid of needless, in these cases, saliva samples are always a better option.

Chronic infection produced by Hepatitis C virus (HCV) and Hepatitis B virus (HBV), could lead us to serious liver complications as cirrhosis and hepatic carcinoma. Recently investigations have shown that HCV and HBV DNAs, antibodies and viral antigens exist in saliva samples and can be accurately detected because the correlation of these particles between blood and saliva samples are quite precise.

PCR saliva based diagnosis is required due to the efficacy of this method in the virus detection but the development of new ELISA based diagnosis tests can also rapidly identify HCV antibodies in saliva samples [43]. The development of new emerging tools into the market based on this suitable ELISA technology could cause a deep impact in the early detection and management of these patients.

Salivary Tests for Forensics Sciences and Drugs Abuse

The use of PCR methods to detect DNA molecules is the based for forensic investigations. The salivary samples can be easily obtained from glasses, cigarettes, food products, envelopes and other sources. An useful biological marker which can be used for the identification of crime suspects and for paternity suit is blood group antigens obtained from saliva secretions [44]. DNA is relatively stable in the dry state; thus DNA testing can be done from the salivary samples collected from different scenarios [45].

The use of saliva has gained importance as drug monitoring and detection of illicit drugs as an alternative to urine samples. The application of these rapid methods have allowed *in situ* applications tests in quite different scenarios, as antidrug police control, rehabilitation centres, prisons, work environment, among others [7]. Moreover, drug particles have the same detection window that they have in blood samples so the mere presence of these components are satisfactory for forensic purposes.

Salivary Diagnostics for Autoimmune Diseases

Sjogren's syndrome (SS) is a chronic autoimmune disorder characterized by salivary and lacrimal gland dysfunction, serologic abnormalities, and multiple organ system changes [46]. To assess the righteous diagnosis of Sjogren's disease, a biopsy of the minor salivary glands is required, therefore the research of different procedures to assess the proper diagnosis has been required. Procedures as sialography, salivary scintigraphy, sialometry and serological tests have shown uneven results [46]. As an example, some authors proposed a panel performed on whole saliva samples that included flow rate, pH, buffer capacity, lactobacillus and yeast concentration that can provide compelling evidence for the presence of Sjogren's Syndrome [47].

The aim of several studies was to seek a positive correlation among cytokines and others specific molecules found in saliva with a diagnostic and clinical utility in Sjogren's syndrome patients. The results of these studies suggest that increased concentration of sodium, chloride, IgA, IgG, lactoferrin, β2-microglobulin, lysozyme C, cystatin C, cystatin S, albumin, α-2 microglobulin, IL-2, IL-6, fatty acids protein binding, and inflammatory mediators such as prostaglandin E2 (PGE2), thromboxane B2, anti-transglutaminase, anti-histone, anti-SSA and anti-SSB can be found in the saliva of Sjogren's disease patients [48 - 50]. On the other hand, controversial data have been reported, so further investigation must be done to correctly assess the role of these biomarkers in the diagnosis of the syndrome.

The presence of Ss-antiLa antibodies can be primarily found in saliva secretions due to a local increased production of these antibodies from salivary glands. Any levels alterations of these antibodies can be used as an indicator for Sjogren's syndrome and a checkpoint for disease progression [6]. That is the reason why analysis of salivary proteins in patients with SS is a promising area but the usefulness of their routine assessment in clinical practice remains limited.

Biomolecules as cathepsin D, α-enolase, and β2-microglobulin have been reported and clinically validated in patients suffering from Systemic Lupus Erythematosus (SLE) [51]. In addition some scientific have identified 24 antibodies which can differentiate patients with primary SS from patients with Systemic Lupus Erythematosus and healthy patients [50].

These biomarkers could become an important clinical tool in the detection of primary SS and other autoimmune diseases to the contrary other diagnosis methods already established that appear to be more expensive, invasive and with non-conclusive results, most of the times.

Salivary Biomarkers for Endocrine Diseases

Most of the hormones enter saliva by passive diffusion across the acinar cells. The lipid-soluble hormones with lower molecular weights can be precisely detected in saliva, but protein-bound hormones (such as gonadotropins, prolactin and thyrotropin) cannot be accurately monitored by salivary analysis methods due to the biochemical and the specific structure of these hormones.

A group of hormones normally measured in plasma, such as steroids, non-steroids, peptides and protein hormones can be found in saliva but the most important fact is to know if the correlation between levels of the compounds in serum and saliva samples [6] are comparable.

Cortisol, testosterone, dehydroepiandrosterone (DHEA), hydroxyprogesterone, estriol, progesterone and aldosterone, are the most frequently measured hormones. Many of them can be accurately determined by commercial assays. Salivary cortisol is in equilibrium with free cortisol in plasma and unaffected by the salivary flow due to diffusion transport through the lipid membranes of salivary glands [52]. The measurement of salivary cortisol is useful as a key indicator to stress situations similar to other stress biomarker as salivary alpha-amylase, and it is highly recommended for Cushing ´s Syndrome screening [53].

Salivary cortisol tests provide the following benefits compared to serum cortisol tests: these tests are easier, simply, non-invasive and can be performed anywhere, without the need of going to the health centre. Taking into account those facilities it is hoped that monitored salivary cortisol will soon become a routine diagnostic method for Cushing's Syndrome diagnosis [54].

Salivary testosterone and its serum concentrations match perfectly thus, salivary testosterone has become an useful tool to assess serum testosterone concentrations when blood extractions are not available [55]. The currently use of saliva testosterone measurements concentrations is to evaluate endocrine disorders, testosterone deficiency and the assessment of androgenic function in prostatic carcinoma patients or after surgical orchiectomy. Among female sex hormones, the detection of progesterone and estradiol in saliva secretions are the most valuable parameters to determine menstrual cycles profiles and in pregnancy monitoring [56].

Salivary Diagnosis of Diabetes Mellitus

There is a lack of correlation between capillary blood glucose and salivary glucose due to insulin hormone effects. This hormone is released into the blood stream to maintain glucose blood levels but this effect does not occur in parallel in salivary secretions. Diabetic patients present higher glucose blood concentration than salivary secretions and blood and salivary glucose levels do not correlate at all [57]. Considering that periodontal inflammation is usual in diabetic patients, Gingival Crevicular Fluid (GCF) that contains blood traces can be used for the detection of capillary glucose levels. GFC can be retrieved with a device similar to dental floss, avoiding other invasive methods. Only a few reports showed a positive correlation among salivary biomarkers (as salivary IL-6 and 1,5 anhydroglycitol) and HbA1C, used as an average measurement of serum glucose concentration.

Salivary Diagnosis on Stress Assessment

The sympathetic nervous system has an important role in alfa-amylase secretions from parotid glands. Anxiety situations are responsible for increments of salivary alpha-amylase. It has been registered an elevated number of patients suffering from anxiety before undergoing surgery. These stressful situations are related to adverse recovering and worse postoperative outcomes. As alpha-amylase levels can vary rapidly, it may be worth to have a specific point of care available to detect these rapid changes. This POCT (Point Of Care Test) method will provide real-time results to assess the anxiety state of patients.

Salivary Diagnostics in Oncology

Analysis of saliva samples could be an important tool in the early diagnosis of different type of tumours because many of them are discovered when the cancer has been finally established in patients. The first salivary biomarker for cancer to be discovered was the HER2/neu, used for breast cancer [58]. In other study, elevated levels of c-erbB2, a tumour marker, and the cancer antigen 15-3 (CA 15-3) were diagnosed in the saliva of breast cancer patients [59]. A positive correlation between CA 15-3 and CA125 serum and salivary levels have been established.

Given the fact that saliva is in direct contact with the oral mucous and cancerous lesions, the early diagnosis and screening of Oral Squamous Cell Carcinoma (OSCC), remains an incredibly useful method in contrast to current more invasive procedures. The p53 is a tumour suppressor protein which is produced in cells exposed to various types of DNA damaging situations of stress. Antibodies acting against p53 gene can be detected in the saliva of Oral Squamous Cell Carcinoma patients [60]. Some of the recent discoveries included some specific bacteria of the oral cavity [61]. Aberrant expressions of CD44, IL-1β, IL-8 S100, the DNA polymorphism of IL-1β, IL-6, IL-8, TNF-α and Vascular Endothelial Growth Factor (VEGF) have been associated with the development of OSCC [58]. Some of these molecules, as cytokines, can be found also elevated in other systemic diseases. However, these concentrations are not as high as those found in patients with carcinoma [62]. To the contrary, other authors did not find a significant statistical increase of TNFα, IL-1α nor IL-8 in the saliva of OSCC patients. Regarding EGF and endotelin-1 (ET-1) molecules, the role that may play in the development of OSCC remain unclear [63, 64]. In addition, human mRNA has shown promising results in the diagnosis of oral cancer but this methods should be validated [65].

Traditionally serum tumour markers as Cyfra 21-1, TPS, CEA, SCC, CA125 and CA19-9 have been recently investigated in OSCC patients. The investigators have found an important rise in salivary concentrations of Cyfra 21-1, TPS and CA125. However the presence of elevated salivary levels of CA19-9, SCC and CEA lack of statistical significance [66]. These increments support the use of salivary tumour markers as a diagnostic tool but the variability of the levels of potential OSCC salivary biomarkers in both cancerous and non-cancerous patients suggests confounding factors still unknown, such as oral inflammation, systemic diseases, and infection diseases states which encourage further investigations. Current developments in proteomic technologies, such as mass spectrometry, liquid chromatography and protein-peptide labelling technologies, will allow the detection of low abundance molecules in saliva proteome [67]. The proteomic profile of saliva of OSCC patients is completely different from saliva samples of control patients according to the latest publications.

Thanks to the advances in molecular biology applied to protein and genomic techniques, several markers of OSCC have been identified. However, further studies are needed to confirm the specificity of these markers, in larger sample research sizes, would definitely be more convenient.

Salivary Diagnostics of Common Oral Diseases

In a healthy oral cavity, there is no correlation between salivary secretion rate and dental caries [68]. On the contrary, when saliva secretion decreases, saliva buffer mechanism capacity, decreases as well. Saliva contains some components such as bicarbonate, phosphate, and some proteins systems (MUC5B and MUC7 type mucins) which not only have a buffer effect but also provide ideal conditions that automatically eliminate certain bacterial components.

That is the reason why the amount of caries is increased when saliva rates falls dramatically [69]. The absence of these mucins is associated with increased susceptibility for dental caries to the fact that they play an important role in the protection of oral surfaces to avoid the colonization of microorganism. Gram-negative and Gram-positive bacteria may attack the oral surface and produce a chronic periodontal disease, causing a persistent inflammation which may alter and destroy the connective tissue and the alveolar bone [70]. Different kind of molecules have been associated with periodontal disease. To check the list of periodontal disease markers see Table **5** [29].

Table 5. Salivary biomarkers of periodontal disease (adapted from Miller *et al*).

Biomarker	Biological role	Compound levels in disease
β-glucuronidase	Marker of neutrophil influx [32]	Elevated
CRP	Acute-phase reactant [71]	Elevated
IL-1β	Proinflammatory cytokine [31]	Elevated
IL-6	Acute-phase protein induction, osteoclast generation and activation [72]	Elevated
MIP1α	CC chemokine stimulates osteoclast progenitors to become active osteoclasts [73]	Elevated in aggressive periodontitis
TNF-α	Proinflammatory cytokine stimulates IL-1, inhibits collagen synthesis and induces collagenases [73, 74]	Elevated
α2-macroglobulin	Proteinases inactivator [32]	Decreased
MMP-8	Neutrophil collagenase degrades type I and III collagens [75]	Elevated
MMP-9	Gelatinase degrades type IV and V collagens [75]	Elevated
AST	Aminotransferase, marker of cell injury [76]	Elevated
ALT	Aminotransferase, marker of cell injury [76]	Elevated
TIMPs	Metalloproteinase inhibitors [75]	Decreased
Alkaline phosphatase	Hydrolase enzyme involved in bone remodelling [76]	Elevated
Osteoprotegerin	Interferes with RANKL binding [31]	Elevated
Osteocalcin	Secreted by osteoblasts. Play a role in mineralization and calcium ion homeostasis [74]	Reduced
SPARC/osteonectin	Secreted by osteoblasts [77]	Reduced
HGF	Osteoclast-activating factor [77]	Elevated

ALT: Alanine aminotransferase; AST: Aspartate aminotransferase; CC: Cysteine-cysteine; CRP: C-reactive protein; MIP: Macrophage inflammatory protein; MMP: Matrix metalloproteinase; RANKL: Receptor activator of NF-κB ligand; SPARC: Secreted protein, acidic, rich in cysteine; TIMP: Tissue inhibitors of metalloproteinase.

Currently, there is a debate about which biomarkers might hold diagnosis relevance to different phases of periodontal disease within salivary diagnosis of whole saliva samples. Nevertheless, recent approaches on the potential role of periodontal disease as a risk factor for cardiovascular and cerebrovascular diseases have provide new important data to this aspect of salivary analysis [78].

To summarize, in the table below, there is a list of different biomarkers used for the diagnosis of different illnesses (see Table **6**).

Table 6. A brief summary of salivary biomarkers in several systemic diseases (adapted from Malathi *et al*).

Biomarker	Disease	Compound levels
	AUTOINMUNE DISEASES	
Immunoglobulins	Sjogren´s syndrome	Elevated
Mediator inflammatory	Sjogren´s syndrome	Elevated
Albumin	Sjogren´s syndrome	Elevated
Sodium, Chloride	Sjogren´s syndrome	Elevated
Phosphate	Sjogren´s syndrome	Decreased
Lactoferrin	Sjogren´s syndrome	Elevated
Lysozime C	Sjogren´s syndrome	Elevated
Cystatin C, S	Sjogren´s syndrome	Elevated
Ss-antiLantibodie	Sjogren´s syndrome	Elevated
Carbonic anhydrase	Sjogren´s syndrome	Decreased
Alpha-amylase	Sjogren´s syndrome/Sarcoidosis	Decreased
Kallikrein	Sarcoidosis	Decreased
Cathepsin -D	Systemic Lupus Erythematosus (SLE)	Elevated
α- enolase	Systemic Lupus Erythematosus (SLE)	Elevated
β-2 microglobulin	Sjogren´s Syndrome/SLE	Elevated
IgA	Multiple Sclerosis	Decreased
	GENETICS DISORDERS	
Total protein	Ectodermal Dysplasia	Elevated
Alpha-amylase	Ectodermal Dysplasia	Decreased
Inorganics Constituents	Ectodermal Dysplasia	Elevated
Lactate dehydrogenase	Cystic Fibrosis	Elevated
Cathepsin -D	Cystic Fibrosis	Elevated
Sodium, Potassium, Chloride, Calcium, Magnesium	Cystic Fibrosis	Elevated

(Table 6) contd.....

Biomarker	Disease	Compound levels
INFECTION DISEASES		
IgA, IgG e IgM antibodies	Viral infections	Elevated
Measles virus-specifics IgM	Viral infections	Elevated
P24 antigens	Viral infections HIV-1 y HIV-2	Elevated
Mycobacterium tuberculosis	Bacterial infection	Present
MUC-5B mucins	Bacterial infection (*Helicobacter pylori*)	Elevated
MUC 7 mucins	Bacterial infection (*Helicobacter pylori*)	Elevated
Immunoglobulins	Fungal infection	Elevated
Hsp 70	Fungal infection	Elevated
Calprotectin	Fungal infection	Elevated
Histatins	Fungal infection	Elevated
Mucins	Fungal infection	Elevated
Basic proline rich proteins	Fungal infection	Elevated
Peroxidases	Fungal infection	Elevated
Candidiasis immunoglobulins	Fungal infection	Elevated
ONCOLOGY DISEASES		
ncRNAs	Carcinoma lung, breast, prostate, Oral Squamous Cell Carcinoma (OSCC)	Aberrant expressions
miRNAs	OSCC	Dysregulation
mRNA: CCNI, EGFR, FGF19, FRS2, GREB2	Lung cancer	Elevated
mRNA: AGPAT1, B2M, BASP2, IER3, IL1B	Ovarian cancer	Elevated
P53 antibodies	OSCC	Detection
Ca 15-3	Breast cancer	Elevated
c-erb-2	Breast cancer	Elevated
HER2/neu	Breast cancer	Elevated
Ca125	Breast cancer, oral cancer and ovarian tumour	Elevated
FGF2	Gland salivary tumour	Elevated
Cyfra 21.1	OSCC	Elevated
TPS	OSCC	Elevated
Cortisol	OSCC	Elevated
Lactate dehydrogenase	OSCC	Elevated
PSA	Prostate adenocarcinoma	Elevated

(Table 6) contd.....

Biomarker	Disease	Compound levels
STRESS ASSESSMENT		
Cortisol	Chronic stress	Elevated
IgA	Chronic stress	Decreased
Lysozyme	Chronic stress	Decreased
Chromogranin A	Acute stress	Elevated
Alpha amylase	Acute stress	Elevated
RENAL DISEASES		
Nitrite	End stage renal	Elevated
Uric acid	End stage renal	Elevated
Sodium chloride	End stage renal	Elevated
pH	End stage renal	Elevated
Alpha-amylase	End stage renal	Elevated
Lactoferrin	End stage renal	Elevated
Cortisol	End stage renal	Elevated
Phosphate	Chronic renal failure	Elevated

PSA: Prostate specific antigen; TPS: Tissue polypeptide specific antigen; FGF: Fibroblast growing factor; Hsp: Heat shock proteins.

SALIVARY PROTEOME

The importance of salivary secretions is not only as a reflexion of body health and physiological status; it may play a role in the detection of diseases. This is the reason why the analyses and cataloguing of the human salivary proteome and the knowledge of the complete protein salivary composition are so important [65]. When saliva and blood protein compositions are compared, the results showed that approximately 27% of the whole saliva proteins are found in blood plasma [79]. Despite this apparently low degree of overlapping, nearly 40% of the proteins that have been suggested to be candidate markers for cancer and cardiovascular disease can be found in whole saliva.

According to their gland origin, (parotid, submandibular and sublingual) a complete catalogue of the salivary proteome has been generated using state-of-the-art, sensitive and high-throughput Mass Spectrometry (MS), a proteomic technology that combined with different protein separation methods, is capable of the detection of this kind of molecules [79, 80].

The entire description of all salivary secretory proteome components, post-translational modifications and protein complexes has been listed and freely available to the general public at: (http://www.hspp.ucla.edu). At this website, the

Saliva Proteome Knowledge Base (SPKB) and the Salivary Proteome Wiki (SPW) are linked. Thanks to the great efforts of investigators and the National Institute of Dental and Craniofacial Research (NIDCR), that encourages the project and makes a collaborative environment, all this knowledge is available to share and annotate information related to salivary proteomics data.

Thanks to these kind of available resources, as new and powerful technologies and bioinformatics tools, it should be interesting to encourage researchers to discover new proteins as biomarkers to the early detection and monitoring of disease status in a non-invasive use of saliva. However, the detection of salivary protein biomarkers for disease detection has its own challenges. Every protein has different properties that could modify detection methodologies and lead special requirements for processing and storing of saliva samples. Besides, the oral cavity environment and the presence of bacterial microorganism impose protein degradation, as well as the presence of proteolytic proteins that could affect the stability of protein analytes over time. Fortunately, it has been developed a safe stabilization method to keep the integrity of the whole saliva sample. This achievement support the promising future of saliva as a wide range non-invasive diagnostic tool [18].

Salivary Transcriptome

The second step in the study of the new "omic" molecular technologies related to salivary analysis is the saliva transcriptome diagnosis. Messenger RNA (mRNA) is the direct precursor of proteins and it can be found in cells and tissue samples [81]. It is already known the important functions related to mRNA, as cellular division, apoptosis, differentiation and cellular development as well as, immune response.

The study of nucleic acids such as DNA and RNA have become nowadays much more affordable due to molecular technologies and because its nature, the new molecular candidates to be used as disease markers, can be verified by sensitive and specific Polymerase Chain Reaction (PCR) methods, the only ones capable of detecting these singular molecules in the past. On the other hand, the often-scarce amount of saliva and the presence of partially fragmented and degraded RNAs, add more difficulties to detect the biomarkers with the currently technology.

Presently, the more useful strategy to identify salivary transcriptomic biomarkers is through microarray technology. After profiling the transcriptomic biomarkers by microarray, they are validated by quantitative (q)PCR, the gold standard for quantification of nucleic acids [82]. To overcome some difficulties in molecular amplification, some investigators have developed methodologies using multiplex pre-amplification techniques [83]. The first studies of the salivary transcriptome showed that the normal salivary transcriptome consists of ~3000 mRNAs in normal populations. The complete profile of mRNAs expression pointed out 185 mRNAs out of 3000 mRNAs initially checked. Those were consistently detected in healthy subjects and they were determined as the Normal Salivary Core Transcripts (NSCT) [84].

Recent studies have demonstrated the practicability of using salivary mRNA as an aid to the diagnosis of oral cancers [85, 86]. The combination of specific RNAs biomarkers yielded a high sensibility and specificity for the detection of OSCC patients. In addition, some mRNAs have been used in forensic fields to identify body fluids [87]. Furthermore, inflammatory mRNAs markers were detected in whole saliva sample to monitor the status of periodontal disease in type II diabetic patients [88]. Also, a highly diagnostic RNAs signatures have been identified for head & neck cancer and Sjögren's syndrome [89].

A complex compositional RNA profile of saliva has raised due to biomolecular technologies by using high-throughput RNA sequencing (RNA-Seq) and powerful bioinformatics platforms [90]. The detection and knowledge of non-coding RNAs (ncRNAs) as microRNAs (miRNAs), small nucleolar RNAs (snoRNAs), and other non-coding RNAs has yielded a large amount of new data which represents a new source of interesting biomarkers information [91].

The nature of ncRNAs lead them to the first position of ideal and suitable salivary biomarkers because of its short size, body fluid stability and their main location inside the exosomes, so they can avoid degradation from proteolytic proteins. On the contrary, complete mRNAs are more prone to be degraded by enzymes [91].

Micro-RNAs (miRNAs) are regarded as important regulators of mRNAs and protein expression and are predicted to regulate the expression of almost one-third of all human transcripts [92]. Recent numerous studies have described the potential of miRNAs as cancer biomarkers for oral cancer [93]. In another study, where the expression of miR-21 and miR-141 were investigated, they found that the two miRNAs were over-expressed in early and advanced prostate cancer patients [94].

While further analyses need to be performed, these results indicate an important role for the salivary transcriptome as a viable and non-invasive source of disease-specific biomarkers. Accuracy study designs and the development of a newest analytical techniques which support this not-stopping advances, may be required to avoid that this incredible work and the effort spent on it, be easily useless. However, the significance of the entire spectrum of saliva RNAs has not been fully elucidated, thus warranting the need for further and comprehensive analysis.

Salivary Metabolome

Metabolomics is the study and description of the wide variety of small molecules present in cells, tissues, organs and biological fluids, known as metabolome [95]. Metabolomics can be considered as the third step in the "omics" new biomarkers research. Just like transcriptome and proteome, metabolome is a field in a process of constant change. A dynamic mixture of complex functions reflexing the changes produced in organs, genes, proteins and even body environment.

Recently, Sugimoto *et al.*, using capillary electrophoresis with time-of-flight mass spectrometry, were able to identify 57 principal metabolites that can accurately predict the probability of being affected by oral cancer, breast cancer, pancreatic cancer, and periodontal disease [96]. Additionally, Wei *et al.*, using ultra-performance liquid chromatography coupled with quadruple time-of-flight mass spectrometry, designed a study where a combination of three salivary metabolic biomarkers (valine, lactic acid and phenylalanine) could accurately discriminate OSCC patients from health controls [97]. Although it could be a promising future, further investigation must be done in order to elucidate the biological importance of these novels biomarkers.

Salivary Microbiome

Bacterial infections have been previously related to malignancies because of their ability to promote chronic inflammation [98]. The aim of a recent study was to compare the microbial species present in tissues from tumour and non-tumour sites of patients with OSCC. *Peptostreptococcus stomatis*, *Streptococcus gordonii* and six other bacterial species were found to be highly associated with tumour sites, whereas *Granulicatella adiacens*, a known factor of endocarditis was associated with non-tumour sites [99].

The Human Oral Microbe Identification Microarray (HOMIM) is a recent development based on the detection of 16s-rRNA, using an oligonucleotide-microarray. This technique has been developed for the profiling and monitoring of changes in the oral microbiota [100]. Regarding to different diseases as pancreatic cancer, oral cancer, lung cancer, colonic neoplasia and extra colonic malignancy, cardiovascular disease and cerebrovascular disease, some investigators have found specific changes in the bacterial profile of the oral cavity with a positive correlation to those diseases.

Human Papilloma virus (HPV) has been associated with the beginning of oropharyngeal squamous cell cancers (OSCC). Current tests for HPV detection in saliva are available utilizing polymerase chain reaction (PCR) methods. This association between HPV and oral cancer can be used for the early diagnosis of

OSCC, with the setting up of the point of care technologies to help the rapid screening of pre-tumour patients [101].

All the collecting data about microorganism profile present in either patients or healthy people could be a reflection of systemic diseases, leading us to another valuable item on salivary diagnosis.

CONCLUSION

Salivary samples have emerged as an ideal biofluid., easy to obtain, safe and ready to collect by mainly all population. Consequently, salivary secretions are a non-invasive collection method that reduces the use of more aggressive procedures.

In addition, salivary secretion is an incredible pool of molecules from every tissue and organ and may reflect and communicate the individual´s current health status, as a real alternative to blood-based analyses.

However, the lack of correlation between the levels of saliva constituents and blood-based constituents, makes more difficult to discern the accuracy of the results. In order to overcome these issues, great efforts have been made towards sample collection standardization, pre-treatment procedures and wide-range protein analysis. Besides, the search of validated and standardised biomarkers as a useful tool to disease assessment with high sensitivity and specificity have lead us to an ultimate technological develop and POCT devices, as well, that have allowed us modern, safe and accuracy platforms to retrieve results.

The most promising results have been made in the salivary proteomics broad field. This has been made possible due to the spectacular contribution of modern biomolecular techniques as qPCR, mass spectrometry, massive sequencing and bioinformatics, among others. A great number of possible biomarkers has been revealed, so a lot of work is needed to elucidate the role of this molecules as diseases biomarkers.

CONFLICT OF INTEREST

The author (editor) declares no conflict of interest, financial or otherwise.

ACKNOWLEDGEMENTS

Declared none.

REFERENCES

[1] Silberring J, Ciborowski P. Biomarker discovery and clinical proteomics. Trends Analyt Chem 2010;

29: 128.
[http://dx.doi.org/10.1016/j.trac.2009.11.007]

[2] Chiappin S, Antonelli G, Gatti R, De-Palo EF. Saliva specimen: a new laboratory tool for diagnostic and basic investigation. Clin Chim Acta 2007; 383: 30-40.
[http://dx.doi.org/10.1016/j.cca.2007.04.011]

[3] Yoshizawa JM, Schafer CA, Schafer JJ, Farrell JJ, Paster BJ, Wong DT. Salivary biomarkers: toward future clinical and diagnostic utilities. Clin Microbiol Rev 2013; 26: 781-91.
[http://dx.doi.org/10.1128/CMR.00021-13]

[4] Humphrey SP, Williamson RT. A review of saliva: normal composition, flow and function. J Prosthet Dent 2001; 85: 162-9.
[http://dx.doi.org/10.1067/mpr.2001.113778]

[5] Tiwari M. Science behind human saliva. J Nat Sci Biol Med 2011; 2: 53-8.
[http://dx.doi.org/10.4103/0976-9668.82322]

[6] Deepa T, Thirrunavukkarasu N. Saliva as a potential diagnostic tool. Indian J Med Sci 2010; 64: 293-306.
[http://dx.doi.org/10.4103/0019-5359.99854]

[7] Aps JK, Martens LC. Review: The physiology of saliva and transfer of drugs into saliva. Forensic Sci Int 2005; 150: 119-31.
[http://dx.doi.org/10.1016/j.forsciint.2004.10.026]

[8] Pfaffe T, Cooper-White J, Beyerlein P, Kostner K, Punyadeera C. Diagnostic potential of saliva: current state and future applications. Clin Chem 2011; 57: 675-87.
[http://dx.doi.org/10.1373/clinchem.2010.153767]

[9] Nieuw Amerongen AV, Ligtenberg AJ, Veerman EC. Implications for diagnostics in the biochemistry and physiology of saliva. Ann N Y Acad Sci 2007; 1098: 1-6.
[http://dx.doi.org/10.1196/annals.1384.033]

[10] Bennick A. Salivary proline-rich proteins. Mol Cell Biochem 1982; 45: 83-99.
[http://dx.doi.org/10.1007/BF00223503]

[11] Edgar WM. Saliva: its secretion, composition and functions. Br Dent J 1992; 172: 305-12.
[http://dx.doi.org/10.1038/sj.bdj.4807861]

[12] Walz A, Stuhler K, Wattenberg A, *et al.* Proteome analysis of glandular parotid and submandibular-sublingual saliva in comparison to whole human saliva by two-dimensional gel electrophoresis. Proteomics 2006; 6: 1631-9.
[http://dx.doi.org/10.1002/pmic.200500125]

[13] Vining RF, McGinley RA, Symons RG. Hormones in saliva: mode of entry and consequent implications for clinical interpretation. Clin Chem 1983; 29: 1752-6.

[14] Korsrud FR, Brandtzaeg P. Characterization of epithelial elements in human major salivary glands by functional markers: localization of amylase, lactoferrin, lysozyme, secretory component, and secretory immunoglobulins by paired immunofluorescence staining. J Histochemistry Cytochem 1982; 30: 657-66.
[http://dx.doi.org/10.1177/30.7.6179983]

[15] Yao Y, Berg EA, Costello CE, Troxler RF, Oppenheim FG. Identification of protein components in human acquired enamel pellicle and whole saliva using novel proteomics approaches. J Biol Chem 2003; 278: 5300-8.
[http://dx.doi.org/10.1074/jbc.M206333200]

[16] Hardt M, Thomas LR, Dixon SE, *et al.* Toward defining the human parotid gland salivary proteome and peptidome: identification and characterization using 2D SDS-PAGE, ultrafiltration, HPLC, and mass spectrometry. Biochemistry 2005; 44: 2885-99.
[http://dx.doi.org/10.1021/bi048176r]

[17] Carpenter GH. The secretion, components, and properties of saliva. Annu Rev Food Sci Technol 2013; 4: 267-76.
[http://dx.doi.org/10.1146/annurev-food-030212-182700]

[18] Wong DT. Salivary diagnostics. Oper Dent 2012; 37: 562-70.
[http://dx.doi.org/10.2341/12-143-BL]

[19] Mittal S, Bansal V, Garg S, Atreja G. The diagnostic role of saliva-A review. J Clin Exp Dent 2011; 3: 314-20.
[http://dx.doi.org/10.4317/jced.3.e314]

[20] Campo J, Perea MA, Del-Romero J, Cano J, Hernando J, Bascones A. Oral transmission of HIV, reality or fiction? An update. Oral Dis 2006; 12: 219-28.
[http://dx.doi.org/10.1111/j.1601-0825.2005.01187.x]

[21] Navazesh M. Methods for collecting saliva. Ann N Y Acad Sci 1993; 694: 72-7.
[http://dx.doi.org/10.1111/j.1749-6632.1993.tb18343.x]

[22] Baldini C, Gallo A, Perez P, Mosca M, Alevizos I, Bombardieri S. Saliva as an ideal milieu for emerging diagnostic approaches in primary Sjogren's syndrome. Clin Exp Rheumatol 2012; 30: 785-90.

[23] Rudney JD, Kajander KC, Smith QT. Correlations between human salivary levels of lysozyme, lactoferrin, salivary peroxidase and secretory immunoglobulin A with different stimulatory states and over time. Arch Oral Biol 1985; 30: 765-71.
[http://dx.doi.org/10.1016/0003-9969(85)90129-3]

[24] Kozaki T, Lee S, Nishimura T, Katsuura T, Yasukouchi A. Effects of saliva collection using cotton swabs on melatonin enzyme immunoassay. J Circadian Rhythms 2011; 9: 1-4.
[http://dx.doi.org/10.1186/1740-3391-9-1]

[25] Schenkels LC, Veerman EC, Nieuw Amerongen AV. Biochemical composition of human saliva in relation to other mucosal fluids. Crit Rev Oral Biol Med 1995; 6: 161-75.
[http://dx.doi.org/10.1177/10454411950060020501]

[26] Ilyin SE, Belkowski SM, Plata-Salaman CR. Biomarker discovery and validation: technologies and integrative approaches. Trends Biotechnol 2004; 22: 411-6.
[http://dx.doi.org/10.1016/j.tibtech.2004.06.005]

[27] Wagner PD, Verma M, Srivastava S. Challenges for biomarkers in cancer detection. Ann N Y Acad Sci 2004; 1022: 9-16.
[http://dx.doi.org/10.1196/annals.1318.003]

[28] Pepe MS, Feng Z, Janes H, Bossuyt PM, Potter JD. Pivotal evaluation of the accuracy of a biomarker used for classification or prediction: standards for study design. J Natl Cancer Inst 2008; 100: 1432-8.
[http://dx.doi.org/10.1093/jnci/djn326]

[29] Miller CS, Foley JD, Bailey AL, *et al.* Current developments in salivary diagnostics. Biomarkers Med 2010; 4: 171-89.
[http://dx.doi.org/10.2217/bmm.09.68]

[30] Floriano PN, Christodoulides N, Miller CS, *et al.* Use of saliva-based nano-biochip tests for acute myocardial infarction at the point of care: a feasibility study. Clin Chem 2009; 55: 1530-8.
[http://dx.doi.org/10.1373/clinchem.2008.117713]

[31] Miller CS, King CP, Langub MC, Kryscio RJ, Thomas MV. Salivary biomarkers of existing periodontal disease: a cross-sectional study. J Am Dent Assoc 2006; 137: 322-9.
[http://dx.doi.org/10.14219/jada.archive.2006.0181]

[32] Aurer A, Jorgic-Srdjak K, Plancak D, Stavljenic-Rukavina A, Aurer-Kozelj J. Proinflammatory factors in saliva as possible markers for periodontal disease. Coll Antropol 2005; 29: 435-9.

[33] Vasan RS. Biomarkers of cardiovascular disease: molecular basis and practical considerations.

Circulation 2006; 113: 2335-62.
[http://dx.doi.org/10.1161/CIRCULATIONAHA.104.482570]

[34] Wu T, Trevisan M, Genco RJ, Dorn JP, Falkner KL, Sempos CT. Periodontal disease and risk of cerebrovascular disease: the first national health and nutrition examination survey and its follow-up study. Arch Intern Med 2000; 160: 2749-55.
[http://dx.doi.org/10.1001/archinte.160.18.2749]

[35] Apple FS, Wu AH, Mair J, *et al.* Future biomarkers for detection of ischemia and risk stratification in acute coronary syndrome. Clin Chem 2005; 51: 810-24.
[http://dx.doi.org/10.1373/clinchem.2004.046292]

[36] Lloyd JE, Broughton A, Selby C. Salivary creatinine assays as a potential screen for renal disease. Ann Clin Biochem 1996; 33(Pt 5): 428-31.
[http://dx.doi.org/10.1177/000456329603300505]

[37] Medina M-L, Medina M-G, Martin G-T, Picon S-O, Bancalari A, Merino L-A. Molecular detection of Helicobacter pylori in oral samples from patients suffering digestive pathologies. Med Oral Patol Oral Cir Bucal 2010; 15: e38-42.

[38] Berroteran A, Perrone M, Correnti M, *et al.* Detection of Helicobacter pylori DNA in the oral cavity and gastroduodenal system of a Venezuelan population. J Med Microbiol 2002; 51: 764-70.
[http://dx.doi.org/10.1099/0022-1317-51-9-764]

[39] Silva DG, Stevens RH, Macedo JM, *et al.* Higher levels of salivary MUC5B and MUC7 in individuals with gastric diseases who harbor Helicobacter pylori. Arch Oral Biol 2009; 54: 86-90.
[http://dx.doi.org/10.1016/j.archoralbio.2008.08.003]

[40] Correia-Silva JF, Victoria JM, Guimaraes AL, *et al.* Cytomegalovirus shedding in the oral cavity of allogeneic haematopoietic stem cell transplant patients. Oral Dis 2007; 13: 163-9.
[http://dx.doi.org/10.1111/j.1601-0825.2006.01240.x]

[41] Holm-Hansen C, Constantine NT. Saliva and HIV testing. Lancet 1993; 341: 382-3.
[http://dx.doi.org/10.1016/0140-6736(93)90196-N]

[42] Corstjens PL, Abrams WR, Malamud D. Detecting viruses by using salivary diagnostics. J Am Dent Assoc 2012; 143: 12S-8S.
[http://dx.doi.org/10.14219/jada.archive.2012.0338]

[43] Drobnik A, Judd C, Banach D, Egger J, Konty K, Rude E. Public health implications of rapid hepatitis C screening with an oral swab for community-based organizations serving high-risk populations. Am J Public Health 2011; 101: 2151-5.
[http://dx.doi.org/10.2105/AJPH.2011.300251]

[44] Sweet D, Hildebrand D. Saliva from cheese bite yields DNA profile of burglar: a case report. Int J Legal Med 1999; 112: 201-3.
[http://dx.doi.org/10.1007/s004140050234]

[45] Kamodyova N, Durdiakova J, Celec P, *et al.* Prevalence and persistence of male DNA identified in mixed saliva samples after intense kissing. Forensic Sci Int Genet 2013; 7: 124-8.
[http://dx.doi.org/10.1016/j.fsigen.2012.07.007]

[46] Daniels TE. Sjogren's Syndrome: clinical spectrum and current diagnostic controversies. Adv Dent Res 1996; 10: 3-8.
[http://dx.doi.org/10.1177/08959374960100010501]

[47] Sreebny LM, Zhu WX. The use of whole saliva in the differential diagnosis of Sjogren's syndrome. Adv Dent Res 1996; 10: 17-24.
[http://dx.doi.org/10.1177/08959374960100010201]

[48] Van der Reijden WA, Van der Kwaak JS, Veerman EC, Nieuw Amerongen AV. Analysis of the concentration and output of whole salivary constituents in patients with Sjogren's syndrome. Eur J Oral Sci 1996; 104: 335-40.

[http://dx.doi.org/10.1111/j.1600-0722.1996.tb00088.x]

[49] Ryu OH, Atkinson JC, Hoehn GT, Illei GG, Hart TC. Identification of parotid salivary biomarkers in Sjogren's syndrome by surface-enhanced laser desorption/ionization time-of-flight mass spectrometry and two-dimensional difference gel electrophoresis. Rheumatology 2006; 45: 1077-86. [http://dx.doi.org/10.1093/rheumatology/kei212]

[50] Hu S, Vissink A, Arellano M, *et al.* Identification of autoantibody biomarkers for primary Sjogren's syndrome using protein microarrays. Proteomics 2011; 11: 1499-507. [http://dx.doi.org/10.1002/pmic.201000206]

[51] Hu S, Gao K, Pollard R, *et al.* Preclinical validation of salivary biomarkers for primary Sjogren's Syndrome. Arthritis Care Res (Hoboken) 2010; 62: 1633-8. [http://dx.doi.org/10.1002/acr.20289]

[52] Mitchell JS, Lowe TE, Ingram JR. Rapid ultrasensitive measurement of salivary cortisol using nano-linker chemistry coupled with surface plasmon resonance detection. Analyst (Lond) 2009; 134: 380-6. [http://dx.doi.org/10.1039/B817083P]

[53] Trilck M, Flitsch J, Ludecke DK, Jung R, Petersenn S. Salivary cortisol measurement-a reliable method for the diagnosis of Cushing's syndrome. Exp Clin Endocrinol Diabetes 2005; 113: 225-30. [http://dx.doi.org/10.1055/s-2005-837667]

[54] Touitou Y, Auzeby A, Camus F, Djeridane Y. Daily profiles of salivary and urinary melatonin and steroids in healthy prepubertal boys. J Pediatr Endocrinol Metab 2009; 22: 1009-15. [http://dx.doi.org/10.1515/JPEM.2009.22.11.1009]

[55] Arregger AL, Contreras LN, Tumilasci OR, Aquilano DR, Cardoso EM. Salivary testosterone: a reliable approach to the diagnosis of male hypogonadism. Clin Endocrinol (Oxf) 2007; 67: 656-62. [http://dx.doi.org/10.1111/j.1365-2265.2007.02937.x]

[56] Gandara BK, Leresche L, Mancl L. Patterns of salivary estradiol and progesterone across the menstrual cycle. Ann N Y Acad Sci 2007; 1098: 446-50. [http://dx.doi.org/10.1196/annals.1384.022]

[57] Malon RSP, Sadir S, Balakrishnan M, Córcoles EP. Saliva-based biosensors: noninvasive monitoring tool for clinical diagnosis Biomed Res Int 2014; 2014: 1-20. [http://dx.doi.org/10.1155/2014/962903]

[58] Wu J-Y, Yi C, Chung H-R, *et al.* Potential biomarkers in saliva for oral squamous cell carcinoma. Oral Oncol 2014; 46: 226-31. [http://dx.doi.org/10.1016/j.oraloncology.2010.01.007]

[59] Streckfus C, Bigler L, Tucci M, Thigpen JT. A preliminary study of CA15-3, c-erbB-2, epidermal growth factor receptor, cathepsin-D, and p53 in saliva among women with breast carcinoma. Cancer Invest 2000; 18: 101-9. [http://dx.doi.org/10.3109/07357900009038240]

[60] Tavassoli M, Brunel N, Maher R, Johnson NW, Soussi T. p53 antibodies in the saliva of patients with squamous cell carcinoma of the oral cavity. Int J Cancer 1998; 78: 390-1. [http://dx.doi.org/10.1002/(SICI)1097-0215(19981029)78:3<390::AID-IJC23>3.0.CO;2-9]

[61] Mager DL, Haffajee AD, Devlin PM, Norris CM, Posner MR, Goodson JM. The salivary microbiota as a diagnostic indicator of oral cancer: a descriptive, non-randomized study of cancer-free and oral squamous cell carcinoma subjects. J Transl Med 2005; 3: 1-8. [http://dx.doi.org/10.1186/1479-5876-3-27]

[62] Arellano-Garcia ME, Hu S, Wang J, *et al.* Multiplexed immunobead-based assay for detection of oral cancer protein biomarkers in saliva. Oral Dis 2008; 14: 705-12. [http://dx.doi.org/10.1111/j.1601-0825.2008.01488.x]

[63] Balicki R, Grabowska SZ, Citko A. Salivary epidermal growth factor in oral cavity cancer. Oral Oncol 2005; 41: 48-55.

[http://dx.doi.org/10.1016/j.oraloncology.2004.06.004]

[64] Pickering V, Jordan RC, Schmidt BL. Elevated salivary endothelin levels in oral cancer patients-a pilot study. Oral Oncol 2007; 43: 37-41.
[http://dx.doi.org/10.1016/j.oraloncology.2005.12.027]

[65] Hu S, Arellano M, Boontheung P, *et al.* Salivary proteomics for oral cancer biomarker discovery. Clin Cancer Res 2008; 14: 6246-52.
[http://dx.doi.org/10.1158/1078-0432.CCR-07-5037]

[66] Nagler RM. Saliva as a tool for oral cancer diagnosis and prognosis. Oral Oncol 2009; 45: 1006-10.
[http://dx.doi.org/10.1016/j.oraloncology.2009.07.005]

[67] Majem B, Rigau M, Reventós J, Wong D. Non-coding RNAs in saliva: emerging biomarkers for molecular diagnostics. Int J Mol Sci 2015; 16: 8676-98.
[http://dx.doi.org/10.3390/ijms16048676]

[68] Lenander-Lumikari M, Loimaranta V. Saliva and dental caries. Adv Dent Res 2000; 14: 40-7.
[http://dx.doi.org/10.1177/08959374000140010601]

[69] Llena-Puy C. The role of saliva in maintaining oral health and as an aid to diagnosis. Med Oral Patol Oral Cir Bucal 2006; 11: E449-55.

[70] Armitage GC. Diagnosis of periodontal diseases. J Periodontol 2003; 74: 1237-47.
[http://dx.doi.org/10.1902/jop.2003.74.8.1237]

[71] Christodoulides N, Mohanty S, Miller CS, *et al.* Application of microchip assay system for the measurement of C-reactive protein in human saliva. Lab Chip 2005; 5: 261-9.
[http://dx.doi.org/10.1039/b414194f]

[72] Janeway CA, Yagi J, Rojo J, *et al.* Immune recognition and effector function in subsets of CD4 T cells. Princess Takamatsu Symp 1988; 19: 193-208.

[73] Fine DH, Markowitz K, Furgang D, *et al.* Macrophage inflammatory protein-1alpha: a salivary biomarker of bone loss in a longitudinal cohort study of children at risk for aggressive periodontal disease? J Periodontol 2009; 80: 106-13.
[http://dx.doi.org/10.1902/jop.2009.080296]

[74] Ng PY, Donley M, Hausmann E, Hutson AD, Rossomando EF, Scannapieco FA. Candidate salivary biomarkers associated with alveolar bone loss: cross-sectional and *in vitro* studies. FEMS Immunol Med Microbiol 2007; 49: 252-60.
[http://dx.doi.org/10.1111/j.1574-695X.2006.00187.x]

[75] Gorska R, Nedzi-Gora M. The effects of the initial treatment phase and of adjunctive low-dose doxycycline therapy on clinical parameters and MMP-8, MMP-9, and TIMP-1 levels in the saliva and peripheral blood of patients with chronic periodontitis. Arch Immunol Ther Exp (Warsz) 2006; 54: 419-26.
[http://dx.doi.org/10.1007/s00005-006-0047-6]

[76] Totan A, Greabu M, Totan C, Spinu T. Salivary aspartate aminotransferase, alanine aminotransferase and alkaline phosphatase: possible markers in periodontal diseases? Clin Chem Lab Med 2006; 44: 612-5.
[http://dx.doi.org/10.1515/CCLM.2006.096]

[77] Scannapieco FA, Ng P, Hovey K, Hausmann E, Hutson A, Wactawski-Wende J. Salivary biomarkers associated with alveolar bone loss. Ann N Y Acad Sci 2007; 1098: 496-7.
[http://dx.doi.org/10.1196/annals.1384.034]

[78] Khader YS, Albashaireh ZS, Alomari MA. Periodontal diseases and the risk of coronary heart and cerebrovascular diseases: a meta-analysis. J Periodontol 2004; 75: 1046-53.
[http://dx.doi.org/10.1902/jop.2004.75.8.1046]

[79] Yan W, Apweiler R, Balgley BM, *et al.* Systematic comparison of the human saliva and plasma

proteomes. Proteomics Clin Appl 2009; 3: 116-34.
[http://dx.doi.org/10.1002/prca.200800140]

[80] Hu S, Loo JA, Wong DT. Human saliva proteome analysis and disease biomarker discovery. Expert Rev Proteomics 2007; 4: 531-8.
[http://dx.doi.org/10.1586/14789450.4.4.531]

[81] Zimmermann BG, Wong DT. Salivary mRNA targets for cancer diagnostics. Oral Oncol 2008; 44: 425-9.
[http://dx.doi.org/10.1016/j.oraloncology.2007.09.009]

[82] Hu Z, Zimmermann BG, Zhou H, *et al.* Exon-level expression profiling: a comprehensive transcriptome analysis of oral fluids. Clin Chem 2008; 54: 824-32.
[http://dx.doi.org/10.1373/clinchem.2007.096164]

[83] Lee JM, Garon E, Wong DT. Salivary diagnostics. Orthod Craniofac Res 2009; 12: 206-11.
[http://dx.doi.org/10.1111/j.1601-6343.2009.01454.x]

[84] Li Y, Zhou X, St John MA, Wong DT. RNA profiling of cell-free saliva using microarray technology. J Dent Res 2004; 83: 199-203.
[http://dx.doi.org/10.1177/154405910408300303]

[85] Li Y, St-John MA, Zhou X, *et al.* Salivary transcriptome diagnostics for oral cancer detection. Clin Cancer Res 2004; 10: 8442-50.
[http://dx.doi.org/10.1158/1078-0432.CCR-04-1167]

[86] St John MA, Li Y, Zhou X, *et al.* Interleukin 6 and interleukin 8 as potential biomarkers for oral cavity and oropharyngeal squamous cell carcinoma. Arch Otolaryngol Head Neck Surg 2004; 130: 929-35.
[http://dx.doi.org/10.1001/archotol.130.8.929]

[87] Nussbaumer C, Gharehbaghi-Schnell E, Korschineck I. Messenger RNA profiling: a novel method for body fluid identification by real-time PCR. Forensic Sci Int 2006; 157: 181-6.
[http://dx.doi.org/10.1016/j.forsciint.2005.10.009]

[88] Gomes MA, Rodrigues FH, Afonso-Cardoso SR, *et al.* Levels of immunoglobulin A1 and messenger RNA for interferon gamma and tumor necrosis factor alpha in total saliva from patients with diabetes mellitus type 2 with chronic periodontal disease. J Periodontal Res 2006; 41: 177-83.
[http://dx.doi.org/10.1111/j.1600-0765.2005.00851.x]

[89] Hu S, Wang J, Meijer J, *et al.* Salivary proteomic and genomic biomarkers for primary Sjogren's syndrome. Arthritis Rheum 2007; 56: 3588-600.
[http://dx.doi.org/10.1002/art.22954]

[90] Bahn JH, Zhang Q, Li F, *et al.* The landscape of microRNA, Piwi-interacting RNA, and circular RNA in human saliva. Clin Chem 2015; 61: 221-30.
[http://dx.doi.org/10.1373/clinchem.2014.230433]

[91] Spielmann N, Ilsley D, Gu J, *et al.* The human salivary RNA transcriptome revealed by massively parallel sequencing. Clin Chem 2012; 58: 1314-21.
[http://dx.doi.org/10.1373/clinchem.2011.176941]

[92] Chang T-C, Mendell JT. MicroRNAs in vertebrate physiology and human disease. Annu Rev Genomics Hum Genet 2007; 8: 215-39.
[http://dx.doi.org/10.1146/annurev.genom.8.080706.092351]

[93] Chang KW, Liu CJ, Chu TH, *et al.* Association between high miR-211 microRNA expression and the poor prognosis of oral carcinoma. J Dent Res 2008; 87: 1063-8.
[http://dx.doi.org/10.1177/154405910808701116]

[94] Hizir MS, Balcioglu M, Rana M, Robertson NM, Yigit MV. Simultaneous detection of circulating oncomiRs from body fluids for prostate cancer staging using nanographene oxide. ACS Appl Mater Interfaces 2014; 6: 14772-8.

[95] Dettmer K, Hammock BD. Metabolomics-a new exciting field within the "omics" sciences. Environ Health Perspect 2004; 112: A396-7.

[96] Sugimoto M, Wong DT, Hirayama A, Soga T, Tomita M. Capillary electrophoresis mass spectrometry-based saliva metabolomics identified oral, breast and pancreatic cancer-specific profiles. Metabolomics 2010; 6: 78-95.
[http://dx.doi.org/10.1007/s11306-009-0178-y]

[97] Wei J, Xie G, Zhou Z, *et al.* Salivary metabolite signatures of oral cancer and leukoplakia. Int J Cancer 2011; 129: 2207-17.
[http://dx.doi.org/10.1002/ijc.25881]

[98] Dewhirst FE, Chen T, Izard J, *et al.* The human oral microbiome. J Bacteriol 2010; 192: 5002-17.
[http://dx.doi.org/10.1128/JB.00542-10]

[99] Pushalkar S, Ji X, Li Y, *et al.* Comparison of oral microbiota in tumor and non-tumor tissues of patients with oral squamous cell carcinoma. BMC Microbiol 2012; 12: 1-35.
[http://dx.doi.org/10.1186/1471-2180-12-144]

[100] Huyghe A, Francois P, Charbonnier Y, *et al.* Novel microarray design strategy to study complex bacterial communities. Appl Environ Microbiol 2008; 74: 1876-85.
[http://dx.doi.org/10.1128/AEM.01722-07]

[101] Marur S, D'Souza G, Westra WH, Forastiere AA. HPV-associated head and neck cancer: a virus-related cancer epidemic. Lancet 2010; 11: 781-9.
[http://dx.doi.org/10.1016/S1470-2045(10)70017-6]

CHAPTER 16

Endocrine Disruptors: What Do We Know about the Effects and Risk Factors in Humans?

Africa de Béjar-Almira[1], **Alice Charlotte Viney**[2] and **Marta M. Castañeda San Cirilo**[1,*]

[1] *Clinical Analysis Department, Santa Lucia General University Hospital, Cartagena, Spain*

[2] *Pharmacy Department, Santa Lucia General University Hospital, Cartagena, Spain*

Abstract: In the last few decades, the theme of endocrine disruption and endocrine disrupting chemicals has become more and more relevant, raising alarm in the scientific community and amongst public health officials, as well as questioning the traditional concepts of toxicology. This chapter is a revision and summary of the latest information and articles concerning this subject. It discusses the terms and definitions of endocrine disruption, the sources and mechanisms of action of these chemicals both naturally and artificially made, reviewing the potential adverse development, reproductive, neurological, and immune effects that they can have on the different systems in the body, and how they can relate to the different epidemics of cancer, diabetes and obesity. It also provides the latest up-to-date information on the tests and methods used to detect and assess these entities, offering ideas and lines of future investigation to attempt to better understand the knowledge in this field and reduce the potential threat of these environmental hazards to mankind.

Keywords: Agonist, Antagonist, BPA, Cancer, DDT, Diabetes, Dioxins, Endocrine disruptor, Environment, Estrogen receptor, Half-live, Health, Hormone system, Obesity, PCBs, Pesticides, Phthalates, Reproductive system, Stability, Thyroid.

INTRODUCTION

In 1991, at a Wingspread work session on endocrine disrupting chemicals (EDCs), a group of expert scientists concluded that "Many compounds introduced into the environment by human activity are capable of disrupting the endocrine system of animal kingdom. These compounds play a crucial role in controlling development" [1]. From that moment, many studies have been conducted assessing the effects of these compounds on human health. There are many side

* **Corresponding author Marta M. Castañeda San Cirilo:** Clinical Analysis Department, Santa Lucia General University Hospital, Cartagena, Spain; Tel/Fax: (+34) 968128600; E-mail: martacastasan@yahoo.es

Juan Antonio Vílchez (Ed.)

effects attributed to EDCs, including alterations of the reproductive system, obesity, type II diabetes and cancer. The catalogue of endocrine disruptors is very broad, varying from synthetic chemicals to natural substances found in the environment, and is growing every day.

For this reason, the study of endocrine disruptors, their effects, mechanisms of action and forms of detection, are becoming increasingly important subjects on which many expert groups are conducting research in order to reduce the adverse health effects.

What are Endocrine Disruptors?

The term endocrine disruptor and the English acronym "Endocrine Disruptor Chemical (EDC)", were coined during the Wingspread Conference in 1991. A group of experts in the field met to assess the causes of the adverse effects observed in epidemiological studies of people and wildlife, including reproductive harm, immune system and hormone-dependent cancers among others. In this moment a novel category of potentially hazardous substances were emerged, although the exact definitions giving the meaning of these terms is still debated and somewhat unclear [2]. The most commonly agreed are the following: An EDC is "an exogenous substance that causes adverse health effects in the intact organism or its progeny, secondary/consequent to changes in endocrine function" [3]. After sorting, many others came later with minor modifications. The most recent definition issued by the EU is that of EDC is "an exogenous substance that causes adverse health effects in an intact organism, or its progeny, secondary to changes in endocrine function" [4]. Therefore, only a substance that produces toxicity in an intact organism *via* a hormonal or hormone-like mechanism represents a genuine EDC [5].

EDCs are highly heterogeneous [6, 7] and can be classified in the following ways: Natural or synthetic compounds [8], or according to their origin [9]. But the most recent classification was developed by Gore [10], who categorized EDCs into three groups based on occurrence:

Pesticides

Dichlorodiphenyltrichloroethane (DDT) and chlorpyrifos are the most commonly used pesticides. These compounds interfere in the neural and reproductive systems of animals, but can also affect human body due to physiological similarities [10]. It has been found that exposure to DDT can interfere with many organs and systems of the human body. Hormonal, metabolic, reproductive systems among others. However, endocrine disruptor are still in use in some countries, for controlling insects that transmit human diseases such as malaria, leishmaniosis, dengue and Chagas disease, frecuently in Africa and India. Illegal diversion to agriculture also continues to occur in other countries [11 - 14]. Studies in South Africa show that the adult population living in homes sprayed with DDT have an average blood DDT concentration of just under 100 µg/g lipid, compared to less than 10 µg/g lipid in people living in nearby communities without DDT spraying [15, 16].

Chemicals in Products

We can find EDCs in many personal and everyday products that are in contact with our body and in our home. For example, children's products, electronics, food contact materials, personal care products, textile/clothing and building products. Generally, consumers have no information about the presence of these compounds. Is important to know that some of these chemicals are released into the atmosphere and can stay in the environment for a long time [10]. For example, plumb is a natural compound that is being used regularly in mining, refining and painting as well as in the making of jewellery and children's products. Young children represent another vulnerable subpopulation because in children:

1. His weight is lower, therefore it accumulates more lead.
2. More dust may be ingested.
3. Lead absorption in the gastrointestinal tract is higher.
4. The blood–brain barrier is not yet fully developed.
5. Lower level of lead are needed to cause neurological alterations [17].

In addition, in women, exposure to lead may cause alterations in the hormones of the reproductive system [18, 19].

Food Contact Materials

Bisphenol A (BPA) is the most commonly used. BPA was found in plastic based containers and also in the epoxy based lining of canned food. BPA is one of the highest-volume chemicals produced and global production is predicted to exceed 5.4 million metric tons by 2015. The exposure appears to be universal; a study conducted by the Centers for Disease Control estimated that between 2003-2004 more than 96% of Americans had BPA in their bodies [20].

Table 1. Common EDCs and their uses [10, 21].

Common EDCs used in our daily life	Uses
DDT, chlorpyrifos, atrazine, 2,4-dichlorophenoxyacetic acid, glyphosate	Pesticides

(Table 1) contd.....

Common EDCs used in our daily life	Uses
Lead, phthalates, cadmium	Children's products
BPA, phthalates, phenol	Food contact materials
Brominated flame retardants, Polychlorinated biphenyls (PCBs)	Electronics and building materials
Phthalates	Personal care products, medical tubing
Triclosan	Antibacterials
Perfluorochemicals	Textiles, clothing
Parabens, phthalates, glycol ethers, fragrances, cyclosiloxanes	Cosmetics, personal care products, cleaners
Tributyltin	Antifoulants used to paint the bottom of ships
Nonylphenol (alkylphenols)	Surfactant detergents, paints, pesticides, personal care products, plastics
Ethinyl estradiol (Synthetic steroid)	Contraceptive

History of Endocrine Disruption

The ability of synthetic chemical substances interfering with the human hormone system has been known since the 40s, when the drug Diethylstilbestrol (DES) was used to prevent spontaneous abortions. In 1971, DES was shown to cause clear-cell-carcinoma, a rare vaginal tumour in girls and women who had been exposed to this drug in the uterus. The United States Food and Drug Administration subsequently withdrew DES from use in pregnant women [22]. In the United States, the production, export and import of PBCs, stopped in the decade of 70, on suspicion of its harmful effects. The general use of DDT was banned by the United States Environmental Protection Agency (U.S. EPA) in 1972 because it had posed unacceptable risks to the environment and potential harm to human health [23].

In 1976, Lindner *et al.* [24] confirmed the existence of over 40 species of plants that contained substances showing active for estrogenic activity in biological assays. For example, sheep infertility caused by grazing on pastures with clover. Today, we know that this is due to coumestrol and daidzein, natural constituents of clover and natural contraceptives.

Gradually, the interest of the effects of these substances on public health is increasing. However, it would be at the Wingspread Conference Center, Racine, Wisconsin in July 1991 organized by Theo Colborn and co-workers that the study of endocrine disruptors would begin. From that moment on, the number of publications related to endocrine disrupters has been constantly rising.

The interest of both public and private institutions for the study of EDCs is increasing significantly. Currently, tests are being conducted to recognize endocrine disruptors. On November 4th, 2010, the US EPA´s Endocrine Disruptor Screening Program released its draft Guidance Document, "Weight-of-Evidence" (WoE). The Agency stated in this document that it would use WoE to determine whether a chemical has the potential to interact with the estrogen, androgen or thyroid hormone components of the endocrine system [25]. Many other organizations like the US EPA are also trying to research and identify endocrine disruptors.

Role of Edcs in the Environment

In our daily life, we are in contact with a wide range of EDCs. Endocrine disruptors are known to work mainly by altering hormonal and homeostatic systems of a living organism. These systems are involved in the regulation of several significant processes like metabolism, sexual development, growth, stress response, gender behaviour, reproduction and even in the fetal development of a living body [23].

In 2009, Somm *et al.* [26] studied prenatal exposure of mice to BPA. They found that such exposure caused accelerated puberty in the descendants, increased body weight, altered mammary glands and altered the female genital tract in male mice. Later in 2013, another group also observed in similar studies decreased fertility and fecundity, masculization of behavior and brain structures in female mice and decreased fertility in male offspring [27].

Some EDCs are able to persist in the environment for a long time being very difficult their degradation, while others, sunlight or bacterial action are enough for remove them [28]. This fact makes the study of these entities even more complicated and increases their association with health effects. In addition, the exposure in many cases is not only to one compound, but usually to a mixture of these.

There are many complications in the study of endocrine disruptors, but this field of investigation is gradually broadening. There are various groups working on this subject, with the aim of discovering what they are and how they work to minimize their effects.

Source and Stability of EDCs

Hundreds of individual anthropogenic and natural chemicals are known or suspected to interact with endocrine systems in humans and wildlife [28]. The sources of EDC exposure are usually diverse and widely distributed all over the world. But the situation is complicated since there is a significant variation in the use of these substances among the different countries. Some of these chemicals have been banned in few countries while others have not. In the case of some EDCs such as steroid hormones, synthetic steroids, and polychlorinated biphenyls their toxicities are obvious, but others like alkyl phenol ethoxylate or gonadotropin compounds have not been discussed as much in detail.

The sources of EDCs are many and varied, and can be found naturally or industrially made. An example of a source is the insufficient treatment of sewage waters. Incomplete elimination of these compounds may result in their release into waterways such as lakes or streams, that can have toxic effects on the environment even though they are present in low concentrations [29 - 32]. Researchers have documented that chemical mixtures in some waste waster effluents can result in feminization of fish populations downstream of treatment plants [23]. Another source is the sex steroids like estradiol, progesterone and testosterone used in agricultural practice for many years. These have growth promoting effects in animals and humans, but they also have significant polluting effects on soil, water and air. Kim *et al.* [33] conducted studies on rats with a synthetic insecticide used worldwide in agriculture, tetramethrin. The results obtained suggested that tetramethrin might exert endocrine disrupting effects on female rats through antiestrogenic actions.

Different studies show an alarming risk of EDC exposure mainly from industries and raw sewage. Fig. (**1**) shows some of these routes of exposure.

Endocrine disruption has been a cause for concern for many years because of the hazardous nature of these entities on the environment. EDCs can persist over long periods of time [11], and low amounts of these can be potentially dangerous for us human beings. For this reason, it is very important to know the stability of these compounds and most importantly, aim to completely remove endocrine disruptors from our surroundings. The stability of EDCs largely depends on their persistent nature in the environment. Some are degraded rapidly by sunlight, bacteria and other chemical processes, whilst others can persevere for months or even for years [28]. For example:

Dioxins such as dibenzodioxins (PCDDs) and dibenzofurans (PCDFs): These compounds are poorly biodegradable. Due to their high stability and lipid solubility they can be found in the food chain. Dioxins are present at trace levels in a variety of foods of animal origin.

Polychlorinated biphenyls (PCBs): They are very persistent compounds that are also incorporated into the food chain. Production is now banned in most developed countries. Special conditions for their destruction have also been set up.

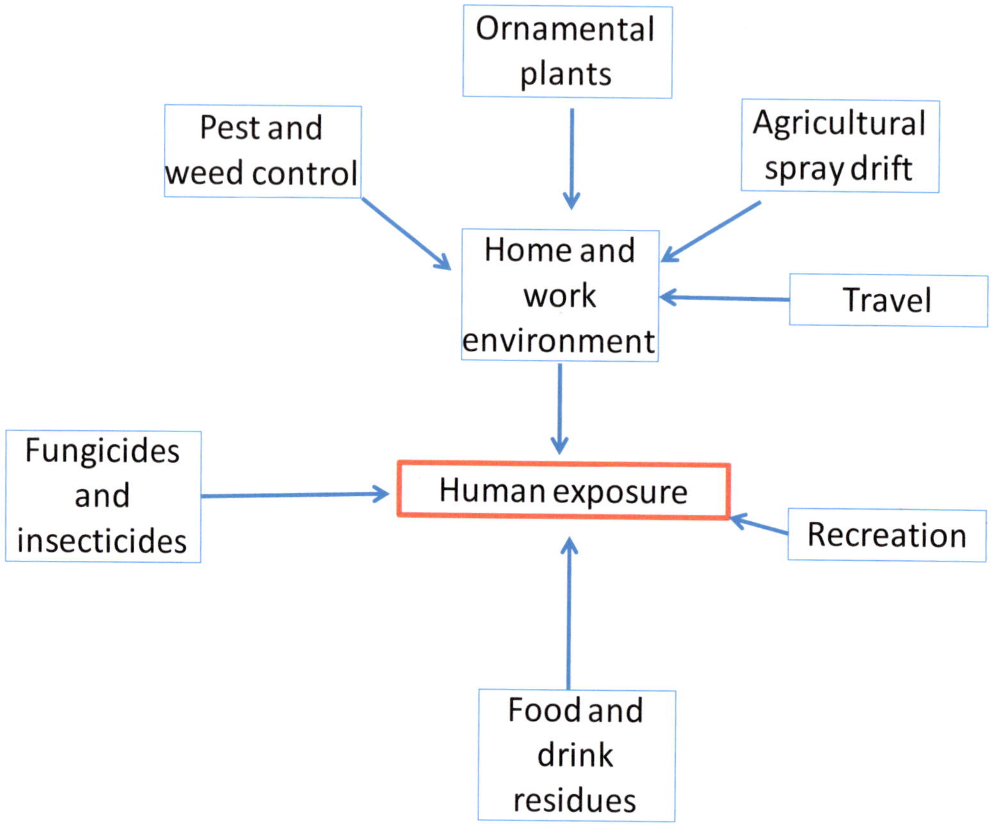

Fig. (1). Different exposure routes for human EDCs. Modified from Cristina *et al.* (2012) [34].

Di (2-ethylhexyl) phthalate (DEHP): These compounds are quickly transformed by biota undergoing metabolic processes.

Atrazine: It is not easily absorbed into the ground, but is very stable in cold environments. It undergoes metabolic processes such as hydroxylation and dialkylation.

The stability of EDCs in the environment remains a major concern. As a result, new techniques are being investigated to solve this problem.

Rosenfeldt *et al.* [35] have investigated the degradation of three EDCs; bisphenol A, ethinyl estradiol and estradiol *via* ultraviolet (UV) radiation photolysis and the UV/hydrogen peroxide advanced oxidation process (AOP).

Important Issues in Endocrine Disruption

A number of issues must be taken into account to analyze the possible effects of EDCs.

Firstly, the age of exposure is very important. The surroundings of a developing organism, including exposure to EDCs, interact with the genes of the individual to determine the individual's propensity to develop a disease or dysfunction later in life [23]. The effects of EDCs can be very different in adults, infants or growing fetus. Exposure at an early age can be potentially more harmful, being that the tissues and organs are not yet fully developed. Research from NIEHS (National Institute of Environmental Health Sciences) shown that the alterations caused by EDCs go from one generation to the next, even though they were not directly exposed to the endocrine disruptors. The mechanisms might involve altering gene function without altering DNA sequence [36].

Another aspect to consider is the latency period. The time from exposure to EDCs until the appearance of clinical symptoms can vary considerably. This makes the association between exposure to endocrine disruptors and different diseases even more difficult [8].

Low-Dose effects of EDCs have also been defined. In studies aiming to identify the levels needed to produce disease, the concentrations in the human body were found to be a lot lower than expected. In the biomonitoring campaign conducted between 2007-2011, it was demonstrated that low levels of cadmium and organochlorine were associated with detectable changes in serum hormone levels, sex maturation, height or in body mass index [37, 38].

It is also important to note the routes by which endocrine disruptors can enter the human body. Polyzos *et al.* [39] saw that adults were primarily in contact with EDCs through ingestion of drinking water, meat, fat-dairy products and by the inhalation of contaminated air. Babies are contaminated with EDCs through breastfeeding, contact with baby products and by inhaling polluted air [39]. Table **2** summarizes the various routes of entry of endocrine disruptors in the human body [10].

Table 2. Examples of EDC routes of exposure in humans.

Pathways of exposure to EDCs.	Sources of EDCs.	EDC example(s)
Oral consumption of contaminated food and/or water	Industrial waste or pesticides contaminating soil or ground- water	PCBs, dioxins, per fluorinated compounds, DDT

(Table 2) contd.....

Pathways of exposure to EDCs.	Sources of EDCs.	EDC example(s)
Oral consumption of contaminated food and/or water	Leaching of chemicals from food or beverage containers; pesticide residues in food or beverages	BPA, phthalates, chlorpyrifos, DDT
Contact with skin and/ or inhalation	Household furniture treated with flame retardants	Brominated flame retardants (BFRs)
Contact with skin and/ or inhalation	Pesticides used in agriculture, homes or for public disease vector control	DDT, chlorpyrifos, vinclozolin, pyrethroids
Intravenous	Intravenous tubing	Phthalates
Skin application	Cosmetics, personal care products, antibacterial agents, sunscreens, *etc.*	Phthalates, triclosan, parabens, insect repellents.
Biological transfer from placenta	Maternal body burden due to prior/current exposures	Numerous EDCs can cross the placenta
Biological transfer from mother's milk	Maternal body burden due to prior/current exposures	Numerous EDCs are detected in milk

Another point to consider is the lack of standardization in the determination of levels of endocrine disruptors. This makes it impossible to do detailed comparisons between studies or assessments of effects.

Currently there is great demand in developing new techniques that allow us to analyze the broad spectrum of endocrine disruptors. Chromatography is only capable of identifying a low number of these compounds. The use of effects directed analysis (EDA) and quantitative structure-activity relationship (QSAR) directed non-target analyses, are two techniques that are currently being used to address this challenge [40].

ENDOCRINE DISRUPTION IN THE HUMAN BODY

Hormones are synthesized, secreted and transported *via* the bloodstream to their target cells or sites of action. They then interact with specific receptors initiating a series of signaling processes that determine subsequent biological effects. In addition, processes related to hormones can also be regulated. In each of these steps; production, action and regulation, EDCs can interfere and mimic hormones, resulting in conformational changes and differences in the functional activity and regulation of gene expression [41].

A decade ago, the focus of concern was the action of EDCs as agonists and antagonists of hormone receptors, in particular, estrogen receptor agonists (ER). Many compounds with estrogenic activity were detected by the transfection of ER in different cell lines [42].

However, in recent years, the scope of EDCs has expanded considerably. There is more concern about EDCs affecting other hormonal systems (such as the thyroid) and EDCs with new modes of action. There is also new evidence claiming that other common environmental chemicals have multiple potential hormone-modifying effects (*e.g.* PCBs) [42].

But lately the main issue is to try to clarify the effects caused by the mixture of endocrine disruptors. Different studies have shown that these products have more significant effects when appearing together than they do individually. These finding raise the question of how to evaluate the effects of mixtures of different EDCs [42].

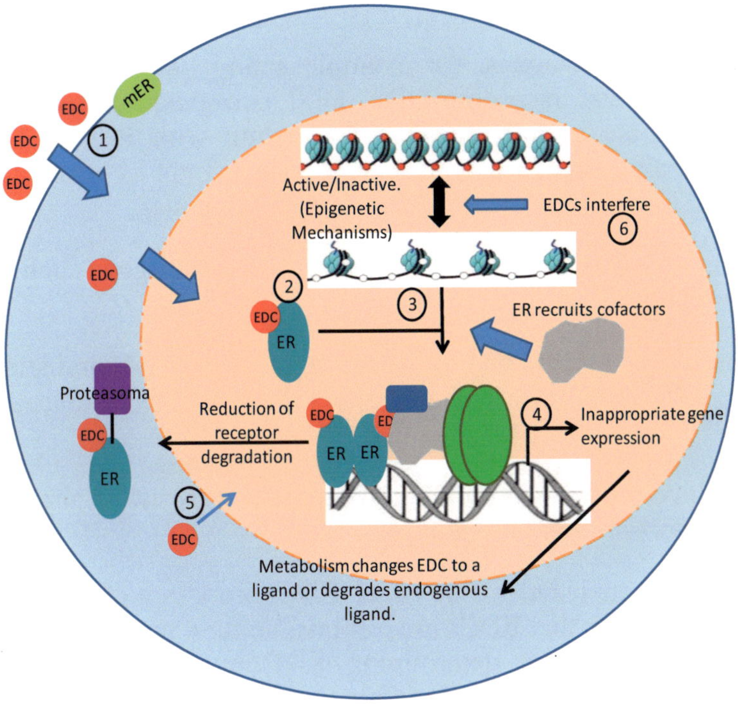

Fig. (2). Shows a variety of mechanisms of action of EDCs. EDCs arrive at a cell membrane, and they may bind to a membrane estrogen receptor (mER) (1). Or pass through the membrane and bind to a nuclear estrogen receptor (ER) in the nucleus (2). When the complex of EDC and ER binds to a gene containing an estrogen-responsive element (3). It recruits molecules that help cause gene expression by boosting gene transcription by the RNA polymerase complex, and so may cause gene expression at inappropriate times (4). EDC/ER complexes can also bind to proteasomes, which can lead to a reduction of the normal process of degradation of ER (5). EDCs may cause methylation of DNA or deacetylation of histone both of which reduce gene expression. Alternatively, EDCs may cause demethylation of DNA or, by inhibiting the enzyme histone deacetylase. The second two effects both induce gene expression (6). Modified from Iguchi T *et al.* (2008) [43].

Mechanisms of Action

To understand endocrine disruption, we must have a clear concept about the endocrine system. It consists of a set of organs and tissues that interact with each other, through hormones. It is responsible for controlling a number of processes in the body. EDCs exhibit the same characteristics as hormones. They interfere in different ways with the production and activity of hormones, and in doing so, can alter endocrine function leading to adverse effects on human health [23].

Understanding the mechanisms and modes of action of different endocrine disruptors is one of the priorities of research in this field (Fig. **2**).

Thus, there are substances able to:

Mimic the action of hormones, for example acting like estrogens, termed as environmental estrogens, including DDT. DDT is organochlorine pesticide that can exhibit estrogenic activity through interaction with ERα and ERβ. This interaction adversely affects the female reproductive tract by stimulating uterine proliferation and impairing normal follicle development [44].

Antagonize the action of hormones, for example antiestrogens such as PCBs or vinclozolin fungicide.

Alter the pattern of synthesis and metabolism, such as polybrominated diphenyl ethers (PBDE-99) (flame retardant) that alters the synthesis of thyroid hormones (THs).

Modulate the levels of corresponding receptors, such as bisphenol A which interferes with the estrogen receptor [45].

Though we know the mechanisms by which EDCs work in biological systems, we cannot say with certainty that EDCs are responsible for a particular disease. There are numerous challenges in determining EDC involvement in a particular disorder; unique exposure rates that different people have to a variety of known and unknown EDCs, individual differences in metabolism and body composition, as well as differences in genetic polymorphisms. Chronic exposure to EDCs is likely to be the cause of disorders in the human body. But this is difficult to study, since the time from exposure until the appearance of the symptomatology is very long, which makes is difficult to associated in many cases [8].

Biotransformation

The exact process of biotransformation that endocrine disruptors go through in the human body is still unknown today. However, we do know that during this process, the intermediary metabolites have stronger disruption properties than the originals compounds, and can persist over longer periods of time in different tissues [28].

Several studies on biotransformation of various endocrine disruptors have been extensively conducted in rats and mice, both *in vivo* and *in vitro*. For example, it was observed that the major metabolite of 1,2-dichloroethane is glutathione conjugate, which is ultimately eliminated from the body as a nonvolatile urinary metabolite. A study with mice demonstrated another transformation route of 1,2-dicloroethane through microsomal hydroxylation obtaining cloroacetaldehyde. This pathway shows the involvement of two enzymes in the oxidation process, microsomal cytochrome P-450 and microsomal mixed-function oxygenase (MFO). Chloroacetaldehyde can then be reduced further to chloroethanol or oxidized to chloroacetic acid [46].

Excretion

Endocrine disruptors can be persistent or non persistent. If the substance is non persistent, it is usually predicted to be metabolized by the liver, then finally eliminated from the body *via* urine and feces. On the other hand, persistent endocrine disruptors are accumulated in different parts of the body, especially in fats and are released slowly. Scientists have observed that breastfeeding can also be a form of excretion of endocrine disruptors from mother to child [28].

Effects on the Human Body

Exposure to endocrine disruptors has been seen to affect many parts of the human body. Epidemiological studies, clinical observations and other experiments conducted on animals have indicated the potential role of endocrine disruptors in affecting the prostate, breast, lung, liver, thyroid and reproductive system as well as metabolism and causing obesity [39]. Epidemiological data shows increased evidence in the incidence and prevalence of some diseases associated with endocrine disrupting chemicals, but no proof of being directly caused by exposure to EDCs. The study of association between exposure to EDCs and the risk of disease, as well as the assessment of exposure is very difficult.

The Thyroid Hormone System

The thyroid hormone has many essential roles in human physiology. Therefore, adverse effects in development, metabolism or adult physiology may occur if there are changes in the function of the thyroid gland or interference with the ability of the hormone to exert its actions [8]. Some environmental chemicals identified as thyroid disruptors include PCBs, bisphenol A, perchlorate, tetrachlorodibenzo-p-dioxin (TCDD) and poly-chlorinated dibenzofuran (PCDF). There is also existing evidence that parabens used in cosmetics and pesticides like DDT, also have thyroid disrupting effects in animals and humans [47]. Thyroid disruptors can affect thyroid physiology in many phases of its regulation. (Table **3**).

Table 3. Mechanisms and effects of thyroid disruptors [47].

Thyroid disruptors	Mechanism	Effect
Perchlorate, thiocyanate, nitrate, bromates, phthalates	Blocking uptake of iodide into thyroid cell	Decreased synthesis of T3 and T4
Methimazole, amitrole, isoflavones, benzophenone	Blocking production of TPO in thyroid follicles.	Decreased synthesis of T3 and T4.
PCBs, pentachlorophenols, flame retardants, phthalates	Competitive binding to thyroid transport protein (TTR)	Possible effect on fetal brain T4 production
Dioxin, Polybrominated diphenyl ethers (PBDE), chlordane	Altering transport across cell membrane	Increased biliary elimination of T3 and T4
Acetochlor (herbicide), PCBs	Enhanced hepatic metabolism	Increased biliary metabolism of T3 and T4
PCBs, triclosan, pentachlorophenol, dioxin, difuran	Inhibition of sulfation	Decreased sulfation of thyroid hormones leading to possible decrease of peripheral T3 synthesis
PCBs, bisphenol A, hexachlorobenzene, flame retardants	Altering binding to thyroid receptor	Altered thyroid hormone directed gene transcription
DDT, PCBs	Inhibiting TSH receptor	Decreased production of T3 and T4

Chemicals that interfere with the sodium/iodide symporter (NIS) can also affect the thyroid hormone synthesis and exacerbate the iodine deficiency problem. A good example would be the case of perchlorate. As a chemical, it is stable in the environment and has become a worldwide contaminant of drinking and irrigating waters and a food contaminant. It has been observed in experimental studies in humans that the serum half life of perchlorate is about 8 h and that a dose of about 5.2 g/kg is sufficient to begin to reduce iodide uptake into the thyroid gland [8]. Thyroperoxidase (TPO) was also observed to be blocked by a number of compounds. An example is 6-propyl-2-thiouracil (PTU), a methylmercapto imidazole used therapeutically to treat patients with Graves disease.

The Nervous system

EDCs are able to cross the placental and blood brain barriers, but little is known about their deleterious actions in the early stages of neural development. Most experimental and epidemiological studies have demonstrated that EDCs affect the nervous system by interacting with the hypothalamus-pituitary-thyroid gland axis, which is essential for proper brain development [48].

It has been observed that the processes of neural progenitor proliferation, differentiation and migration as well as synaptogenesis and myelination are sensitive to EDCs. It has been demonstrated that neural progenitor cells express robust levels of the aryl hydrocarbon receptor (AhR), which is known to be responsible for dioxin and dioxin-like PCB intoxications [49]. Several studies have shown that embryonic stem cells are vulnerable to tetrachloro-dibenzo-p-dioxin (TCDD), benzo(a)pyrene [B(a)P], mono-(2-ethylhexyl) phthalate and bisphenol A [50, 51].

The fetal basis of adult-onset disease could be a result of epigenetic factors. Recently, EDCs and DNA methylation have been implicated in the etiology of neurological disorders [48]. In some circumstances, the epigenetic effects are exerted during in utero-exposure, while in others circumstances, the effects are transmitted across generations *via* incorporation into the germline cells. For example, exposure to the fungicide vinclozolin early in pregnancy is imprinted in the male lineage, resulting in anxiety behavior and unique patterns of gene expression in relevant brain regions [52].

It has been demonstrated that chronic exposure to pesticides, such as paraquat or rotenone, is associated with alterations in dopaminergic neurotransmission, which may result in neurodegenerative disorders. Exposure to pesticides has been cited as a potential risk factor for amyotrophic lateral sclerosis and Alzheimer's and Parkinson's diseases. Post-mortem studies showed a strong correlation between high concentrations of PCB congeners (PCB153, PCB180) and an increased ratio of Parkinson's disease-related pathologies, especially in females [53]. EDCs including bisphenol A, chlorpyrifos, rote-none and TCDD were found to stimulate oxidative stress and induce apoptosis and excitotoxicity [54 - 58].

Numerous other classes of EDCs have been linked to neural degenerations, including organochlorines and organophosphates, but epidemiological evidence is lacking.

The Reproductive System

The ability of chemicals to affect reproduction and development has garnered significant attention [59]. Studies suggest that one of the causes of declining fertility is exposure to these chemicals products [60]. The list of compounds involved is very long, ranging from pesticides, PCBs (polychlorinated biphenyls), dioxin, and the other types of anthropogenic chemicals.

There is epidemiological evidence that endocrine effects may be occurring in humans as a result of DDT exposure. DDT exposure damages the reproductive system, reduces reproductive success, semen quality, menstruation, gestational length and duration of lactation [61 - 63]. Given the long half-life of this molecule, up to 16 years, it would be important to study girls going into puberty and throughout the rest of their lifetime, who may be exposed to these chemicals regularly, to see the possible effects.

In newborn boys exposed to PCB, dioxin, furan and DDT, micropenis was observed, but no effects were reported in testicular size or development stages of Tanner [64].

Between 1949 and 1990, the sperm volume has decreased from 3.4 to 2.75 mL per ejaculate, and in that same period, from 113 to 66 million sperm/mL [65]. In addition, persistent organic pollutants have been seen to induce significant changes in sperm function. In studies conducted in 61 countries, organic pollutants were associated with a decreased sperm volume. Fig. (**3**) shows the mechanisms underlying the development of reproductive disorders in human male fetuses. Different classes of endocrine-disrupting chemicals (EDCs) inhibit the production of androgens and insulin-like factor 3 (Insl3) by fetal Leydig cells (FLCs) in the sensitive androgen-dependent period of male reproductive organ development. Under masculinization disturbs the proper formation and growth of the penis, increasing the risk of developing hypospadias and micropenis. A lack of Insl3 alters the normal testis descent, leading to cryptorchidism. EDCs can impair Sertoli cell (SC) differentiation from their progenitors and alter the proliferation of gonocytes that may lead to poor semen quality in postnatal life. EDCs may also directly impair the development of the reproductive organs by antiandrogenic or estrogen-mediated mechanisms and affect the methylation and/or acetylation status of the fetal testicular cells, which may create an abnormal genetic background and an increased risk of testicular cancer in postnatal life [66].

Side effects from exposure are not always necessarily negative. In birth control pills, a stable derivate of estradiol called Ethinyl estradiol (EE) can be found, that alters hormonal signaling through estrogen receptor binding effectively prevents ovulation by inhibiting the release of gonadotropins. Tamoxifen and Flutamide are receptors antagonist that are routinely used in the treatment of breast and prostate cancer, respectively. Phytoestrogens have been shown to have some beneficial effects in humans [67, 68]. People living in regions where traditional diets contain high quantities of plant estrogens, are reported to have lower incidences of atherosclerotic cardiovascular disease as well as breast and prostate cancer than people who consume a "Western" diet. However, more recent studies have demonstrated that Genistein, an isoflavone estrogen, can have a negative impact oocyte maturation and subsequent embryonic development [69]. The use of soya meal in infant formula is seen to be somewhat controversial with expert panels concluding that more detailed studies are required to evaluate the long-term effects of phytoestrogen exposure and uptake [70].

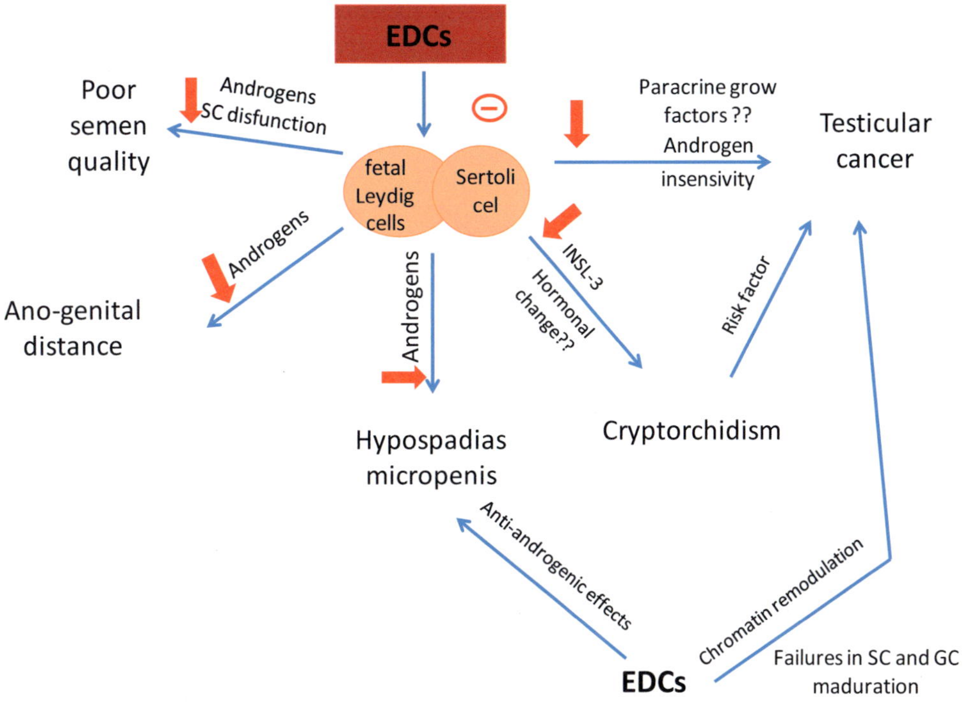

Fig. (3). Summary diagram of the putative mechanisms underlying the development of reproductive disorders in human male fetuses. Modified from Svechnikov *et al*. [66].

The exposure to EDCs during pregnancy can drastically modify fetus developmental programming cues resulting in permanent long-term consequences. Exposure to EDCs during development may result in different effects than exposure during adulthood. Low-dose exposure during fetal development can result in long-standing adverse effects after the EDC is eliminated from the body. Such interactions between the developing organism *in utero* and the toxic components of the environment may determine the predisposition of that individual to develop disease later in life [8]. The term *"fragile fetus"* was created by Howard Bern in 1992 to express the extreme vulnerability of the developing organism to alterations that environmental chemicals can make, especially those with hormone-like activity. Exposure to environmental chemicals during development can result in structural malformations and/or functional alterations in the embryo or fetus and even death in the most severe cases. In individuals that have been associated with antiandrogenic action of certain EDCs, effects such as reduced anogenital distance (AGD) in newborn boys [71] and disorders of puberty [72] have been observed.

Prenatal exposure to thalidomide, previously used to treat maternal anxiety and depression, resulted in limb deformities in the exposed offspring [73].

Different adverse consequences from EDCs that have been identified in various maturing organ systems include, but are not limited to, reproductive tract tissues. This data shows the extreme sensitivity of the growing organism and emphasizes the need for identification and avoidance of EDCs, especially during the critical periods of prenatal and neonatal development [73].

Cancer

The proceedings at the Wingspread Conference helped generate a hypothesis that stated the fetal exposure to agents that were hormonally active may explain epidemiological trends observed in the last half of the 20th century in European and North American populations. The epidemiological observations indicated the following including an increased incidence of congenital malformations of the male genital tract, for example undescended testis and hypospadia, decreased sperm quality and an increased incidence of tumors-uterine leiomyoma, testicular cancer and breast cancer [74, 75].

The focus point is mainly on the carcinogenic potential of exposure to diethylstilbestrol, BPA and dioxins in the male and female sexual organs (genital tract) and mammary gland during organogenesis.

The hypothesis that prenatal exposure to EDCs could possibly cause cancer, began in the 1990s when two entrenched notions were challenged. These notions stated that mammalian growth and maturity were was merely the unfolding of a genetic program [76] and that only mutagenic agents (biological, physical and chemical agents that induce DNA mutations) could cause cancer [77].

In the last decade, carcinogenesis has been believed to be a cell-based resulting in DNA mutations in a single founder cell. This theory has been represented by the somatic mutation theory (SMT). Probably owing to the defects of the SMT, and to the fact that other nonmutagenic agents can also cause cancer, various scientifics have proposed a review and rectification for the SMT, particularly, that epigenetic changes play a central part in carcinogenesis [78, 79]. But these cell-based theories don´t explain the long latency period and the regression of hormone-dependent tumors when there is no longer exposure to the hormone.

Diethylstilbestrol is considered an EDC as it was released into the environment while simultaneously being consumed by cattle and poultry to accelerate weight gain. Scientific experiments involving mice showed that prenatal and neonatal diethylstilbestrol blocked stromal induction of stratification of the vaginal epithelium, resulting in the development of simple columnar epithelium, usually present in the uterus. The majority of the vaginal epithelium, however, becomes stratified upon discontinuation of diethylstilbestrol exposure, but islands of noninduced epithelium eventually develop glands similar to the ones found in the endometrium which persist throughout adult life. These ectopic glands are thought to be the origin of clear-cell carcinoma of the vagina [80]. This discovery is yet another example of how altered tissue interactions lead to carcinogenesis. Another finding that was noted was the development of uterine benign neoplasms (leiomyomas) in female rodents that had been exposed during fetal life and neonatally to diethylstilbestrol when they reached adulthood [81, 82].

The xenoestrogen BPA is one of the EDCs that has been most thoroughly investigated over the years. Exposure to BPA in the human fetus occurs through maternal exposure and in the neonate through ingestion of maternal milk, tinned food and infant formula [83 - 85]. Rats exposed prenatally to environmentally relevant doses of BPA show an increased number of intraductal hyperplasias (precancerous lesions) that appear during adulthood, while high doses induce the development of carcinomas *in situ* [86, 87]. Animals exposed during fetal life to BPA develop palpable tumors during early adulthood when treated at 50 days of age with nitrosomethyurea, a chemical carcinogen [86]. Exposure of mouse dams to environmentally relevant levels of BPA during organogenesis results in considerable alterations in the mammary gland. At embryonic day 18, BPA accelerates maturation of the fat pads and increases the density of collagen fibers directly abutting the epithelium. Within the epithelium, BPA exposure leads to a decreased cell size, delayed lumen formation, increased ductal area and ductal extension. At this stage of development, the estrogen receptors are present only in the mesenchyme, and mammary gland development is dependent on reciprocal interactions between the mesenchyme and the epithelium [88]. At puberty, an increased sensitivity to estradiol was observed in the mammary glands of animals fetally exposed to BPA, which led to the induction of progesterone receptors in epithelial cells and to increased lateral branching of ducts [89]. Thus, perinatal exposure to low doses of BPA results in altered mammary gland morphogenesis, induction of precancerous lesions, and carcinoma *in situ*. It is worth noting that neoplasias that result from exposures to BPA and diethylstilbestrol during embryogenesis and organogenesis usually appear after sexual maturity is complete. Investigators have hypothesized that this phenomenon is due to the effects of sex hormones on the proliferation and remodeling of these organs [90].

Dioxins are by-products of combustion and of multiple industrial processes; they have been identified as a class 1 carcinogen by the International Agency for Research on Cancer [91] and have also been identified as reproductive and endocrine toxicants [92]. The TCDD is the most potent dioxin of the series. In cell culture studies, TCDD, by binding to the aryl hydrocarbon receptor, interacts with oestrogen receptors and thus behaves either as an oestrogen agonist or antagonist. Exposure of rats to the carcinogen dimethylbenzantracene at puberty increases their tumor incidence and shortens the latency period as compared to animals not exposed to TCCD [93]. In males, EDC exposure has been linked to the occurrence of testicular dysgenesis syndrome, as well as prostate development and cancer.

Skakkebaek *et al.* [94] hypothesized that fetal exposure to EDCs may have lead to the triad of diminished semen quality, male genital tract malformations and testicular germ line cancer, through a mechanism involving altered Leydig and Sertoli cell function and impaired germ cell development. In regard to testicular cancer, a small study has found that blood organochlorine levels in mothers, measured decades after they gave birth to sons, correlated with the sons' increased risk of testicular germ cell cancer [95].

The fetal development of the prostate is affected by exposure to estrogens and TCDD [96, 97]. Animal experiments have confirmed this concept, also revealing that estrogens have a significant role in the induction of prostate cancer [98]. The developing prostate has been shown to be sensitive to minute doses of estrogens, which suggests that altered morphogenesis may predispose the prostate to neoplastic development [99, 100].

In conclusion, the consequences of exposure to a single EDC are dependent on the sex and gonadal status of the individual, the dose level and the time and length of exposure. Humans and wildlife are exposed to a mixture of EDCs. To explain this mindboggling complexity will require the design of novel experimental approaches that integrate the effects of different doses of widely structurally different chemicals acting at different ages on different target tissues having different susceptibilities to those diverse EDCs [88].

Diabetes and Obesity

The current prevalence of diabetes and obesity is unprecedented. In 2013, the World Health Organization (WHO) reported that 347 million people all over the world suffer from diabetes (90% of them with type 2 diabetes) [101, 102].

Furthermore, based on WHO statistics in 2011, the prevalence of obesity worldwide had doubled since 1980, and tripled in children and adolescents between 2 and 19 years of age. This trend was also apparent in preschool children between 2 and 5 years of age [103] and more recently, could also be seen in developing countries too. The reasons for this rapid increase in diabetes and obesity remain unclear. Excess caloric consumption and a sedentary lifestyle are undoubtedly key causal factors for obesity and diabetes. However, there is growing interest in the contribution of "non-traditional" risk factors, such as environmental chemicals, micronutrients and the gut microbiome, to the etiology of these health conditions [104].

Being that obesity is a multifactorial and complex endocrine disease, its etiology involves interactions between genes and the environment. Interestingly, the current increase in obesity and other metabolic diseases correlates with substantial increases in environmental chemical production and exposures over the past few decades [105, 106]. A subclass of EDCs have been identified that can disrupt sensitive metabolic processes if exposure occurs during early development, which leads to obesity, type 2 diabetes mellitus and the metabolic syndrome. These chemicals, so called 'obesogens', are thought to predispose individuals to weight gain owing to changes in metabolic 'set-points', particularly if exposure occurs during sensitive periods [107 - 109]. Early in life, EDCs can affect fetal adipose tissue by increasing the number and size of adipocytes [110]. Fat cells are generated from mesenchymal stem cells, which are also capable of differentiating into bone cells, cartilage cells and cells of other tissues [111]. Obesogenic chemicals can artificially direct mesenchymal stem cells to differentiate into adipocytes and promote the accumulation of triglycerides in mature adipocytes. These effects can lead to alterations in the 'set point' for gaining weight and thus contribute both to weight gain and the problems associated with weight loss. EDCs can also cause changes in the hypothalamus. Disruptions in hypothalamic programming might result in altered metabolic 'set points' in adolescents and adult individuals. Exposure to EDCs can also alter the organization and function of dopaminergic pathways. For example, early-life exposure to bisphenol A has been shown to alter both presynaptic and postsynaptic dopamine activity in brain regions associated with addictive and impulsive behaviour [112]. Thus, obsessive eating patterns observed in adult individuals with obesity might be, in part, due to chemically-induced alterations in neural programming early in life. A variety of prescription drugs have adverse effects that result in weight gain, such as thiazolidinediones (anti-diabetic drugs), tricyclic antidepressants, selective 5-hydroxytryptamine uptake inhibitors and atypical antipsychotic drugs [113 - 115], which provides a proof-of-principle that chemicals with similar structures and modes of action might have a role in the obesity epidemic (Table **4**).

An old perception persists that oestrogens have a harmful effect on glucose metabolism, however, studies have shown that oestrogens, when used alone, in healthy and postmenopausal women with type 2 diabetes, can lower blood glucose levels as well as improve insulin sensitivity [116, 117].

Both human and rodent cells express the estrogen receptor ERα, in addition ERβ. More recently, Nadal *et al.* [118] acknowledged the presence of GPR30, a non-classical ER also known as GPER (G protein-coupled oestrogen receptor), at the membrane of β cells. Most of the ER actions related to glucose homoeostasis include extra nuclear pathways such as ERα and GPER. Activation of ERα by physiological concentrations of E2 (estradiol) induces to the stimulation of proinsulin gene transcription [119]. On the contrary, activation of ERβ and GPER has no effect on insulin biosynthesis, but predominantly enhances glucose-induced Ca^{2+} signals and also glucose-stimulated insulin release through membrane pathways [120]. Interestingly, most rodent models of diabetes due to β-cell failure, exhibit sexual dimorphism, suggesting that oestrogens can prevent β-cell failure [121]. There are many theories about the role of E2 on β-cell survival, one supports that E2 is responsible for the inhibition of lipogenesis and therefore, prevents islet lipotoxicity induced by ER activation [122].

Table 4. Environmental chemicals associated with obesogenic properties. Modified from Heindel *et al.* (2015) [109].

Chemical	Source/commercial use	Potential mechanism
Cigarette smoke	First-hand and second-hand smoke	Prenatal nicotine exposure alters neurological development and exposures ↑ rates of preterm and low-weight births
Air pollution Polycyclic aromatic hydrocarbons	Incomplete combustion of fossil fuels	↑ Accumulation of visceral fat Inflammation
Tributyltin	Fungicide in paints and components of polyvinyl chlorides	Activation of peroxisome proliferator-activated receptor γ and increased fat cell differentiation
Bisphenol A	Plastics and epoxy resins	Estrogenic Inhibition of proliferation of neural progenitor cells
Flame retardants	Chemicals applied to furniture and electronics	↑ Rate of adipogenesis ↑ Glucose intolerance
Polychlorinated biphenyls	Coolants, plasticizers and flame retardants	Altered thyroid function Altered metabolism Bioaccumulation in fat cells

(Table 4) contd.....

Chemical	Source/commercial use	Potential mechanism
Phthalates	Plasticizers, adhesives and personal care products	↑ Rate of adipocyte differentiation
Perfluorooctanoic acid Perfluorooctanoate sulphonate	Components of lubricants, non-stick coatings and stain-resistant compounds	↑ Serum levels of insulin ↑ Serum levels of leptin

Many studies have reported that organic pollutants could enhance insulin resistance *in vitro* by a direct action on adipocytes. For example, cultured and differentiated mouse adipocyte 3T3-L1 cells exposed to a mixture of persistent organic pollutants showed impaired responses to insulin (especially with organochlorine pesticides) and downregulation of insulin-induced gene 1 (INSIG1) and lipin 1 (LPIN1), two key regulators of lipid metabolism [123]. Exposure of mouse and human cultured cells to phthalates,especially mono-(2 -ethylhexyl)-phthalate (MEHP) and monobenzyl-phthalate (MBzP), led to activation of peroxisome proliferator-activated receptor alpha (PPAR-α) and PPAR-γ, then to fatty acid oxidation and strong adipocyte differentiation [124]. Exposure to a single low dose of BPA decreased blood glucose levels while simultaneously increasing serum insulin levels, leading to postprandial hyperinsulinaemia favouring insulin resistance, especially in peripheral tissues, such as liver, adipocytes and skeletal muscle [125]. Augmentation in β-cell insulin content and release after contact with BPA appears to be a direct result of its oestrogenic properties, as these effects were not observed in ERα-knockout animals [126].

Several epidemiological studies from around the world have reported suggestive-to-strong links between diabetes and exposure to common EDCs, especially the organic pollutants. Although extrapolation of animal results to human health may be fraught with dangers, animal studies using doses similar to those found in humans should be considered good approximations. Furthermore, there is now considerable molecular evidence indicating that ERs are involved in the induction of insulin resistance and β-cell failure. However, further studies are needed to definitively confirm or refute the link between EDC exposure and type 2 diabetes [127].

IMPORTANCE OF EDCS IN MEDICAL LABORATORIES AND MECHANISMs OF DETECTION

The need to obtain reliable information on the state of particular environmental compartments and processes occurring has motivated many scientific centers to investigated new analytical tools. Both instrumental and biological tools are being used in analytical practice in the field of bioanalytics and biomonitoring.

Since the early 1970s, immunological techniques have been applied in environmental studies with the very basics of these methods consisting in the use of the affinity between the antibody and antigen (analyte) [128]. There are numerous variations of immunoanalysis (EIA). One of the most popular is ELISA (Enzyme linked immunoabsorbent assay). Table **5** summarizes the essential characteristics of the different immunoanalytical tests used.

In literature we can find information on using chromatographic techniques for detection, identification, and quantification of an extensive range of chemicals that have endocrine affecting properties (both naturally occurring or anthropogenic). Cellular biotests are a good alternative to traditional analytical procedures as well as to immunotechniques and methods using living organisms as biomarkers of exposure to EDCs [129]. Yeast or human cells (*e.g.*, cells of breast or kidney cancer) are used in these types of biotest to determine disturbances in the run of hormonal signaling [130]. Out of all the different cellular tests, these three are the ones that are most frequently used to determine endocrine potency.

Table 5. Different immunoanalytical test. Modified from Kudlak *et al.* (2015) [128].

Test Acronym	Full name
RIA	Radioimmunoassay
IRMA	Immunoradiometric assay
ELISA	Enzyme-linked immunosorbent assay
EMIT	Enzymatic multiplied immunoassay technique
FPIA	Fluorescence polarization immunoassay
SLFIA	Substrate-labeled fluorescent immunoassay

E-Screen: The presence of estrogen induces a response in the cell proliferation rate. The number of cells being determined is proportional to the estrogen concentration in the sample studied. Estrogen-dependent cell lines are used *e.g.*, cell lines of human breast cancer MCF-7 [131].

YES/YAS Assay: The human estrogen receptor gene is introduced into genetically modified yeast cells of *Saccharomyces cerevisiae* (in the case of the YES test; the androgenic gene in the case of the YAS test) and coupled to the LacZ reporter gene. When taking in account a substance's estrogenic potential, it binds the estrogen receptor and signals the presence of an estrogenic chemical, initiating the reporting gene activity. The reporting gene codes the synthesis of β-galactosidase, which subsequently participates converting the dye present in the sample solution from yellow to red [132]. The sample´s estrogenic activity is directly related on the intensity of the red color.

CALUX Assay: In this test mammalian cells are used. The specific type of cell is proliferated under standard conditions in multi-well plates. These cells show sensitivity to certain types of chemicals, which may potentially be estrogenic, androgenic or xenoestrogenic substances. As in the YES/YAS assay, the main receptor gene is coupled with the reporting gene, in this case the luciferase gene [128].

There are loads of methods for detecting these compounds, but in detection and quantification in biological fluids, the most commonly used method is liquid chromatography mass spectrometry. Zimmers *et al.* [133] conducted a study whose purpose was to elaborate a sensitive method to detect free BPA in human breast milk. Using solid-phase extraction, BPA was isolated from the milk of 21 nursing mothers in the U.S. It was then derivatized to improve sensitivity and subsequently analyzed by ultra high performance liquid chromatography–tandem mass spectrometry.

Other similar studies carried out by Yi *et al.* [134] compare two methods of detection of BPA in breast milk. The analytical methods include single trace chromatographic separations such as high performance liquid chromatography with fluorescence detection (HPLC/FLD), and lipid chromatography mass spectrometry (LC/MS/MS). Even though there was a strong positive association between BPA levels, which were analyzed with the two different methods, the BPA levels in the HPLC/FLD were lower than those in the LC/MS/MS. Hence, to avoid possible error in biological monitoring of BPA, we recommend severe guidelines for identification and confirmation of BPA with the LC/MS/MS method.

Another interesting study was done by the group Capiello *et al.* [135]. In the last decade, environmental pollution has been contemplated as the possible cause for Sudden Infant Death Syndrome (SIDS) and Sudden Intrauterine Unexplained Death Syndrome (SIUDS). In this study, they developed a rapid and sensitive analytical method to measure the levels of organochlorine pesticides (OCPs) and organophosphate pesticides (OPPs), carbamates and phenols in human fetal and newborn tissues (brain and liver) and unveiled the potential existence of non-targeted compounds. A liquid-solid extraction method was used on human and animal tissues and the extracts, after a solid phase extraction (SPE) clean-up procedure, were analyzed by gas chromatography coupled to a quadrupole mass spectrometric detector (GC-qMS). A GC-TOF MS (time-of-flight) instrument was applied for a parallel detection of non-targeted compounds because of its higher full-scan sensitivity.

The interest in EDCs is becoming increasingly more significant. Knowing the levels in humans, half-life and accumulation in tissues, would help to relate their presence with certain diseases.

CONCLUSIONS

The methods for monitoring and evaluating metals for instance lead and mercury were developed early on. In recent times, on other compounds such as PBDEs, PFCs, phthalates and BPA are being focused on by scientists. Due to the difficulty of finding new compounds, work on the development of new human biological monitoring methods for modern compounds is rarely found. Unusual and unknown exposures to different compounds can be determined through scientific case reports. The easiest way to identify new substances would be if the industry was more transparent with the introduction of these in society.

Human epidemiological studies on the association between exposure to EDCs and thyroid obesity, cancer, diabetes and/or metabolic syndrome are needed, as the rates of all of these diseases and disorders are steeply rising.

In Europe, the new Legislation on Registration, Evaluation, Authorisation and Restriction of Chemicals (REACH) legislation was adopted on the first of June 2007. According to REACH, it is compulsory that the industry test all chemicals substances made within or imported to Europe that exceed one metric ton concerning health and environmental aspects. These chemicals must then be registered by the European Chemicals Agency (ECHA). This information provides scientists with the possibility to detect new chemicals that have endocrine disrupting characteristics.

Another question is the exposure time window, which is extremely important in EDCs. The first thing to consider when using exposure strategies in a scientific investigation is the half-lives of the monitored compounds. There are several EDCs, there are several compounds such as PCBs and PFCs, that have a very long half-life. However, the majority of EDCs for example phthalates, parabens and bisphenol A have very short half-lives, with some consisting of just hours. This makes it even more challenging to develop human biological monitoring methods when trying to estimate accurately the possible effects of lifelong exposure to these EDCs, considering that several samples are needed to obtain a good estimate of the exposure over a short period.

Lately, techniques such as direct injection of protein and urine depleted serum using analyses by LC-MS/MS and the use of more rapid work-up procedures, have been introduced.

Biobanked samples are increasingly being used to obtain accurate estimations of lifelong exposure. More recently, the government has been more accepting of this process of work and many new biobanks have now been created all over the world.

However, the creation of biobanks has not been able to solve the problem of existing chemicals that do not remain stable during storage.

In conclusion, many more studies are needed to allow us to correlate different diseases with endocrine disruptors to which we are exposed, thus enabling us to foresee the emergence of such diseases and prevent them in their early stages.

CONFLICT OF INTEREST

The author (editor) declares no conflict of interest, financial or otherwise.

ACKNOWLEDGEMENTS

Declared none.

ABBREVIATONS

AGD	Anogenital distance.
AhR	Aryl hydrocarbon receptor.
B(a)P	Benzo(a)pyrene.
BFRs	Brominated flame retardants.
BPA	Bisphenol A.
DDT	Dichlorodiphenyltrichloroethane.
DEHP	Di (2-ethylhexyl) phthalate.
DES	Diethylstilbestrol.
E2	Estradiol.
ECHA	European chemicals agency.
EDA	Effects directed analysis.
EDC	Endocrine disrupting chemical.
EE	Ethinyl estradiol.
EIA	Enzyme immunoanalysis.
ELISA	Enzyme linked immunoabsorbent assay.
EPA	Environmental protection agency.
ER	Estrogen receptor.
FLCs	Fetal Leydig cells.
GPER	G protein-coupled oestrogen receptor.

GC-qMS	Gas chromatography coupled to a quadrupole mass spectrometric detector.
HPLC-FLD	High performance liquid chromatography with fluorescence detection.
INSIG1	Insulin-induced gene 1.
Insl3	Insulin-like factor 3.
LC/MS/MS	Lipid chromatography mass spectrometry.
LPIN1	Lipin 1.
MBzP	Monobenzyl-phthalate.
MEHP	Mono-(2-ethylhexyl)-phthalate.
mER	Membrane estrogen receptor.
MFO	Microsomal mixed-function oxygenase.
NIEHS	National Institute of Environmental Health Sciences.
NIS	Sodium/iodide symporter.
OCPs	Organochlorine pesticides.
OPPs	Organophosphate pesticides.
PBDE-99	Polybrominated diphenyl ethers.
PCBs	Polychlorinated biphenyl.
PCDDs	Dibenzodioxins.
PCDF	Poly-chlorinated dibenzofuran.
PPAR	Peroxisome proliferator-activated receptor.
PTU	6-propyl-2-thiouracil.
QSAR	Quantitative structure-activity relationship.
REACH	Registration, evaluation, authorisation and restriction of chemicals.
SC	Sertoli cell.
SIDS	Sudden infant death syndrome.
SIUDS	Sudden intrauterine unexplained death syndrome.
SMT	Somatic mutation theory.
SPE	Solid phase extraction.
TCDD	Tetrachlorodibenzo-p-dioxin.
THs	Thyroid hormones.
TPO	Thyroperoxidase.
TTR	Thyroid transport protein.
WHO	World Health Organization.
WoE	Weight-of-Evidence.

REFERENCES

[1] Hotchkiss AK, Rider CV, Blystone CR, *et al.* Fifteen years after "Wingspread"--environmental endocrine disrupters and human and wildlife health: where we are today and where we need to go. Toxicol Sci 2008; 105: 235-59.
[http://dx.doi.org/10.1093/toxsci/kfn030]

[2] Fosters WG, Agzarian J. Towards less confusing terminology in endocrine disruptor research. Environ Health 2008; 11: 152-61.

[3] Weybridge European Workshop on the impact of endocrine disrupters on human health and wildlife Rep Proceedings 1996.

[4] European UnionEndocrine Disruptor Site 4 December 2012.

[5] Nohynek GJ, Borgert CJ, Dietrich D, Rozman KK. Endocrine disruption: fact or urban legend? Toxicol Lett 2013; 223: 295-305.
[http://dx.doi.org/10.1016/j.toxlet.2013.10.022]

[6] Blumberg B. Endocrine disruptors as obesogens. Mol Cell Endocrinol 2009; 304: 19-29.
[http://dx.doi.org/10.1016/j.mce.2009.02.018]

[7] Grün F. Obesogens. Curr Opin Endocrinol Diabetes Obes 2010; 17(5): 453-9.
[http://dx.doi.org/10.1097/MED.0b013e32833ddea0]

[8] Diamanti-Kandarakis E, Bourguignon JP, Giudice LC, Hauser R. Endocrine-disrupting chemicals: an endocrine society scientific statement. Endocr Rev 2009; 30: 293-342.
[http://dx.doi.org/10.1210/er.2009-0002]

[9] Caliman FA, Gavrilescu M. Pharmaceuticals, personal care products and endocrine disrupting agents in the environment – a review. Clean – Soil AirWater 2009; 277-303.

[10] Gore AC, Crews D, Doau LL, *et al.* Introduction to Endocrine disrupting Chemicals (EDCs)-A Guide for Public interest Organizations and Policy-markers. Endocrine Society 2014.

[11] Amoako PK, Kumah P. Pesticides usage in cabbage (Brassica oleracea) cultivation in the forest ecozone of Ghana. Int J Res Chem Env 2012; 2: 26-31.

[12] Sharma BM, Bharat GK, Tayal S, Nizetto L, Cupr PL. Environment and human exposure to persistent organic pollutants (POPs) in India: A systematic review of recent and historical data. Environ Int 2014; 66: 48-64.
[http://dx.doi.org/10.1016/j.envint.2014.01.022]

[13] Mekonen S, Ambelu SP. Pesticide residue evaluation in major staple food items of Ethiopia using the Quechers method: A case study from the Jimma zone. Environ Toxicol Chem 2014; 33: 1294-302.
[http://dx.doi.org/10.1002/etc.2554]

[14] Gebremichael S, Birhanu Tessema DA. Organochlorine pesticide residues in human and cow's milk in the towns of Asendabo, Serbo and Jimma in South-Western Ethiopia. Chemosphere 2013; 90: 1652-7.
[http://dx.doi.org/10.1016/j.chemosphere.2012.09.008]

[15] Aneck-Hahn NH, Schulenburg GW, Bornman MS, Farias P. Impaired semen quality associated with environmental DDT exposure in young men living in a malaria area in the Limpopo Province, South Africa. J Androl 2007; 28: 423-34.
[http://dx.doi.org/10.2164/jandrol.106.001701]

[16] Bouwman H. Of sibship on DDT residue levels in human serum from a malaria endemic area in northern Kwazulu. Bull Environ Contam Toxicol 1993; 50: 300-7.
[http://dx.doi.org/10.1007/BF00191737]

[17] World Health Organization. Exposure to lead: a major public health concern. Geneva: International Programme on Chemical Safety 2010.

[18] Gollenberg AL, Hediger ML, Lee PA, Himes JH. LC between lead and cadmium and reproductive

hormones in peripubertal US girls. Environ Health Perspect 2010; 118: 1156.

[19] Pollack AZ, Schisterman EF, Goldman LR, Mumford SL, Albert PS, Jones RL. Cadmium, lead, and mercury in relation to reproductive hormones and anovulation in premenopausal women. Environ Health Perspect 2011; 119: 1156.
[http://dx.doi.org/10.1289/ehp.1003284]

[20] Calafat AM, Ye X, Wong LY, Reidy JA. Exposure of the US population to bisphenol A and 4-tertiar--octylphenol: 2003-2004. Environ Health Perspect 2008; 116: 39-44.
[http://dx.doi.org/10.1289/ehp.10753]

[21] Dodson RE, Nishioka M, Standley LJ, Perovich LJ, Brody JG, Rudel R. Endocrine disruptors and asthma-associated chemicals in consumer products. Environ Health Perspect 2012; 120: 935-43.
[http://dx.doi.org/10.1289/ehp.1104052]

[22] Update DE. For ConsumersCenters for Disease Control and Prevention http://wwwcdcgov/de (Accessed Gener 11, 2016) 2016.

[23] Kabir ER, Rahman MS, Rahman I. A review on endocrine disruptors and their possible impacts on human health. Environ Toxicol Pharmacol 2015; 40: 241-58.
[http://dx.doi.org/10.1016/j.etap.2015.06.009]

[24] Lindner HR. Occurrence of anabolic agents in plants and their importance. Environ Qual Saf Suppl 1976; 1: 151-8.

[25] Borgert CJ, Mihaich EM, Ortego LS, *et al.* Hypothesis-driven weight of evidence framework for evaluating data within the US EPA's Endocrine Disruptor Screening Program 2011; 185-91.

[26] Somm E, Schwitzgebe VM, Toulotte A, Cederroth CR. Perinatal exposure to bisphenol A alters early adipogenesis in the rat. Environ Health Perspect 2009; 117: 1549-55.
[http://dx.doi.org/10.1289/ehp.11342]

[27] Nicolas JC, Cecile C, Perinnaaz RW, *et al.* Effects of low doses of bisphenol a on the metabolome of perinatally exposed CD-1 mice. Environ Health Perspect 2013; 121: 586-93.

[28] Kidd KA, Becher G, Muir DC, Woodruff TJ. Human and wildlife exposures to EDCs.Chapter 3. State of the science of endocrine disrupting chemicals UNEP 2012; 189-250.

[29] Westerhoff P, Yoon Y, Snyder S, Wert E. Fate of endocrine-disruptor, pharmaceutical, and personal care product chemicals during simulated drinking water treatment processes. Environ Sci Technol 2005; 39: 6649-63.
[http://dx.doi.org/10.1021/es0484799]

[30] Gibs J, Stackelberg PE, Furlong ET, Meyer M, Zaugg SD, Lippincott RL. Persistence of pharmaceuticals and other organic compounds in chlorinated drinking water as a function of time. Sci Total Environ 2007; 373: 240-9.
[http://dx.doi.org/10.1016/j.scitotenv.2006.11.003]

[31] Kim SD, Cho J, Kim IS, Vanderford BJ, Snyder SA. Occurrence and removal of pharmaceuticals and endocrine disruptors in South Korean surface, drinking, and waste waters. Water Res 2007; 41: 1013-21.
[http://dx.doi.org/10.1016/j.watres.2006.06.034]

[32] Schenck K, Rosenblum L, Wiese TE, *et al.* Removal of estrogens and estrogenicity through drinking water treatment. J Water Health 2012; 1(10): 43-55.
[http://dx.doi.org/10.2166/wh.2011.135]

[33] Kim SS, Kwack SJ, Lee RD, *et al.* Assessment of estrogenic and androgenic activities of tetramethrin *in vitro* and *in vivo* assays. J Toxicol Environ Health A 2005; 68: 2277-89.
[http://dx.doi.org/10.1080/15287390500182453]

[34] Preda C, Ungureanu MC. Endocrine disruptors in the environment and their impact on human health. Enviromental Eng Manag J 2012; pp. 1697-706.

[35] Rosenfeldt EJ, Linden K. Degradation of endocrine disrupting chemicals bisphenol A, ethinyl estradiol, and estradiol during UV photolysis and advancedoxidation processes. Environ Sci Technol 2004; 38: 5476-83.
[http://dx.doi.org/10.1021/es035413p]

[36] Submission HE. National Institute of Environmental Health Sciences. J Consum Health Internet 2008; 174.

[37] Dhooge W, Hond ED, Koppen G, Bruckers L, Nelen V. Internal exposure to pollutants and sex hormone levels in Flemish male adolescents in a cross-sectional study: Associations and dose–response relationships. J Expo Sci Environ Epidemiol 2011; 21: 106-13.
[http://dx.doi.org/10.1038/jes.2009.63]

[38] Den HE, Roels HA, Hoppenbrouwers K, *et al.* Sexual maturation in relation to polychlorinated aromatic hydrocarbons: Sharpe and Skakkebaek's hypothesis revisited. Environ Health Perspect 2002; 110: 771-6.
[http://dx.doi.org/10.1289/ehp.02110771]

[39] Polyzos SA, Kountouras J, Deretzi G, Zavos C, Mantzoros CS. The emerging role of endocrine disruptors in pathogenesis of insulin resistant: a concept implicating nonalcoholic fatty liver disease. Curr Mol Med 2012; 12: 68-82.
[http://dx.doi.org/10.2174/156652412798376161]

[40] WHO U and United Nations Environment Programme and the World Health Organization. State of the Science of Endocrine Disrupting Chemicals 2013; 2012: 2013.

[41] Annamalai J, Namasivayam V. Endocrine disrupting chemicals in the atmosphere: their effects on humans and wildlife. Environ Int 2015; 76: 78-97.
[http://dx.doi.org/10.1016/j.envint.2014.12.006]

[42] Weybridge T. European environment Agency.. The impacts of endocrine disrupters on wildlife, people and their environments. The Weybridge+15 (1996–2011) report 2012.

[43] Iguchi T, Katsu Y. Commonality in signaling of endocrine disruption from snail to human. Bioscience 2008; 58: 1061-7.
[http://dx.doi.org/10.1641/B581109]

[44] Kiyama R, Wada-Kiyama Y. Estrogenic endocrine disruptors: molecular mechanisms of action. Environ Int 2015; 83: 11-40.
[http://dx.doi.org/10.1016/j.envint.2015.05.012]

[45] Romano Mozo MD. Disruptores Endocrinos. Nuevas respuestas para nuevos retos Instituto sindical de trabajo, ambiente y salud (ISTAS) 2013; 1-6.

[46] Guengerich FP, Crawford WM, Domoradzki JY, Macdonald TL, Watanabe PG. *In vitro* activation of 1,2-dichloroethane by microsomal and cytosolic enzymes. Toxicol Appl Pharmacol 1980; 55: 303-17.
[http://dx.doi.org/10.1016/0041-008X(80)90092-7]

[47] Patrick L. Thyroid disruption: mechanism and clinical implications in human health. Altern Med Rev 2009; 14: 326-46.

[48] Kajta M, Wójtowicz AK. Impact of endocrine-disrupting chemicals on neural development and the onset of neurological disorders. Pharmacol Rep 2013; 65: 1632-9.
[http://dx.doi.org/10.1016/S1734-1140(13)71524-X]

[49] Latchney SE, Lioy DT, Henry EC, *et al.* Neural precursor cell proliferation is disrupted through activation of the aryl hydrocarbon receptor by 2,3,7,8-tetrachlorodibenzo-p-dioxin. Stem Cells Dev 2011; 20: 313-26.
[http://dx.doi.org/10.1089/scd.2009.0529]

[50] Lim CK, Kim SK, Ko DS, *et al.* Differential cytotoxic effects of mono-(2- ethylhexyl) phthalate on blastomere-derived embryonic stem cells and differentiating neurons. Toxicology 2009; 29(264): 145-

54.
[http://dx.doi.org/10.1016/j.tox.2009.08.015]

[51] Okada M, Makino A, Nakajima M, Okuyama S, Furukawa S, Furukawa Y. Estrogen stimulates proliferation and differentiation of neural stem/progenitor cells through different signal transduction pathways. Int J Mol Sci 2010; 11: 4114-23.
[http://dx.doi.org/10.3390/ijms11104114]

[52] Anway MD. Epigenetic Transgenerational Actions of Endocrine Disruptors and Male Fertility. Science 2005; 3(308): 1466-9.
[http://dx.doi.org/10.1126/science.1108190]

[53] Hatcher-Martin JM, Gearing M, Steenland K, Levey AI, Miller GW, Pennell KD. Association between polychlorinated biphenyls and Parkinson's disease neuropathology. Neurotoxicology 2012; 33: 1298-304.
[http://dx.doi.org/10.1016/j.neuro.2012.08.002]

[54] Huang SC, Giordano G, Costa LG. Comparative cytotoxicity and intracellular accumulation of five polybrominated diphenyl ether congeners in mouse cerebellar granule neurons. Toxicol Sci 2010; 9(114): 124-32.
[http://dx.doi.org/10.1093/toxsci/kfp296]

[55] Ishido M, Yonemoto J, Morita M. Mesencephalic neurodegeneration in the orally administered bisphenol A-caused hyperactive rats. Toxicol Lett 2007; 173(1): 66-72.
[http://dx.doi.org/10.1016/j.toxlet.2007.06.014]

[56] Ishido M, Suzuki J. Inhibition by rotenone of mesencephalic neural stem-cell migration in a neurosphere assay *in vitro*. Toxicol In Vitro 2010; 24: 552-7.
[http://dx.doi.org/10.1016/j.tiv.2009.11.005]

[57] Ki YW, Park JH, Lee JE, Shin IC, Koh HC. JNK and p38 MAPK regulate oxidative stress and the inflammatory response in chlorpyrifos-induced apoptosis. Toxicol Lett 2013; 26(218): 235-45.
[http://dx.doi.org/10.1016/j.toxlet.2013.02.003]

[58] Nayyar T, Zawia NH, Hood DB. Transplacental effects of 2,3,7,8-tetrachlorodibenzo-p-dioxin on the temporal modulation of Sp1 DNA binding in the developing cerebral cortex and cerebellum. Exp Toxicol Pathol Off J Gesellschaft fur Toxikologische Pathol 2002; 53: 461-8.
[http://dx.doi.org/10.1078/0940-2993-00219]

[59] Campion S, Catlin N, Heger N, *et al.* Male reprotoxicity and endocrine disruption. EXS 2012; 101: 315-60.
[http://dx.doi.org/10.1007/978-3-7643-8340-4_11]

[60] Colborn T, vom Saal FS, Soto AM. Developmental effects of endocrine-disrupting chemicals in wildlife and humans. Environ Health Perspect 1993; 1(101): 378-84.
[http://dx.doi.org/10.1289/ehp.93101378]

[61] Chen A, Rogan WJ. Nonmalarial infant deaths and DDT use for malaria control. Emerg Infect Dis 2003; 9: 960-4.
[http://dx.doi.org/10.3201/eid0908.030082]

[62] Roberts D, Curtis C, Tren R, Sharp B, Shiff C, Bate R. Malaria control and public health. Emerg Infect Dis 2004; 10: 1170-1.
[http://dx.doi.org/10.3201/eid1006.030787]

[63] Rogan WJ, Ragan NB. Evidence of effects of environmental chemicals on the endocrine system in children. Pediatrics 2003; 112: 247-52.

[64] GUo YL. Hse CC LG. Effects of environmental chemicals (EC) on sexual maduration. Pediatr Res 1996; 39: 74-8.
[http://dx.doi.org/10.1203/00006450-199604001-00447]

[65] Hanser R, Altshul L, Chen Z. Environmental organochlorines and semen quality: results of a pilot

study. Environmental Health Perspectives 2002; 110(3): 229-33.

[66] Svechnikov K, Stukenborg JB, Savchuck I, Söder O. Similar causes of various reproductive disorders in early life. Asian J Androl 2014; 16: 50-9. [http://dx.doi.org/10.4103/1008-682X.122199]

[67] Cheek AO, Vonier PM, Oberdörster E, Burow BC, McLachlan JA. Environmental signaling: a biological context for endocrine disruption. Environ Health Perspect 1998; 106 (Suppl.): 5-10. [http://dx.doi.org/10.2307/3433910]

[68] Cheek AO, McLachlan JA. Environmental hormones and the male reproductive system. J Androl 1998; 19: 5-10.

[69] Chan WH. Impact of genistein on maturation of mouse oocytes, fertilization, and fetal development. Reprod Toxicol 2009; 28: 52-8. [http://dx.doi.org/10.1016/j.reprotox.2009.03.014]

[70] Rozman KK, Bhatia J, Calafat AM, *et al.* NTP-CERHR expert panel report on the reproductive and developmental toxicity of soy formula. Birth Defects Res B Dev Reprod Toxicol 2006; 77: 280-397. [http://dx.doi.org/10.1002/bdrb.20086]

[71] Swan SH, Main KM, Liu F, *et al.* Decrease in anogenital distance among male infants with prenatal phthalate exposure. Environ Health Perspect 2005; 113: 1056-61. [http://dx.doi.org/10.1289/ehp.8100]

[72] Buck Louis GM, Gray LE, Marcus M, Ojeda SR, Pescovitz OH, Witchel SF, *et al.* Environmental factors and puberty timing: expert panel research needs. Pediatrics 2008; 121 (Suppl.): S192-207. [http://dx.doi.org/10.1542/peds.1813E]

[73] Newbold RR. Developmental exposure to endocrine-disrupting chemicals programs for reproductive tract alterations and obesity later in life. Am J Clin Nutr 2011; 94: 1939S-42S. [http://dx.doi.org/10.3945/ajcn.110.001057]

[74] Sharpe RM, Skakkebaek NE. Are oestrogens involved in falling sperm counts and disorders of the male reproductive tract? Lancet 1993; 341: 1392-5. [http://dx.doi.org/10.1016/0140-6736(93)90953-E]

[75] Markey CM, Rubin BS, Soto AM, Sonnenschein C. Endocrine disruptors: from Wingspread to environmental developmental biology. J Steroid Biochem Mol Biol 2002; 83: 235-44. [http://dx.doi.org/10.1016/S0960-0760(02)00272-8]

[76] Gilbert SF, Opitz JM, Raff RA. Resynthesizing evolutionary and developmental biology. Dev Biol 1996; 173: 357-72. [http://dx.doi.org/10.1006/dbio.1996.0032]

[77] Baker SG, Kramer BS. Paradoxes in carcinogenesis: new opportunities for research directions. BMC Cancer 2007; 7: 151. [http://dx.doi.org/10.1186/1471-2407-7-151]

[78] Ho SM. Developmental exposure to estradiol and bisphenol a increases susceptibility to prostate carcinogenesis and epigenetically regulates phosphodiesterase type 4 variant 4. Cancer Res 2006; 66: 5624-32. [http://dx.doi.org/10.1158/0008-5472.CAN-06-0516]

[79] Keri RA, Ho SM, Hunt PA, Knudsen KE, Soto AM, Prins GS. An evaluation of evidence for the carcinogenic activity of bisphenol A. Reprod Toxicol 2007; 24: 240-52. [http://dx.doi.org/10.1016/j.reprotox.2007.06.008]

[80] Kurita T, Mills AA, Cunha GR. Roles of p63 in the diethylstilbestrol-induced cervicovaginal adenosis. Development 2004; 131: 1639-49. [http://dx.doi.org/10.1242/dev.01038]

[81] Newbold RR, Moore AB, Dixon D. Characterization of uterine leiomyomas in CD-1 mice following

developmental exposure to diethylstilbestrol (DES). Toxicol Pathol 2011; 30: 611-6. [http://dx.doi.org/10.1080/01926230290105839]

[82] Burroughs KD, Fuchs-Young R, Davis B, Walker CL. Altered hormonal responsiveness of proliferation and apoptosis during myometrial maturation and the development of uterine leiomyomas in the rat. Biol Reprod 2000; 63: 1322-30. [http://dx.doi.org/10.1095/biolreprod63.5.1322]

[83] Schönfelder G, Wittfoht W, Hopp H, Talsness CE, Paul M, Chahoud I. Parent bisphenol A accumulation in the human maternal-fetal-placental unit. Environ Health Perspect 2002; 110: A703-7. [http://dx.doi.org/10.1289/ehp.021100703]

[84] Sun Y, Irie M, Kishikawa N, Wada M, Kuroda N, Nakashima K. Determination of bisphenol A in human breast milk by HPLC with column-switching and fluorescence detection. Biomed Chromatogr 2004; 18: 501-7. [http://dx.doi.org/10.1002/bmc.345]

[85] Onn Wong K, Woon Leo L, Leng Seah H. Dietary exposure assessment of infants to bisphenol A from the use of polycarbonate baby milk bottles. Food Addit Contam 2005; 22: 280-8. [http://dx.doi.org/10.1080/02652030500077502]

[86] Durando M, Kass L, Piva J, *et al.* Prenatal bisphenol A exposure induces preneoplastic lesions in the mammary gland in Wistar rats. Environ Health Perspect 2007; 115: 80-6. [http://dx.doi.org/10.1289/ehp.9282]

[87] Murray TJ, Maffini MV, Ucci AA, Sonnenschein C, Soto AM. Induction of mammary gland ductal hyperplasias and carcinoma *in situ* following fetal bisphenol A exposure. Reprod Toxicol 2007; 23: 383-90. [http://dx.doi.org/10.1016/j.reprotox.2006.10.002]

[88] Soto AM, Sonnenschein C. Environmental causes of cancer: endocrine disruptors as carcinogens. Nat Rev Endocrinol 2010; 6: 363-70. [http://dx.doi.org/10.1038/nrendo.2010.87]

[89] Muñoz-de-Toro M, Markey CM, Wadia PR, *et al.* Perinatal exposure to bisphenol-A alters peripubertal mammary gland development in mice. Endocrinology 2005; 146: 4138-47. [http://dx.doi.org/10.1210/en.2005-0340]

[90] Hatch EE, Palmer JR, Titus-Ernstoff L, *et al.* Cancer risk in women exposed to diethylstilbestrol in utero. JAMA 1998; 280: 630-4. [http://dx.doi.org/10.1001/jama.280.7.630]

[91] Steenland K, Bertazzi P, Baccarelli A, Kogevinas M. Dioxin revisited: developments since the 1997 IARC classification of dioxin as a human carcinogen. Environ Health Perspect 2004; 112: 1265-8. [http://dx.doi.org/10.1289/ehp.7219]

[92] Kogevinas M. Human health effects of dioxins: cancer, reproductive and endocrine system effects. Hum Reprod Update 2001; 7: 331-9. [http://dx.doi.org/10.1093/humupd/7.3.331]

[93] Brown NM, Manzolillo PA, Zhang JX, Wang J, Lamartiniere CA. Prenatal TCDD and predisposition to mammary cancer in the rat. Carcinogenesis 1998; 19: 1623-9. [http://dx.doi.org/10.1093/carcin/19.9.1623]

[94] Skakkebaek NE, Rajpert-De Meyts E, Main KM. Testicular dysgenesis syndrome: An increasingly common developmental disorder with environmental aspects. Hum Reprod 2001; 16: 972-8. [http://dx.doi.org/10.1093/humrep/16.5.972]

[95] Hardell L, Bavel B, Lindström G, Eriksson M, Carlberg M. *In utero* exposure to persistent organic pollutants in relation to testicular cancer risk. Int J Androl 2006; 29: 228-34. [http://dx.doi.org/10.1111/j.1365-2605.2005.00622.x]

[96] Timms BG, Peterson RE, vom Saal FS. 2,3,7,8-tetrachlorodibenzo-p-dioxin interacts with endogenous

estradiol to disrupt prostate gland morphogenesis in male rat fetuses. Toxicol Sci 2002; 67: 264-74. [http://dx.doi.org/10.1093/toxsci/67.2.264]

[97] Timms BG, Howdeshell KL, Barton L, Bradley S, Richter CA, vom Saal FS. Estrogenic chemicals in plastic and oral contraceptives disrupt development of the fetal mouse prostate and urethra. Proc Natl Acad Sci USA 2005; 102: 7014-9. [http://dx.doi.org/10.1073/pnas.0502544102]

[98] Bosland MC, Dreef-Van Der Meulen HC, Sukumar S, *et al.* Multistage prostate carcinogenesis: The role of hormones. Princess Takamatsu Symp 1991; 22: 109-23.

[99] Huang L, Pu Y, Alam S, Birch L, Prins GS. Estrogenic regulation of signaling pathways and homeobox genes during rat prostate development. J Androl 2004; 25: 330-7. [http://dx.doi.org/10.1002/j.1939-4640.2004.tb02796.x]

[100] Prins GS, Birch L, Tang WY, Ho SM. Developmental estrogen exposures predispose to prostate carcinogenesis with aging. Reprod Toxicol 2003; 23: 374-82. [http://dx.doi.org/10.1016/j.reprotox.2006.10.001]

[101] World Health Organization. http://wwwwhoint/mediacentre/factsheets/fs312/en/ Diabetes 2013; 2013.

[102] Danaei G, Finucane MM, Lu Y, *et al.* National, regional, and global trends in fasting plasma glucose and dia- betes prevalence since 1980: Systematic analysis of health examination surveys and epidemiological studies with 370 country-years and 2.7 million participants. Lancet 2011; 378: 31-40. [http://dx.doi.org/10.1016/S0140-6736(11)60679-X]

[103] Thayer KA, Heindel JJ, Bucher JR. Role of environmental chemicals in diabetes and obesity: a National Toxicology Program work-shop review. Environ Health Perspect 2012; 120: 779-89. [http://dx.doi.org/10.1289/ehp.1104597]

[104] Chevalier N, Fenichel P. Endocrine disruptors: New players in the pathophysiology of type 2 diabetes? Diabetes Metab 2015; 41: 107-15. [http://dx.doi.org/10.1016/j.diabet.2014.09.005]

[105] Casals-Casas C. Endocrine disruptors: from endocrine to metabolic disruption Annu Rev Physiol 2011; 173: 135-62. [http://dx.doi.org/10.1146/annurev-physiol-012110-142200]

[106] Baillie-Hamilton PF. Chemical toxins: a hypothesis to explain the global obesity epidemic. J Altern Complement Med 2002; 8: 185-92. [http://dx.doi.org/10.1089/107555302317371479]

[107] Regnier SM, Sargis RM. Adipocytes under assault: environmental disruption of adipose physiology. Biochim Biophys Acta 2014; 1842: 520-33.

[108] Grün F, Blumberg B. Endocrine disrupters as obesogens. Mol Cell Endocrinol 2009; 304: 19-29. [http://dx.doi.org/10.1016/j.mce.2009.02.018]

[109] Heindel JJ. Parma consensus statement on metabolic disruptors. Environ Health 2015; 14: 52. [http://dx.doi.org/10.1186/s12940-015-0042-7]

[110] Janesick A, Blumberg B. Endocrine disrupting chemicals and the developmental programming of adipogenesis and obesity. Birth Defects Res C Embryo Today 2011; 93: 34-50. [http://dx.doi.org/10.1002/bdrc.20197]

[111] Pittenger MF. Multilineage potential of adult human mesenchymal stem cells. Science 1999; 284: 143-7. [http://dx.doi.org/10.1126/science.284.5411.143]

[112] Adriani W, Seta DD, Dessi-Fulgheri F, Farabollini F, Laviola G. Altered profiles of spontaneous novelty seeking, impulsive behavior, and response to D-amphetamine in rats perinatally exposed to bisphenol A. Environ Health Perspect 2003; 111: 395-401. [http://dx.doi.org/10.1289/ehp.5856]

[113] Kupfer DJ, Coble PA, Rubinstein D. Changes in weight during treatment for depression. Psychosom Med 1979; 41: 535-44.
[http://dx.doi.org/10.1097/00006842-197911000-00004]

[114] Shim WS. The long-term effects of rosiglitazone on serum lipid concentrations and body weight. Clin Endocrinol (Oxf) 2006; 65: 453-9.
[http://dx.doi.org/10.1111/j.1365-2265.2006.02614.x]

[115] Serretti A, Mandelli L. Antidepressants and body weight: a comprehensive review and meta-analysis. J Clin Psychiatry 2010; 71: 1259-72.
[http://dx.doi.org/10.4088/JCP.09r05346blu]

[116] Danaei G, Finucane MM, Lu Y, Singh GM, Cowan MJ. National, regional, and global trends in fasting plasma glucose and diabetes prevalence since 1980: systematic analysis of health examination surveys and epidemiological studies with 370 country-years and 2.7 million participants. Lancet 2011; 378: 31-40.
[http://dx.doi.org/10.1016/S0140-6736(11)60679-X]

[117] World Health Organization. Diabetes 2013;http:// wwwwhoint/ 2013.

[118] Nadal A, Rovira JM, Laribi O, *et al.* Rapid insulinotropic effect of 17β-estradiol *via* a plasma membrane receptor. FASEB J 1998; 12: 1341-8.

[119] Wong WP, Tiano JP, Liu S, *et al.* Extranuclear estrogen receptor-alpha stimulates NeuroD1 binding to the insulin promoter and favors insulin synthesis. Proc Natl Acad Sci USA 2010; 107: 13057-62.
[http://dx.doi.org/10.1073/pnas.0914501107]

[120] Sharma G, Prossnitz ER. Mechanisms of estradiol-induced insulin secretion by the G protein-coupled estrogen receptor GPR30/GPER in pancreatic beta-cells. Endocrinology 2011; 152: 3030-9.
[http://dx.doi.org/10.1210/en.2011-0091]

[121] Liu S, Mauvais-Jarvis F. Rapid, nongenomic estrogen actions protect pancreatic islet survival. Islets 2009; 1: 273-5.
[http://dx.doi.org/10.4161/isl.1.3.9781]

[122] Tiano JP, Delghingaro-Augusto V, Le May C, *et al.* Estrogen receptor activation reduces lipid synthesis in pancreatic islets and prevents β cell failure in rodent models of type 2 diabetes. J Clin Invest 2011; 121: 3331-42.
[http://dx.doi.org/10.1172/JCI44564]

[123] Ruzzin J, Petersen R, Meugnier E, *et al.* Persistent organic pollutant exposure leads to insulin resistance syndrome. Environ Health Perspect 2010; 118: 465-71.
[http://dx.doi.org/10.1289/ehp.0901321]

[124] Hurst CH, Waxman DJ. Activation of PPARalpha and PPARgamma by environmental phthalate monoesters. Toxicol Sci 2003; 74: 297-308.
[http://dx.doi.org/10.1093/toxsci/kfg145]

[125] Alonso-Magdalena P, Morimoto S, Ripoll C, Fuentes E, Nadal A. The estrogenic effect of bisphenol a disrupts pancreatic β-cell function *in vivo* and induces insulin resistance. Environ Health Perspect 2005; 114: 106-12.
[http://dx.doi.org/10.1289/ehp.8451]

[126] Alonso-Magdalena P, Ropero AB, Carrera MP, *et al.* Pancreatic insulin content regulation by the estrogen receptor ER alpha. PLoS One 2008; 3: e2069.
[http://dx.doi.org/10.1371/journal.pone.0002069]

[127] Chevalier N, Fénichel P. Endocrine disruptors: new players in the pathophysiology of type 2 diabetes? Diabetes Metab 2015; 41: 107-15.
[http://dx.doi.org/10.1016/j.diabet.2014.09.005]

[128] Kudłak B, Szczepańska N, Owczarek K, Mazerska Z, Namieśnik J. Revision of biological methods for

determination of EDC presence and their endocrine potential. Crit Rev Anal Chem / CRC 2015; 45: 191-200.

[129] Chang HS, Choo KH, Lee B, Choi S-J. The methods of identification, analysis, and removal of endocrine disrupting compounds (EDCs) in water. J Hazard Mater 2009; 172: 1-12. [http://dx.doi.org/10.1016/j.jhazmat.2009.06.135]

[130] Albayrak N, Yang S. Immobilization of aspergillus oryzae b-Galacto-sidase on tosylated cotton cloth. Enzyme Microb Technol 2002; 31: 371-83. [http://dx.doi.org/10.1016/S0141-0229(02)00115-1]

[131] Riffle BW, Henderson WM, Laws SC. Measurement of steroids in rats after exposure to an endocrine disruptor: mass spectrometry and radioimmunoassay demonstrate similar results. J Pharmacol Toxicol Methods 2013; 68: 314-22. [http://dx.doi.org/10.1016/j.vascn.2013.07.003]

[132] Barouki R, Gluckman PD, Grandjean P, Hanson M, Heindel JJ. Developmental origins of non-communicable disease: implications for research and public health. Environ Health 2012; 11: 42. [http://dx.doi.org/10.1186/1476-069X-11-42]

[133] Zimmers SM, Browne EP, O'Keefe PW, *et al.* Determination of free Bisphenol A (BPA) concentrations in breast milk of U.S. women using a sensitive LC/MS/MS method. Chemosphere 2014; 104: 237-43. [http://dx.doi.org/10.1016/j.chemosphere.2013.12.085]

[134] Yi B, Kim C, Yang M. Biological monitoring of bisphenol A with HLPC/FLD and LC/MS/MS assays. J Chromatogr B Analyt Technol Biomed Life Sci 2010; 878: 2606-10. [http://dx.doi.org/10.1016/j.jchromb.2010.02.008]

[135] Cappiello A, Famiglini G, Palma P, Termopoli V, Maria Lavezzi A, Matturri L. Determination of selected endocrine disrupting compounds in human fetal and newborn tissues by GC-MS. Anal Bioanal Chem 2014; 406: 2779-88. [http://dx.doi.org/10.1007/s00216-014-7692-0]

CHAPTER 17

Monitoring of Monoclonal Antibodies, A New Window in the Use of Biomarkers in Pharmacology

Enrique Jiménez-Santos[1], **Iris Muñoz-García**[2], **Maria S García-Simón**[2], **Jose Pedregosa-Díaz**[1] and **Juan Antonio Vílchez**[1,*]

[1] *Clinical Analysis Department, Santa Lucia General University Hospital, Cartagena, Spain*

[2] *Pharmacy Department, Santa Lucia General University Hospital, Cartagena, Spain*

Abstract: Pharmacology has been a major field of clinical laboratory. Firstly in the emergency measures for the diagnosis of possible poisoning, and secondly in the monitoring of various treatments, mainly antibiotics and anticonvulsants, but also others in such as antidepressants, digitalis, anxiolytics, *etc.*

The discovery of biological treatments, based on the use of monoclonal antibodies, and their use for the treatment of various chronic diseases, whether the digestive tract or the rheumatic diseases, is considered a big step in favor of patients and their quality of life.

However, these treatments have had a worse response than they were supposed to, sometimes due to their own immunogenicity of these monoclonal antibodies in the organism, triggering responses to the drug itself, which reduces their effectiveness and also involves high costs and secondary effects to patients. That is why the biological monitoring of drugs has been proven to be a way to improve treatment adherence, and cost-effectiveness. In turn, it is also important to determine the presence of antibodies generated by the patient in response to the drug, because these cases determine inefficient and highly expensive treatments and the injury to the patient.

Keywords: ADA, Adalimumab, Ankylosing Spondylitis, Anti-TNF alpha Therapy, Antibodies, Biologic Treatment, Chronic Inflammation, Etanercept, HACA, Inflammatory Bowel Diseases, Infliximab, Monitoring, Monoclonal, Pharmacodynamics, Pharmacokinetics, Pharmacology, Rheumatoid Arthritis, TNF alpha.

INTRODUCTION

The determination of drug concentrations in the clinical laboratory is constant since a long time; it is very useful as it can provide a lot of information to the

[*] **Corresponding author Juan Antonio Vílchez:** Clinical Analysis Department, Santa Lucia General University Hospital, Cartagena, Spain; Tel: 968128600; E-mail: juananv81@hotmail.com

requesting physician [1]. For example, drug concentrations such as salicylates or barbiturates are determined in suspected poisoning, often involuntary, such as in children, or deliberated in other cases, such as cases of autolytic actions [2]. In these cases, to know the drug level in blood is vital to the clinician to take decisions.

Meanwhile in other drugs, such as aminoglycoside antibiotics, the importance of drug monitoring is important, on one hand to optimize performance in the treatment of infection and on the other hand, to prevent complications due to the nephrotoxic properties [3]. In case of drugs, such as antidepressants (lithium), anticonvulsants (phenytoin, phenobarbital), digitalis (digoxin), *etc*., is necessary to know the concentration of the drug to adjust the exact dose based on renal elimination, absorption, response and effectiveness of treatment, as well as follow-up or failure in treatment [4].

Therefore, the clinical laboratory should be updated in recent new treatments and therapeutic applications in pharmacology. This could suppose a benefit to the patient, and could be useful in clinical practice due to the information provided.

In the search for solutions from the pharmaceutical industry to different pathologies, the discovery of treatment with antibodies, and specifically those generated by a single cell clone and its large-scale production, has been a qualitative step in the treatment of many diseases. Highlighting the field against cancer, has led to the finding of most of the chemotherapies based on this principle [5]. These monoclonal antibodies recognize different antigens, receptors or proteins produced by tumor cells and their binding neutralizing the activity of those proteins.

In oncology as well as in many other fields the biological therapies have shown interesting progress, which is important to determine the behavior of viral infections treatment, for example of hepatitis C virus [6] or the human immunodeficiency virus [7]. The biological therapies have allowed a better control in remission, getting chronic infections and reducing the high mortality rates that have been associated with this kind of infection in the recent past.

The usefulness of immunotherapies or therapies with biological drugs is also significant in the treatment of diseases associated with an autoimmune disorder, many of which are chronic, such as multiple rheumatic diseases (arthritis, spondylitis, osteomyelitis, *etc*.), inflammatory bowel diseases, for example Crohn's disease or ulcerative colitis, or topical diseases such as psoriasis. Commonly to this variety of pathologies appears an immunological baseline of the disease with an inflammatory exacerbated response, which triggers a response of the immune system against the body, resulting in damage of the affected tissue.

This inflammatory response involves cytokines, messengers and proinflammatory mediators, being tumor necrosis factor (TNF) alpha the major player in the global response. That is why, most biological therapies with monoclonal antibodies have been focused to directly block the inflammatory response of this cytokine.

TNF ALPHA: A MAIN ROLE IN INFLAMMATION

TNF alpha superfamily consists of 20 proteins, including TNF alpha, whose main action takes place in the immune system, thus modulating innate and acquired immunity.

TNF alpha is a glycopeptidic hormone formed by 185 amino acids, which comes from a propeptide consisting of 212 amino acids that is hydrolyzed by the TNF alpha conversion enzyme (TACE). TNF alpha plays a role as proinflammatory and immunomodulatory cytokine that is produced by a wide spectrum of cells such as monocytes, macrophages, B and T lymphocytes, NK cells and non-immune system cells, as fibroblasts. This cytokine modulates a great number of biological processes such as mediation of host defense against neoplastic cell growth (hence its name) [8] and increased expression of antigens of the major histocompatibility complex class I [9]. It has also been implicated in the development of cachexy (in fact, there are several authors who have named it as cachectin) [10] in states of shock caused by sustained vasodilatation and vasopressor resistant (due to increased production of nitric oxide, the stimulate increased expression of the inducible form of nitric oxide synthase) [11], and also could act as an effector molecule in various inflammatory processes.

First, TNF alpha is synthetized as a 26-kDa transmembrane molecule (memTNF alpha) which is cut by TACE to produce a soluble molecule of 17 kDa (sTNF alpha). MemTNF alpha is a type II membrane protein that needs a cytoplasmic tail for signal transduction. As TNF alpha, memTNF is activated as homo-trimers that can bind in two cell surface receptors, TNF alpha-R1 (p55) and TNF alpha-R2 (p75) [12].

sTNF alpha is usually attached by the receptor TNF alpha-R1; this receptor presents a death domain that mediates the apoptosis process. In contrast, TNF alpha-R2 has preference for memTNF alpha and while it lacks a death domain, TNF alpha-R2 may also initiate cell death by enhancing TNF alpha-R1 signaling [13]. Both receptors activate the Rel/NF-kappaB and MAP kinase signal transduction pathways [14]. These multiple options in signaling let TNF alpha promote a great diversity of cellular responses, which include the clearance of intracellular bacterial infections, apoptosis, cell proliferation and regulation of leucocyte movement through chemokine and adhesion molecule expression [15]. But it should behave as a defense mechanism of immune system of the host against external aggression, such as infection or tumor development and can turn against the same individual, generating varied immune diseases. In these scenarios it has been found that excessive expression of TNF alpha plays a pivotal role.

Rheumatoid arthritis (RA), ankylosing spondylitis (AS) and psoriatic arthritis (PsA) are all associated with an overproduction of proinflammatory cytokines, particularly TNF alpha, IL-6 and IL-1 [16]. So, these pathologies are characterized by distinct immune-mediated pathogenesis that is related to the pathophysiology of each disease and finally leads to a chronic inflammatory response through a final common pathway [17].

TNF alpha induces macrophages and other cells to secrete other proinflammatory cytokines (for example, IL-1, IL-6, IL-8), leading to T-cell activation and inducing endothelial cells to express both adhesion molecules that lead to an increase in T-cell infiltration and vascular growth factors which promote angiogenesis and keratinocyte proliferation. TNF alpha also participates in the differentiation and maturation of osteoclasts, the main cells engaged in bone destruction in arthritis [18].

TNF alpha also participates in the inflammatory bowel diseases (IBD), that are a group of chronic systemic diseases involving inflammation of the gastrointestinal tract, and in which ulcerative colitis is included, which affects only the large bowel; Crohn's disease, which can affect the entire gastrointestinal tract; and indeterminate colitis, which consists of large-bowel inflammation that shows features of both Crohn's disease, ulcerative colitis, and microscopic colitis [19].

PATHOPHYSIOLOGY

Rheumatoid Arthritis

This disease is characterized by progressive bone destruction. Under physiological conditions, bone remodeling is a balanced process that persistently occurs by the formation and degradation of bone. Bone formation is regulated by osteoblasts, and bone resorption is mediated by osteoclasts, and this combination ensures bone homeostasis. In pathological conditions, such as rheumatoid arthritis (RA), an uncoordinated osteoclast formation is produced, due to an alteration of bone homeostasis [20].

Osteoclasts are derivate from monocyte-macrophage lineage. Several systemic and local factors influence the process of osteoclastogenesis. Proinflammatory cytokines such as interleukin IL-6, IL-1, TNF alpha, IL-11, IL-17 and IL-8 have been frequently reported to be osteoclastogenic, and on the other hand, non-osteoclastogenic cytokines as interferon (IFN)-γ, IFN-β, IL-4, IL-13, IL-13 and IL-10 are listed [21].

Increased production of RANKL and macrophage colony-stimulating factor (M-CSF) from osteoblasts and stromal cells is promoted by TNF alpha and bone wreckage as well as the differentiation into osteoclasts independently of RANK-RANKL signaling [22]. In addition, TNF alpha and RANKL synergistically upregulate the expression of RANK [23]. TNF alpha is associated with other inflammatory cytokines, like IL-1 and M-CSF combining osteoclastogenic effect [24].

Although osteoclastogenesis is a more dominant mechanism in the bone erosion of inflammatory disease, osteoblast formation is also affected by TNF alpha. The mechanism through which TNF alpha modulates the osteoblast differentiation is primarily TNF alpha receptor 1 signaling [25].

Ankylosing Spondylitis

Ankylosing spondylitis (AS) is a chronic inflammatory arthritis characterized by inflammation in spine that results in pain, stiffness, disability, and fusion of the joints in the spine, and in many patients, due to new bone formation. AS characteristically affects the sacroiliac joints at the base of the spine, with radiographic evidence of bilateral sacroiliac damage that defines the modified New York criteria for AS [26].

Although AS is pathogenetically different from RA, a biopsy study indicates that TNF alpha was expressed in inflamed sacroiliac joints in patients with AS [27]. As well in AR, an increasing evidence occurs that immune cells and cytokines are critically responsible for bone resorption and formation changes and *vice versa*, resulting in changes in bone quality and chronic inflammatory conditions [28].

AS is more prevalent in men than in women, and usually appears around the third decade of life [29]. Moreover, extra-articular manifestations such as uveitis, psoriasis, or osteoporosis are frequently associated with this rheumatologic disease [30].

A special feature of this disease is its strong genetic association. The association of HLA-B27 and AS, was the first described for HLA alleles related to inflammatory diseases. This association with AS is one of the strongest associations known between HLA and disease, with 96% of patients carrying at least one HLA-B27 allele compared with <10% of healthy people [31, 32].

Psoriatic Arthritis

After RA, Psoriatic arthritis (PsA) is the second most common inflammatory arthropathy. Most of the patients, firstly present established psoriasis (PsO), usually for years, and then they develop a joint pain and swelling; in addition, associated features of nail disease, dactylitis, enthesitis, spondylitis or uveitis are usual [33]. It is presented as a symmetrical polyarthritis, similar to RA, or an asymmetrical oligoarthritis with a predilection for the distal interphalangeal joints. Spinal involvement is similar, although not identical, to AS [33].

In PsA, a convoluted combination of genetic and environmental factors appears leading to a self sustained and constant immune response. However, the relationship between skin disease and joint disease is not clear yet, and also an independent behaviour appears rather than shared genetic associations.

In a recent report, the RUNX3 gene, which is an important transcriptional regulator of CD8 T-cell differentiation, was reported as an overexpressed gene in patients with PsO and PsA. This fact supports the potential role of CD8 T-cell-dependent pathways in PsA [34].

The association with HLA B27 and CD8 T-lymphocyte function also supported a potential pathogenic role as the major histocompatibility complex class I molecules, may present arthritogenic peptides to CD8+ T lymphocytes or select an auto reactive T-cell repertoire [35].

In PsO skin lesions, IFN-γ and TNF alpha appear to be the main cytokines but the genomic data implicate T cells and also dendritic cells as a key in the inflammatory process [33].

Many angiogenic cytokines including TNF alpha, transforming growth factor beta, platelet-derived growth factor, and vascular endothelial growth factor are overexpressed in PsO skin and they have been postulated as likely to be responsible for the induction of changes in effector cell function as well as in endothelial cell function and growth. Similar morphological changes have been described in the synovial blood vessels in the PsA joint, and recently strong associations have been reported with significant functional changes in infiltrating immune cells at a molecular level [36]. This approach further suggests that vascular endotelial cells actively participate in the immune response, rather than acting as a passive conduit for lymphocytes and monocytes entering inflamed tissues. Indeed, there is a close link between angiogenesis, inflammation, tissue hypoxia and the production of cytokines and matrix degrading enzymes, such metalloproteinases, that facilitate the effector cell activation and migration in the synovial tissue *in vivo* [37].

Many of the destructive matrix metalloproteinase enzymes are upregulated in the synovium, usually in response to cytokines, principally IL-1 and TNF alpha. Increased osteoclast activity at the cartilage–pannus junction also appears to be associated with increased receptor activator of nuclear factor-κB ligand expression and reduced osteoprotegerin. It has shown that the receptor activator of nuclear factor-κB ligand/osteoprotegerin can be altered *via* histamine acting through an orphan receptor in human chondrocytes [38].

Ulcerative Colitis

It is manifested by inflammation in the colon mucosa and in most cases, it is extended to the rectum. Similar to other IBDs, common symptoms are abdominal pain, diarrhea, and rectal bleeding as well as weight loss, fever, and fatigue. Ulcerative colitis (UC) is an autoimmune disorder of unknown etiology that results from excessive immune responses to intestinal microbiota which is triggered by increased activity of effector T cells and/or decreased activity of regulatory T cells, changes in the composition of intestinal microflora, and/or damage in the epithelial barrier [39].

In UC, the type of immune response is considered to be generally a T helper 2- (Th2-) mediated condition [40]. However, expression of a typical Th2 cytokine, IL-4, was not elevated in intestinal mucosa of UC patients. Instead, it has been suggested that the central role in the UC pathogenesis is played by IL-13 [41], which not only acts as a Th2 effector cytokine but also disrupts the continuity of colonic epithelium by inducing apoptosis of epithelial cells and upregulating expression of claudin-2, a pore-forming tight junction protein. The level of TNF alpha mRNA is up regulated in involved and uninvolved colonic tissue of UC patients compared to healthy subjects [42, 43]. The production of TNF alpha in the colon mucosa of UC patients was localized in the lamina propria macrophages [44]. Several studies showed that TNF alpha contributes to the disruption of intestinal epithelial barrier which allows the intestinal penetration of luminal antigens and promotes intestinal inflammation [45 - 47].

Crohn's Disease

It mainly affects small intestine and colon, however any other segment of the gastrointestinal tract could also be involved. Crohn's disease (CD) is characterized by discontinuous ulcerations and bowel wall inflammation. Furthermore, CD patients often develop strictures between segments of the bowel or between the bowel and other organs [39, 48].

There are several genetic factors that contribute to the pathogenesis of the IBD which include genetic mutations leading to enhanced inflammatory response [49 -

51], deficient elimination of intracellular bacteria [52], or disruption of the intestinal epithelial barrier [53].

Unlike UC, in CD, the inflammation is mainly driven by T helper 1 (Th1) or T helper 17 (Th17) cells [54]. In a recent report [43], elevated levels of TNF alpha protein which correlates with the activity of the disease, were observed in blood serum of CD patients while other groups found increased levels of TNF alpha protein [55, 56] in the intestinal lamina propria of both CD and UC patients as well as the intestinal submucosa of CD patients.

Pathogenesis of IBD is also associated with altered expression of TNF alpha receptors since both CD and UC patients showed elevated expression of TNF alphaR2 on colonic epitelial cells [57]. Moreover, a positive correlation was observed between CD and UC activity and serum concentration of soluble forms of TNF alphaR1 and TNF alphaR2 [43]. Furthermore, upregulated expression of TNF alphaR2 (but not TNF alphaR1) was found on intestinal lamina propria CD4+ cells as well as peripheral blood T cells of CD patients [58].

Intestinal epithelium integrity is provided by the presence of tight junctions (TJ) located in the apical region of intestinal epithelial cells. TNF alpha induced redistribution of several TJ and adhered junction proteins, including zonula occludens-1, occludins, claudins, and E-cadherin, as well as myosin light chain kinase, to basolateral membranes of intestinal epithelial cells. Furthermore, administration of TNF alpha resulted not only in the rearrangement of junctional proteins but also in the shedding of whole cells from intestinal epithelium. These events were preceded by caspase-3 activation due to the TNF alpha induced activation of NF-kB-dependent signaling pathway and proapoptotic pathways. These data suggest that TNF alpha triggered loss of intestinal epithelium integrity is a complex process which involves not only rearrangement of cytoskeletal elements but also direct induction of intestinal cells' apoptosis by TNF alpha [59].

THERAPY WITH MONOCLONAL ANTIBODIES

Since the development of hybridoma technology by Kohler and Milstein in 1975, the necessary tool was provided to get a cloned hybrid cell line which could generate unlimited quantities of a single antibody of specific and well-characterized binding specificity, avidity and isotype. From there, the production of most monoclonal antibody products is easily amenable. So, monoclonal antibodies represent a potent therapy due to their high specificity, safety and generally well-tolerance, as well as monoclonal antibodies often provide the most rapid route to a new clinical target, for activating, inhibiting, or blocking.

Following the approval of the first monoclonal antibody product in 1986, sales growth and approval of additional products were slow until the late 1990s when the first chimeric monoclonal antibodies were approved. With the approval of these products, followed by the approval of humanized and then fully human monoclonal antibodies, the rate of monoclonal antibody products has increased dramatically. So, in 2013, global sales revenue for all monoclonal antibody products was nearly $75 billion representing approximately half of the total sales of all biopharmaceutical products [60].

Due to the mentioned properties of the monoclonal antibodies, an increasing knowledge is observed on the role of TNF alpha in multiple autoimmune diseases, as rheumatic and intestinal, in which its involvement has been demonstrated in a sustained inflammatory response in time, and in consequence with serious complications in the affected tissue. That is why, the pharmaceutical industry has developed various therapies based on monoclonal antibodies focusing the TNF alpha as target.

Therefore, the inhibition of inflammatory cytokines, and TNF alpha in particular became the major focus of clinical research in RA in the 1990s, which has continued to this day. The first open-label trial using a TNF alpha blocking agent was conducted in the United Kingdom in 1992, in which 20 patients with RA were given infliximab, a chimeric antibody specific for TNF alpha, resulting in a significant decrease in the signs and symptoms of RA, accompanied by a decrease in inflammatory markers [61]. The early promise of anti-TNF alpha therapy in RA was confirmed in several recent multicenter studies.

Since the introduction of infliximab, other anti-TNF alpha agents and other cytokine inhibitors have been developed and reached clinic use. Although these agents have largely been developed and tested in RA, their use has extended to other chronic inflammatory conditions, occasionally with unexpected and apparently divergent results. For the treatment of RA, there are currently five TNF alpha inhibitors. Of these, three are full-length monoclonal antibodies (infliximab, adalimumab, and golimumab), one is a humanized fragment antigen-binding fragment (Fab) conjugated to polyethylene glycol (certolizumab), and one is a soluble fusion protein (etanercept) (Table **1**) [62].

In the case of AS, various TNF alpha inhibitors seem to have remarkably similar response rates in clinical trials, although there have been no direct head-to-head comparisons [62].

Etanercept was approved for the treatment of PsA in the United States and Europe in 2002. The efficacy of infliximab (5 mg/kg) in PsA was demonstrated in IMPACT 1 and 2 (Infliximab Multinational Psoriatic Arthritis Controlled Trials)

[63, 64]. Similarly, the efficacy of adalimumab in PsA was demonstrated in the ADEPT study (Adalimumab Effectiveness in Psoriatic Arthritis Trial) [65].

Table 1. Cytokine inhibitors licensed for the treatment of rheumatoid arthritis.

Target	Active substance	TradeName	Half-Life
TNFalpha	Infliximab	Remicade®	8-10 days
TNFalpha	Adalimumab	Humira®	2 weeks
TNFalpha	Etanercept	Enbrel®	70 hours
TNFalpha	Golimumab	Simponi®	9-15 days
TNFalpha	Certolizumab	Cimzia®	14 days
IL-6R	Tocilizumab	Roactemra®	13 days
IL-1R	Anakinra	Kineret®	3-6 hours

TNF alpha inhibitor studies in IBD are more complex and difficult to interpret due to the variable doses induction and maintenance regimens employed. Three anti-TNF alpha molecules are used currently in IBD therapy: infliximab, adalimumab, and certolizumab. Etanercept has not shown to be effective in Crohn's disease treatment [19]. Future prospects are encouraging, with multiple antibodies in advanced stages of studies (Table **2**).

Infliximab

Infliximab is a recombinant chimeric human–murine monoclonal antibody that binds soluble and membrane-bound TNF alpha. Stable complexes are formed, binding TNF alpha to TNF alpha receptors. The TNF alpha binding region presents a mouse origin and comprises 30% of the amino acid sequence of infliximab. The remainder is a human IgG1 heavy-chain and kappa-chain constant region.

Infliximab was allowed for use in RA with methotrexate, although in clinical practice, it is used without methotrexate or with other disease-modifying antirheumatic drug (DMARD) if patients are intolerant to methotrexate. The recommended dose of infliximab for RA is 3 mg kg^{-1} (body weight given) as an intravenous infusion, followed by further infusion, at the same dose, 2 and 6 weeks later. Thereafter, infusions are given at 8-week intervals. An interval between infusions more than 16 weeks is not recommended due to an increased risk of hypersensitivity reactions, although infusions after longer gaps are administered safely [66]. Infliximab is also licensed for use in severe Crohn's disease (5 mg kg^{-1}), including disease complicated by fistulae, ankylosing spondylitis (5 mg kg^{-1}) and psoriatic arthritis (5 mg kg^{-1}). Use of higher doses of infliximab in trials has encouraged the use of higher doses or a shorter interval between infusions in RA.

Table 2. Main antibodies approved for treatment of IBD, or also in advanced stages. Adapted from Rutgeerts P. *et al.* [19].

Target	Compound	Stage	Disease
TNFalpha	Infliximab	Approved	CD, UC
TNF alpha	Adalimumab	Approved	CD, UC
TNF alpha	Certolizumab	Approved	CD
TNF alpha	Golimumab	Phase III	CD, UC
TNF alpha	Etanercept	Stopped	
TNF alpha	Onercept	Stopped	
α4-integrin	Natalizumab	Approved	CD, UC
α4β7-integrin	Vedolizumab	Phase III	CD, UC
IFN-γ	Fontolizumab	Stopped	CD
IL-2	Daclizumab	Stopped	UC
IL-2	Basiliximab	Phase III	UC
IL-6	Tocilizumab	Phase II	CD
CTLA-4	Abatacept	Phase III	CD, UC
CD20	Rituximab	Phase III	CD, UC
IL-17	AIN457	Phase II	CD

Adalimumab

Adalimumab is a recombinant monoclonal antibody, made from human peptide sequences, using phage display technology resulting in an antibody with human derived heavy and a light chain with variable regions and human IgG1 constant regions.

It binds specifically to TNF alpha and neutralizes its biological functions by blocking interactions with the p55 and p75 cell-surface TNF alpha receptors. Treatment is currently recommended in patients with moderate or severe RA who have not responded to one or more DMARDs, including methotrexate.

Concomitant treatment with methotrexate is recommended for optimum efficacy, but adalimumab may be used alone if methotrexate is not tolerated or is contraindicated. Clearance of adalimumab from the body is decreased with age and by concomitant methotrexate administration, whereas adalimumab increases methotrexate clearance [66]. Patients normally self-administer adalimumab by subcutaneous injections, after training. The standard dose used is 40 mg in alternate weeks; but the dose may be increased to 40 mg weekly if disease is poorly controlled.

Etanercept

Etanercept is a combinated protein consisting of the extracellular portion of two of the 75-kDa TNF alpha receptors (TNFalpha-R2) for TNF alpha combined with a human Fc portion of human IgG class 1 (IgG1).

Etanercept binds soluble and cell-bound TNF alpha with high affinity competing with TNF alpha receptors. Etanercept is administrated as a twice-weekly subcutaneous injection of 25 mg or a once-weekly injection of 50 mg. Patients or caregivers normally administer etanercept, after suitable training. No dose changes are necessary for patients with renal or hepatic failure or in elderly subjects.

Etanercept may be used in combination with methotrexate or alone for the treatment of active RA in adults when the response to DMARDs, including methotrexate (unless contraindicated), has been inadequate, and for the treatment of severe, active and progressive RA not previously treated with methotrexate. Etanercept is also licensed for use in juvenile idiopathic arthritis, psoriatic arthritis, ankylosing spondylitis and severe psoriasis [66].

MONITORING AND IMMUNOGENICITY OF ANTI-TNF ALPHA THERAPY

Nowadays, Infliximab, Adalimumab and Etanercept are the most common antibodies used in treatments, and the most studied of all. The main problem with the monoclonal antibodies cited above is their capacity to generate anti-drug antibodies (ADA). These anti-drug antibodies could have potential repercussions on the efficacy and safety of treatments, but their clinical relevance is not yet fully understood and not all antibodies are neutralized. However, it is clear that the TNF alpha inhibitors differ in their immunogenicity, which is influenced by factors that include the structure of the drug; route, dose, and schedule of administration; concomitant medications; underlying disease; age; immune status; and genetics [67].

Infliximab is the most immunogenic drug of them. It is caused by a consistent chimeric structure that contains 25% murine sequences, leading to the development of human anti-chimeric antibodies [68]. The reported prevalence of anti-infliximab antibodies greatly varies depending on the assay and the disease, but, it appears around 30% in RA [69] and may be as high as 61% in Crohn's disease [70], with this latter high prevalence due to the episodic dosing schedule used in CD. Another finding is that higher anti-infliximab antibodies levels are found in patients with earlier drug clearances [71].

Despite its humanized structure, adalimumab antibodies have been detected in 28% of RA patients in a 3-year follow-up study, with two-thirds of these patients developing anti-adalimumab antibodies within the first 28 weeks of treatment. Anti-adalimumab antibody development was associated with lower adalimumab levels and reduced clinical response to adalimumab, but this fact was not universal with anti-adalimumab antibodies detected in 4% of patients with sustained remission [72].

Etanercept appears to be the least immunogenic of the current anti-TNF alpha agents. In a 5-year extension study to evaluate the long-term efficacy and safety of etanercept in RA, Klareskog *et al.* (2011) found that anti-etanercept antibodies were present in <5% of patients and all those antibodies were non-neutralized [73]. There was also no statistically significant relationship between autoantibody titers and adverse clinical events in this study.

Several groups have shown that the development of antibodies to infliximab and adalimumab correlates with a decreased clinical response, not only with decreased drug concentrations. The clinical management of patients who responded to treatment and then lost the response remains largely empiric. Management strategies include escalation of dose or shortening the infusion interval, switching to another anti-TNF alpha agent, or switching to another therapeutic class [74].

Nowadays, a new tool from clinical laboratory has been developed. It is the immunoassays which are able to determine drug levels in serum, as well as ADA. In the case of Infliximab, it is logical that the incorporation of routine measurement of human anti-chimeric antibody (HACA) and drug levels could lead to a better approach in the follow up of the pathology [75]. Cut-off levels are considered a prerequisite [76] to further investigations of the clinical usefulness of measurements of monoclonal antibodies and HACA in patients who failed infliximab therapy.

The presence of HACAs to Infliximab provides clear evidence that immunogenicity to anti-TNF alpha was developed and that is why, further treatment would result in a decreased clinical response or possible infusion reactions [77, 78]. In patients with subtherapeutic concentrations, drug dose escalation was associated with a significantly increased clinical response compared with changing to another anti-TNF alpha [75].

Current clinical practice in monitoring response, dose regimens and planning decision making after primary or secondary failure of anti-TNF alpha therapy, still does not include routine monitoring of HACA (about infliximab) or ADA, neither to drug serum concentrations in patients. Thus, therapeutic intensification and/or drug switching, especially in patients with sub therapeutic concentration [75] is carried out without the knowledge of factors that could explain the failure.

The knowledge of the predictive power of serum drug determination and ADA appearing for responses to dose escalation, alternative TNF alpha inhibitors or drugs with different mechanisms of action could radically change the way of patients with anti-TNF alpha drugs management [79]. Here the importance role that clinical laboratory could play in decision-making can be highlighted, to abort or to change the treatment, avoiding unnecessary costs and complications for patients. Due to the high cost of TNF alpha inhibitors drugs, and the consequent costs associated with the use of such drugs when they are destined to fail, the potential benefit of monitoring drug concentrations and ADA in patients on anti-TNF alpha therapy could be economically justified.

Currently, there are many authors that have postulated the inclusion of updated clinical guidelines in the anti-TNF alpha therapy. So they have suggested to include algorithms to evaluate change treatment or anti-TNF alpha antibody dose escalation, as well as suspend therapy based on laboratory parameters tested [75, 80 - 82] (Fig. **1**).

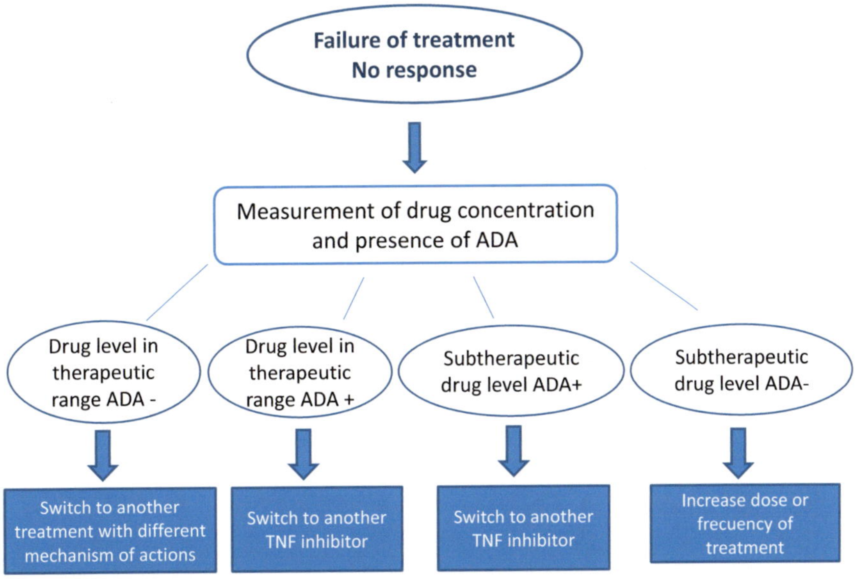

Fig. (1). Algorithm to make decisions in anti-TNF alpha therapy basis in drug levels and presence or not of ADAs. Adapted from FB Vincent *et al.* [79].

LABORATORY METHODS AND TESTING

There is no standardized test for measuring a monoclonal antibody or antidrug antibodies in this field. Various laboratories use different assays, which ultimately limit the ability to directly compare concentrations between assays [83].

We could classify different methods to measure anti-TNF alpha and ADA in 3 categories. Standard direct and indirect enzyme-linked immunosorbent assay (ELISAs), is the most overall-used methodology. Due to ELISAs, a high throughput and relatively inexpensive costs present, with very low detection limits, common to the variety of ELISAs; the detection limit and assay range can switch between commercial kits available in market, but these differences are residual [84].

Maybe, these methodologies are more prone to false-positive results and non-specific binding, and also these have the inconvenience that is not possible to measure ADA in the same assay through which we measure the drug, because Infliximab or Adalimumab levels interfere with the ADA testing. Another limitation is that this method could not discriminate yet between non-neutralizing/ neutralizing antibodies.

To improve assay methods, the two-site (or bridging) ELISA has been proposed. The two-site ELISA involves the binding of the drug to a solid support matrix, adding patient sera and then a detection using a biotinylated form of the drug of interest. This type of assay is both highly specific and sensitive.

Still there are limitations, such as the presence of rheumatoid proteins, complement factors or heterophilic antibodies which are responsible for false positive in the assays [85]. Another problem is that most commercially available ELISA fail to detect IgG4 antibodies which are prevalent after sustained immunizations [86]. More recently, an ECLIA (electrochemiluminescence immunoassay) has been developed, with a lower limit of detection than ELISA (0.4 mg/mL) and it can measure ADA in the presence of serum drug. However, no clinical validation data are available with this assay yet, and so its current use in clinical practice is uncertain [87].

Second methodology is radio-immunoassay (RIA). It is similar to ELISA, and is a sensitive, specific (less sensitive to artefacts than ELISAs), and inexpensive form of testing, but it uses a radioactive reagent, which is the main problem, considerably limiting its practical use [88].

The last methodology is an assay which uses high-performance liquid chromatography to measure both, anti-TNF alpha concentration and ADA [89].

This test was developed as an alternative to ELISA and RIA, as neither of them could measure ADA when drug is detectable in the serum, a big disadvantage in comparison with thechromatographic method is that, however, the cost of this assay may limit its widespread use.

DISCUSSION

Pharmacology has played a key role in the clinical laboratory, as the professionals of clinical laboratory have updated their knowledge of new treatments and therapies.

Although, the pathophysiology of the immune diseases is complicated, there are some aspects well-established in the schemes of these diseases.

The main role that TNF alpha plays in the mechanism of the inflammatory response is well-known which generates a sustained response over time, causing tissue damage [18].

Due to the main role of TNF alpha therapies focused on blocking this cytokine, the present use of monoclonal antibodies therapy has offered benefits for patients in diverse immune pathologies [19]. However, these therapies have shown minor effectiveness than was expected, principally due to the ability of these biological treatments to generate antibodies against them or their own immunogenicity.

The clinical laboratory may contribute to avoid ineffective costs in those treatments, as well as secondary effects to patients, measuring levels of monoclonal antibodies, as well as the presence or absence of ADA. So, the role of TNF alpha in the treatment of inflammatory autoimmune diseases has recently appeared as a new target in the use of therapy based on monoclonal antibodies that block cytokines, messenger molecules and proteins [62 - 65]. These new treatments have notably improved the quality of life of patients.

In this area, investigations focus their efforts trying to evaluate those treatments that are most effective, with low immunogenicity by "humanizing" monoclonal antibodies, as well as the introduction of new similar biomolecules, called "biosimilars".

Biosimilars are biological products, a version of original "innovator" products, that are supposed to be a copy of the reference product in terms of "quality, safety and efficacy", according to European Medicines Agency, or "safety, purity and potency", according to Food and Drug Administration. These are always characterized by the absence of relevant differences of each parameter: quality/purity, safety, efficacy/potency [90].

To the appearance of biosimilars, the next expiration of biological products patents is added and several pharmaceutical industries are developing new versions of their patents or biosimilars. It is expected that an increased offer will broaden the range of therapeutic options.

From the analytical point of view, the introduction of new monoclonal antibodies, new genetic and molecular biology techniques are being developed, such as microarrays and molecular probes. These techniques aim to study the involvement of genetic polymorphisms associated with disease progression and response to treatment.

In this context and with a future perspective, the role of clinical laboratory specialist is based on an updated and adapted knowledge of the technology to improve the performance of new treatments as well as the existing ones.

Clinical laboratory specialists, as a part of healthcare team, will provide essential information and could become a tool to understand an optimize treatments based on monoclonal antibodies.

CONFLICT OF INTEREST

The author (editor) declares no conflict of interest, financial or otherwise.

ACKNOWLEDGEMENTS

Declared none.

REFERENCES

[1] Dasgupta A. Clinical utility of free drug monitoring. Clin Chem Lab Med 2002; 40: 986-93. [http://dx.doi.org/10.1515/CCLM.2002.172]

[2] Kwong TC. Salicylate measurement: clinical usefulness and methodology. Crit Rev Clin Lab Sci 1987; 25: 137-59. [http://dx.doi.org/10.3109/10408368709105880]

[3] Wenk M, Vozeh S, Follath F. Serum level monitoring of antibacterial drugs: A review. Clin Pharmacokinet 1984; 9: 475-92. [http://dx.doi.org/10.2165/00003088-198409060-00001]

[4] Perucca E, Grimaldi R, Crema A. Interpretation of drug levels in acute and chronic disease states. Clin Pharmacokinet 1985; 10: 498-513. [http://dx.doi.org/10.2165/00003088-198510060-00003]

[5] Levy R, Miller RA. Biological and clinical implications of lymphocyte hybridomas: tumor therapy with monoclonal antibodies. Annu Rev Med 1983; 34: 107-16. [http://dx.doi.org/10.1146/annurev.me.34.020183.000543]

[6] Sautto G, Tarr AW, Mancini N, Clementi M. Structural and antigenic definition of hepatitis C virus E2 glycoprotein epitopes targeted by monoclonal antibodies. Clin Dev Immunol 2013; 2013: 450963. [http://dx.doi.org/10.1155/2013/450963]

[7] Barouch DH, Deeks SG. Immunologic strategies for HIV-1 remission and eradication. Science 2014; 345(80-): 169-74.

[8] Maury CP. Tumour necrosis factor - an overview. Acta Med Scand 1986; 220: 387-94. [http://dx.doi.org/10.1111/j.0954-6820.1986.tb02785.x]

[9] Le J, Vilcek J. Tumor necrosis factor and interleukin 1: cytokines with multiple overlapping biological activities. Lab Invest 1987; 56: 234-48.

[10] Levine B, Kalman J, Mayer L, Fillit HM, Packer M. Elevated circulating levels of tumor necrosis factor in severe chronic heart failure. N Engl J Med 1990; 323: 236-41. [http://dx.doi.org/10.1056/NEJM199007263230405]

[11] Landry DW, Oliver JA. Mechanisms of disease: The pathogenesis of vasodilatory shock. N Engl J Med 2001; 345: 588-95. [http://dx.doi.org/10.1056/NEJMra002709]

[12] Campbell IK, Roberts LJ, Wicks IP. Molecular targets in immune-mediated diseases: The case of tumour necrosis factor and rheumatoid arthritis. Immunol Cell Biol 2003; 81: 354-66. [http://dx.doi.org/10.1046/j.0818-9641.2003.01185.x]

[13] Chan FK, Siegel RM, Lenardo MJ. Signaling by the TNF receptor superfamily and T cell homeostasis. Immunity 2000; 13: 419-22. [http://dx.doi.org/10.1016/S1074-7613(00)00041-8]

[14] Grell M, Douni E, Wajant H, *et al.* The transmembrane form of tumor necrosis factor is the prime activating ligand of the 80 kDa tumor necrosis factor receptor. Cell 1995; 83: 793-802. [http://dx.doi.org/10.1016/0092-8674(95)90192-2]

[15] Bazzoni F, Beutler B. The tumor necrosis factor ligand and receptor families. N Engl J Med 1996; 334: 1717-25. [http://dx.doi.org/10.1056/NEJM199606273342607]

[16] Smolen JS, Emery P. Infliximab: 12 years of experience. Arthritis Res Ther 2011; 13(Suppl 1): S2. [http://dx.doi.org/10.1186/1478-6354-13-S1-S2]

[17] Smolen JS, Steiner G. Therapeutic strategies for rheumatoid arthritis. Nat Rev Drug Discov 2003; 2: 473-88. [http://dx.doi.org/10.1038/nrd1109]

[18] Redlich K, Hayer S, Ricci R, *et al.* Osteoclasts are essential for TNF-α-mediated joint destruction. J Clin Invest 2002; 110: 1419-27. [http://dx.doi.org/10.1172/JCI0215582]

[19] Rutgeerts P, Vermeire S, Van Assche G. Biological therapies for inflammatory bowel diseases. Gastroenterology 2009; 136: 1182-97. [http://dx.doi.org/10.1053/j.gastro.2009.02.001]

[20] Jung SM, Kim KW, Yang C-W, Park S-H, Ju JH, Mamura M. Cytokine-mediated bone destruction in rheumatoid arthritis. J Immunol Res 2014; 2014: 263625. [http://dx.doi.org/10.1155/2014/263625]

[21] Nakashima T, Kobayashi Y, Yamasaki S, *et al.* Protein expression and functional difference of membrane-bound and soluble receptor activator of NF-κB ligand: Modulation of the expression by osteotropic factors and cytokines. Biochem Biophys Res Commun 2000; 275: 768-75. [http://dx.doi.org/10.1006/bbrc.2000.3379]

[22] Kawai VK, Stein CM, Perrien DS, Griffin MR. Effects of anti-tumor necrosis factor α agents on bone. Curr Opin Rheumatol 2012; 24: 576-85. [http://dx.doi.org/10.1097/BOR.0b013e328356d212]

[23] Komine M, Kukita A, Kukita T, Ogata Y, Hotokebuchi T, Kohashi O. Tumor necrosis factor-α cooperates with receptor activator of nuclear factor κB ligand in generation of osteoclasts in stromal

cell-depleted rat bone marrow cell culture. Bone 2001; 28: 474-83.
[http://dx.doi.org/10.1016/S8756-3282(01)00420-3]

[24] Kudo O, Fujikawa Y, Itonaga I, Sabokbar A, Torisu T, Athanasou NA. Proinflammatory cytokine (TNFα/IL-Iα) induction of human osteoclast formation. J Pathol 2002; 198: 220-7.
[http://dx.doi.org/10.1002/path.1190]

[25] Abbas S, Zhang Y-H, Clohisy JC, Abu-Amer Y. Tumor necrosis factor-α inhibits pre-osteoblast differentiation through its type-1 receptor. Cytokine 2003; 22: 33-41.
[http://dx.doi.org/10.1016/S1043-4666(03)00106-6]

[26] Van Der Linden S, Valkenburg HA, Cats A. Evaluation of diagnostic concept in patients with rheumatoid arthritis and an inaduquate response to methotrexate: results of a phase II, randomized, placebo-controlled trial. Arthritis Rheumatol 1984; 63: 1782-92.

[27] Braun J, Bollow M, Neure L, *et al.* Use of immunohistologic and *in situ* hybridization techniques in the examination of sacroiliac joint biopsy specimens from patients with ankylosing spondylitis. Arthritis Rheum 1995; 38: 499-505.
[http://dx.doi.org/10.1002/art.1780380407]

[28] Geusens P, Lems WF. Osteoimmunology and osteoporosis. Arthritis Res Ther 2011; 13(5): 242.
[http://dx.doi.org/10.1186/ar3375]

[29] El Maghraoui A. Extra-articular manifestations of ankylosing spondylitis: Prevalence, characteristics and therapeutic implications. Eur J Intern Med 2011; 22: 554-60.
[http://dx.doi.org/10.1016/j.ejim.2011.06.006]

[30] Braun J, Van Den Berg R, Baraliakos X, *et al.* 2010 update of the ASAS/EULAR recommendations for the management of ankylosing spondylitis. Ann Rheum Dis 2011; 70: 896-904.
[http://dx.doi.org/10.1136/ard.2011.151027]

[31] Brewerton DA, Hart FD, Nicholls A, Caffrey M, James DC, Sturrock RD. Ankylosing spondylitis and HL-A 27. Lancet 1973; 301: 904-7.
[http://dx.doi.org/10.1016/S0140-6736(73)91360-3]

[32] Schlosstein L, Terasaki PI, Bluestone R, Pearson CM. High association of an HL-A antigen, W27, with ankylosing spondylitis. N Engl J Med 1973; 288: 704-6.
[http://dx.doi.org/10.1056/NEJM197304052881403]

[33] Veale DJ. Psoriatic arthritis: Recent progress in pathophysiology and drug development. Arthritis Res Ther 2013; 15(6): 224.
[http://dx.doi.org/10.1186/ar4414]

[34] Nair RP, Duffin KC, Helms C, *et al.* Genome-wide scan reveals association of psoriasis with IL-23 and NF-κB pathways. Nat Genet 2009; 41: 199-204.
[http://dx.doi.org/10.1038/ng.311]

[35] Bluett J, Barton A. What have genome-wide studies told us about psoriatic arthritis? Curr Rheumatol Rep 2012; 14: 364-8.
[http://dx.doi.org/10.1007/s11926-012-0255-5]

[36] Gao W, Sweeney C, Walsh C, *et al.* Notch signalling pathways mediate synovial angiogenesis in response to vascular endothelial growth factor and angiopoietin 2. Ann Rheum Dis 2013; 72: 1080-8.
[http://dx.doi.org/10.1136/annrheumdis-2012-201978]

[37] Ng CT, Biniecka M, Kennedy A, *et al.* Synovial tissue hypoxia and inflammation *in vivo*. Ann Rheum Dis 2010; 69: 1389-95.
[http://dx.doi.org/10.1136/ard.2009.119776]

[38] Marzaioli V, McMorrow JP, Angerer H, *et al.* Histamine contributes to increased RANKL to osteoprotegerin ratio through altered nuclear receptor 4A activity in human chondrocytes. Arthritis Rheum 2012; 64: 3290-301.
[http://dx.doi.org/10.1002/art.34554]

[39] Strober W, Fuss I, Mannon P. The fundamental basis of inflammatory bowel disease. J Clin Invest 2007; 117: 514-21. [http://dx.doi.org/10.1172/JCI30587]

[40] Bamias G, Kaltsa G, Ladas SD. Cytokines in the pathogenesis of ulcerative colitis. Discov Med 2011; 11: 459-67.

[41] Fuss IJ, Strober W. The role of IL-13 and NK T cells in experimental and human ulcerative colitis. Mucosal Immunol 2008; 1: S31-3. [http://dx.doi.org/10.1038/mi.2008.40]

[42] Ślebioda TJ, Kmieć Z. Tumour necrosis factor superfamily members in the pathogenesis of inflammatory bowel disease. Mediators Inflamm 2014; 2014: 325129. [http://dx.doi.org/10.1155/2014/325129]

[43] Owczarek D, Cibor D, Głowacki MK, Cieśla A, Mach P. TNF-α and soluble forms of TNF receptors 1 and 2 in the serum of patients with Crohn's disease and ulcerative colitis. Pol Arch Med Wewn 2012; 122: 616-23.

[44] Murch SH, Braegger CP, Walker-Smith JA, MacDonald TT. Location of tumour necrosis factor a by immunohistochemistry in chronic inflammatory bowel disease. Gut 1993; 34: 1705-9. [http://dx.doi.org/10.1136/gut.34.12.1705]

[45] Ma TY, Iwamoto GK, Hoa NT, *et al.* TNF-α-induced increase in intestinal epithelial tight junction permeability requires NF-κB activation. Am J Physiol Gastrointest Liver Physiol 2004; 286: G367-76. [http://dx.doi.org/10.1152/ajpgi.00173.2003]

[46] Fréour T, Jarry A, Bach-Ngohou K, *et al.* TACE inhibition amplifies TNF-α-mediated colonic epithelial barrier disruption. Int J Mol Med 2009; 23: 41-8.

[47] Wang F, Schwarz BT, Graham WV, *et al.* IFN-γ-Induced TNFR2 Expression Is Required for TNF-Dependent Intestinal Epithelial Barrier Dysfunction. Gastroenterology 2006; 131: 1153-63. [http://dx.doi.org/10.1053/j.gastro.2006.08.022]

[48] Molodecky NA, Kaplan GG. Environmental risk factors for inflammatory bowel disease. Gastroenterol Hepatol 2010; 6: 339-46.

[49] Glocker E-O, Kotlarz D, Boztug K, *et al.* Inflammatory bowel disease and mutations affecting the interleukin-10 receptor. N Engl J Med 2009; 361: 2033-45. [http://dx.doi.org/10.1056/NEJMoa0907206]

[50] Yamazaki K, McGovern D, Ragoussis J, *et al.* Single nucleotide polymorphisms in TNFSF15 confer susceptibility to Crohn's disease. Hum Mol Genet 2005; 14: 3499-506. [http://dx.doi.org/10.1093/hmg/ddi379]

[51] Maeda S, Hsu L-C, Liu H, *et al.* Nod2 mutation in Crohn's disease potentiates NF-κB activity and IL-1β processing Science 2005; 307(80-): 734-8.

[52] Parkes M, Barrett JC, Prescott NJ, *et al.* Sequence variants in the autophagy gene IRGM and multiple other replicating loci contribute to Crohn's disease susceptibility. Nat Genet 2007; 39: 830-2. [http://dx.doi.org/10.1038/ng2061]

[53] Lin Z, Poritz L, Franke A, *et al.* Genetic association of DLG5 R30Q with familial and sporadic inflammatory bowel disease in men. Dis Markers 2009; 27: 193-201. [http://dx.doi.org/10.1155/2009/167534]

[54] Brand S. Crohn's disease: Th1, Th17 or both? The change of a paradigm: new immunological and genetic insights implicate Th17 cells in the pathogenesis of Crohn's disease. Gut 2009; 58: 1152-67. [http://dx.doi.org/10.1136/gut.2008.163667]

[55] Murch SH, Lamkin VA, Savage MO, Walker-Smith JA, MacDonald TT. Serum concentrations of tumour necrosis factor in childhood chronic inflammatory bowel disease. Gut 1991; 32: 913-7. [http://dx.doi.org/10.1136/gut.32.8.913]

[56] Maeda M, Watanabe N, Neda H, *et al.* Serum tumor necrosis factor activity in inflammatory bowel disease. Immunopharmacol Immunotoxicol 1992; 14: 451-61. [http://dx.doi.org/10.3109/08923979209005404]

[57] Mizoguchi E, Mizoguchi A, Takedatsu H, *et al.* Role of tumor necrosis factor receptor 2 (TNFR2) in colonic epithelial hyperplasia and chronic intestinal inflammation in mice. Gastroenterology 2002; 122: 134-44. [http://dx.doi.org/10.1053/gast.2002.30347]

[58] Holtmann MH, Douni E, Schütz M, *et al.* Tumor necrosis factor-receptor 2 is up-regulated on lamina propria T cells in Crohn's disease and promotes experimental colitis *in vivo.* Eur J Immunol 2002; 32: 3142-51. [http://dx.doi.org/10.1002/1521-4141(200211)32:11<3142::AID-IMMU3142>3.0.CO;2-4]

[59] Marchiando AM, Shen L, Graham WV, *et al.* The epithelial barrier is maintained by *in vivo* tight junction expansion during pathologic intestinal epithelial shedding. Gastroenterology 2011; 140: 1208-18. [http://dx.doi.org/10.1053/j.gastro.2011.01.004]

[60] Ecker DM, Jones SD, Levine HL. The therapeutic monoclonal antibody market. MAbs 2015; 7: 9-14. [http://dx.doi.org/10.4161/19420862.2015.989042]

[61] Elliott MJ, Maini RN, Feldmann M, *et al.* Treatment of rheumatoid arthritis with chimeric monoclonal antibodies to tumor necrosis factor α. Arthritis Rheum 1993; 36: 1681-90. [http://dx.doi.org/10.1002/art.1780361206]

[62] Siebert S, Tsoukas A, Robertson J, McInnes I. Cytokines as therapeutic targets in rheumatoid arthritis and other inflammatory diseases. Pharmacol Rev 2015; 67: 280-309. [http://dx.doi.org/10.1124/pr.114.009639]

[63] Antoni C, Krueger GG, De Vlam K, *et al.* Infliximab improves signs and symptoms of psoriatic arthritis: Results of the IMPACT 2 trial. Ann Rheum Dis 2005; 64: 1150-7. [http://dx.doi.org/10.1136/ard.2004.032268]

[64] Antoni CE, Kavanaugh A, Kirkham B, *et al.* Sustained benefits of infliximab therapy for dermatologic and articular manifestations of psoriatic arthritis: Results from the Infliximab Multinational Psoriatic Arthritis Controlled Trial (IMPACT). Arthritis Rheum 2005; 52: 1227-36. [http://dx.doi.org/10.1002/art.20967]

[65] Mease PJ, Gladman DD, Ritchlin CT, *et al.* Adalimumab for the treatment of patients with moderately to severely active psoriatic arthritis: Results of a double-blind, randomized, placebo-controlled trial. Arthritis Rheum 2005; 52: 3279-89. [http://dx.doi.org/10.1002/art.21306]

[66] Chen Y-F, Jobanputra P, Barton P, *et al.* A systematic review of the effectiveness of adalimumab, etanercept and infliximab for the treatment of rheumatoid arthritis in adults and an economic evaluation of their cost-effectiveness. Health Technol Assess (Rockv) 2006; 10: 1-129.

[67] Garcês S, Demengeot J, Benito-Garcia E. The immunogenicity of anti-TNF therapy in immune-mediated inflammatory diseases: a systematic review of the literature with a meta-analysis. Ann Rheum Dis 2013; 72: 1947-55. [http://dx.doi.org/10.1136/annrheumdis-2012-202220]

[68] Atzeni F, Talotta R, Salaffi F, *et al.* Immunogenicity and autoimmunity during anti-TNF therapy. Autoimmun Rev 2013; 12: 703-8. [http://dx.doi.org/10.1016/j.autrev.2012.10.021]

[69] Pascual-Salcedo D, Plasencia C, Ramiro S, *et al.* Influence of immunogenicity on the efficacy of long-term treatment with infliximab in rheumatoid arthritis. Rheumatology (Oxford) 2011; 50: 1445-52. [http://dx.doi.org/10.1093/rheumatology/ker124]

[70] Baert F, Noman M, Vermeire S, *et al.* Influence of immunogenicity on the long-term efficacy of

infliximab in Crohn's disease. N Engl J Med 2003; 348: 601-8.
[http://dx.doi.org/10.1056/NEJMoa020888]

[71] Plasencia C, Pascual-Salcedo D, Alcocer P, *et al.* The timing of serum infliximab loss, or the appearance of antibodies to infliximab (ATI), is related with the clinical activity in ATI-positive patients with rheumatoid arthritis treated with infliximab. Ann Rheum Dis 2013; 72: 1888-90.
[http://dx.doi.org/10.1136/annrheumdis-2013-203353]

[72] Bartelds GM, Krieckaert CL, Nurmohamed MT, *et al.* Development of antidrug antibodies against adalimumab and association with disease activity and treatment failure during longterm follow-up. JAMA -. JAMA 2011; 305: 1460-8.
[http://dx.doi.org/10.1001/jama.2011.406]

[73] Klareskog L, Gaubitz M, Rodríguez-Valverde V, Malaise M, Dougados M, Wajdula J. Assessment of long-term safety and efficacy of etanercept in a 5-year extension study in patients with rheumatoid arthritis. Clin Exp Rheumatol 2011; 29: 238-47.

[74] Rutgeerts P, Van Assche G, Vermeire S. Optimizing Anti-TNF treatment in inflammatory bowel disease. Gastroenterology 2004; 126: 1593-610.
[http://dx.doi.org/10.1053/j.gastro.2004.02.070]

[75] Afif W, Loftus EV Jr, Faubion WA, *et al.* Clinical utility of measuring infliximab and human anti-chimeric antibody concentrations in patients with inflammatory bowel disease. Am J Gastroenterol 2010; 105: 1133-9.
[http://dx.doi.org/10.1038/ajg.2010.9]

[76] Steenholdt C, Bendtzen K, Brynskov J, Thomsen OØ, Ainsworth MA. Cut-off levels and diagnostic accuracy of infliximab trough levels and anti-infliximab antibodies in Crohn's disease. Scand J Gastroenterol 2011; 46: 310-8.
[http://dx.doi.org/10.3109/00365521.2010.536254]

[77] Maser EA, Villela R, Silverberg MS, Greenberg GR. Association of trough serum infliximab to clinical outcome after scheduled maintenance treatment for Crohn's disease. Clin Gastroenterol Hepatol 2006; 4: 1248-54.
[http://dx.doi.org/10.1016/j.cgh.2006.06.025]

[78] Hanauer SB, Wagner CL, Bala M, *et al.* Incidence and importance of antibody responses to infliximab after maintenance or episodic treatment in Crohn's disease. Clin Gastroenterol Hepatol 2004; 2: 542-53.
[http://dx.doi.org/10.1016/S1542-3565(04)00238-1]

[79] Vincent FB, Morand EF, Murphy K, Mackay F, Mariette X, Marcelli C. Antidrug antibodies (ADAb) to tumour necrosis factor (TNF)-specific neutralising agents in chronic inflammatory diseases: A real issue, a clinical perspective. Ann Rheum Dis 2013; 72: 165-78.
[http://dx.doi.org/10.1136/annrheumdis-2012-202545]

[80] Pallagi-Kunstár É, Farkas K, Szepes Z, *et al.* Utility of serum TNF-α, infliximab trough level, and antibody titers in inflammatory bowel disease. World J Gastroenterol 2014; 20: 5031-5.
[http://dx.doi.org/10.3748/wjg.v20.i17.5031]

[81] Krieckaert CL, Nair SC, Nurmohamed MT, *et al.* Personalised treatment using serum drug levels of adalimumab in patients with rheumatoid arthritis: An evaluation of costs and effects. Ann Rheum Dis 2015; 74: 361-8.
[http://dx.doi.org/10.1136/annrheumdis-2013-204101]

[82] Chen D-Y, Chen Y-M, Tsai W-C, *et al.* Significant associations of antidrug antibody levels with serum drug trough levels and therapeutic response of adalimumab and etanercept treatment in rheumatoid arthritis. Ann Rheum Dis 2015; 74
[http://dx.doi.org/10.1136/annrheumdis-2015-207978]

[83] Vaughn BP, Sandborn WJ, Cheifetz AS. Biologic concentration testing in inflammatory bowel disease. Inflamm Bowel Dis 2015; 21: 1435-42.

[84] Malíčková K, Ďuricová D, Bortlík M, *et al.* Serum trough infliximab levels: A comparison of three different immunoassays for the monitoring of CT-P13 (infliximab) treatment in patients with inflammatory bowel disease. Biologicals 2016; 44: 33-6. [http://dx.doi.org/10.1016/j.biologicals.2015.09.005]

[85] Bendtzen K. Immunogenicity of anti-TNF-α biotherapies: II. Clinical relevance of methods used for anti-drug antibody detection. Front Immunol 2015; 6 [http://dx.doi.org/10.3389/fimmu.2015.00109]

[86] Svenson M, Geborek P, Saxne T, Bendtzen K. Monitoring patients treated with anti-TN--αbiopharmaceuticals: Assessing serum infliximab and anti-infliximab antibodies. Rheumatology 2007; 46: 1828-34. [http://dx.doi.org/10.1093/rheumatology/kem261]

[87] Bendtzen K. Personalized medicine: Theranostics (therapeutics diagnostics) essential for rational use of tumor necrosis factor-alpha antagonists. Discov Med 2013; 15: 201-11.

[88] Aarden L, Ruuls SR, Wolbink G. Immunogenicity of anti-tumor necrosis factor antibodies-toward improved methods of anti-antibody measurement. Curr Opin Immunol 2008; 20: 431-5. [http://dx.doi.org/10.1016/j.coi.2008.06.011]

[89] Wang S-L, Ohrmund L, Hauenstein S, *et al.* Development and validation of a homogeneous mobility shift assay for the measurement of infliximab and antibodies-to-infliximab levels in patient serum. J Immunol Methods 2012; 382: 177-88. [http://dx.doi.org/10.1016/j.jim.2012.06.002]

[90] U.S. Food and Drug Administration. Inf Healthc Prof Available at: http://www.fda.gov/Drugs/DevelopmentApprovalProcess/HowDrugsareDevelopedandApproved/ApprovalApplications/TherapeuticBiologicApplications/Biosimilars/ucm241719.htm

SUBJECT INDEX

A

Juan Antonio Vílchez (Ed.)

B

C

D

E

I

K

L

M

N

Q

R

W

Made in the USA
Las Vegas, NV
03 November 2026

57425093R00335